Fundamentals of MOS Digital Integrated Circuits

JOHN P. UYEMURA

GEORGIA INSTITUTE OF TECHNOLOGY

ADDISON-WESLEY PUBLISHING COMPANY
Reading, Massachusetts • Menlo Park, California • New York
Don Mills, Ontario • Wokingham, England • Amsterdam • Bonn
Sydney • Singapore • Tokyo • Madrid • Bogotá
Santiago • San Juan

This book is in the **Addison-Wesley Series in Electrical and Computer Engineering**

Library of Congress Cataloging-in-Publication Data
Uyemura, John P. (John Paul), 1952–
  Fundamentals of MOS digital integrated circuits.

  Includes index.
  1. Digital integrated circuits.   2. Metal oxide
semiconductors.   I. Title
TK7874.U94   1988   621.381'73   87-11458
ISBN 0-201-13318-0

Reprinted with corrections July, 1988

4 5 6 7 8 9 10 HA 9594939291

My wife Melba Valerie and I dedicate this book to our parents
Reverend George and Mrs. Ruby Uyemura
and to the loving memory of our mother
Mrs. Rebecca V. Byram
who have always provided unlimited support and encouragement.

Preface

Microelectronics has experienced explosive growth in recent years. The advent of VLSI (very large scale integration) has made possible digital systems that were inconceivable only a short time ago. New developments continue to advance the field. Although the marketplace and discipline are showing signs of maturity, work continues at a rapid pace and should continue well into the future. This applies to all aspects of microelectronics: device physics, fabrication, circuit design, and system development.

PHILOSOPHY

This book addresses digital chip design in nMOS and CMOS technologies. It is intended for use as a text and should also be of interest to practicing engineers. The presentation deals with basics and fundamentals. It assumes a background equivalent to an undergraduate course in electronics and devices. Logic theory is kept to a minimum, with emphasis instead placed on calculating currents and voltages. Since this diverges from recent trends in VLSI books, I should first mention what is not covered before listing the topics that are included.

First, this book is not about VLSI layout or patterning techniques. Layout drawings are used as needed in the analysis, but the topology is not a central theme. Device physics is studied at a basic level, and fabrication techniques are only summarized. Since all else is eliminated, it is easily deduced that emphasis must be on circuit design and analysis. This is indeed the case, but the subject cannot be decoupled from the other disciplines that make up microelectronics. This book deals with the interaction of circuit design with device physics, fabrication, and system logic. It is intended to provide a clearer understanding of the *foundations* of VLSI system design. As such, it directly complements VLSI system courses, and strengthens the background of both circuit and system designers.

The approach to circuit analysis presented here is based on square-law MOSFET models. Although this automatically introduces limitations on the accuracy of the results, it allows the discussion to progress in an analytic manner that helps clarify circuit operation. Mathematizing the circuits provides a stepping stone for developing a design approach and philosophy. Numbers can be calculated, and design estimates are possible. This provides the reader with a measure of confidence and a manner for checking understanding. Worked examples are provided to illustrate important principles, equations, and computer algorithms.

It should be mentioned that nMOS and CMOS circuits are given roughly equal weighting. This may seem puzzling, since CMOS is rapidly becoming the dominant technology. The decision to pursue this approach was actually quite simple to make. nMOS logic structuring is straightforward to learn and forms the basis for most emerging CMOS design styles. The former consideration is pedagogically important, while the latter is more practical.

USE OF COMPUTERS
One relatively unique feature of the treatment is the acceptance of personal computers as an everyday tool. This allows for problem solving at a level that would have been prohibitive only a few years ago. Transcendental equations are found throughout the treatment, requiring numerical calculations. Most of the algorithms used in the book can be solved with a scientific calculator. However, I encourage my students to write their own programs, and simple programming examples are provided in the text. The programs are particularly useful for varying parameters to study the changes in the circuit characteristics. Using an interpreted language on a personal computer simplifies this procedure.

The accuracy of the analysis is another important point. Since square-law $I-V$ MOSFET models are used, care must be taken when interpreting the results. Computer simulation is the best safeguard. Examples in this text are referenced to SPICE (for Simulation Program with Integrated Circuit Emphasis) as a standard, for two reasons. First, the program was developed at my alma mater, and I was one of the hundreds of beta-site testers (then called "guinea pigs") when it was introduced into the curriculum. Over the years I have developed a certain measure of trust in the results. The second reason is more pragmatic. SPICE is in public domain and can be obtained in a variety of computer system formats for the cost of a tape. Moreover, commercial versions of SPICE for desktop computers are available. These range from complete programs with pre- and postprocessors down to free programs that provide full capabilities but limit circuit size. Students at Georgia Tech have their choice of SPICE systems to use. Many choose mainframe- or minicomputer-based code, but an increasing number are purchasing PCs and using a free or scaled-down classroom version of SPICE.

This book structures a design philosophy by incorporating the above considerations into the presentation. Understanding of the circuits is attained through the derivations and examples. The interplay between circuit parameters and performance is stressed. Results are used to develop design equations. Computer simulation allows for verification of an analysis or "fine-tuning" of a design.

USE OF THIS BOOK

This book was written around a two-quarter graduate level sequence at Georgia Tech entitled Digital MOS Integrated Circuits. The audience is made up of first-year graduate students and senior undergraduates. The book has also been used in self-study groups and in teaching courses of varying duration. It is possible to skip sections without losing continuity, which helps adapt it to other schedules. In classroom usage, the material is greatly enhanced by design projects. These can be either individual or group efforts.

The subject flow is organized to allow flexibility for study. Sections that may be review material for the more seasoned readers can be skipped entirely. This includes the chapters on MOSFETs and chip fabrication in particular. In addition, more advanced topics, such as VLSI device corrections in Chapter 2, can also be omitted in a first reading. Other variations are possible, as seen by the following short summaries.

Chapter 1 introduces MOSFETs by developing the gradual channel approximation. Sections 1.1–1.4 develop the square-law MOSFET equations, which are used for the rest of the book. The last two sections are optional reading. Section 1.5 examines the errors involved in modeling the MOSFETs with the simplified equations, while Section 1.6 introduces the depletion-mode MOSFET. (D-mode MOSFETs are introduced again in Chapter 3 using simplified modeling).

VLSI device corrections are examined in **Chapter 2**. The topics of scaling theory, geometry-induced threshold voltage shifts, and basic two-dimensional effects are introduced. This chapter can be skipped without loss of continuity.

The circuit development starts in **Chapter 3**, which is devoted to analyzing the DC characteristics of nMOS and CMOS inverters. It is recommended that this chapter be read in its entirety, since it develops equations that are used for the remainder of the text. It might be tempting to skip Section 3.3 on enhancement-MOSFET load configurations, since these are relatively rare in state-of-the-art design. However, the analysis directly applies to pass transistor logic and synchronous circuits and is referenced extensively in Chapter 8.

Chapter 4 centers around the transient switching behavior of MOS inverters. Although the material is structured to be read after the DC analysis in Chapter 3, this order is not mandatory. An alternative is to simultaneously discuss the DC and transient characteristics of each circuit by drawing from both Chapters 3 and 4. Either approach works in practice. MOSFET capacitance modeling completes the connection between DC design, transient response, and layout. In addition, the overall design aspects become clear.

Silicon fabrication is covered in **Chapter 5**. Some reviewers felt this was out of place. They may be correct, but in my mind the actual location is not that important. The chapter should be accessed as needed. If the material is review, then its placement is accurate. It can be skimmed through or skipped entirely. For those with less background in the material, this marks the farthest one can progress in the book without learning some aspects of fabrication. I did not place it at the beginning of the book (as might seem logical) since I have found that students interested in learning digital circuit design are not necessarily the same group that enjoy the rigors of silicon process analysis. Rather, they want to get into circuits as soon as possible. Fabrication considerations tend to be more meaningful after the circuit aspects are developed; motivation for learning the material increases accordingly.

Chapter 5 will probably seem either excessively complicated or much too simplified, depending on one's outlook. The emphasis is on the physical processes and the resulting *electrical* properties for circuit design. As such, layout considerations are discussed only in terms of circuit parameters. The reader is referred to the references for more detailed treatments.

Combinational nMOS and CMOS logic is the subject of **Chapter 6**. This takes the basic inverter in Chapters 3 and 4 and examines the building of logic gates. AOI (and-or-invert) nMOS and transmission gate CMOS techniques are emphasized.

Chapter 7 is a short discussion of bistable logic gates, including flip-flops, latches, and Schmitt triggers. Both nMOS and CMOS circuits are covered.

Synchronous nMOS logic is the subject of **Chapter 8**. The analysis is directed toward pass transistor circuits, with discussions of charge transfer and charge leakage in dynamic circuits. Various clocking schemes are introduced, and charge sharing problems are analyzed. Bootstrapping and precharging complete the presentation.

Chapter 9 deals with synchronous CMOS logic. The treatment examines the charge transfer properties of transmission gates and charge retention on CMOS soft nodes. Different circuit design approaches are covered, and the system design techniques of domino, NORA, and zipper structuring are included.

Circuit aspects of structured logic are introduced in **Chapter 10**. This forms the interface with VLSI system design considerations. Switch logic capacitances and transmission line effects are studied. Weinberger arrays and programmable logic arrays round out the discussion.

References are provided at the end of each chapter. The lists are mostly books that the reader may find useful. No attempt was made to even begin to list journal articles. However, the background provided by the treatment should be sufficient to allow access into current journals, such as the *IEEE Journal of Solid-State Circuits*.

Problems are also provided at the end of each chapter. The degree of difficulty varies. Some are straightforward applications of the material, while others seem to go on for pages. Numerical calculations are emphasized throughout. Problems marked with an asterisk (*) require access to a computer; this can imply a SPICE simulation or writing a program.

The book emphasizes circuit design and analysis in a chip environment. However, only the fundamentals are covered. It is my hope that this book will provide insight into the subject and will lead to a deeper understanding of a fast-moving field that directly affects the world we live in.

ACKNOWLEDGMENTS

Early in my academic career my interest in the subject was greatly enhanced during discussions with A. A. Alvarez. The manuscript itself had its roots in a course requested by a group of ambitious students, including M. Ash, D. Carter, K. Jackoski, C. Morgan, R. Perry, D. Schaeffer, and D. Zyriek. Subsequent feedback from these persons as employees of AMD, Cypress, IBM, MCC, and Motorola has helped to steer the presentation to its present form.

A great deal of thanks is due those who have suffered through the different versions of the manuscript while students enrolled in classes at Georgia Tech. Many useful comments were received from K. Banach (AT & T; U.S. Navy), S. Bartling (Intel), J. DiChristina (GTE), J. Eckhardt (IBM), D. Norris (AMD), H. T. T. Nguyen (GE), S. Park (Intel), J. Pena-Finol (Harris), and D. Reginold (GE). R. Perry (IBM) provided comments, corrections, and problem suggestions. N. Mansour spent countless hours finding errors in the final manuscript. L. Clendenning and E. Snyder gave insight into using the text for self study.

My teachers, friends, and colleagues in academia have had a tremendous amount of influence on the content and approach contained in the book. I was introduced to the subject in the outstanding lectures of Professors D. A. Hodges and R. G. Meyer. I am grateful to the reviewers at Arizona, Columbia, Illinois, MIT, Purdue, Washington, and Waterloo for their critical commentary of the manuscript. Special thanks are due Professors R. B. Darling, M. I. Elmasry, S. M. Kang, G. W. Neudeck, and C. Zukowski for their careful analyses and suggestions.

I have known Tom Robbins of Addison-Wesley since I first started teaching and have always enjoyed and profited from our conversations. His work as a publisher has increased my respect even more. Barbara Rifkind, my editor on this project, has continuously demonstrated her expertise in all aspects of publishing. Her professional attitude combined with a refreshing sense of humor provided an ideal atmosphere for completing the work. Bette Aaronson, production manager for the book, has done a spectacular job in starting with a stack of manuscript papers and then organizing production details to yield the final product with speed and efficiency.

In closing, it is unfortunate that tradition dictates only my name appears on the title page. While it is true that I wrote the words, the love and support of my wife Melba Valerie really made this a joint effort. The countless hours spent on the writing and the drawings were tolerated with patience and understanding. Her engineering background (EECS/NE, U.C. Berkeley) made her proofreading of the manuscript, by providing technical comments and suggestions, more than just another check in spelling and grammar. Sharing both personal and professional aspects of our lives has given added meaning and strength to our partnership.

Atlanta   J. P. U.

Note on Program Listings

The program examples written in BASIC should work without modification using any standard interpreter. Some implementations recognize only the first two characters of a variable, so that VT0D and VT0L, for example, would represent the same quantity. In such a case, the variables should be changed—e.g., to VD and VL. Programs have been kept simple to aid in understanding. They can be refined by using higher-level commands such as function (FN) statements and WHILE . . . WEND loops if available.

SPICE listings include only the important statements; for simplicity, commands such as .OPTIONS are not shown. In addition, the implementation of SPICE2.G used to generate the plots contains an internal incrementing routine for .DC and .TRAN analysis. It may be necessary to modify the voltage and time increments when running the code without a postprocessor. This applies to all SPICE runs that yield plots using standard line printer output.

MOSFET SPICE models are defaulted at LEVEL = 1 for simplicity. This also correlates with the analytic modeling developed in the text. A good exercise is to increase the modeling to LEVEL = 2 and compare the results. This requires additional parameters, as described in any SPICE user guide.

Commercial equation solving programs can also be used to implement the algorithms in the book. Codes that can iteratively solve for unknowns in coupled equation sets are especially useful. Various programs of this type have been used to check the examples.

Contents

Preface   vii
Note on Program Listings   xiii

CHAPTER 1:   MOSFETs   1

   1.1 Introduction to the MOS System   *1*
   1.2 MOSFET Structure and Operation   *12*
   1.3 The Gradual Channel Approximation   *18*
   1.4 MOSFET Circuit Equations   *22*
   1.5 The Complete MOSFET GCA Analysis   *30*
   1.6 Depletion-Mode MOSFETs   *35*

CHAPTER 2:   Small-Geometry MOSFETs for VLSI   48

   2.1 MOSFET Scaling Theory   *49*
      2.1.1 Full Scaling   *51*
      2.1.2 Constant-Voltage Scaling   *54*
      2.1.3 Practical Limitations on the Use of Scaling Theory   *55*

2.2 Threshold Voltage Corrections for Small MOSFETs   *57*
    2.2.1 Short-Channel Effects   *58*
    2.2.2 Narrow-Width Effects   *63*
    2.2.3 Minimum-Size Effects   *69*
2.3 Two-Dimensional MOSFET Effects   *71*

CHAPTER 3:   DC Characteristics of MOS Inverters   **79**

3.1 Basic Digital Concepts   *80*
3.2 Linear Resistors as Load Elements   *91*
3.3 Enhancement-Mode MOSFETs as Loads   *101*
    3.3.1 Saturated E-Mode MOSFET Loads   *101*
    3.3.2 Nonsaturated Enhancement MOSFET Loads   *113*
3.4 Depletion-Mode MOSFET Loads   *119*
3.5 The CMOS Inverter   *137*
    3.5.1 The p-Channel MOSFET   *139*
    3.5.2 CMOS Inverter Characteristics   *142*

CHAPTER 4:   Switching of MOS Inverters   **158**

4.1 The Output High-to-Low Time   *159*
4.2 nMOS Rise Time   *165*
4.3 nMOS Propagation Delay Times   *175*
4.4 CMOS Transient Response   *182*
4.5 The Power-Delay Product   *188*
4.6 MOSFET Capacitances   *197*
4.7 Inverter Output Capacitance   *208*
4.8 Scaled Inverter Performance   *218*

CHAPTER 5:   MOS Integrated Circuit Fabrication   **228**

5.1 Overview of Basic Silicon Processing Steps   *229*
    5.1.1 Thermal Oxidation   *229*
    5.1.2 Impurity Diffusion   *233*
    5.1.3 Ion Implantation   *243*
    5.1.4 Chemical Vapor Deposition   *247*
    5.1.5 Metallization and Silicides   *249*
5.2 Lithography   *250*
    5.2.1 Basic Pattern Transfer   *250*
    5.2.2 Lateral Doping Effects   *256*
    5.2.3 Etching   *258*
5.3 Isolation Techniques   *263*
    5.3.1 Etched Field Oxide Isolation   *263*
    5.3.2 LOCOS   *267*
    5.3.3 Trench Isolation   *269*

5.4 nMOS LOCOS Process Flow   *273*
5.5 Threshold Voltage Adjustment   *282*
5.6 nMOS Design Rules and Layout   *284*
    5.6.1 Physical Basis of Design Rules   *284*
    5.6.2 Mask Drawings   *288*
    5.6.3 Design Rule Sets   *289*
    5.6.4 Some Comments on Layout   *291*
5.7 Processing Variations   *292*
    5.7.1 Generalized Modeling   *292*
    5.7.2 Analysis of Device Parameters   *294*
    5.7.3 Circuit Design Criteria   *296*
5.8 CMOS Technologies   *296*
5.9 CMOS Design Rules   *304*

CHAPTER 6: Combinational MOS Logic Circuits   **311**

6.1 nMOS NOR Gate   *311*
6.2 nMOS NAND Gate   *317*
6.3 Complex Static nMOS Logic   *325*
6.4 nMOS XOR, XNOR, and Associated Circuits   *331*
6.5 CMOS NAND and NOR Gates   *344*
    6.5.1 2-Input NAND Analysis   *346*
    6.5.2 2-Input NOR Analysis   *352*
    6.5.3 Comparison of CMOS NAND and NOR Gates   *357*
6.6 Complex CMOS Logic   *358*
6.7 Transmission Gate CMOS Logic   *359*
    6.7.1 CMOS Transmission Gate Characteristics   *360*
    6.7.2 TG Logic Implementation   *366*
6.8 Nonstandard CMOS Gates   *371*

CHAPTER 7: Bistable Logic Elements   **386**

7.1 nMOS SR Flip-Flop   *386*
7.2 Clocked Flip-Flops   *392*
7.3 The Schmitt Trigger   *398*
7.4 CMOS SR Flip-Flops   *408*
7.5 Clocked CMOS Flip-Flops   *413*
7.6 CMOS Schmitt Trigger   *420*

CHAPTER 8: Synchronous nMOS Logic   **428**

8.1 nMOS Pass Transistors   *428*
    8.1.1 Charging Analysis: Logic 1 Transfer   *431*
    8.1.2 Discharging Analysis: Logic 0 Transfer   *435*
    8.1.3 Charge Leakage from Soft Nodes   *438*

8.2 2-Phase Clock Timing   *451*
8.3 Synchronous Depletion Load Logic   *454*
    8.3.1 Basic Shift Register   *454*
    8.3.2 Synchronous Complex Logic   *459*
    8.3.3 Clocked Static Register Circuits   *465*
8.4 Dynamic Charge Sharing   *467*
8.5 2-Phase Enhancement Load Logic   *473*
    8.5.1 2-Phase Ratioed Dynamic Logic   *475*
    8.5.2 Ratioless 2ϕ Logic   *480*
8.6 Voltage Bootstrapping Techniques   *484*
8.7 Precharging   *491*

CHAPTER 9: Synchronous CMOS Logic   503

9.1 Switching Properties of CMOS Transmission Gates   *503*
    9.1.1 Discharging Analysis: Logic 0 Transfer   *506*
    9.1.2 Charging Analysis: Logic 1 Transfer   *508*
    9.1.3 Charge Leakage   *513*
9.2 Synchronized TG CMOS Logic   *523*
    9.2.1 Single-Clock Logic   *523*
    9.2.2 Pseudo 2-Phase Shift Register   *532*
    9.2.3 Complex Logic   *534*
9.3 Single-Clock Dynamic Logic   *541*
    9.3.1 Basic Operation   *542*
    9.3.2 Generalized Combinational Logic   *545*
9.4 Domino CMOS Logic   *554*
9.5 NORA Logic   *562*
9.6 Zipper CMOS   *566*
9.7 Dynamic Pseudo-2ϕ Logic   *573*

CHAPTER 10: Structured MOS Logic   584

10.1 Concepts in Structured Logic   *585*
10.2 Switch Logic   *585*
10.3 Weinberger Structuring   *601*
10.4 Programmable Logic Arrays   *605*

APPENDIX: Summary of *pn* Junction Properties   615

INDEX   619

CHAPTER 1

MOSFETs

Integrated circuits are designed using the current-voltage (I-V) characteristics of the devices. This chapter introduces MOSFETs by first examining the physics of the MOS system and then applying the results to form transistors. Sections 1.1–1.4 are referenced throughout the book. However, the remaining two sections are optional and may be omitted in a first reading.

1.1 Introduction to the MOS System

The MOS (metal oxide semiconductor) system is illustrated in Fig. 1.1. The acronym MOS describes the layering scheme for the three different materials used in the structure: metal on top of oxide on top of semiconductor. This arrangement forms the basis for MOS integrated circuits (MOS ICs). It has the important property of exhibiting the *field effect*, where the charge carriers in the semiconductor are controlled by electric fields in the structure; the fields

1

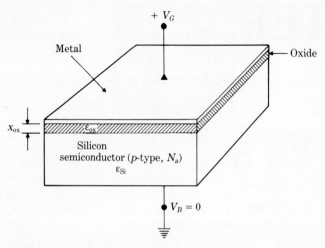

FIGURE 1.1 The basic MOS structure.

are established by externally applied voltages. This phenomenon allows for the construction of MOSFETs (MOS field effect transistors), the switching and amplifying devices in MOS circuits.

The physics involved in describing the operational characteristics of the MOS system is complicated. However, it is possible to gain a basic understanding of the system properties using arguments from electrostatics and semiconductor theory. This section will examine the MOS field effect to illustrate the most important properties. The equations will be constructed by combining physical reasoning with basic device physics. It is assumed that the reader is familiar with the static analysis of *pn* junctions, which is summarized in the Appendix.

Consider first the electrical characteristics of the layers used to construct the MOS system. The semiconductor is taken to be single-crystal *p*-type silicon, which will give the basic structure required to form an *n*-channel MOSFET (nMOS). The *p*-type wafer is uniformly doped with an acceptor (e.g., boron) concentration N_a. Typically, N_a is on the order of $10^{15}\,[\text{cm}^{-3}]$, which gives a resistivity ρ of a few $[\Omega\text{-cm}]$. The bottom of the wafer is coated with a metal (e.g., gold) to define the *bulk* or *substrate* electrode to which an external bulk (or *body*) voltage V_B may be applied. In Fig. 1.1, this voltage has been chosen to be at ground potential ($0\,[\text{V}]$). Close to the bulk electrode, the majority and minority thermal equilibrium carrier concentrations are approximated by

$$p_{po} \simeq N_a \quad \text{and} \quad n_{po} \simeq \frac{n_i^2}{N_a}, \tag{1.1-1}$$

where the extra o subscript denotes equilibrium quantities and n_i is the intrinsic carrier density, which is a function of temperature T. At room temperature ($T = 300$ [°K]), $n_i \simeq 1.45 \times 10^{10}$ [cm^{-3}]. The top of the substrate (at the oxide-semiconductor interface) is termed the *surface* of the silicon and will be the region of greatest interest in the discussion of the MOS behavior.

The oxide layer used in the silicon technology is silicon dioxide (SiO$_2$), which occurs in an amorphous state. Silicon dioxide is more commonly known as quartz glass. This layer is grown on top of the substrate by passing an oxygen-rich gas over the surface of the wafer at an elevated temperature; the silicon reacts with the oxygen to form the SiO$_2$ molecules. The silicon dioxide layer in the MOS system is used as an insulating dielectric between the metal and semiconductor layers. Ideally, the oxide would be a perfect insulator with an infinite resistivity and would be free of stray charge. In reality, the silicon dioxide layer is a very good insulator, typically having a resistivity greater than 10^{15} [Ω-cm]. The grown oxide layer is also found to contain unwanted positive charges. Some originate from alkali contaminants, with positively charged sodium ions being particularly troublesome in this respect. The charges are found throughout the oxide layer and can also accumulate at the oxide-semiconductor interface.

The top layer of the MOS system is the metal, and it is used to form the *gate* electrode. The external voltage V_G applied to this layer is defined to be positive, as shown in Fig. 1.1, and is referenced to the system ground. At one time, aluminum was the most common gate metal used to form the MOS structure. However, other conducting materials have replaced Al as the gate substance in modern technologies. The descriptive term MOS will be used to denote any conductor-oxide-semiconductor arrangement, even though the top layer may not be a metal. This is sometimes referred to as an "insulated gate" structure, with the resulting transistor denoted as an IGFET; however, the MOS terminology will be maintained here.

Polycrystalline silicon (polysilicon or poly) is of particular importance to state-of-the-art MOS ICs for use as a gate material. Polysilicon is macroscopically classified as an amorphous material. At the microscopic level, it consists of small regions of silicon crystal called *crystalites*. The term *polycrystalline* arises from viewing the material as being made up of randomly distributed crystalites. One of the advantages of using polysilicon instead of a metal is that it has the same melting temperature as crystal Si. This allows the wafer to be subjected to processing steps that require heat treatments after the gate has been formed. Another useful property is that the poly layer can be doped either *n*-type or *p*-type.

Now that the basic electrical properties of the individual layers have been discussed, we can explore the electrical behavior of the MOS system. In the simplest analysis, the structure may be viewed as a basic capacitor. The *gate capacitance* exists because of the insulating oxide layer between the gate electrode and the semiconductor. This is usually described in terms of an oxide gate capacitance per unit area C_{ox} to keep the treatment independent of any specific geometry. If the permittivity of the oxide is denoted by $\varepsilon_{\mathrm{ox}}$, the parallel-plate capacitance formula from electrostatics can be used to write

$$C_{\mathrm{ox}} = \frac{\varepsilon_{\mathrm{ox}}}{x_{\mathrm{ox}}} \; [\mathrm{F/cm^2}], \tag{1.1-2}$$

where x_{ox} is the thickness of the gate oxide in [cm]. Silicon dioxide has a (DC) dielectric constant of approximately 3.9, so that $\varepsilon_{\mathrm{ox}} = (3.9)\varepsilon_o$, with ε_o the free space permittivity ($\varepsilon_o = 8.854 \times 10^{-14}\,[\mathrm{F/cm}]$). In state-of-the-art structures, x_{ox} is less than about $0.05\,[\mu\mathrm{m}]$ ($= 500\,[\text{Å}]$), which gives C_{ox} values on the order of $10^{-8}\,[\mathrm{F/cm^2}]$. For a gate having an area $A\,[\mathrm{cm^2}]$, the total gate capacitance is computed from

$$C_g = C_{\mathrm{ox}}A \; [\mathrm{F}]. \tag{1.1-3}$$

The field effect in the MOS system arises from the application of external voltages. In terms of Fig. 1.1, the gate voltage V_G is viewed as the basic parameter that controls the electric fields within the structure; V_G may be either positive or negative in polarity. Varying the gate voltage gives three primary regions of operation for the MOS system. These are respectively termed *accumulation, depletion,* and *inversion.*

Suppose that V_G is negative, as shown in Fig. 1.2. With this choice of polarity, the oxide electric field $\mathscr{E}_{\mathrm{ox}}$ in the MOS capacitor points toward the gate electrode. Positively charged majority carrier holes then "accumulate" at the oxide-semiconductor interface (the bottom "plate" of the capacitor), thus defining the operational region of *accumulation*. When biased into this mode, the MOS system behaves like a capacitor with a value given by eqn. (1.1-2). The field effect is manifest in the fact that the electric field within the semiconductor induces the accumulation of holes at the surface. Although this bias arrangement is extremely useful in measuring some basic MOS properties,

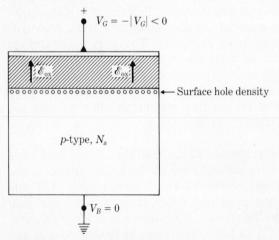

FIGURE 1.2 The accumulation region of operation in the MOS structure.

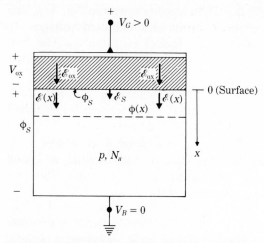

FIGURE 1.3 Fields and potentials in the MOS system for positive gate voltages.

it is not the operational region used to create MOSFETs. Rather, the *enhancement-mode n*-channel MOSFETs of immediate interest here use positive values of the gate voltage.

Consider then the case where V_G is positive, as illustrated in Fig. 1.3. The magnitude of V_G is initially assumed to be "small"; the exact meaning of this will become clear as the discussion progresses. The oxide electric field $\mathscr{E}_{ox}$ for this gate voltage is seen to point from the gate toward the substrate. The electric field $\mathscr{E}(x)$, which penetrates into the semiconductor, is the basic quantity used to control the carrier densities. Note that the x-coordinate system is defined with $x = 0$ at the semiconductor surface; x increases with increasing depth into the substrate.

The MOS field effect results from the fact that the semiconductor electric field is "controlled" by the externally applied gate voltage V_G. The relationship between the two may be established in a straightforward manner. First, the electrostatic semiconductor potential $\phi(x)$ is related to the field by the usual gradient equation

$$\mathscr{E}(x) = -\frac{d\phi(x)}{dx}. \tag{1.1-4}$$

The potential $\phi(x)$ has the boundary condition that $\phi(x) \to V_B = 0$ as x increases toward the bulk electrode. Consequently, the total voltage across the semiconductor is

$$\phi_S = \phi(x = 0), \tag{1.1-5}$$

with ϕ_S being the *surface potential.* Applying Kirchhoff's voltage law (KVL) to the system then gives

$$V_G = V_{ox} + \phi_S, \tag{1.1-6}$$

where V_{ox} is the voltage drop across the oxide layer as shown in Fig. 1.3. This equation describes the field properties in terms of the system voltages. In essence, the value of V_G sets the value of the surface potential ϕ_S. The final connection between V_G and $\mathscr{E}(x)$ is obtained by introducing the *surface electric field*

$$\mathscr{E}_S = \mathscr{E}(x = 0) = -\left.\frac{d\phi}{dx}\right|_{x=0}. \tag{1.1-7}$$

This represents the largest value of the semiconductor field. Once ϕ_S is established by V_G as the $x = 0$ boundary condition on the potential, $\phi(x)$ and hence $d\phi(x)/dx$ may be computed by solving the Poisson equation. Thus, V_G may be viewed as directly controlling the value of the surface field $\mathscr{E}_s$.

The MOS field effect centers around controlling the silicon carrier densities with the semiconductor electric field $\mathscr{E}(x)$. Since the surface field $\mathscr{E}_s$ constitutes the maximum value of $\mathscr{E}(x)$, the effects will be most pronounced around $x = 0$. The *surface carrier concentrations*

$$p_S = p_p(x = 0), \qquad n_S = n_p(x = 0) \tag{1.1-8}$$

are the hole and electron densities that exhibit the greatest variations as V_G is changed.

For small positive values of V_G, the semiconductor electric field points toward the bulk electrode (i.e., in the $+x$ direction). The field repels the majority carrier holes away from the surface, which reduces the surface hole density to the point where $p_S \ll N_a$ is satisfied. This defines the *depletion* condition in the MOS system. The basic features of this operational mode are shown in Fig. 1.4. The depletion region is assumed to extend from $x = 0$ to $x = x_d$. Since this region is almost free of mobile carriers, the dominant

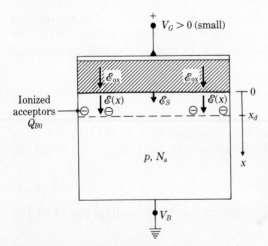

FIGURE 1.4 Depletion in the MOS system.

charges are the negatively charged acceptor ions embedded in the crystal lattice. These serve as the termination points for the electric field lines.

The depletion phenomenon in the MOS system is analogous to that found on the p-side of a one-sided n^+p step profile junction. The only major difference between the two is that the MOS depletion region has a voltage ϕ_S across it. The n^+p static junction results may be used to describe the MOS depletion region by simply replacing the built-in voltage ϕ_o by the surface potential ϕ_S. The total depletion width (or depth) in the system is then found to be

$$x_d = \sqrt{\frac{2\varepsilon_{Si}}{qN_a}}\,\phi_S \qquad (1.1\text{-}9)$$

where ε_{Si} is the silicon permittivity. The dielectric constant for silicon is approximately 11.8, so $\varepsilon_{Si} = (11.8)\varepsilon_o$.

The *bulk depletion charge* per unit area is constructed by writing

$$Q_{B0} = -qN_a x_d \;[C/cm^2], \qquad (1.1\text{-}10)$$

where

$$Q_{B0} = Q_B|_{V_B=0} \qquad (1.1\text{-}11)$$

specifies the $V_B = 0$ value of the bulk charge density. Using eqn. (1.1-9) for x_d, eqn. (1.1-10) may be rewritten in the form

$$Q_{B0} = -\sqrt{2q\varepsilon_{Si}N_a\phi_S}. \qquad (1.1\text{-}12)$$

When the MOS system is biased into the depletion mode of operation, the bulk charge constitutes the total *surface charge density* around $x = 0$:

$$Q_S = Q_{B0} < 0. \qquad (1.1\text{-}13)$$

Since the system may be viewed as a basic MOS capacitor, the oxide voltage V_{ox} across the oxide can be expressed as

$$V_{ox} = -\frac{Q_S}{C_{ox}}, \qquad (1.1\text{-}14)$$

where the negative sign is required because the surface charge Q_S resides on the semiconductor surface (not on the positively charged gate). This equation assumes that there are no stray charges trapped in the oxide and is valid only for an ideal MOS capacitor.

Now suppose that the gate voltage V_G is increased. Equation (1.1-6) implies that ϕ_S must also increase, which in turn drives the depletion edge x_d deeper, as seen from eqn. (1.1-9). This is justified on physical grounds by noting that additional bulk charge is needed to support the larger electric field strengths. When V_G reaches a critical value, called the *threshold voltage* V_{T0}, the MOS system changes operational regions from depletion to inversion. V_{T0} denotes the threshold voltage when $V_B = 0$.

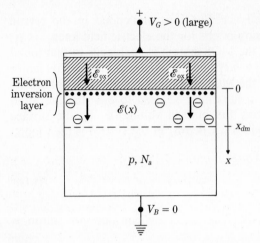

FIGURE 1.5 Surface inversion in the MOS system.

The basic features of the inversion phenomenon are shown in Fig. 1.5. The semiconductor has a depletion region that extends from $x = 0$ to $x = x_{dm}$. In addition, a layer of minority carriers has accumulated at $x = 0$, giving an *inversion layer* of free electrons. The creation of the inversion layer occurs because the semiconductor electric field is oriented such that the electrons are drawn from the substrate to the surface. Once the surface is inverted, increasing V_G enhances the electron density in the inversion layer. The maximum depletion depth x_{dm} remains constant, since the inversion layer electrons provide the charge needed to support the surface field. Figure 1.5 shows that the inversion layer "shields" the bulk charge from the electric field.

Surface inversion occurs when the surface potential reaches a particular value, denoted by ϕ_{si}. This is found to depend on the level of the substrate doping N_a. It is convenient to introduce the *bulk Fermi potential*

$$|\phi_F| = \left(\frac{kT}{q}\right) \ln\left(\frac{N_a}{n_i}\right) \tag{1.1-15}$$

as a reference voltage for the *p*-type substrate. The factor (kT/q) is the thermal voltage, approximately 0.026 [V] at room temperature. Note that only the magnitude $|\phi_F|$ will be used here. The condition for creating an inversion layer is that

$$\phi_{si} \simeq 2|\phi_F|. \tag{1.1-16}$$

This is somewhat arbitrary, but it can be justified from a detailed analysis of the system. The maximum depletion width x_{dm} is then computed by

$$x_{dm} = \sqrt{\frac{2\varepsilon_{Si}}{qN_a}(2|\phi_F|)}, \tag{1.1-17}$$

which corresponds to a maximum bulk depletion charge density of

$$Q_{B0} = -\sqrt{2q\varepsilon_{Si}N_a(2\,|\phi_F|)}.$$ (1.1-18)

The total surface charge density Q_S must now include the presence of the electron inversion layer charge Q_I [C/cm^2]. Thus,

$$Q_S = Q_{B0} + Q_I$$ (1.1-19)

accounts for both contributions. Note that Q_S is a function of V_G since $Q_I = Q_I(V_G)$.

Consider now the threshold voltage V_{T0}, which is the gate voltage needed to initiate the surface inversion. At the onset of inversion, $Q_I \ll Q_{B0}$, so $Q_S \simeq Q_{B0}$. If the oxide capacitor were ideal, then V_{ox} would be described by eqn. (1.1-14). Combining this with the KVL equation (1.1-6) and Q_{B0} in eqn. (1.1-18) gives the *ideal threshold voltage*

$$V_{T0}^{ideal} = \frac{\sqrt{2q\varepsilon_{Si}N_a(2\,|\phi_F|)}}{C_{ox}} + 2\,|\phi_F|.$$ (1.1-20)

This correlates with physical reasoning: the first term represents the voltage required to support the bulk charge, while the second is the condition for strong inversion. In a realistic MOS system, this expression must be modified to account for deviations from the ideal behavior. Additional terms are found in the threshold voltage expression such that

$$V_{T0} = V_{FB} + \frac{\sqrt{2q\varepsilon_{Si}N_a(2\,|\phi_F|)}}{C_{ox}} + 2\,|\phi_F|,$$ (1.1-21)

where V_{FB} is called the *flatband voltage* and arises from an analysis of the energy band diagram of a realistic system having oxide charges. Explicitly, it assumes the form

$$V_{FB} = \Phi_{GS} - \frac{1}{C_{ox}}(Q_{ox} + Q_{ss}).$$ (1.1-22)

In this equation,

$$\Phi_{GS} = \Phi_G - \Phi_S$$ (1.1-23)

represents the difference in work functions Φ between the gate and the substrate materials. Q_{ox} and Q_{ss} respectively give the oxide and surface state charge densities in [C/cm^2]; these terms must be included to account for the stray charges in the oxide. In general, both Q_{ox} and Q_{ss} are positive.

The enhancement-mode n-channel MOSFETs used to construct digital circuits require a positive threshold voltage for proper switching. Since Q_{ox} and Q_{ss} are both positive, a problem arises in that V_{FB} may be a negative number. (The magnitude and sign of Φ_{GS} depend on the gate and substrate materials.) This in turn may give a negative value for V_{T0}. To ensure that $V_{T0} > 0$, an additional processing step is included in the wafer fabrication process. This takes the form of an acceptor ion implantation into the substrate and is termed the *threshold adjustment* implant. The implant is characterized by an ion *dose* D_I [ions/cm^2]. The final working value of the threshold voltage is then given by

$$V_{T0} = V_{FB} + \frac{\sqrt{2q\varepsilon_{Si}N_a(2|\phi_F|)}}{C_{ox}} + 2|\phi_F| + \frac{qD_I}{C_{ox}}, \tag{1.1-24}$$

where V_{T0} can be "adjusted" to the desired value by varying the implant dose.

As stated above, Q_I is a function of V_G. To find this dependence, suppose that $V_G > V_{T0}$ is applied to the system. The effective voltage across the MOS capacitor is $(V_G - V_{T0})$, since V_{T0} is needed to start the inversion layer formation. Thus,

$$Q_I = -C_{ox}(V_G - V_{T0}) \tag{1.1-25}$$

gives the electron charge density in the inversion layer.

The discussion in this section has assumed that the bulk electrode was set at $V_B = 0$. When MOSFETs are used in circuits, it is very common to find nonzero values of the bulk voltage. The effects of this bulk, or *body bias*, can be accounted for by modifying the threshold voltage expression. To analyze this situation, consider the drawing in Fig. 1.6, where a nonzero V_B is applied; the MOS system is assumed to be in a state of inversion. A useful viewpoint to introduce here is that the inversion layer of electrons acts like an *n*-type material, so that a *pn* junction is formed with the *p*-type substrate. The negative terminal of V_B is applied to the substrate, which then reverse-biases the *pn* junction. The reverse bias changes the bulk depletion charge to a value

$$Q_B = -\sqrt{2q\varepsilon_{Si}N_a(2|\phi_F| + V_B)}. \tag{1.1-26}$$

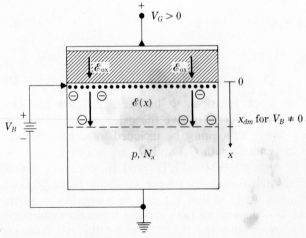

FIGURE 1.6   Increase in depletion charge from body bias V_B.

The change in the threshold voltage is computed by means of

$$\Delta V_T = V_T - V_{T0}$$

$$= \frac{\sqrt{2q\varepsilon_{Si}N_a}}{C_{ox}}(\sqrt{2|\phi_F| + V_B} - \sqrt{2|\phi_F|}), \tag{1.1-27}$$

where V_T is the threshold voltage with the applied body bias. The general expression for V_T then assumes the form

$$V_T = V_{T0} + \gamma(\sqrt{2|\phi_F| + V_B} - \sqrt{2|\phi_F|}), \tag{1.1-28}$$

where

$$\gamma = \frac{\sqrt{2q\varepsilon_{Si}N_a}}{C_{ox}} [V^{1/2}] \tag{1.1-29}$$

is the *body bias coefficient*. The inversion charge density is now given by

$$Q_I = -C_{ox}(V_G - V_T), \tag{1.1-30}$$

which is valid for arbitrary bias arrangements. This important expression will form the basis for the MOSFET analysis.

EXAMPLE 1.1-1

Consider a silicon MOS system with the following parameters:

$$x_{ox} = 500\,[\text{Å}], \qquad \Phi_{GS} = -0.85\,[\text{V}], \qquad D_I = 5.26 \times 10^{11}\,[\text{cm}^{-2}],$$

$$N_a = 10^{15}\,[\text{cm}^{-3}], \qquad Q_{ss} = q(1.5 \times 10^{11})\,[\text{C/cm}^2] \gg Q_{ox}, \qquad T = 27\,[\text{°C}].$$

Compute V_{T0}.

Solution

From eqn. (1.1-24),

$$V_{T0} = 2|\phi_F| + \frac{1}{C_{ox}}\sqrt{2q\varepsilon_{Si}N_a(2|\phi_F|)} + \Phi_{GS} - \frac{(Q_{ss} + Q_{ox})}{C_{ox}} + \frac{qD_I}{C_{ox}}.$$

Computing each term gives

$$2|\phi_F| = 2\left(\frac{kT}{q}\right)\ln\left(\frac{N_a}{n_i}\right) \simeq 2(0.026)\ln\left(\frac{10^{15}}{1.45 \times 10^{10}}\right) \simeq 0.579\,[\text{V}],$$

$$C_{ox} = \frac{\varepsilon_{ox}}{x_{ox}} = \frac{(3.9)(8.854 \times 10^{-14})}{0.05 \times 10^{-4}} \simeq 6.91 \times 10^{-8}\,[\text{F/cm}^2],$$

$$\frac{1}{C_{ox}}\sqrt{2q\varepsilon_{Si}N_a(2|\phi_F|)} \simeq \frac{[2(1.6 \times 10^{-19})(11.8)(8.854 \times 10^{-14})(10^{15})(0.579)]^{1/2}}{6.91 \times 10^{-8}}$$

$$\simeq 0.201\,[\text{V}],$$

$$\frac{(Q_{ss} + Q_{ox})}{C_{ox}} \simeq \frac{Q_{ss}}{C_{ox}} \simeq \frac{(1.6 \times 10^{-19})(1.5 \times 10^{11})}{6.91 \times 10^{-8}} \simeq 0.348 \,[\text{V}],$$

$$\frac{qD_I}{C_{ox}} \simeq \frac{(1.6 \times 10^{-19})(5.26 \times 10^{11})}{6.91 \times 10^{-8}} \simeq 1.218 \,[\text{V}].$$

Adding gives

$$V_{T0} \simeq 0.579 + 0.201 - 0.85 - 0.348 + 1.218,$$

so $V_{T0} \simeq +0.80 \,[\text{V}]$. ∎

EXAMPLE 1.1-2

Compute γ for the MOS system described in Example 1.1-1. Then find V_T for body bias voltages of $V_B = 1 \,[\text{V}]$ and $V_B = 3 \,[\text{V}]$.

Solution
Equation (1.1-29) gives

$$\gamma = \frac{\sqrt{2q\varepsilon_{Si}N_a}}{C_{ox}} \simeq \frac{[2(1.6 \times 10^{-19})(11.8)(8.854 \times 10^{-14})(10^{15})]^{1/2}}{6.91 \times 10^{-8}}$$

$$\simeq 0.265 \,[\text{V}^{1/2}].$$

To compute V_T, use eqn. (1.1-28):

$$V_T = V_{T0} + \gamma(\sqrt{2|\phi_F| + V_B} - \sqrt{2|\phi_F|}).$$

With $V_B = 1 \,[\text{V}]$,

$$V_T \simeq 0.80 + 0.265(\sqrt{0.579 + 1.0} - \sqrt{0.579}) \simeq 0.931 \,[\text{V}].$$

With $V_B = 3 \,[\text{V}]$,

$$V_T \simeq 0.80 + 0.265(\sqrt{0.579 + 3} - \sqrt{0.579}) \simeq 1.100 \,[\text{V}].$$ ∎

1.2 MOSFET Structure and Operation

The structure of a typical n-channel MOSFET is shown in Fig. 1.7(a). The central portion of the device consists of a poly gate MOS arrangement that has a gate oxide with thickness x_{ox}. The 4-terminal MOSFET is formed by adding n^+ *source* and *drain* regions to the p-type substrate. Contacting these n^+ wells with an interconnect layer (e.g., polysilicon or metal) gives the source and drain electrodes shown in the drawing. The symmetry of the structure does not allow one to distinguish between the source and drain in a zero-bias state; the roles of the two n^+ regions are defined only after the terminal voltages are applied.

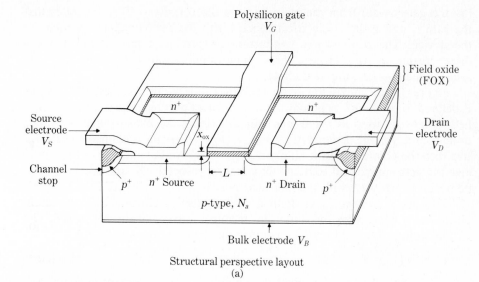

Structural perspective layout
(a)

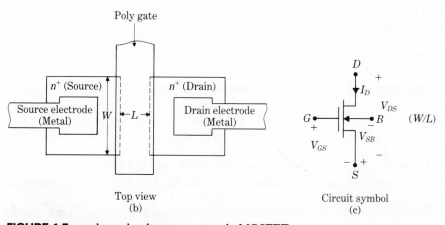

Top view
(b)

Circuit symbol
(c)

FIGURE 1.7 *n*-channel enhancement-mode MOSFET.

L is called the *channel length* of the MOSFET and is the distance between the n^+ source and drain edges. Figure 1.7(b) provides a top view of the transistor and shows both L and the *channel width* W, which is the extent of the device in the lateral direction perpendicular to L. The *aspect ratio* (W/L) is a geometrical factor used to describe the relative dimensions of a MOSFET. As will be seen later, (W/L) may be used as a design parameter that can be varied to set the desired drain-source conduction properties of the transistor.

The circuit symbol for an *n*-channel enhancement-mode (E-mode) MOSFET is shown in Fig. 1.7(c); an E-mode MOSFET is defined to have a positive threshold voltage $V_{T0} > 0$. The circuit symbol reflects the basic structural features of the device. In particular, the gate is physically isolated (by the gate oxide) from the main body of the transistor. This illustrates that the gate is not in direct electrical contact with the semiconductor portions of the device. The drain-source conduction properties of the MOSFET are a direct result of the electric field penetrating from the oxide into the semiconductor—i.e., the field effect.

Figure 1.7(c) shows the basic MOSFET operating voltages. The source is defined to be the n^+ region at the lower potential. Once the source is identified, it may be used as a reference for describing the bias arrangement. The drain-source voltage is denoted by $V_{DS} = (V_D - V_S)$ and is defined to be positive with the polarity shown. Similarly, $V_{GS} = (V_G - V_S)$ and $V_{SB} = (V_S - V_B)$ give the gate-source and source-bulk voltages, respectively, and are defined to be positive quantities with the indicated polarities.

The *drain current* I_D is defined to be positive flowing into the drain electrode; it constitutes the important terminal current of the MOSFET. The current-voltage $(I - V)$ relation assumes the general functional form

$$I_D = I_D(V_{GS}; V_{DS}; V_{SB}), \tag{1.2-1}$$

indicating that all of the (relative) device voltages are important in controlling the drain current. The remaining sections in this chapter examine the device physics involved in describing the operational characteristics of the MOSFET. This will lead to explicit analytical relationships between the device voltages and the drain current. These equations may then be used in the analysis and design of MOSFET-based integrated circuits.

The simplest bias arrangement that can be used to illustrate the operation of a MOSFET is one in which both the source and bulk electrodes are grounded: $V_S = 0 = V_B = V_{SB}$. The gate-source voltage V_{GS} is used to control the creation of an electron inversion layer underneath the gate oxide; this constitutes the *channel* in a MOSFET. When a channel exists in the transistor, a conduction path is established between the drain and the source, allowing drain current I_D to flow. The value of I_D is set by V_{GS} and V_{DS} in addition to device-specific parameters such as V_{T0} and (W/L).

Consider first the case where the drain is grounded, so that $V_{DS} = 0$. When $V_{GS} < V_{T0}$, the p-type bulk semiconductor is depleted, as shown in Fig. 1.8(a). The depletion actually originates from two distinct phenomena. The surface underneath the gate oxide depletes because of the electric field penetration from the MOS field effect. On the other hand, the depletion around the n^+ source and drain regions arises from the pn^+ junctions formed with the p-type bulk. This bias gives $I_D = 0$, as may be seen by noting the absence of both a conducting path and a driving voltage.

If the gate-source voltage is increased to a value $V_{GS} > V_{T0}$, an inversion layer is established, as shown in Fig. 1.8(b). The presence of the conducting channel electrically connects the drain and source regions. However, since there is no voltage difference between the two n^+ regions ($V_{DS} = 0$), the current remains at a value $I_D = 0$.

A more common situation is one in which a nonzero drain-source voltage

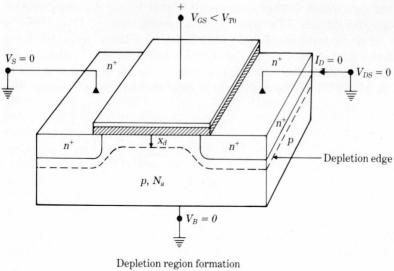

Depletion region formation
(a)

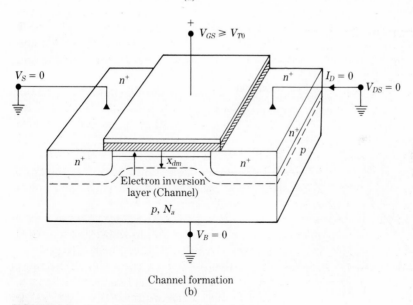

Channel formation
(b)

FIGURE 1.8 Basic MOSFET channel formation.

is established on the MOSFET. This is shown in Fig. 1.9(a) for the case where $V_{GS} < V_{T0}$. Since the value of V_{GS} is not sufficient to create an inversion layer, no channel exists. If leakage currents are ignored, then $I_D \simeq 0$ even though $V_{DS} \neq 0$. This describes the *cutoff* mode of MOSFET operation. Cutoff is crucially important for digital switching networks since it allows the drain current to be "switched off" by setting $V_{GS} < V_{T0}$.

When the gate-source voltage is increased so that $V_{GS} > V_{T0}$, drain current flows through the device. This arises from a combination of (1) creating a channel by inverting the surface and (2) establishing a *lateral* (parallel to the surface) electric field in the channel through the difference in n^+ potentials set by V_{DS}. When I_D flows, the MOSFET is said to be in the *active* mode of operation. A MOSFET in the active mode exhibits two distinct current flow

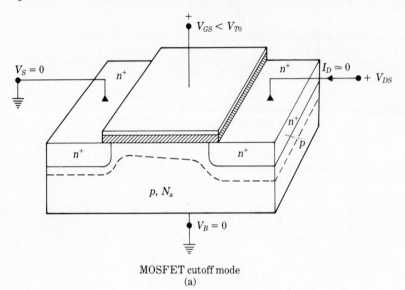

MOSFET cutoff mode
(a)

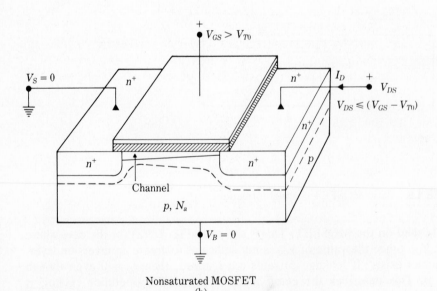

Nonsaturated MOSFET
(b)

FIGURE 1.9 Basic operational regions of a MOSFET.

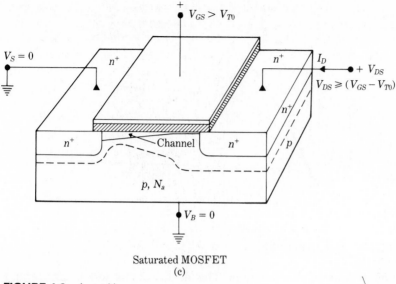

Saturated MOSFET
(c)

FIGURE 1.9 (*contd.*)

characteristics, depending on the relative value of V_{DS} with respect to $(V_{GS} - V_{T0})$.

For "small" values of V_{DS} such that $V_{DS} < (V_{GS} - V_{T0})$ is approximately satisfied, the channel geometry assumes a form similar to that shown in Fig. 1.9(b). This describes a MOSFET in the *nonsaturated* mode of active operation. Note that the small drain-source voltage gives a slight "tilt" to the bottom edge of the channel. This results from the presence of the *channel voltage V* that exists in the inversion layer due to V_{DS}. The channel voltage assumes a maximum value of $V = V_{DS}$ at the drain n^+ region and goes to $V_S = 0$ at the source. Its polarity is such that it opposes the applied gate-source voltage used to set up the inversion layer. As such, the channel is modified as it is traced from source to drain.

If the transistor bias is set so that $V_{DS} > (V_{GS} - V_{T0})$, the MOSFET is said to be *saturated*. A typical channel geometry for a saturated MOSFET is shown in Fig. 1.9(c). Saturation is characterized by the fact that the channel undergoes a type of "pinch-off" in which the thickness of the channel decreases to a minimum value before reaching the drain n^+ region. This minimum channel thickness is extremely small compared with the overall device dimensions. Consequently, the channel is usually modeled as undergoing a complete pinch-off in which the thickness of the channel is essentially zero beyond the pinch-off point.

The basic MOSFET $I - V$ characteristics are shown in graphical form in Fig. 1.10. Generally, the plot of I_D vs. V_{DS} is the most useful in understanding digital circuits. Varying V_{GS} gives the family of curves illustrated in the plot. The curves show the important features of both $I_D(V_{DS})$ and $I_D(V_{GS})$ in

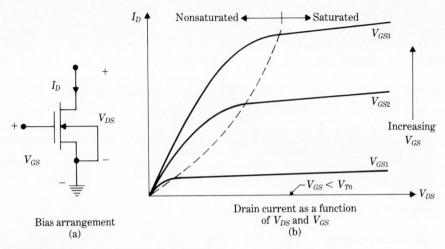

Bias arrangement
(a)

Drain current as a function
of V_{DS} and V_{GS}
(b)

FIGURE 1.10   MOSFET characteristics.

saturated and nonsaturated operation. The next sections will be concerned with the derivation of these curves.

1.3 The Gradual Channel Approximation

To deduce the analytical expressions for I_D as a function of the MOSFET terminal voltages, it is necessary to introduce approximations for both the channel and electric field characteristics. This requirement arises because the complexity of the realistic three-dimensional problem prohibits the derivation of closed-form equations. The easiest approach to studying MOSFET pro-perties is to employ the *gradual channel approximation* (GCA) in which the system is reduced to a one-dimensional current flow problem. Although this may sound like a drastic simplification, the GCA is capable of yielding results that agree fairly well with experimental measurements. These equations are restricted to describing what are generally termed "large" devices and cannot be directly applied to the "small" MOSFETs required for VLSI-level chip structures without modifications. However, the equations derived in this chapter may be used as a basis for the initial design and analysis of MOS circuits while still maintaining a reasonable degree of accuracy.

The basic geometry used to describe a nonsaturated MOSFET within the limits of the gradual channel approximation is illustrated in Fig. 1.11. The bias voltages have been set with $V_S = 0 = V_B$, so that V_{GS} and V_{DS} constitute the important controlling parameters. This will be extended to arbitrary bias arrangements later. Note that the x-coordinate system has been defined to be positive into the substrate, which is consistent with the treatment in the previous sections. The y-coordinate system is defined to be parallel to the surface of the semiconductor such that the channel extends from $y = 0$ to $y = L$.

The mechanism of drain-source current flow is easily understood by noting that $V_{GS} > V_{T0}$ creates an electron inversion layer that acts as the MOSFET channel. The channel provides an electrical conduction path between the source and the drain. The *channel electric field* $\mathscr{E}_y(y)$ is established by the drain-source voltage V_{DS}. The field is related to the channel voltage $V(y)$ by means of

$$\mathscr{E}_y(y) = -\frac{dV(y)}{dy} \tag{1.3-1}$$

such that the channel voltage is constrained to satisfy the boundary conditions

$$V(y = 0) = V_S = 0, \qquad V(y = L) = V_{DS}. \tag{1.3-2}$$

Note that these imply $\mathscr{E}_y < 0$, indicating that the channel field points in the $-y$ direction. The GCA assumes that this is the dominant electric field component in the channel and therefore ignores the vertical (x) electric field used to invert the surface. With this modeling in mind, it is seen that I_D is created by the electrons that originate at the source (hence the name source) and travel toward the drain by field-aided transport through the channel.

The shape of the depletion edge shown in Fig. 1.11 deserves some explanation. The gradual channel approximation assumes that the depletion charge is supported entirely by the vertical electric field $\mathscr{E}_x(y)$. The depletion depth is greater at the drain side of the channel because of the channel voltage $V(y)$. Recall from Section 1.1 that the inversion layer may be viewed as forming a pn^+ junction with the p-type substrate. $V(y)$ acts as a reverse-bias

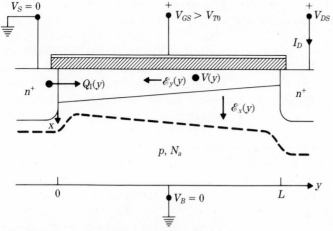

FIGURE 1.11 MOSFET geometry used in the gradual channel approximation.

voltage across this junction such that the maximum depletion depth becomes a function of y in the form

$$x_{dm}(y) \simeq \sqrt{\frac{2\varepsilon_{Si}}{qN_a}[2|\phi_F| + V(y)]}. \tag{1.3-3}$$

Since $V(y)$ has the boundary conditions stated in eqn. (1.3-2), x_{dm} varies along the channel, with its maximum value at the drain.

Consider now the electron inversion charge density $Q_I(y)$ [C/cm²] in the channel. At the source end of the device, the channel voltage is zero. Consequently, the MOS analysis of Section 1.1 may be used to write

$$Q_I(y = 0) = -C_{ox}[V_{GS} - V_{T0}]. \tag{1.3-4}$$

For arbitrary values of y, this must be modified to account for the presence of the channel voltage. Since $V(y) > 0$, it acts against V_{GS} in inverting the surface. Thus,

$$Q_I(y) = -C_{ox}[V_{GS} - V_{T0} - V(y)] \tag{1.3-5}$$

describes the inversion charge as a function of position. This also explains the differences in the channel thickness portrayed by the figure.

With the basic channel and field characteristics established, it is now possible to analyze the current flow. A simple viewpoint of this problem is illustrated in Fig. 1.12. Using Kirchhoff's current law (KCL), we see that the total channel current is simply the terminal current I_D flowing into the device.

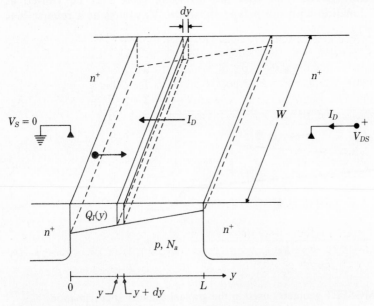

FIGURE 1.12 Geometry for GCA current analysis.

The resistance presented to the current by the differential channel increment with length dy is

$$dR = -\frac{dy}{\mu_n W Q_I(y)} \, [\Omega], \tag{1.3-6}$$

as may be verified by noting that the product of the channel conductivity σ and the cross-sectional area A of the channel is given by

$$\sigma A = -\mu_n W Q_I(y). \tag{1.3-7}$$

(The minus sign appears in these equations to compensate for the negative value of Q_I.) μ_n is the electron *surface mobility* and must be kept distinct from the *bulk mobility* found deep inside the semiconductor. The surface mobility describes the electron motion along the surface of the silicon and is smaller than the bulk mobility because of surface scattering and other processes that inhibit the electron motion. The surface mobility is a function of the substrate doping N_a; a reasonable approximation is to take the surface mobility to be about one-half the bulk value.

The differential voltage dV that appears across dy is approximated by Ohm's law in the form

$$dV = I_D \, dR = -\frac{I_D \, dy}{\mu_n W Q_I(y)}. \tag{1.3-8}$$

Rearranging and integrating along the channel then gives

$$I_D \int_0^L dy = -\mu_n W \int_0^{V_{DS}} Q_I(V) \, dV. \tag{1.3-9}$$

Note that the right-hand side of the expression has been cast into a voltage integral, so that Q_I has been written explicitly as a function of the channel voltage V. The limits of integration correspond to the values of V at $y = 0$ and $y = L$, respectively. I_D has been assumed constant in the channel, so it may be factored out of the y integration. Using $Q_I(V)$ from eqn. (1.3-5) and integrating the left side of eqn. (1.3-9) then gives the basic GCA current equation

$$I_D = (\mu_n C_{\text{ox}})\left(\frac{W}{L}\right)\int_0^{V_{DS}} (V_{GS} - V_{TO} - V) \, dV. \tag{1.3-10}$$

This will serve as the starting point for the current-voltage calculations in the next two sections.

As a final point in the discussion, it should be noted that the gradual channel approximation deals only with the drift currents in the channel. To qualify this comment, combine eqns. (1.3-1) and (1.3-8) to write

$$I_D = \mu_n W Q_I \mathscr{E}_y, \tag{1.3-11}$$

which is equivalent to the basic form used to describe drift currents. If diffusion currents were included in the analysis, the generalized drain current expression would read

$$I_D = W\left(\mu_n Q_I \mathscr{E}_y + D_n \frac{dQ_I}{dy}\right), \tag{1.3-12}$$

where $D_n = (kT/q)\mu_n$ is the surface diffusion coefficient for electrons. The GCA current flow analysis assumes that $Q_I(y)$ is a slowly varying function of y, so that (dQ_I/dy) is negligible. This corresponds to a "gradual" change in the channel properties and provides the name for this approximation technique. In a very small MOSFET, the channel lengths may be short enough to require that diffusion currents be added to the analysis.

1.4 MOSFET Circuit Equations

The gradual channel approximation formulated in the previous section is capable of providing different levels of MOSFET modeling. The simplest current-voltage relationships that can be obtained from the GCA are extremely useful for performing basic circuit analysis. They may also be used to derive *design equations*, which are approximate relations between the desired circuit performance specifications and the design parameters. This level of modeling yields what will be termed the MOSFET "circuit equations" since they permit circuits to be analyzed in a straightforward manner. This section is devoted to the derivation and study of the simplified $I - V$ equation set.

The circuit equations for the MOSFET are ideal for "hand calculations" (as opposed to computer simulations) because of their structural simplicity. However, they ignore some important information contained within the full GCA model. Section 1.5 will deal with the complete MOSFET $I - V$ equations as developed from the gradual channel approximation. These are much more complicated than those found in the present discussion. The two approaches will then be correlated to conform to the level of device modeling required for specific situations.

The starting point for the MOSFET circuit equation derivation is the general integral

$$I_D = (\mu_n C_{\text{ox}})\left(\frac{W}{L}\right)\int_0^{V_{DS}} (V_{GS} - V_{T0} - V)\, dV, \tag{1.4-1}$$

which is valid for a nonsaturated device. To effect the integration, it is assumed that V_{T0} is a constant along the channel. This is clearly an approximation, since the value of V_{T0} depends on the bulk depletion charge density Q_{B0} and the channel voltage $V(y)$. However, proceeding with this assumption allows the integral to be easily evaluated. Then

$$I_D = k'\left(\frac{W}{L}\right)\left[(V_{GS} - V_{T0})V_{DS} - \frac{1}{2}V_{DS}^2\right], \tag{1.4-2}$$

where the *process transconductance parameter k'* is defined by

$$k' = \mu_n C_{ox} \ [A/V^2] \tag{1.4-3}$$

and accounts for the basic fabrication-dependent factors in the device equations. An important point implied by this analysis is that V_{T0} is interpreted as the threshold voltage required to invert the source end of the device where $V = 0$.

It is convenient to rewrite the nonsaturated current in the form

$$I_D = \frac{\beta}{2} [2(V_{GS} - V_{T0})V_{DS} - V_{DS}^2], \tag{1.4-4}$$

where β is termed the *device transconductance parameter* such that

$$\beta = k'\left(\frac{W}{L}\right) \ [A/V^2]. \tag{1.4-5}$$

β is introduced to allow for the description of specific device geometries. This equation provides the basic nonsaturated functional dependence of $I_D(V_{GS}; V_{DS})$.

The simplest way to understand the MOSFET characteristics predicted by this equation is to hold V_{GS} constant and then plot I_D as a function of V_{DS}. The family of curves generated by different choices for V_{GS} is shown in Fig. 1.13. By inspection, I_D is seen to be a parabolic function of V_{DS}. It is also noted that I_D is directly proportional to V_{GS}, which sometimes leads to the nonsaturated operation being termed the *linear* region. Since the treatment here is concerned with I_D as a function of V_{DS}, this terminology will not be adopted.

The curves shown in Fig. 1.13 have been plotted from the origin to the peak current values. If the curves were continued beyond their respective maximas, then the behavior shown by the dashed lines would result. The dashed-line characteristics are not observed in practice. This might be

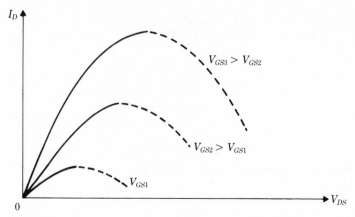

FIGURE 1.13 I_D vs. V_{DS} as predicted by eqn. (1.4–4) (nonsaturated region).

suspected on physical grounds by noting that these would lead to negative resistances for the drain-source current flow. Instead, measurements on MOSFETs have shown that the current tends to reach a peak value and then remain almost constant with increasing V_{DS}.

The peak points in I_D represent the onset of the *saturation* mode of MOSFET operation. The value of the *saturation voltage* $V_{DS,\text{sat}}$, which defines the starting point of the saturated mode, may be computed by finding the maximum current points using

$$\frac{\partial I_D}{\partial V_{DS}} = 0 = \beta(V_{GS} - V_{T0} - V_{DS}). \tag{1.4-6}$$

Thus,

$$V_{DS,\text{sat}} = V_{GS} - V_{T0} \tag{1.4-7}$$

is the desired expression. Note that the value of $V_{DS,\text{sat}}$ is a function of the applied gate-source voltage. The drain current for a saturated MOSFET is predicted to be

$$I_{D,\text{sat}} = I_D(V_{DS} = V_{DS,\text{sat}}) = \frac{\beta}{2}(V_{GS} - V_{T0})^2. \tag{1.4-8}$$

This describes a parabolic border between the saturated and non-saturated regions of operation on the $I_D - V_{DS}$ curve. The basic MOSFET characteristics found up to this point are illustrated in Fig. 1.14. The saturated current has been extrapolated to be a constant for $V_{DS} > V_{DS,\text{sat}} = (V_{GS} - V_{T0})$.

To understand the phenomenon of saturation, note that the inversion charge density at the drain end of the channel is given by

$$Q_I(L) = -C_{\text{ox}}[V_{GS} - V_{T0} - V_{DS}]. \tag{1.4-9}$$

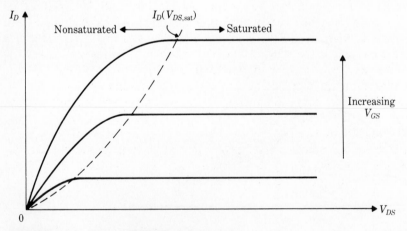

FIGURE 1.14 Basic MOSFET characteristics.

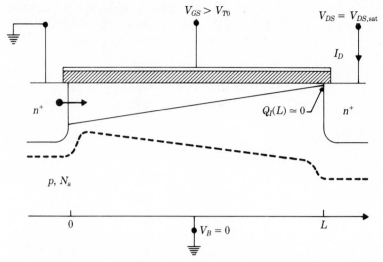

FIGURE 1.15 Start of saturation in a MOSFET.

When $V_{DS} = V_{DS,\text{sat}}$, the gradual channel approximation predicts that

$$Q_I(L) = 0. \tag{1.4-10}$$

This, of course, is incorrect; it arises from the approximations contained within the depletion analysis for the MOS system. A more correct statement would be that

$$Q_I(L) \ll Q_I(0), \tag{1.4-11}$$

which indicates that the inversion charge density at the drain side of the channel is reduced by the presence of the channel voltage $V(y)$. A MOSFET at the onset of saturation can be visualized as in Fig. 1.15. The channel at $y = L$ has been "pinched" to a minimum thickness, which then limits the electron flow through the region. The basic GCA assumes that the current flow through the channel at the start of saturation represents the maximum value of I_D that the device can support for a given V_{GS}. In particular, as the drain-source voltage is increased beyond $V_{DS,\text{sat}}$, I_D is taken to be a constant, as mentioned earlier in connection with Fig. 1.14. This approximation will be used extensively in the circuit analysis and design presentations later.

It is possible to extend this analysis beyond the basic characteristics discussed above to include the phenomenon of *channel-length modulation*, which occurs when $V_{DS} > V_{DS,\text{sat}}$. This is illustrated in Fig. 1.16, which shows that the effective length of the channel is reduced to $L' < L$ such that

$$L' = L - \Delta L, \tag{1.4-12}$$

where ΔL represents the portion of the channel that is effectively pinched off.

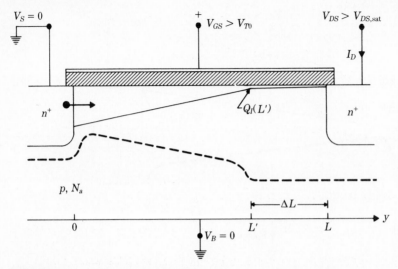

FIGURE 1.16 Channel-length modulation.

To model this effect analytically, note that the GCA now gives

$$Q_I(L') = 0 \tag{1.4-13}$$

to this order of approximation. This in turn implies that the channel voltage at L' is

$$V(L') \simeq V_{DS,\text{sat}} \tag{1.4-14}$$

since this is the condition required for saturation. The length ΔL may be approximated as a depletion region with a voltage $(V_{DS} - V_{DS,\text{sat}})$ across it. Using the one-sided step profile pn junction analysis gives the estimate

$$\Delta L \simeq \sqrt{\frac{2\varepsilon_{\text{Si}}}{qN_a}(V_{DS} - V_{DS,\text{sat}})}. \tag{1.4-15}$$

This is found to be a reasonable approximation so long as $x_{\text{ox}} \ll \Delta L$.

The main effect of channel-length modulation is to modify the saturated MOSFET current to

$$I_D \simeq \frac{k'}{2}\left(\frac{W}{L'}\right)(V_{GS} - V_{T0})^2, \tag{1.4-16}$$

i.e., L is replaced by L'. This may be rewritten in the form

$$I_D \simeq \frac{I_{D0}}{\left(1 - \dfrac{\Delta L}{L}\right)}, \tag{1.4-17}$$

where I_{D0} represents the saturated current in eqn. (1.4-8). Since ΔL is a function of $V_{DS} > V_{DS,\text{sat}}$, channel-length modulation gives a "tilt" to the

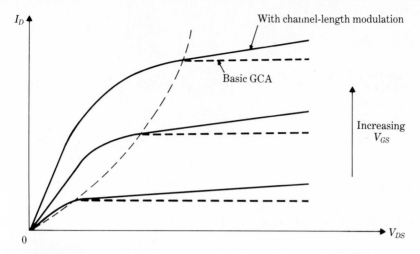

FIGURE 1.17 Channel-length modulation effects in the MOSFET characteristics.

current-voltage curves in the MOSFET saturation region. This is shown in Fig. 1.17.

Owing to the complicated form of the approximation for ΔL given in eqn. (1.4-15), this expression is much too cumbersome for use in hand calculations. Instead, the results are simplified by writing the empirical relation

$$1 - \frac{\Delta L}{L} \simeq 1 - \lambda V_{DS}, \tag{1.4-18}$$

where $\lambda \, [\text{V}^{-1}]$ is called the *channel-length modulation factor*. Then, performing a binomial expansion by assuming that $\lambda V_{DS} \ll 1$ gives

$$I_D \simeq \frac{\beta}{2}(V_{GS} - V_{T0})^2[1 + \lambda V_{DS}]. \tag{1.4-19}$$

This level of simplification predicts that the saturated drain current increases linearly with V_{DS}, which is distinctly different from the dependence found using the depletion analysis. However, it does provide the basic feature of a nonzero slope for the saturated drain current. For this reason, eqn. (1.4-19) is commonly used as a first-order approximation when channel-length modulation effects must be accounted for. Note that λ may be varied to obtain the closest fit to the actual characteristics of a given MOSFET. As might be suspected, channel-length modulation becomes important in MOSFETs with short channel lengths.

The basic MOSFET characteristics described above may be extended to the case of arbitrary bias arrangements. Consider the generalized situation shown in Fig. 1.18. Since the source voltage may still be used as a reference, only the relative voltages V_{GS}, V_{DS}, and V_{SB} are important in determining the current flow. The current equations must then be modified to account for the

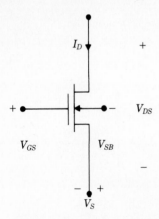

FIGURE 1.18   General MOSFET bias.

body bias applied to the MOSFET through V_{SB}. To this end, recall that V_{T0} is the $V_{SB} = 0$ value of the threshold voltage, which was interpreted to be the voltage needed to create the channel at the source. From Section 1.1, body bias may be included by writing

$$V_T = V_{T0} + \gamma(\sqrt{2|\phi_F| + V_{SB}} - \sqrt{2|\phi_F|}) \qquad \text{(1.4-20)}$$

since V_{SB} now takes the place of V_B in a MOSFET. The resulting equations are then modified by simply replacing V_{T0} by the more general value V_T. Thus, in cutoff,

$$I_D \simeq 0 \qquad (V_{GS} < V_T). \qquad \text{(1.4-21)}$$

The nonsaturated current is modified to read

$$I_D = \frac{\beta}{2}[2(V_{GS} - V_T)V_{DS} - V_{DS}^2] \qquad (V_{GS} > V_T, \; V_{DS} < V_{DS,\text{sat}}), \qquad \text{(1.4-22)}$$

where the saturation voltage is now

$$V_{DS,\text{sat}} = V_{GS} - V_T. \qquad \text{(1.4-23)}$$

Finally, the circuit-level approximation to the saturated current is given by

$$I_D = \frac{\beta}{2}(V_{GS} - V_T)^2(1 + \lambda V_{DS}) \qquad (V_{GS} > V_{T0}, \; V_{DS} \geq V_{DS,\text{sat}}). \qquad \text{(1.4-24)}$$

Digital nMOS circuit analysis generally requires inclusion of body bias effects, since significant errors can arise in the calculations if they are ignored. On the other hand, basic CMOS logic can be biased so that $V_T = V_{T0}$. To simplify the notation, V_T will be used to denote the threshold voltage even if $V_T = V_{T0}$.

The effects of body bias are easily understood by referring to the circuit arrangement shown in Fig. 1.19(a). Since the gate and drain electrodes are

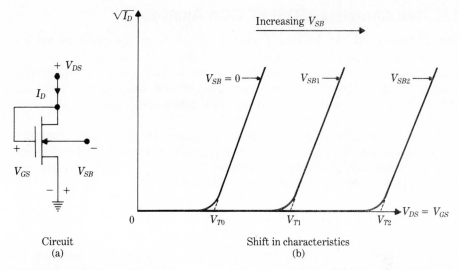

FIGURE 1.19 Body bias effects.

connected,

$$V_{DS} = V_{GS},\qquad(1.4\text{-}25)$$

which implies that

$$V_{DS} > (V_{GS} - V_T),\qquad(1.4\text{-}26)$$

i.e., the MOSFET is saturated. Assuming that $\lambda = 0$ for simplicity, the saturated drain current is then given by

$$I_D = \frac{\beta}{2}[V_{GS} - V_T(V_{SB})]^2\qquad(1.4\text{-}27)$$

where $V_T(V_{SB})$ is described by eqn. (1.4-20). This is plotted in Fig. 1.19(b) for various body bias levels. The "shift" in the I_D curve results from the fact that application of a positive V_{SB} increases the threshold voltage of the transistor. This then requires a larger gate-source voltage to invert the surface for the creation of a channel. It is noted in passing that this bias arrangement is useful for experimentally determining the value of MOSFET threshold voltages.

The remaining sections of the chapter are directed toward more detailed studies of MOSFET characteristics. In particular, the full modeling provided by the GCA is discussed in the next section. If the reader is eager to move quickly into circuit analysis and design, the rest of this chapter may be omitted in a first reading. In this track, the choice must be made whether to go to Chapter 2 for an introduction to the physics of small devices or to jump directly to Chapter 3, which initiates the circuit discussions. The presentation has been structured so that either choice is viable.

1.5   The Complete MOSFET GCA Analysis

The derivation of the MOSFET circuit equations in the previous section assumed that V_{T0} was a constant along the channel. As mentioned earlier, this is an approximation that ignores the increase in bulk depletion charge Q_{B0} created by the channel voltage $V(y)$. This section will center around the MOSFET analysis that results when this additional depletion charge is included.

To understand the change that must be made in the derivation, recall that $V(y)$ gives a reverse bias across the n^+p junction created at the channel-substrate boundary. Assuming for simplicity that $V_S = 0 = V_B$, the reverse-bias voltage may be included by writing (see eqn. 1.1-24)

$$V_{T0}(V) = V_{FB} + 2\,|\phi_F| + \frac{qD_I}{C_{ox}} + \frac{1}{C_{ox}}\sqrt{2q\varepsilon_{Si}N_a(2\,|\phi_F| + V)}, \qquad (1.5\text{-}1)$$

where a threshold-adjustment ion implantation term has been included for completeness. This explicitly shows the change in Q_{B0} through the additional dependence on V in the last term.

The basic GCA integral

$$I_D = \beta \int_0^{V_{DS}} [V_{GS} - V_{T0}(V) - V]\, dV \qquad (1.5\text{-}2)$$

is now modified to read

$$I_D = \beta \int_0^{V_{DS}} \left\{ \left[V_{GS} - V_{FB} - 2\,|\phi_F| - \frac{qD_I}{C_{ox}} \right] - V \right.$$
$$\left. - \frac{1}{C_{ox}}\sqrt{2q\varepsilon_{Si}N_a(2\,|\phi_F| + V)} \right\} dV. \qquad (1.5\text{-}3)$$

Although the integral may appear somewhat complicated, it is really quite easy to evaluate. The first group of terms is a constant to the integration, the second term in the integrand is simply V, and the last term merely requires integrating the square root of V. Consequently, the integration gives

$$I_D = \beta \left\{ \left(V_{GS} - V_{FB} - 2\,|\phi_F| - \frac{qD_I}{C_{ox}} \right) V_{DS} - \frac{1}{2} V_{DS}^2 \right.$$
$$\left. - \frac{2}{3C_{ox}}\sqrt{2q\varepsilon_{Si}N_a}\, [(2\,|\phi_F| + V_{DS})^{3/2} - (2\,|\phi_F|)^{3/2}] \right\} \qquad (1.5\text{-}4)$$

as the nonsaturated drain current.

The inclusion of the bulk charge variations in the channel has led to an expression for the nonsaturated current that is much more complicated than that given in eqn. (1.4-4) for the circuit equation set. The analysis still gives terms that are proportional to V_{DS} and V_{DS}^2, however. Thus, the basic shape of the curves predicted by this equation are similar to those shown in Fig. 1.13.

The actual numerical values will be different when applied to a given device, but this will be accounted for when the two theories are correlated.

The peak value of the nonsaturated current is again computed by means of

$$\frac{\partial I_D}{\partial V_{DS}} = 0. \tag{1.5-5}$$

Evaluating the derivatives then gives the saturation voltage of the MOSFET as

$$V_{DS,\text{sat}} = V_{GS} - V_{FB} - 2|\phi_F|$$
$$- \frac{qD_I}{C_{\text{ox}}} - \frac{q\varepsilon_{\text{Si}}N_a}{C_{\text{ox}}^2}\left[\sqrt{1 + \frac{2C_{\text{ox}}^2}{q\varepsilon_{\text{Si}}N_a}(V_{GS} - V_{FB})} - 1\right], \tag{1.5-6}$$

which is considerably more complicated than the expression found for the circuit equation treatment. Since this derivation is based on the depletion approximation, the inversion charge density still satisfies

$$Q_I(L) = -C_{\text{ox}}[V_{GS} - V_{T0}(V_{DS,\text{sat}}) - V_{DS,\text{sat}}] = 0, \tag{1.5-7}$$

as can be verified by direct substitution.

The saturated MOSFET current is estimated to lowest order as being a constant for $V_{DS} > V_{DS,\text{sat}}$ such that

$$I_{D,\text{sat}} = I_D(V_{DS} = V_{DS,\text{sat}}). \tag{1.5-8}$$

Substituting eqn. (1.5-6) into eqn. (1.5-4) thus gives the desired equation for $I_{D,\text{sat}}$. Although a closed-form expression can be obtained, it is quite messy and will not be reproduced here. The overall shape of the I_D vs. V_{DS} curve will be similar to that shown in Fig. 1.14. Channel-length modulation effects may be treated using the theory developed in the previous section, since it modifies only the channel length L.

The level of modeling obtained by including variations of the bulk depletion charge along the channel is much more accurate than that provided by the circuit equations. In particular, the equations typified by (1.5-4) yield numerical results that agree quite well with experimental measurements on large MOSFETs. The price paid for this increased precision is an equation set that is significantly more complicated to work with. This generally restricts the full MOSFET modeling to use in computer-based circuit simulation codes, as the hand calculations quickly become intractable.

The digital circuit discussions beginning in Chapter 3 are based on the simplified MOSFET circuit equations found in the previous section. Since the equations represent only a basic level of device modeling, they should be examined further to understand the accuracy of any subsequent circuit analysis. It is therefore useful to compare the two MOSFET equation sets. As will be seen below, the simplified circuit equations may be modified to the point where they are accurate enough for initial circuit calculations.

Consider first the case where a MOSFET is nonsaturated with a small drain-source voltage such that

$$V_{DS} \ll 2\,|\phi_F|. \tag{1.5-9}$$

The last term in eqn. (1.5-4) may be simplified by using the binomial expansion

$$(2\,|\phi_F| + V_{DS})^{3/2} = (2\,|\phi_F|)^{3/2} + \frac{3}{2}(2\,|\phi_F|)^{1/2}V_{DS} + \frac{3}{8}(2\,|\phi_F|)^{-1/2}V_{DS}^2 + \cdots. \tag{1.5-10}$$

Keeping only terms to first order in V_{DS} gives

$$I_D \simeq \beta(V_{GS} - V_{T0})V_{DS}, \tag{1.5-11}$$

where V_{T0} from eqn. (1.5-1) has been used with $V = 0$ (i.e., at the source end of the channel). If the nonsaturated circuit equation (1.4-4) is reduced for $V_{DS} \ll (V_{GS} - V_{T0})$, the same expression results. Consequently, the two models agree for small values of drain-source voltages.

Increasing V_{DS} requires that the second order terms be included. The current then assumes the form

$$I_D \simeq \frac{\beta}{2}\left[2(V_{GS} - V_{T0})V_{DS} - \left(1 + \frac{1}{2C_{ox}}\sqrt{\frac{q\varepsilon_{Si}N_a}{|\phi_F|}}\right)V_{DS}^2\right]. \tag{1.5-12}$$

Comparing this with the nonsaturated circuit equation,

$$I_D = \frac{\beta}{2}[2(V_{GS} - V_{T0})V_{DS} - V_{DS}^2], \tag{1.5-13}$$

shows that the simplified treatment will overestimate the current flow through the device. The difference between the two models may be understood on physical grounds by noting that the circuit equations assumed a constant bulk charge of

$$|Q_{B0}| = \sqrt{2q\varepsilon_{Si}N_a(2\,|\phi_F|)}, \tag{1.5-14}$$

while the more precise treatment used

$$|Q_{B0}| = \sqrt{2q\varepsilon_{Si}N_a(2\,|\phi_F| + V)}. \tag{1.5-15}$$

Since the GCA current integral given in eqn. (1.5-2) sums the differential depletion charge contributions, the sign of V_{T0} indicates that the increased bulk charge reduced I_D in the full treatment. This also gives rise to the fact that the circuit equations overshoot the actual value of $V_{DS,\text{sat}}$, i.e., eqn. (1.4-7) predicts a larger saturation voltage than eqn. (1.5-6).

An example of the differences between the two levels of modeling is illustrated by the curves in Fig. 1.20. The parameters given in the drawing were used to generate the (room temperature) I_D vs. V_{DS} characteristics as predicted by the two MOSFET equation sets. The complete GCA treatment, which includes bulk charge variations in the channel, is shown in the solid-line plot, while the dashed-line behavior results from using the simplified circuit

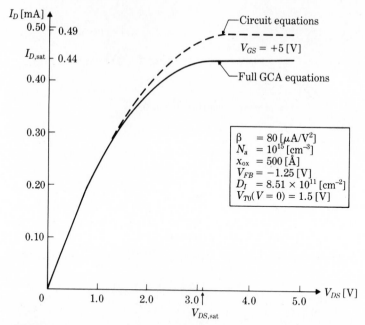

FIGURE 1.20 Comparison of the circuit equations with the complete GCA model.

equations. The fact that the circuit equations overestimate the current is easily seen in the comparison. Of particular interest are the saturation quantities. The complete model predicts $I_{D,\text{sat}} \simeq 0.44\,[\text{mA}]$ and $V_{DS,\text{sat}} \simeq 3.07\,[\text{V}]$; these should be compared to $I_{D,\text{sat}} \simeq 0.49\,[\text{mA}]$ and $V_{DS,\text{sat}} \simeq 3.5\,[\text{V}]$ as computed from the simplified equations.

It is obvious that the circuit equations must be modified if they are to be used to describe MOSFETs in a realistic circuit arrangement. This is most easily accomplished by introducing a "reduction factor" $M < 1$ such that the nonsaturated current equation is modified to

$$I_D = M\frac{\beta}{2}[2(V_{GS} - V_{T0}) - V_{DS}^2]. \tag{1.5-16}$$

The saturated current is then given by

$$I_{D,\text{sat}} = M\frac{\beta}{2}(V_{GS} - V_{T0})^2, \tag{1.5-17}$$

such that M is chosen to "reduce" the value of $I_{D,\text{sat}}$ to that found from the more rigorous treatment. Specifically, M can be obtained by the ratio

$$M = \frac{I_{D,\text{sat}}(\text{Full GCA})}{I_{D,\text{sat}}(\text{Circuit equations})}. \tag{1.5-18}$$

When applied to the example of Fig. 1.20, $M \simeq (0.44/0.49) = 0.898$. The

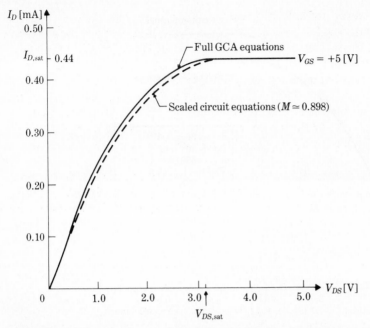

FIGURE 1.21 Comparison of full GCA MOSFET characteristics with the modified circuit equations. The device properties are those listed in Fig. 1.20.

modified circuit equation characteristics are denoted by a dashed-line plot in Fig. 1.21. This technique is capable of approximating the actual current through the MOSFET by a simple modification of the circuit equations. However, the discrepancy between $V_{DS,sat}$ values still remains and sometimes must be included with additional modifications. The reduction factor has the most important functional dependences of

$$M = M(C_{ox}, N_a). \tag{1.5-19}$$

M is also a function of V_{GS}, but it is usually small enough to allow a single value of M to be used for the entire range of V_{GS} values of interest. In the circuit design discussions, M will not be written explicitly in any of the equations. Rather, it will be assumed that the quoted value for k' or β has been appropriately reduced by this type of argument.

The critical reader will undoubtedly feel uneasy about the modification technique introduced above to justify the use of the simplified MOSFET circuit equations. This is understandable, particularly since there is no theoretical basis for M. However, the important point to remember here is that the results obtained from this approach provide enough accuracy for the initial phases of a circuit design. A given circuit must always be checked by a simulation program that contains accurate device modeling (e.g., SPICE) to ensure that the design will function as intended.

Another factor that should be mentioned at this point is that unavoidable *processing variations* are always present. These arise from the inability to precisely control the wafer fabrication steps to an arbitrary degree of accuracy. Process variations prohibit the exact specification of the MOSFET parameters that enter into the drain current equations. Consequently, the circuit design must proceed using average values of the required process-dependent quantities (such as k' and V_{T0}). The resulting circuit must then be able to correctly function over the expected range of MOSFET parameters introduced by the processing variations. With this in mind, the use of M as discussed above becomes more plausible. Processing variations are discussed in detail in Chapter 5.

As a closing comment, the techniques employed in this section may be used to analyze the effects of bulk charge variations in a MOSFET having nonzero values of V_S and V_B. The analysis is straightforward and will not be presented here.

1.6 Depletion-Mode MOSFETs

Enhancement-mode (E-mode) n-channel MOSFETs are defined to have $V_{T0} > 0$. This terminology arises from the fact that increasing V_{GS} above the threshold voltage "enhances" the drain-source conduction. While E-mode transistors constitute the basic building blocks for digital MOS switching networks, *depletion-mode* (D-mode) MOSFETs with $V_{T0} < 0$ have found wide use as "load" devices in nMOS designs. As will be demonstrated in later discussions, this design technique allows for improved switching characteristics and also requires less chip area than circuits that employ only E-mode MOSFETs.

A negative threshold voltage can be obtained by ion implanting donors (instead of acceptors) in a threshold adjustment step. With the donor implant dose denoted by D_I, V_{T0} is modified to

$$V_{T0} = V_{FB} + 2\,|\phi_F| + \frac{1}{C_{\text{ox}}} \sqrt{2q\varepsilon_{\text{Si}}N_a(2\,|\phi_F|)} - \frac{qD_I}{C_{\text{ox}}}. \tag{1.6-1}$$

D_I may be chosen to set the desired working value of $V_{T0} < 0$. Assuming that the implanted donor density on the surface is greater than the substrate acceptor doping N_a, this process gives a MOSFET with an n-type layer connecting the source and drain n^+ regions, as illustrated in Fig. 1.22(a). The circuit symbol for an n-channel D-mode MOSFET is shown in Fig. 1.22(b). This is simply the basic MOSFET symbol with an added line to explicitly show the presence of the implanted n-type layer. The n-type surface region in the D-mode MOSFET gives rise to a different conduction mechanism than that found for the E-mode device. As suggested by the terminology, current flow in the D-mode transistor depends on the properties of depletion regions within the n-type layer. Since a conduction path is automatically established between

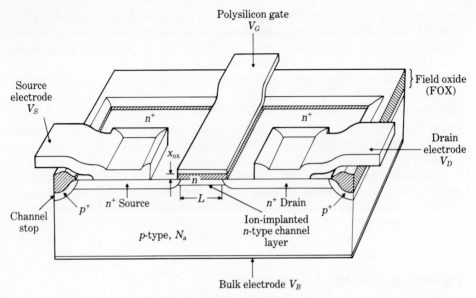

Basic D-mode MOSFET structure
(a)

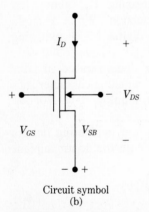

Circuit symbol
(b)

FIGURE 1.22   Depletion-mode MOSFET.

the source and the drain, the operation of this device is expected to be similar
to that of a JFET (junction field effect transistor).

It is possible to formulate a basic charge model to describe the characteristics of a D-mode MOSFET. The structure to be analyzed for this purpose is
shown in Fig. 1.23, where the bias has been chosen with $V_S = 0 = V_B$ for
simplicity. The n-type layer resulting from the donor implant is modeled as
having a net constant donor doping of $(N_d - N_a) > 0$. In this context,
N_d [cm^{-3}] represents the average donor density in the layer. The depletion
edges shown by the dashed lines in the drawing originate from two phenomena. The depletion region at the surface of the semiconductor is due to the

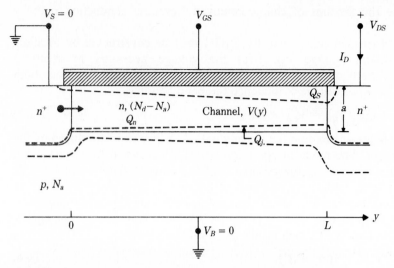

FIGURE 1.23 Simplified depletion-mode MOSFET model.

MOS field effect, such that Q_S [C/cm^2] gives the surface charge density. On the other hand, the depletion at the bottom of the n-type layer comes from the pn junction formed with the substrate. The amount of depletion charge on the n-side of the junction is denoted by Q_j [C/cm^2].

Current flow is established through the undepleted portions of the n-type layer (for the case of the simplest operational mode); this region then defines the channel for the device. Control of I_D is attained by using the voltage dependence of the depletion region depths. This may be modeled with the basic form of the GCA current from eqn. (1.3-9) to write

$$I_D = -\mu_n\left(\frac{W}{L}\right)\int_0^{V_{DS}} Q_C(V)\,dV, \tag{1.6-2}$$

where $Q_C(V)$ has been introduced as the *channel* charge density that describes the region of current flow. Note that this replaces the inversion charge density $Q_I(V)$ needed to describe the enhancement MOSFET.

An analytical expression for I_D may be obtained by writing the channel charge in the form

$$Q_C(V) = -Q_n + Q_S(V) + Q_j(V), \tag{1.6-3}$$

where Q_n represents the total charge density from electrons in the n-type layer. Explicitly,

$$Q_n = -q(N_d - N_a)a. \tag{1.6-4}$$

This gives the free electron charge density in the layer if depletion were not present. The remaining terms on eqn. (1.6-3) are voltage-dependent functions that describe the amount of charge removed from the conduction path by depletion.

The MOS surface charge density $Q_S(V)$ may be constructed by recalling that the flatband voltage V_{FB} gives the voltage necessary to create a charge-neutral flatband state at the surface of the semiconductor. Since both V_{GS} and the channel voltage V influence the band bending,

$$Q_S(V) = -C_{ox}[V_{GS} - V_{FB} - V] \tag{1.6-5}$$

is a reasonable approximation for the amount of surface charge induced by the MOS field effect. The mode of operation portrayed in the figure requires that $(V_{GS} - V_{FB}) < 0$ to obtain the positive depletion charge. This is not the only possibility. For example, an electron accumulation layer may form at the surface of the n-type region, which then gives additional conduction modes.

The last term in eqn. (1.6-3) is the pn junction depletion charge $Q_j(V)$. Employing the results of a step profile junction analysis gives

$$Q_j(V) = \sqrt{2q\varepsilon_{Si}N(\phi_o + V)}, \tag{1.6-6}$$

where ϕ_o is the built-in voltage

$$\phi_o \simeq \left(\frac{kT}{q}\right) \ln\left[\frac{(N_d - N_a)N_a}{n_i^2}\right] \tag{1.6-7}$$

while N is given by

$$N = \frac{(N_d - N_a)N_a}{(N_d - N_a) + N_a} = \frac{N_a}{N_d}(N_d - N_a) \tag{1.6-8}$$

as required to include only the depletion on the n-side of the junction.

Using the charge densities above gives

$$I_D = \mu_n\left(\frac{W}{L}\right)\int_0^{V_{DS}} [q(N_d - N_a)a + C_{ox}(V_{GS} - V_{FB} - V) \\ - \sqrt{2q\varepsilon_{Si}N(\phi_o + V)}]\, dV, \tag{1.6-9}$$

which may be easily integrated. A fine point that should be mentioned here is that μ_n represents the bulk mobility since the conduction is through the center of the n-type layer. This should be contrasted with the surface mobility required for describing electron inversion layers in an E-mode device. Performing the integration gives

$$I_D = \beta\left\{\frac{q(N_d - N_a)a}{C_{ox}}V_{DS} + \left[(V_{GS} - V_{FB})V_{DS} - \frac{1}{2}V_{DS}^2\right] \\ - \frac{2}{3C_{ox}}\sqrt{2q\varepsilon_{Si}N}\left[(\phi_o + V_{DS})^{3/2} - (\phi_o)^{3/2}\right]\right\}. \tag{1.6-10}$$

This has been found to be fairly accurate in describing experimental measurements on D-mode MOSFETs. Note that this equation is valid for the special case where $V_{GS} = 0$ since the n-type layer provides a conduction path for the current flow.

Although the level of analytic device modeling contained in eqn. (1.6-10) may be employed in computer-based circuit simulations, the functional dependence of I_D on V_{GS} and V_{DS} is too complicated for use in hand calculations. This problem is usually dealt with by adopting the simplified viewpoint that the (nonsaturated) D-mode MOSFET current can be described by

$$I_D = \frac{\beta}{2}[2(V_{GS} - V_{T0})V_{DS} - V_{DS}^2], \qquad (1.6\text{-}11)$$

where V_{T0} is a negative number as computed from eqn. (1.6-1). In this level of modeling, the distinction between E-mode and D-mode MOSFETs lies primarily in the sign of V_{T0}. When the simplified equation is compared to (1.6-10), it is seen that this viewpoint tends to ignore the basic current flow mechanisms of the depletion device. Consequently, eqn. (1.6-11) cannot be justified on theoretical grounds. However, it is possible to modify the equation to the point where it provides a reasonable approximation for the behavior of a D-mode MOSFET in a circuit. This consideration tends to support its wide use.

Figure 1.24(a) shows a comparison of the two equations for the device parameters specified in the graph; note that $V_{GS} = 0$ has been chosen. The relationship between the donor implant dose and the doping levels has been approximated by

$$D_I \simeq (N_d - N_a)a. \qquad (1.6\text{-}12)$$

This leads to a value of $V_{T0} \simeq -3.0\,[\text{V}]$, which has been used to plot the simplified equation (1.6-11). The curve shows that this approximation over-estimates both the saturation current $I_{D,\text{sat}}$ and the saturation voltage $V_{DS,\text{sat}}$. A similar situation was found for the E-mode transistor, so that a reduction factor M is introduced as in eqn. (1.5-18). For this particular example, $M \simeq 0.86$ gives the correlation shown in Fig. 1.24(b). Although there is still a noticeable difference between the two curves, the simplified model is usually accurate enough for initial circuit calculations. M is sensitive to changes in N_d and a, in addition to the dependence specified in eqn. (1.5-19). As in the case of E-mode MOSFETs, the reduction factor M will not be written explicitly when the equations are used to analyze circuits. Rather, it will always be assumed that k' has been scaled to the appropriate value. With this in mind, the general equation for the saturated current is easily found as

$$I_{D,\text{sat}} = \frac{\beta}{2}(V_{GS} + |V_{T0}|)^2, \qquad (1.6\text{-}13)$$

which describes the current for V_{DS} greater than the saturation voltage

$$V_{DS,\text{sat}} = V_{GS} + |V_{T0}|. \qquad (1.6\text{-}14)$$

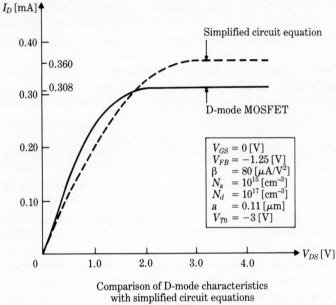

Comparison of D-mode characteristics
with simplified circuit equations
(a)

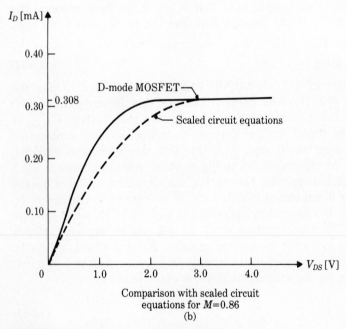

Comparison with scaled circuit
equations for $M=0.86$
(b)

FIGURE 1.24   Depletion-mode MOSFET characteristics.

The magnitude $|V_{T0}|$ has been used to emphasize the fact that $V_{T0} < 0$. Channel-length modulation may be included by the same techniques used to describe E-mode devices.

The example introduced in conjunction with Fig. 1.24 illustrates the important point that a D-mode MOSFET supports current flow when $V_{GS} = 0$. In the simplistic model, this may be understood by noting that since V_{T0} is negative, $V_{GS} = 0$ still satisfies the requirement $V_{GS} > V_{T0}$. Since the case of zero gate-source voltage is of particular importance to nMOS circuits, it warrants additional comments.

First, note that when $V_{GS} = 0$, the nonsaturated current reduces to

$$I_D = \frac{\beta}{2}(2|V_{T0}|V_{DS} - V_{DS}^2) \qquad (V_{GS} = 0). \tag{1.6-15}$$

This should be contrasted with the E-mode MOSFET, which is in cutoff with $I_D = 0$ for this gate-source voltage. The saturated D-mode current is

$$I_{D,\text{sat}} = \frac{\beta}{2}|V_{T0}|^2 \qquad (V_{GS} = 0). \tag{1.6-16}$$

To include the effects of body bias, this may be modified to read

$$I_{D,\text{sat}} = \frac{\beta}{2}|V_T(V_{SB})|^2 \qquad (V_{GS} = 0), \tag{1.6-17}$$

where

$$V_T(V_{SB}) = V_{T0} + \gamma(\sqrt{2|\phi_F| + V_{SB}} - \sqrt{2|\phi_F|}). \tag{1.6-18}$$

Note that $|V_T|$ decreases with increasing V_{SB}.

The difference between D-mode and E-mode characteristics is best illustrated by the bias arrangement shown in Fig. 1.25(a). V_{DS} is assumed to be large enough to produce saturated current flow. The solid-line plot in Fig. 1.25(b) demonstrates the nonzero current flow for $V_{GS} = 0$. The dashed-line curve represents an enhancement MOSFET that will not "turn on" until V_{GS} is positive and greater than the threshold voltage.

As a final point in the discussion, note that the value of $k' = \mu_n C_{ox}$ for a D-mode MOSFET is expected to be different from that used for an enhancement transistor. However, in practice, the two may be very close to each other. To understand this comment, recall that the mobility is a function of the doping density. Letting μ_B represent the bulk electron mobility, the E-mode MOSFET requires that

$$\mu_n \simeq \frac{1}{2}\mu_B(N_a). \tag{1.6-19}$$

On the other hand, the D-mode MOSFET current flow is established by electron motion through the n-type layer, so this device uses

$$\mu_n = \mu_B(N_d + N_a). \tag{1.6-20}$$

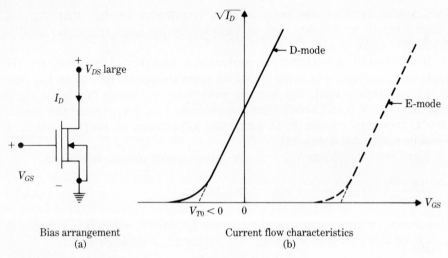

Bias arrangement
(a)

Current flow characteristics
(b)

FIGURE 1.25 $\sqrt{I_D}$ vs. V_{GS} for a depletion-mode MOSFET.

The bulk mobilities satisfy

$$\mu_B(N_d + N_a) < \mu_B(N_a) \tag{1.6-21}$$

because of differences in impurity scattering. When this inequality is used to compare the μ_n values above, it is seen that the two types of MOSFETs may indeed exhibit similar k' values.

REFERENCES

These books deal with general device physics, and all include chapters on the MOS system and MOSFETs.

(1) B. El-Kareh and R. J. Bombard, *Introduction to VLSI Silicon Devices,* Hingham, MA: Kluwer Academic Publishers, 1986.

(2) L. A. Glasser and D. W. Dobberpuhl, *The Design and Analysis of VLSI Circuits,* Reading, MA: Addison-Wesley, 1985.

(3) R. S. Muller and T. I. Kamins, *Device Electronics for Integrated Circuits,* 2nd ed., New York: Wiley, 1986.

(4) R. F. Pierret and G. W. Neudeck, *Modular Series on Solid State Devices,* Vols. 1–4, Reading, MA: Addison-Wesley, 1983.

(5) P. Richman, *MOS Field-Effect Transistors and Integrated Circuits,* New York: Wiley, 1973.

(6) B. G. Streetman, *Solid State Electronic Devices,* 2nd ed., Englewood Cliffs, NJ: Prentice-Hall, 1980.

(7) S. M. Sze, *Physics of Semiconductor Devices*, 2nd ed., New York: Wiley, 1981.

(8) R. M. Warner, Jr., and B. L. Grung, *Transistors*, New York: Wiley, 1983.

(9) E. S. Yang, *Fundamentals of Semiconductor Devices*, New York: McGraw-Hill, 1978.

Depletion-Mode MOSFETs are the subject of the paper

(10) Y. F. El-Mansy, "Analysis and Characterization of the Depletion-Mode IGFET," *IEEE J. Solid-State Circuits*, Vol. SC-15, pp. 331–340, 1980.

The above paper is reprinted in

(11) M. I. Elmasry (ed.), *Digital MOS Integrated Circuits*, New York: IEEE Press (Wiley), 1981.

PROBLEMS

1.1 A silicon MOS system at room temperature has

$$x_{ox} = 600\,[\text{Å}], \quad N_a = 8 \times 10^{14}\,[\text{cm}^{-3}], \quad Q_{ss} = q(10^{11})\,[\text{C/cm}^2],$$
$$\Phi_{GS} = -0.5\,[\text{V}].$$

(a) Compute the value of the acceptor ion implant dose $D_I\,[\text{cm}^{-3}]$ needed to set $V_{T0} = +1\,[\text{V}]$. Assume Q_{ox} is negligible.

(b) Compute the value of the body bias factor $\gamma\,[\text{V}^{1/2}]$.

1.2 A polysilicon gate can be doped n-type, with donor doping $N_{d,\text{poly}}$, or p-type, with acceptor doping $N_{a,\text{poly}}$. This alters the value of $\Phi_{GS} = (\Phi_G - \Phi_S)$ in the threshold voltage.

The work function in an n-type sample may be approximated by

$$\Phi \simeq \Delta\phi - \left(\frac{kT}{q}\right) \ln\left(\frac{N_d}{n_i}\right),$$

while a p-type sample has

$$\Phi \simeq \Delta\phi + \left(\frac{kT}{q}\right) \ln\left(\frac{N_a}{n_i}\right) \qquad (\Delta\phi = \text{Constant}).$$

These can be used for both the poly and substrate regions.

(a) Find a simple expression for Φ_{GS} when applied to a system with a p-poly gate and p-type substrate. Then calculate the room temperature value of $\Phi_{GS}\,[\text{V}]$ if $N_{a,\text{poly}} = 10^{19}\,[\text{cm}^{-3}]$ while the substrate doping is $N_a = 10^{15}\,[\text{cm}^{-3}]$.

(b) Repeat for the case of an n-type polysilicon gate and a p-type substrate. After finding the general equation for Φ_{GS}, calculate the value of Φ_{GS} if $N_{d,\text{poly}} = 3 \times 10^{19}\,[\text{cm}^{-3}]$, $N_a = 10^{15}\,[\text{cm}^{-3}]$.

1.3 A MOSFET is known to have $2|\phi_F| = 0.57\,[\text{V}]$ and $\gamma = 0.45\,[\text{V}^{1/2}]$ at room temperature. The surface mobility is $\mu_n \approx 550\,[\text{cm}^2/\text{V-sec}]$, and the adjusted threshold voltage is set to be $V_{T0} = +0.8\,[\text{V}]$. Assume $\lambda \approx 0$.

(a) Compute the value of $k'\,[\mu\text{A}/\text{V}^2]$ for the transistor.

(b) Suppose that the device aspect ratio is designed to be $(W/L) = 10/2$. Compute the current flow when $V_{GS} = 2\,[\text{V}]$, $V_{SB} = 1\,[\text{V}]$, and $V_{DS} = 4\,[\text{V}]$. Repeat for $V_{GS} = 2\,[\text{V}]$, $V_{SB} = 1\,[\text{V}]$, and $V_{DS} = 2\,[\text{V}]$. Use the circuit equations.

*(c) Use SPICE to plot I_D vs. V_{DS} in the range $V_{DS} = 0$ to $5\,[\text{V}]$, $V_{GS} = 0$ to $5\,[\text{V}]$. Assume that $V_{SB} = 0\,[\text{V}]$ and use $(W/L) = 10/2$. (Use LEVEL 1 modeling and compare it with LEVEL 2 modeling.)

1.4 Consider an enhancement-mode MOSFET characterized by

$$V_{T0} = +0.8\,[\text{V}], \quad \gamma = 0.40\,[\text{V}^{1/2}], \quad \lambda = 0.05\,[\text{V}^{-1}],$$
$$2|\phi_F| = 0.58\,[\text{V}], \quad k' = 20\,[\mu\text{A}/\text{V}^2].$$

(a) The device is biased with $V_{GS} = 2.2\,[\text{V}]$, $V_{DS} = 4[\text{V}]$, and $V_{SB} = 2\,[\text{V}]$. Compute the value of the aspect ratio (W/L) needed to obtain a current flow of $I_D = 100\,[\mu\text{A}]$. Use the simplified circuit equation set.

(b) Suppose that the minimum linewidth for W and L is given as $2.5\,[\mu\text{m}]$. Calculate the *gate capacitance* $C_g = C_{ox}WL$ for the device designed in part (a). Place your answer in units of *femtofarads* [fF] where $1\,[\text{fF}] = 10^{-15}\,[\text{F}]$.

1.5 An enhancement-mode MOSFET process yields a zero body bias threshold voltage of $V_{T0} = 0.70\,[\text{V}]$. The body bias parameter is given as $\gamma = 0.35\,[\text{V}^{1/2}]$, while the oxide capacitance is $C_{ox} = 7 \times 10^{-8}\,[\text{F}/\text{cm}^2]$ (room temperature values). Assume $\lambda = 0$, and use the simplified MOSFET (circuit) equations.

(a) Plot $V_T(V_{SB})$ for $V_{SB} = 0\,[\text{V}]$ to $V_{SB} = 5\,[\text{V}]$.

(b) A MOSFET made in this process is designed with $\beta = 85\,[\mu\text{A}/\text{V}^2]$. The drain-source voltage is set at $V_{DS} = 5\,[\text{V}]$, which is the highest voltage in the circuit. Plot $\sqrt{I_D}$ as a function of V_{GS} in the range 0 to $5\,[\text{V}]$ for body bias voltages of $V_{SB} = 0\,[\text{V}]$ and $2.5\,[\text{V}]$.

1.6 A MOSFET is made with processing parameters of

$$x_{ox} = 600\,[\text{Å}], \quad N_a = 10^{15}\,[\text{cm}^{-3}], \quad \beta = 60\,[\mu\text{A}/\text{V}^2],$$
$$V_{FB} = -1.0\,[\text{V}], \quad Q_{ss} = 1.5 \times 10^{-8}\,[\text{C}/\text{cm}^2] \gg Q_{ox},$$
$$n\text{-poly gate with } N_{d,\text{poly}} = 10^{20}\,[\text{cm}^{-3}].$$

(a) The threshold voltage V_{T0} is set to a value of $+0.75\,[\text{V}]$ using an acceptor ion implant. Compute the dose $D_I\,[\text{cm}^{-2}]$ of the implant step. (Φ_{GS} is computed using the results of Problem 1.2.)

Assume that $V_{GS} = +3\,[\text{V}]$ and $V_{SB} = 0\,[\text{V}]$ for the remainder of the problem.

(b) Compute the value of $V_{DS,\text{sat}}$ using the simplified circuit equations. Then compute $V_{DS,\text{sat}}$ as predicted by the complete GCA analysis equation (1.5–6).

(c) Compute the value of $I_{D,\text{sat}}$ from the simplified circuit equations and then repeat using the complete GCA analysis eqn. (1.5-8).

(d) Calculate the reduction factor M needed to correlate the two models.

1.7 An empirical expression for the *bulk* mobility with doping $N\,[\text{cm}^{-3}]$ is (see Reference [3])

$$\mu = \mu_{\min} \frac{(\mu_{\max} - \mu_{\min})}{1 + (N/N_{\text{ref}})^{\alpha}}\,[\text{cm}^2/\text{V-sec}].$$

For electrons,

$$\mu_{\max} = 1360, \quad \mu_{\min} = 92, \quad N_{\text{ref}} = 1.3 \times 10^{17}\,[\text{cm}^{-3}], \quad \alpha = 0.91.$$

For holes,

$$\mu_{\max} = 495, \quad \mu_{\min} = 48, \quad N_{\text{ref}} = 6.3 \times 10^{16}\,[\text{cm}^{-3}], \quad \alpha = 0.76.$$

(a) Suppose that a MOSFET is made with $x_{\text{ox}} = 750\,[\text{Å}]$, Calculate the value of $k'_n = \mu_n C_{\text{ox}}$ for a substrate doping of $N_a = 10^{15}\,[\text{cm}^{-3}]$ under the assumption that μ_n is 1/2 the bulk value,

(b) Repeat the k'_n calculation for $x_{\text{ox}} = 500\,[\text{Å}]$,

(c) Compare the values of hole and electron surface mobilities for $N_a = 10^{15}\,[\text{cm}^{-3}]$. What does this say about carrier transport through MOSFETs?

1.8 The gradual channel approximation may be used to extract information about the channel electric field $\mathscr{E}_y(y)$ in a MOSFET.

(a) Combine eqns. (1.3-5) and (1.3-8) to obtain an equation for dV in terms of dy. Then integrate the expression from the source ($y = 0$) to an arbitrary channel position y and thus find a quadratic equation for V.

(b) Solve the quadratic equation subject to the boundary condition that $V(y = 0) = V_S = 0$.

(c) Eliminate I_D from the expression found in (b) by using the simplified circuit equations, and thus show that

$$V(y) \simeq (V_{GS} - V_T)\left(1 - \sqrt{1 - \frac{y}{L}}\right).$$

(Assume that the MOSFET is saturated.)

(d) Use $V(y)$ to calculate $\mathscr{E}_y(y)$. Then plot the electric field in the channel for the case of $L = 10\,[\mu\text{m}]$ with $(V_{GS} - V_T)$ values of 1 [V], 2 [V], and 4 [V].

1.9 MOSFET characteristics are temperature sensitive because of the presence of thermal voltage (kT/q) factors and also because the mobility has an approximate dependence of

$$\mu \propto \frac{1}{T^{3/2}},$$

where T is in [°K]. Note that n_i is also a function of T.

Consider a MOSFET with $N_a = 10^{15}$ [cm^{-3}] and $x_{ox} = 500$ [Å]. At $T = 27$ [°C], the MOSFET has $V_{T0} = +0.8$ [V] and $k' = 40$ [μA/V^2].

(a) Calculate V_{T0} at $T = 60$ [°C]. Assume that V_{FB} is not affected by temperature.

(b) Find the percent increase in the body bias coefficient γ as the temperature increases from 27 [°C] to 60 [°C]. Then plot $V_T(V_{SB})$ for $V_{SB} = 0$ [V] to 5 [V] at the two temperatures.

(c) Find k' at 60 [°C].

(d) Assume $V_{SB} = 0$ [V] and $V_{GS} = V_{DS}$. Plot I_D vs. V_{DS} for the device at 27 [°C] and 60 [°C] on the same graph.

*(e) Use SPICE to study the temperature dependence of the MOSFET.

1.10 The simplified equations for the depletion-made MOSFET current flow resulted in a nonsaturated equation of

$$I_D = \frac{\beta}{2}(2|V_{T0}|V_{DS} - V_{DS}^2).$$

Suppose that the transistor is to be used in a circuit with a power supply range of 0 [V] to $V_{DD} = 5$ [V].

Discuss the limitations on setting the value of $|V_{T0}|$ for use in a circuit of this type.

1.11 The approach adopted in this book approximates the body bias factor γ using eqn. (1.1-29):

$$\gamma = \frac{1}{C_{ox}}\sqrt{2q\varepsilon_{Si}N_a} \ [\text{V}^{1/2}].$$

This accounts only for the substrate acceptor doping and ignores the dopants introduced by the threshold adjustment ion implantation step. An approach that can be used to include the extra dopants is to write

$$N_{a,\text{eff}} = N_a + \frac{D_I}{d}$$

where $N_{a,\text{eff}}$ is the effective doping, D_I is the acceptor ion implant dose, and d is the average depth of the ion implanted layer.

Suppose that a MOSFET is made with $x_{ox} = 500$ [Å] and $N_a = 10^{15}$ [cm^{-3}]. The threshold adjustment ion implant dose is $D_I = 2 \times 10^{11}$ [cm^{-2}] with an average implant depth of $d = 0.2$ [μm].

Calculate γ using N_a and $N_{a,\text{eff}}$ and then compare the percentage difference between the two values.

1.12 The mobility in a MOSFET is found to be affected by the electric field. In the simplest level of approximation, this introduces a degradation described by

$$\mu_{\text{eff}} = \frac{\mu}{1 + \theta(V_{GS} - V_T)},$$

where θ is an empirical factor.

Consider an n-channel MOSFET that has a surface mobility of $\mu_n = 550 \, [\text{cm}^2/\text{V-sec}]$, $x_{\text{ox}} = 500 \, [\text{Å}]$, and $V_{T0} = +0.8 \, [\text{V}]$. Assuming that $\theta = 0.08 \, [\text{V}^{-1}]$, plot I_D vs. V_{DS} for drain-source voltages in the range $0 \, [\text{V}]$ to $5 \, [\text{V}]$ with $V_{GS} = 3 \, [\text{V}]$. Compare this with the case where $\theta = 0$.

CHAPTER 2

Small-Geometry MOSFETs for VLSI

Very-large-scale integration (VLSI) requires dense circuit layouts on silicon. The level of integration depends on the smallest-size feature permitted by the fabrication processes. To obtain the highest packing density, the size of the transistors must be made as small as possible. This, however, changes the internal operating physics of the MOFSETs. Phenomena that are negligible in "large" devices become limiting factors as device geometries are reduced.

This chapter discusses some of the important aspects involved in describing small MOSFETs. The level is introductory, with emphasis on parameters that affect circuit design. The remaining chapters in the book have been structured so that this chapter may be bypassed without loss of continuity. Persons interested in digital circuit design at this point are directed to Chapter 3, which initiates the subject.

2.1   MOSFET Scaling Theory

The theory of MOSFET scaling is concerned with the characteristics of a transistor as its overall dimensions are systematically reduced. The basic problem is illustrated in Fig. 2.1, where a "large" device is scaled downward to give a MOSFET that is smaller in size but maintains the same geometric ratios. This type of study is important because it may be used as a guide for the miniaturization of existing devices and circuits. The motivation for such an analysis centers around the observation that small devices allow for higher-density integration. In terms of digital MOS chips, this implies that more complicated logic can be implemented in a given area.

Scaling theory may be viewed as the first step toward understanding the chip structures required for VLSI designs. It is highly idealized and is therefore

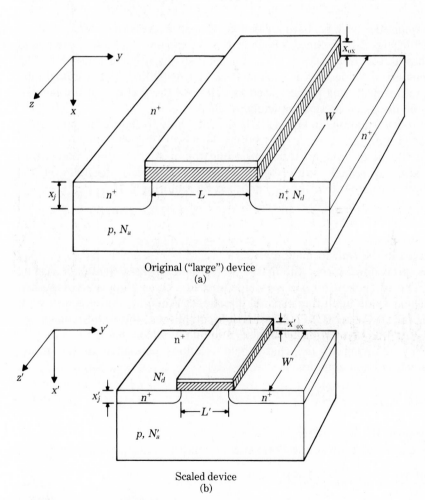

Original ("large") device
(a)

Scaled device
(b)

FIGURE 2.1   Basic MOSFET scaling problem.

limited when applied in a realistic engineering environment. However, it is extremely useful for examining the device modifications that must be made and also provides a foundation for more detailed analyses.

The basic premise of scaling theory centers around reducing the *real estate* (chip surface area) required for a MOSFET while simultaneously maintaining the current-voltage characteristics of the larger device. This would, in principle, allow existing circuits to be directly transformed into smaller chip structures. With regard to the MOSFET in Fig. 2.1(a), L and W constitute the most important parameters that need to be reduced. To effect this transformation, a *scaling factor* $S > 1$ is introduced such that the scaled device is described by

$$L' = \frac{L}{S}, \qquad W' = \frac{W}{S}. \tag{2.1-1}$$

Primed quantities will be used to denote values in the scaled MOSFET, as indicated in Fig. 2.1(b). Since S is defined to be greater than unity, the scaled device has a gate area that is $(1/S^2)$ smaller than the original MOSFET. The actual numerical value of S is restricted by the resolution of the lithography (the pattern-transferring process used to delineate the features of the layers); this will be discussed more completely in Chapter 5.

The transformations in eqn. (2.1-1) describe how to reduce the size of the transistor in the y and z surface directions. However, the MOS field effect is a result of the fields that are oriented along the x-axis, so these dimensions should also be scaled. With regard to the x (vertical) distances in the illustration, this is most easily accomplished by requiring that

$$x'_{\text{ox}} = \frac{x_{\text{ox}}}{S}, \qquad x'_j = \frac{x_j}{S} \tag{2.1-2}$$

be included in the transformation set.

This procedure forms the basis of *isotropic three-dimensional* scaling theory. The adjective *isotropic* is included because the scaling is accomplished by a single S value in all three spatial directions. While the major objective of reducing the size of a MOSFET has been accomplished, there still remains the question of the current-voltage characteristics in the scaled device. This may be studied by referring to the charge-field relations contained in the three-dimensional Poisson equation

$$\nabla^2 \phi(x, y, z) = -\frac{\rho(x, y, z)}{\varepsilon}, \tag{2.1-3}$$

which governs the field effect.

The expanded form of the Poisson equation for the original device is

$$\frac{\partial^2 \phi(x, y, z)}{\partial x^2} + \frac{\partial^2 \phi(x, y, z)}{\partial y^2} + \frac{\partial^2 \phi(x, y, z)}{\partial z^2} = -\frac{\rho(x, y, z)}{\varepsilon}. \tag{2.1-4}$$

When the basic spatial transformations

$$x' = \frac{x}{S}, \qquad y' = \frac{y}{S}, \qquad z' = \frac{z}{S} \tag{2.1-5}$$

are made in the Laplacian, this becomes

$$\frac{1}{S^2} \nabla'^2 \phi(x', y', z') = -\frac{\rho(x', y', z')}{\varepsilon}, \tag{2.1-6}$$

where

$$\nabla'^2 \equiv \frac{\partial^2}{\partial x'^2} + \frac{\partial^2}{\partial y'^2} + \frac{\partial^2}{\partial z'^2} \tag{2.1-7}$$

is the Laplacian operator in the primed coordinate system of the scaled MOSFET. This equation may be used to extrapolate the scaled potential $\phi'(x', y', z')$ and charge density $\rho'(x', y', z')$ for the smaller transistor. The permittivity ε cannot be scaled since it is a property of the material. The fundamental requirement for preserving the properties of the field effect is that

$$\nabla'^2 \phi'(x', y', z') = -\frac{\rho'(x', y', z')}{\varepsilon} \tag{2.1-8}$$

describe the new primed (scaled) system. There are two common choices for the scaling. Each is discussed below for the purpose of scaling the MOSFET circuit equations.

2.1.1 Full Scaling

The first choice that will be studied is scaling the potentials by S such that

$$\phi' = \frac{\phi}{S}. \tag{2.1-9}$$

Equation (2.1-6) may then be written in the form

$$\nabla'^2 \left(\frac{\phi}{S}\right) = -\frac{(S\rho)}{\varepsilon}, \tag{2.1-10}$$

which gives a scaled charge density of

$$\rho' = S\rho, \tag{2.1-11}$$

as seen by comparison with the desired Poisson form in eqn. (2.1-8).

In *full scaling*, the voltages are reduced by a factor $(1/S)$ in the scaled MOSFET. However, the charge densities must be increased by a factor S to maintain the proper form of the charge-field relation. With regard to the MOSFET shown in Fig. 2.1(b), this implies that the doping densities must be scaled upward by means of

$$N'_a = SN_a, \qquad N'_d = SN_d. \tag{2.1-12}$$

Full scaling is obtained by combining the transformations in eqns. (2.1-1), (2.1-2), (2.1-9), and (2.1-12) to form a single equation set.

Interest is now directed toward the scaling of the MOSFET circuit equations. First, note that reducing the gate oxide thickness to (x_{ox}/S) gives a scaled gate oxide capacitance of

$$C'_{ox} = \frac{\varepsilon_{ox}}{x'_{ox}} = SC_{ox}, \tag{2.1-13}$$

which indicates a linear upward scaling.

Now consider the basic threshold voltage V_{T0}, which has the scaled form

$$V'_{T0} = V'_{FB} + 2|\phi'_F| + \frac{1}{C'_{ox}}\sqrt{2q\varepsilon_{Si}N'_a(2|\phi'_F|)} + \frac{qD'_I}{C'_{ox}}, \tag{2.1-14}$$

as may be verified by a direct application of the full scaling set. The scaled flatband voltage is given by

$$V'_{FB} = \Phi'_{GS} - \frac{1}{C'_{ox}}(Q_{ss} + Q_{ox}), \tag{2.1-15}$$

where it has been assumed that the trapped oxide and surface state charges remain invariant. The work function difference Φ_{GS} depends on the gate material. As an example, suppose that an n-type poly gate is used. The unscaled expression in Problem 1.2 may be used to write

$$\Phi'_{GS} \simeq -\left(\frac{kT}{q}\right)\ln\left[\frac{S^2 N_{d,\text{poly}}N_a}{n_i^2}\right] < \Phi_{GS}, \tag{2.1-16}$$

which shows that the scaling required by eqn. (2.1-12) gives a more negative value. The second term in V_{FB} has the opposite behavior, since it is reduced by a factor $(1/S)$.

The surface inversion potential is described by

$$2|\phi'_F| = 2\left(\frac{kT}{q}\right)\ln\left(\frac{SN_a}{n_i}\right) > 2|\phi_F|, \tag{2.1-17}$$

indicating a nonlinear upward scaling. On the other hand, the bulk charge support term assumes the form

$$\frac{1}{SC_{ox}}\sqrt{2q\varepsilon_{Si}SN_a(2|\phi'_F|)}, \tag{2.1-18}$$

which is smaller than the original unscaled contribution.

The term-by-term examination above demonstrates the fact that V_{T0} does not directly yield to simple scaling. The desired value of the scaled threshold voltage is

$$V'_{T0} = \frac{V_{T0}}{S},$$

(2.1-19)

which is needed to satisfy the potential scaling in eqn. (2.1-9). Although this cannot be attained by a brute force application of the scaling rules, V_{T0} can be set by adjusting the acceptor ion implant dose D_I. Consequently, scaling the (zero body bias) threshold voltage does not present any insurmountable problems.

The MOSFET circuit equations may now be scaled in a relatively straightforward manner. The scaled nonsaturated current is given by

$$I'_D = \frac{\beta'}{2}[2(V'_{GS} - V'_{T0})V'_{DS} - V'^2_{DS}].$$

(2.1-20)

The device transconductance parameter scales according to

$$\beta' = \mu_n(SC_{ox})\left(\frac{W/S}{L/S}\right) = S\beta.$$

(2.1-21)

Note that the aspect ratio is invariant under the scaling:

$$\left(\frac{W'}{L'}\right) = \left(\frac{W}{L}\right).$$

(2.1-22)

This, of course, was built in to the initial surface area transformations. A fine point that should be mentioned is that the surface mobility is actually reduced because

$$\mu_n(SN_a) < \mu_n(N_a).$$

(2.1-23)

However, this change is small and constitutes one of the many *second-order* scaling effects that are ignored at this level of analysis.

The operating voltages of the scaled MOSFET are reduced by means of

$$V'_{GS} = \frac{V_{GS}}{S}, \qquad V'_{DS} = \frac{V_{DS}}{S}.$$

(2.1-24)

Substituting these into the drain current expression gives

$$I'_D = \frac{I_D}{S}.$$

(2.1-25)

In a similar manner, the saturated drain current scales according to

$$I'_{D,\text{sat}} = \frac{\beta'}{2}(V'_{GS} - V'_{T0})^2 = \frac{I_{D,\text{sat}}}{S}.$$

(2.1-26)

Consequently, a MOSFET that is subjected to full scaling exhibits a uniform reduction of current and voltage magnitudes while maintaining the basic equation structure of the original device.

Scaling theory was introduced as a means of attaining a lower real estate requirement for a MOSFET. An added advantage of full scaling is that the power consumption is reduced. This can be seen by writing the scaled power as

$$P' = I'_D V'_{DS} = \frac{P}{S^2},$$ (2.1-27)

where

$$P = I_D V_{DS}$$ (2.1-28)

is the power dissipated by the unscaled device. The $(1/S^2)$ power reduction illustrated by this equation is very attractive for high-density chip designs.

The scaling above assumed that body bias effects were not present. If body bias is included, then the threshold voltage must be changed to

$$V'_T = V'_{T0} + \gamma'(\sqrt{2|\phi'_F| + V'_{SB}} - \sqrt{2|\phi'_F|}).$$ (2.1-29)

Since the body bias coefficient becomes

$$\gamma' = \frac{\sqrt{2q\varepsilon_{Si}SN_a}}{SC_{ox}}$$ (2.1-30)

and the bulk Fermi potential does not scale linearly, there is no simple relation for scaling the full threshold voltage, even if $V'_{SB} = (V_{SB}/S)$. Consequently, this must be accounted for when using the scaled circuit equations.

Although the discussion has centered around the scaling of the MOSFET circuit equations, it is worthwhile to mention that all chip dimensions can be reduced in a similar manner. Among the more important parameters that directly yield to the scaling transformations are the n^+ source and drain geometries and the interconnect patternings. These directly affect the transient response of a scaled circuit.

As a final point in the analysis of full scaling, it should be noted that only the simplified circuit equations scale in a direct manner. If the complete GCA MOSFET equations developed in Section 1.5 are subjected to the scaling transformations, it is found that I_D scales only approximately under certain conditions. The details of this analysis are left as an exercise.

2.1.2 Constant-Voltage Scaling

Full scaling requires that the voltages be reduced by a factor of $(1/S)$. An alternate approach to MOSFET scaling is the case where the voltages are left unchanged when applied to the scaled device. This is termed *constant-voltage* scaling and is employed when an *a priori* choice is made for the power supply level. This is a practical consideration that can be understood by examining a system made up of different logic families. For example, it is quite common to use bipolar TTL (transistor-transistor logic) chips to provide gate-level logic that directly interfaces with a MOS integrated circuit. TTL requires a power supply level of $V_{DD} = +5$ [V]. Although it would be possible to design the MOS logic chip around a different supply voltage constraint, this would require adding another power source to the system board. As such, it would be undesirable to the board designer, and the system would cost more to implement.

Under constant-voltage scaling, the terminal MOSFET voltages are given by

$$V'_{GS} = V_{GS}, \qquad V'_{DS} = V_{DS}. \tag{2.1-31}$$

When these are used to scale the MOSFET circuit equations, one obtains

$$I'_D = SI_D, \qquad I'_{D,\text{sat}} = SI_{D,\text{sat}}, \tag{2.1-32}$$

indicating that the currents are increased by a factor S. The power dissipated by the scaled device is given by

$$P' = I'_D V'_{DS} = SP, \tag{2.1-33}$$

so that constant-voltage scaling increases the power consumption of the circuit.

Another important point to consider is that preservation of the Poisson equation structure requires

$$\nabla'^2(\phi') = -\frac{(S^2 \rho)}{\varepsilon}. \tag{2.1-34}$$

Thus, the doping densities must be increased by means of

$$N'_a = S^2 N_a, \qquad N'_d = S^2 N_d. \tag{2.1-35}$$

While constant-voltage scaling is difficult to implement in practice, it is a very realistic situation that is commonly encountered in MOSFET circuit problems.

2.1.3   Practical Limitations on the Use of Scaling Theory

It was mentioned at the onset that scaling theory is used primarily as an initial guide for reducing the size of a MOSFET. In a realistic environment, practical considerations tend to limit the scaling to only a portion of the specified transformation set. This can occur because of limitations in the processing capabilities (beyond the lithographic constraint that establishes the maximum S value). In addition, it is found that full implementation of the scaling rules can lead to the degradation of some important MOSFET parameters that are not included in the scaling treatment.

One important dimension that cannot be arbitrarily scaled is the gate oxide thickness x_{ox}. There are two basic reasons prohibiting the reduction of the thickness to (x_{ox}/S) for arbitrary S. The first is a processing limitation. It is very difficult to grow a uniform oxide layer with a thickness less than about 150 [Å] because of the presence of *pinholes,* localized points where the oxide has failed to grow to the thickness of surrounding regions. When a gate is formed over a pinhole, an electrical short with the substrate may occur. This, of course, eliminates the field effect and renders the device (and chip) inoperative. As processing technology improves, the minimum attainable x_{ox} value may be reduced.

The second limitation on scaling x_{ox} arises because of *oxide breakdown*. The silica layer tends to undergo breakdown for oxide electric fields greater than about $\mathscr{E}_{ox,BD} = 5 \times 10^6$ [V/cm]. This is usually a destructive process. Since the voltage across the gate MOS capacitor is roughly proportional to x_{ox}, very thin oxide layers cannot support large applied voltages. This situation is even worse when it is noted that *weak spots* can exist in the oxide. These regions are characterized by lower breakdown field values that arise from nonuniform structural properties of the silica layer.

Another problem in implementing the scaling transformations originates from the requirement that the substrate doping N_a be increased. This tends to lower the MOSFET drain-source breakdown voltage, as can be seen from a simplified analysis. Consider the drain-substrate *pn* junction shown in Fig. 2.2. Assuming that the results of a one-sided step profile junction can be used, the total depletion width is approximately

$$x_d \simeq \sqrt{\frac{2\varepsilon_{Si}}{qN_a}(\phi_o + V_{DS})}, \tag{2.1-36}$$

where ϕ_o is the built-in potential. The maximum depletion electric field occurs at the junction with

$$\mathscr{E}_{max} \simeq \frac{qN_a x_d}{\varepsilon_{Si}}. \tag{2.1-37}$$

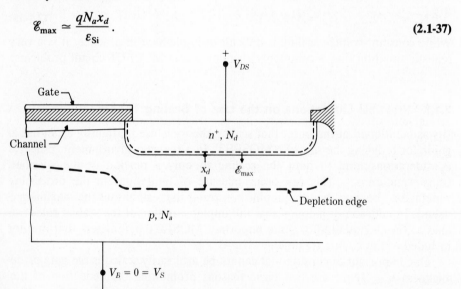

FIGURE 2.2 Simplified MOSFET geometry for computing drain-source breakdown voltage.

When the maximum field intensity reaches a critical value $\mathscr{E}_{BD}$, the reverse-bias junction will break down. For example, avalanching may result. To compute the drain-source breakdown voltage $V_{DS,BD}$, the two equations are combined to write

$$\mathscr{E}_{BD}^2 \simeq \frac{2qN_a}{\varepsilon_{Si}}(\phi_o + V_{DS,BD}) = \text{Constant.} \tag{2.1-38}$$

This demonstrates the general dependence

$$V_{DS,BD} \propto \frac{1}{N_a} \tag{2.1-39}$$

so that increasing N_a lowers the breakdown voltage. This may cause a problem in circuit design, depending on the power supply levels that are available.

The most important limitation to scaling theory arises from the fact that the device physics of small-geometry MOSFETs can be significantly different from the results obtained by scaling the gradual channel equations. The remaining sections of this chapter examine some small device effects that are important to VLSI circuit design but are ignored by scaling theory. They may be omitted in a first reading, in which case one may turn directly to Chapter 3, which initiates the discussion of digital circuits.

2.2 Threshold Voltage Corrections for Small MOSFETs

The threshold voltage

$$V_{T0} = V_{FB} + 2|\phi_F| + \frac{1}{C_{ox}}\sqrt{2q\varepsilon_{Si}N_a(2|\phi_F|)} + \frac{qD_I}{C_{ox}} \tag{2.2-1}$$

used to describe a MOSFET was obtained from the MOS analysis in Chapter 1. As such, it does not account for the structural modifications required to change a basic MOS system into a transistor. For example, this expression ignores the presence of the n^+ source and drain regions in the substrate. While the equation is fairly accurate for describing large devices, it breaks down when applied to small-geometry MOSFETs.

This section centers around the study of various models for the small-device effects that induce changes in the threshold voltage. The approach employed here develops correction terms that must be added to V_{T0} in eqn. (2.2-1). Although the extra terms represent at best only first-order corrections, including them in V_{T0} provides enough accuracy for use in a large number of realistic problems.

The MOSFETs discussed in the remaining portions of this chapter are assumed to be constructed without reference to any larger device. Consequently, the "primed" notation introduced in scaling theory will be abandoned.

2.2.1 Short-Channel Effects

The threshold voltage in eqn. (2.2-1) implicitly assumes that the bulk depletion charge originates solely from the MOS electric field created by the gate voltage. This may be understood by referring to Fig. 2.3, where it is seen that the bulk charge volume is approximated by LWx_{dm} at the onset of strong inversion in the GCA. Although this is a good approximation for large devices, it ignores the fact that the depletion charge near the n^+ source and drain regions is induced by pn junction band bending, not the MOS field. Equation (2.2-1) thus overestimates the amount of bulk charge that must be supported by the gate voltage. Consequently, it yields a value of V_{T0} that is larger than the actual threshold voltage of the MOSFET.

The existence of pn junction depletion charge was neatly sidestepped in the GCA equations by using Q_{B0} $[C/cm^2]$ and C_{ox} $[F/cm^2]$. Since both quantities are defined with regard to a unit area, the ratio $-Q_{B0}/C_{ox}$ $[V]$ employed in constructing V_{T0} automatically implies that the depletion charge volume is rectangular. To illustrate the actual geometric dependence in this term, it is instructive to write

$$-\frac{Q_{B0}WL}{C_{ox}WL} = \frac{1}{C_{ox}}\sqrt{2q\varepsilon_{Si}N_a(2\,|\phi_F|)}\,\frac{WL}{WL},\qquad\text{(2.2-2)}$$

which is a true charge-capacitance ratio.

A short-channel MOSFET is one where the channel length L is very small; more concrete definitions will emerge from the analysis. This type of device exhibits characteristics that differ from those found in a long-channel MOSFET. The deviations from the GCA behavior are collectively termed *short-channel effects* (SCEs). The reduction of the threshold voltage discussed in this subsection is one example of a short-channel effect.

In a short-channel MOSFET a significant fraction of the total bulk depletion charge underneath the gate originates from pn junction depletion. This charge must be subtracted out of the threshold voltage expression since it does not require gate voltage support. Although the actual geometry of the

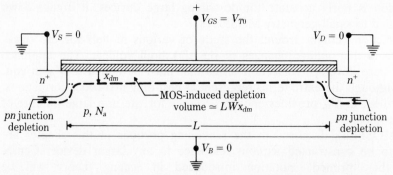

FIGURE 2.3 Bulk depletion charge geometry implied in the gradual channel analysis.

problem is quite complicated, a simple model that exhibits reasonable accuracy may be constructed as shown in Fig. 2.4. The geometric modeling has been structured so that the bulk charge important to V_{T0} exists in a trapezoidal volume. The bottom of the trapezoid has a length L_1 such that

$$L = L_1 + 2(\Delta L), \tag{2.2-3}$$

as verified by the dimensions shown in Figs. 2.4(a) and (b). ΔL is the lateral extent of the depletion charge that is attributed to the pn junctions. With the

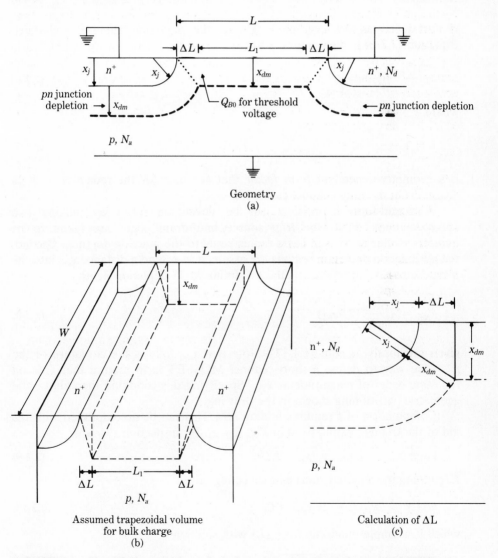

Geometry
(a)

Assumed trapezoidal volume
for bulk charge
(b)

Calculation of ΔL
(c)

FIGURE 2.4   Simplified model for calculation of V_{T0} in a short-channel MOSFET.

choice of bias $V_S = 0 = V_D$, ΔL as given by

$$\Delta L = \frac{L - L_1}{2} \tag{2.2-4}$$

applies to both the source and drain ends of the channel.

To describe the amount of bulk depletion charge that contributes to V_{T0}, the effective channel length

$$L_{\text{eff}} = \frac{L + L_1}{2} = L - \Delta L \tag{2.2-5}$$

is introduced as the average length of the depletion along the channel. Equation (2.2-2) is then modified to read

$$-\frac{Q_{B0} W L_{\text{eff}}}{C_{\text{ox}} W L} = \frac{1}{C_{\text{ox}}} \sqrt{2 q \varepsilon_{\text{Si}} N_a (2 |\phi_F|)} f, \tag{2.2-6}$$

where

$$f = 1 - \frac{\Delta L}{L} < 1 \tag{2.2-7}$$

is a geometry-dependent *form factor* that accounts for the reduction in bulk charge from the long-channel case.

A closed-form expression can be developed for f by making two approximations. First, the n^+p source and drain edges are taken to be quarter-circular arcs; each has a radius equal to the junction depth x_j. Second, the *pn* junction depletion regions are assumed to extend a distance x_{dm} into the *p*-type substrate, where x_{dm} is the maximum MOS depletion depth

$$x_{dm} \simeq \sqrt{\frac{2 \varepsilon_{\text{Si}}}{q N_a} (2 |\phi_F|)} \tag{2.2-8}$$

given previously as eqn. (1.1-17). Note that x_{dm} follows the curvature of the n^+. One way to define a short-channel MOSFET is to require that L is on the same order of magnitude as x_{dm}. Specifying this condition then allows the geometric partitioning shown in the drawing.

Computation of f requires finding ΔL. This can be accomplished with the aid of the triangle emphasized in Fig. 2.4(c). By inspection,

$$(x_j + x_{dm})^2 = x_{dm}^2 + (x_j + \Delta L)^2. \tag{2.2-9}$$

Expanding the squared sums and canceling gives

$$(\Delta L)^2 + 2 x_j \Delta L - 2 x_j x_{dm} = 0, \tag{2.2-10}$$

which is a simple quadratic for (ΔL) with solution

$$\Delta L = -x_j + \sqrt{x_j^2 + 2 x_j x_{dm}}. \tag{2.2-11}$$

Using this in eqn. (2.2-7) then yields

$$f = 1 - \frac{x_j}{L}\left[\sqrt{1 + \frac{2x_{dm}}{x_j}} - 1\right] \tag{2.2-12}$$

as an approximate expression for f within the limits of this model.

The threshold voltage due to the SCE is obtained by simply replacing the third term in eqn. (2.2-1) with eqn. (2.2-6). Thus,

$$(V_{T0})_{\text{SCE}} = V_{FB} + 2|\phi_F| + \frac{qD_I}{C_{\text{ox}}}$$
$$+ \frac{1}{C_{\text{ox}}}\sqrt{2q\varepsilon_{\text{Si}}N_a(2|\phi_F|)}\left[1 - \frac{x_j}{L}\left(\sqrt{1 + \frac{2x_{dm}}{x_j}} - 1\right)\right] \tag{2.2-13}$$

gives the complete expression for the threshold voltage when short-channel effects are included. It is convenient to examine the amount of threshold voltage shift

$$(\Delta V_{T0})_{\text{SCE}} = (V_{T0})_{\text{SCE}} - V_{T0} \tag{2.2-14}$$

induced by the short-channel geometry. This is easily computed to be

$$(\Delta V_{T0})_{\text{SCE}} = -\frac{1}{C_{\text{ox}}}\sqrt{2q\varepsilon_{\text{Si}}N_a(2|\phi_F|)}\left(\frac{x_j}{L}\right)\left(\sqrt{1 + \frac{2x_{dm}}{x_j}} - 1\right) < 0, \tag{2.2-15}$$

which clearly shows the reduction from the GCA value. Since this voltage shift is proportional to (x_j/L), one may alternately define a short-channel MOSFET as one in which L becomes comparable to the junction depth x_j of the n^+ regions.

The plot in Fig. 2.5 provides an example of the threshold voltage variations predicted by eqn. (2.2-13). Note that $(V_{T0})_{\text{SCE}}$ is sensitive to the basic processing parameters N_a, x_j, and x_{ox}. In particular, increasing the substrate doping (as required in scaling theory) enhances the effect.

The threshold voltage of a short-channel MOSFET is also sensitive to any

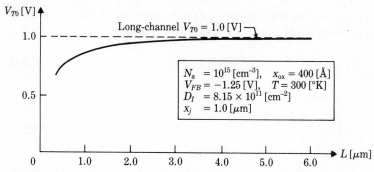

FIGURE 2.5   Short-channel threshold voltage.

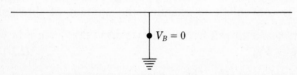

FIGURE 2.6 Model for calculation of SCE threshold voltage with drain-source bias.

drain-source voltage that may exist across the device. This occurs because V_{DS} reverse-biases the drain-substrate junction, increasing the *pn* depletion charge in that region.

The analysis above may be modified to provide an estimate of this effect. Figure 2.6 shows a simplified cross-sectional drawing of the depletion-edge shape when V_{DS} is applied. The trapezoid is asymmetric, with

$$L = L_1 + \Delta L_s + \Delta L_D. \tag{2.2-16}$$

This changes the form factor to

$$f = 1 - \frac{1}{2L}(\Delta L_S + \Delta L_D). \tag{2.2-17}$$

Reference is now made to the depletion depths underneath the n^+ regions. Using the step-profile results, these are given by

$$x_{pS} \simeq \sqrt{\frac{2\varepsilon_{Si}}{qN_a}(\phi_o)}, \qquad x_{pD} \simeq \sqrt{\frac{2\varepsilon_{Si}}{qN_a}(\phi_o + V_{DS})}, \tag{2.2-18}$$

where

$$\phi_o = \left(\frac{kT}{q}\right)\ln\left(\frac{N_dN_a}{n_i^2}\right) \tag{2.2-19}$$

is the junction built-in voltage.

The lowest order of approximation for the bulk charge geometry is obtained by estimating

$$\Delta L_S \simeq x_{pS}, \qquad \Delta L_D \simeq x_{pD}(V_{DS}). \tag{2.2-20}$$

Employing the same techniques as before then yields

$$(\Delta V_{T0})_{\text{SCE}} \simeq -\frac{1}{C_{\text{ox}}} \sqrt{2q\varepsilon_{\text{Si}}N_a(2\,|\phi_F|)}\frac{1}{2}\left[\left(\sqrt{1 + \frac{2\Delta L_S}{x_j}} - 1\right)\right.$$
$$\left. + \left(\sqrt{1 + \frac{2\Delta L_D}{x_j}} - 1\right)\right], \tag{2.2-21}$$

which shows the dependence on V_{DS} through ΔL_D.

2.2.2 Narrow-Width Effects

Narrow-width MOSFETS are defined to have channel widths W on the order of x_{dm}. These devices exhibit *narrow-width effects* (NWEs) that change the characteristics. The most important NWE is that the threshold voltage is larger than predicted from the basic MOS analysis.

The origin of this NWE can be understood by referring to the MOSFET structure shown in Fig. 2.7(a). This is typical of the geometry found in a LOCOS (local oxidation of silicon) technology. The channel width W is defined to be the planar portion of the p-type substrate, which has a gate oxide thickness x_{ox} over it. The GCA analysis assumes that the bulk depletion charge required in V_{T0} is contained within the channel region having width W. However, the geometry of the device shows that additional depletion occurs beyond this limit. Accounting for this extra charge gives a larger value of V_{T0} than predicted by eqn. (2.2-1). Although this contribution is negligible in wide devices, it becomes increasingly important as smaller devices are considered.

Two features shown in the drawing deserve additional comments before proceeding into the analysis. First, note that the MOFSET is surrounded by a *field oxide* (FOX) with a thickness $X_{\text{FOX}} > x_{\text{ox}}$. This is included as a form of device *isolation* in the LOCOS technology. It is required because the poly gate level is also used as an interconnect. The FOX capacitance per unit area,

$$C_{\text{FOX}} = \frac{\varepsilon_{\text{ox}}}{X_{\text{FOX}}} < C_{\text{ox}}, \tag{2.2-22}$$

contributes to the *field threshold voltage* $(V_{T0})_{\text{FOX}}$. It is necessary to keep $(V_{T0})_{\text{FOX}}$ large to avoid forming an unwanted inversion layer under the field oxide. If this region of the silicon surface is inverted, it creates an electrical path to a neighboring device. This, of course, must be avoided. The second feature to notice is the extra p^+ layer under the field oxide. This is known as a *channel stop* (or *field implant*), as it is used to increase $(V_{T0})_{\text{FOX}}$ to a value large enough to avoid the formation of stray channels.

The transition from the gate to the field oxide in a LOCOS technology is commonly known as the *bird's beak* region. An expanded drawing of the bird's

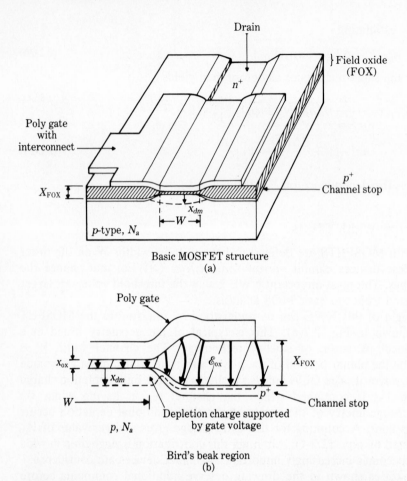

FIGURE 2.7 Narrow-width effects in a LOCOS MOSFET.

beak is shown in Fig. 2.7(b). The terminology arises from the shape of the oxide transition. The emphasized lines in the drawing show the oxide electric field $\mathscr{E}_{ox}$ when a gate voltage is applied. This illustrates the point that the capacitance associated with the bird's beak induces bulk charge that must be supported by the gate voltage. The actual threshold voltage of the MOSFET is larger than the GCA value since both the bird's beak and FOX depletion regions must be accounted for.

The geometric complexity of the LOCOS bird's beak prohibits a closed-form analytic derivation of the NWE threshold voltage. Owing to this fact, the present discussion will center around the simpler MOSFET structure shown in Fig. 2.8. This type of device results from a *thick field oxide* process and is representative of older MOS technologies. The channel width W is defined by the steplike transition from the gate to field oxide regions. Note that the entire silicon surface is taken to be planar, which simplifies the problem considerably.

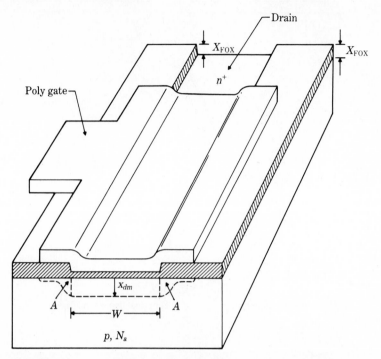

FIGURE 2.8   A thick field oxide MOSFET for NWE calculations.

Although p^+ channel stops are usually included, they will be ignored in the analysis.

Calculation of the threshold voltage increase brought about by NWEs centers around finding the extra bulk charge. This may be accomplished by first writing the numerator in eqn. (2.2-2) as

$$-Q_{B0}LW = qN_a x_{dm}WL \text{ [C]}, \tag{2.2-23}$$

which is valid for wide devices. Letting A denote the cross-sectional area of the additional bulk charge (as shown in Fig. 2.8), inclusion of NWEs requires that the total depletion charge term be modified to

$$-Q_{B0}LW \to qN_a L(x_{dm}W + 2A) \text{ [C]}. \tag{2.2-24}$$

Consequently, the bulk charge voltage support term originally given in eqn. (2.2-2) now reads as

$$-\frac{Q_{B0}}{C_{\text{ox}}} = \frac{qN_a x_{dm}}{C_{\text{ox}}} g, \tag{2.2-25}$$

where g is the NWE form factor

$$g = 1 + \frac{2A}{x_{dm}W} > 1. \tag{2.2-26}$$

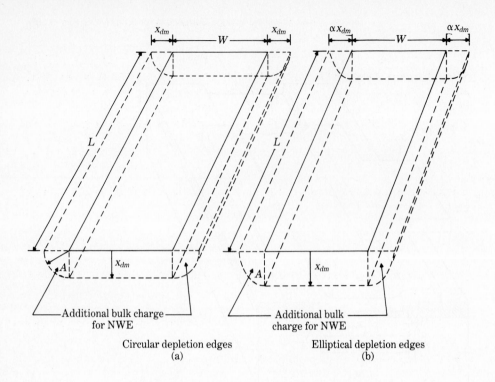

Circular depletion edges
(a)

Elliptical depletion edges
(b)

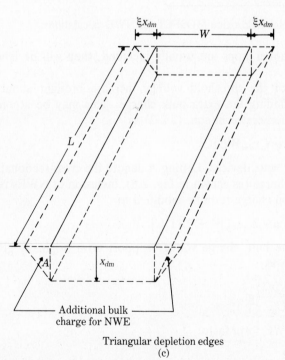

Triangular depletion edges
(c)

FIGURE 2.9 Models for computing depletion contributions.

Even though the thick field oxide MOSFET is simpler than the LOCOS structure, it is still necessary to make approximations for the cross-sectional area A required to compute the form factor. Three simple geometries are illustrated in Fig. 2.9. In Fig. 2.9(a), the depletion edges are modeled as quarter-circular arcs with radius x_{dm}. Thus, $A = (1/4)\pi x_{dm}$, which gives

$$g = 1 + \frac{\pi x_{dm}}{2W}. \tag{2.2-27}$$

Elliptical contours are used in Fig. 2.9(b) such that $A = (1/4)\alpha x_{dm}$. For this case,

$$g = 1 + \frac{\pi \alpha x_{dm}}{2W}. \tag{2.2-28}$$

Finally, Fig. 2.9(c) assumes a triangular cross-section with $A = (1/2)\xi x_{dm}$. This choice sets

$$g = 1 + \frac{\xi x_{dm}}{W}. \tag{2.2-29}$$

The above variations may be summarized by writing

$$g = 1 + \frac{\kappa x_{dm}}{W}, \tag{2.2-30}$$

where κ is chosen according to the assumed depletion edge contour.
   The complete threshold voltage expression now becomes

$$(V_{T0})_{\text{NWE}} = V_{FB} + 2|\phi_F| + \frac{qD_I}{C_{\text{ox}}} + \frac{1}{C_{\text{ox}}}\sqrt{2q\varepsilon_{\text{Si}}N_a(2|\phi_F|)}\left(1 + \frac{\kappa x_{dm}}{W}\right) \tag{2.2-31}$$

so that the change

$$(\Delta V_{T0})_{\text{NWE}} = (V_{T0})_{\text{NWE}} - V_{T0} \tag{2.2-32}$$

induced by the narrow-width effect is given by

$$(\Delta V_{T0})_{\text{NWE}} = +\frac{1}{C_{\text{ox}}}\sqrt{2q\varepsilon_{\text{Si}}N_a(2|\phi_F|)}\left(\frac{\kappa x_{dm}}{W}\right) > 0. \tag{2.2-33}$$

The presence of the factor (x_{dm}/W) justifies the original definition of a narrow-width MOSFET. A typical variation of the threshold voltage as predicted by this analysis is illustrated in Fig. 2.10 for $\kappa = (\pi/2)$, corresponding to an assumed circular depletion contour. Note that $(V_{T0})_{\text{NWE}}$ is sensitive to the basic processing parameters N_a and x_{ox}.
   As a final point in the NWE discussion, it should be emphasized that the actual amount of threshold voltage shift is strongly dependent on the device geometry. Although the LOCOS and thick field oxide processes have been

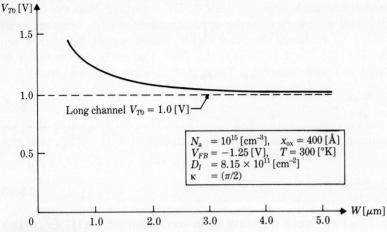

FIGURE 2.10   Narrow-width threshold voltage.

chosen as examples, the general technique can, in principle, be modified for application to any new processing technology. An example of an alternate structure is shown in Fig. 2.11. This represents a *fully recessed* LOCOS MOSFET, with its name arising from the fact that the field oxide has been "buried" in the substrate to give a relatively flat oxide surface. It is seen that NWEs will still be important in this type of device.

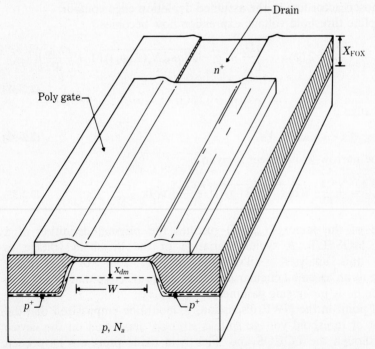

FIGURE 2.11   A full-recessed LOCOS MOSFET showing NWE effects.

2.2.3 Minimum-Size Effects

A minimum-size MOSFET is defined to have the smallest L and W allowed by the processing technology. Such a device is important because it represents the minimum real estate requirement and thus allows for maximum logic integration levels. Note that this type of MOSFET has a set value for β since its aspect ratio is predefined. Consequently, (W/L) cannot be used as a circuit design variable if minimum-size transistors are desired.

Minimum-size effects (MSEs) alter the threshold voltage when both L and W are on the order of x_{dm}. A first-order estimate of the threshold voltage shift induced by MSEs can be obtained by superposing the short-channel and narrow-width corrections to write

$$(\Delta V_{T0})_{\text{MSE}} \simeq (\Delta V_{T0})_{\text{SCE}} + (\Delta V_{T0})_{\text{NWE}} \tag{2.2-34}$$

such that

$$(V_{T0})_{\text{MSE}} = V_{T0} + (\Delta V_{T0})_{\text{MSE}}. \tag{2.2-35}$$

Using eqns. (2.2-15) and (2.2-33) gives

$$(\Delta V_{T0})_{\text{MSE}} \simeq \frac{\sqrt{2q\varepsilon_{\text{Si}}N_a(2\,|\phi_F|)}}{C_{\text{ox}}}\left(\frac{x_j}{L} + \frac{\kappa x_{dm}}{W} - \frac{x_j}{L}\sqrt{1 + \frac{2x_{dm}}{x_j}}\right) \tag{2.2-36}$$

MSEs can either increase or decrease the threshold voltage, depending on the relative values of L, W, x_j, and x_{dm}.

It is interesting to note that this equation predicts $(\Delta V_{T0})_{\text{MSE}} = 0$ if

$$x_j = \frac{(\kappa L/W)^2 x_{dm}}{2\left(1 - \kappa\dfrac{L}{W}\right)}. \tag{2.2-37}$$

Since x_j must be a positive number, satisfaction of this condition requires that

$$\kappa\frac{L}{W} < 1. \tag{2.2-38}$$

In a minimum-size MOSFET, L is approximately equal to W. The analysis is therefore quite sensitive to the choice of κ used to approximate the NWE depletion-edge contour. This compensation can be used to reduce the MSE threshold voltage changes in a realistic device.

Equation (2.2-36) overestimates the positive shift induced by NWEs. It is possible to obtain a more accurate expression by carefully examining the geometry of the MSE bulk charge geometry. In Fig. 2.12(a), the NWE depletion volume is shown for an assumed triangular region, i.e., $\kappa = \xi$. The drawing has been inverted from that shown previously in Fig. 2.9(c) to aid in the visualization. Addition of short-channel effects reduces the bulk charge as shown in Fig. 2.12(b). The resulting geometry indicates that the amount of extra charge induced by NWEs has been decreased because of the volume lost by the short-channel angle.

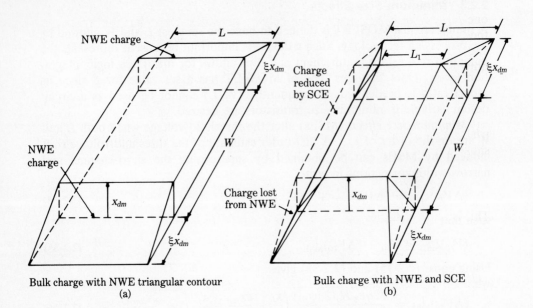

Bulk charge with NWE triangular contour
(a)

Bulk charge with NWE and SCE
(b)

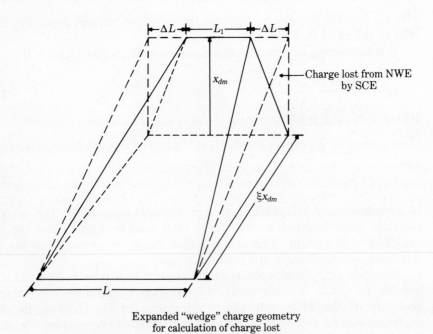

Expanded "wedge" charge geometry
for calculation of charge lost
from NWE by the SCE
(c)

FIGURE 2.12   Geometry for computation of MSE threshold voltage.

Consider the "wedge" geometry illustrated in Fig. 2.12(c). This represents one side of the NWE charge. The volume of the depletion that contributes to the NWE threshold shift may be computed by first finding the wedge volume and then subtracting the volume of the "pyramid-shaped" regions shown in dotted lines. The total volume of importance is thus

$$\mathcal{V} = 2\left[\frac{1}{2}\xi L x_{dm}^2 - \frac{1}{3}\xi(\Delta L)x_{dm}^2\right],$$

(2.2-39)

where the factor of 2 accounts for both wedges in the device. Using the short-channel ΔL from eqn. (2.2-11) gives

$$\mathcal{V} = \xi x_{dm}^2 L\left[1 - \left(\frac{2}{3}\right)\frac{x_j}{L}\left(\sqrt{1 + \frac{2x_{dm}}{x_j}} - 1\right)\right].$$

(2.2-40)

This analysis modifies the NWE form factor to read

$$g = 1 + \frac{\xi x_{dm}}{W} - \frac{2}{3}\left(\frac{\xi x_{dm}}{W}\right)\left(\frac{x_j}{L}\right)\left(\sqrt{1 + \frac{2x_{dm}}{x_j}} - 1\right),$$

(2.2-41)

which is easily verified by writing eqn. (2.2-24) as

$$-Q_{B0}LW \rightarrow qN_a[x_{dm}LW + \mathcal{V}].$$

(2.2-42)

The expression for $(\Delta V_T)_{\text{SCE}}$ remains unchanged by the MSE analysis. Thus, the NWE form factor above gives

$$(\Delta V_{T0})_{\text{MSE}} = \frac{\sqrt{2q\varepsilon_{\text{Si}}N_a(2\,|\phi_F|)}}{C_{\text{ox}}}\left[\frac{x_j}{L} + \frac{\xi x_{dm}}{W} - \left(1 + \frac{2\xi x_{dm}}{3W}\right)\left(\frac{x_j}{L}\right)\sqrt{1 + \frac{2x_{dm}}{x_j}}\right].$$

(2.2-43)

It is seen that the positive threshold shift from NWEs has been reduced by the extra term that originated from the volume reduction calculation.

This concludes the discussion of the most important threshold voltage variations experienced in small-geometry MOSFETs. It must be remembered that the equations developed in this section represent only first-order estimates of the various effects. The accuracy of the calculations depends primarily on the correlation between the actual and assumed geometries of the depletion regions.

2.3 Two-Dimensional MOSFET Effects

The gradual channel approximation reduces the three-dimensional MOSFET analysis down to a one-dimensional current flow problem that can be solved analytically. The simplification works quite well for large devices but breaks down as the size of the MOSFET is reduced. Device characteristics are then influenced by physical effects that are ignored in the one-dimensional treatment.

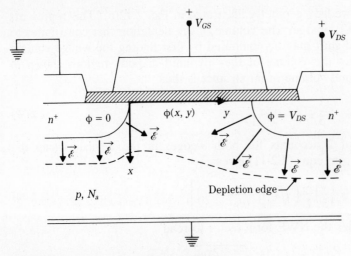

FIGURE 2.13   Two-dimensional potential region.

Extracting device physics in two or three dimensions is generally classified as being analytically intractable. Numerical simulations become mandatory because of the complexity of the coupled differential equation set. Details of this type of analyses can be found in the literature and are beyond the scope of the present discussion. Instead, interest will be directed toward obtaining a qualitative feeling for the physics of small devices by examining some important two-dimensional effects.

The two-dimensional nature of the small-geometry MOSFET problem can be seen in Fig. 2.13. The control voltages are set at V_{GS} and V_{DS}. Inside the transistor, the electrostatic potential $\phi(x, y)$ is described by the Poisson equation

$$\nabla^2\phi(x, y) = -\frac{\rho(x, y)}{\varepsilon_{Si}}, \tag{2.3-1}$$

where $\phi(x, y)$ must satisfy the boundary conditions established by the external voltages and the semiconductor physics. For example, the source is grounded so that

$$\phi(x_S, y_S) = 0, \tag{2.3-2}$$

with (x_S, y_S) denoting the coordinates describing the source n^+ boundary. At the drain, the boundary condition becomes

$$\phi(x_D, y_D) = V_{DS}. \tag{2.3-3}$$

The electric field in the device is obtained from

$$\vec{\mathscr{E}} = -\vec{\nabla}\phi(x, y)$$

$$= -\frac{\partial\phi}{\partial x}\hat{a}_x - \frac{\partial\phi}{\partial y}\hat{a}_y \tag{2.3-4}$$

such that $\vec{\mathscr{E}}$-field lines are perpendicular to lines of constant potential. Electric field lines originate on positively charged surfaces.

The GCA makes two assumptions on the electric fields: (1) the channel electric field is parallel to the surface and points from drain to source, while (2) the depletion electric field is perpendicular to the surface, with field lines terminating on ionized acceptors. The electric field components are

$$\mathscr{E}_{\text{channel}} \simeq \mathscr{E}_y, \qquad \mathscr{E}_{\text{depletion}} \simeq \mathscr{E}_x, \tag{2.3-5}$$

indicating that they are decoupled. The two-dimensional analysis maintains the vector characteristics of $\vec{\mathscr{E}}$. This alters the viewpoint used to describe the MOSFET by directing interest toward the variations in the potential along the surface.

Current flow through the MOSFET is established by creating an inversion layer with $V_{GS} > V_{T0}$. For $V_{GS} < V_{T0}$, the current flow is viewed as being blocked by a *barrier potential* ϕ_B. Increasing the gate-source voltage to $V_{GS} > V_{T0}$ reduces the barrier potential to $\phi_B = 0$, allowing drift current flow. The barrier potential is also affected by V_{DS} in the two-dimensional treatment.

The most important two-dimensional effects in small MOSFETs arise from variations in the barrier potential with $V_{GS} < V_{T0}$, which is assumed for the rest of the section. In the GCA, this would correspond to cutoff. However, the physics of barrier control shows that current flow does exist. This is the result of *weak-inversion* or *subthreshold* conditions in the device. Simple analytic models proposed for ϕ_B allow the physics to be studied without excessive mathematics.

Figure 2.14 schematically illustrates how two-dimensional effects alter the internal potential distribution in a small MOSFET. The electric field lines induced by V_{DS} point away from the n^+ drain region. In a short-channel MOSFET, a significant fraction point toward the source, changing the

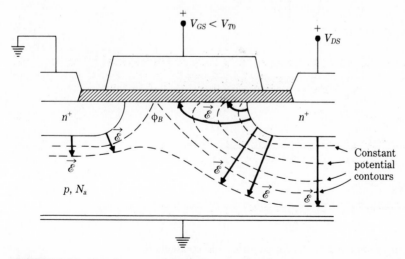

FIGURE 2.14 Barrier control by external voltages.

potential barrier seen by electrons. A simple model for the barrier potential is

$$\phi_B = \phi_r + AV_{GS} + BV_{DS}, \tag{2.3-6}$$

where $\phi_r < 0$ is a reference potential and A, B are structural constants determined by the geometry. A is less than unity, and $B \ll A$ generally holds in this model. The expression for ϕ_B shows a reduction in barrier magnitude with increasing V_{DS}. This effect has been termed *drain-induced barrier lowering* (DIBL), *static induction,* and the *second gate* effect in the literature. If V_{DS} is increased (or the channel length is decreased), the barrier-control equation changes to

$$\phi_B = \phi_r + AV_{GS} + BV_{DS} - CV_{GS}V_{DS}. \tag{2.3-7}$$

This results in the condition classically termed *punch-through.*

Integrated circuit design can be influenced by the current flow created in barrier reduction. Since this occurs for voltages $V_{GS} < V_{T0}$, it is termed *subthreshold* current. The most important point about subthreshold current is that $V_{GS} < V_{T0}$ does not give $I_D = 0$. Rather, both subthreshold and reverse-leakage currents exist. Dynamic circuits are particularly sensitive to this effect.

A simple barrier-control model for subthreshold current is based on the geometry in Fig. 2.15. With a short channel length L, the lateral n^+pn^+ layering resembles a bipolar *npn* transistor! The operation is similar. A reduced barrier allows electrons to diffuse away from the source. The electron diffusion current in the p-region is

$$I = qD_n A \frac{dn(y)}{dy}, \tag{2.3-8}$$

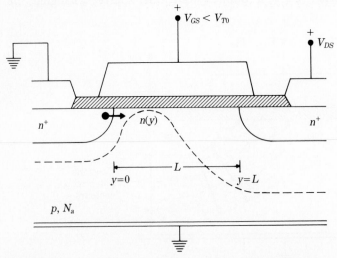

FIGURE 2.15   Simplified model for subthreshold conduction.

where D_n is the electron diffusion coefficient and

$$A = Wx_c \tag{2.3-9}$$

is the cross-sectional current flow area. W is the channel width and x_c is the subthreshold channel thickness. The carrier gradient may be approximated to give

$$I \simeq \frac{qD_nWx_c}{L_B}[n(0) - n(L)], \tag{2.3-10}$$

where L_B is the length of the barrier region $(L_B < L)$. Barrier control is described by the exponential dependence

$$\Delta n \simeq n_o e^{q\phi_B/kT}, \tag{2.3-11}$$

with n_o a constant. This is combined with eqn. (2.3-6) to give the approximate expression

$$I_{\text{sub}} \simeq \frac{qD_nWx_cn_o}{L_B}e^{q(\phi_r+AV_{GS}+BV_{DS})/kT} \tag{2.3-12}$$

for the subthreshold current. Although this is at best a first-order approximation, it displays the exponential voltage dependence

$$I_{\text{sub}} \propto e^{qAV_{GS}/kT}e^{qBV_{DS}/kT}, \tag{2.3-13}$$

which is consistent with more accurate models. Subthreshold current flow is illustrated by the plot in Fig. 2.16.

The two-dimensional properties of MOSFETs are still under study. Many problems exist, particularly in the area of obtaining accurate device models for circuit design. As chip dimensions decrease, the number of unanswered questions seems to grow. Device modeling thus remains a viable research topic.

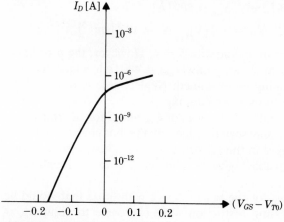

FIGURE 2.16 Typical subthreshold current dependence.

REFERENCES

(1) L. Akers, "Small Geometry Effects," Chap. 6 in D. Ong, *Modern MOS Technology*, New York: McGraw-Hill, 1984.

(2) D. F. Barbe (ed.), *Very Large Scale Integration*, 2nd ed., New York: Springer-Verlag, 1982.

(3) W. L. Engl, H. K. Dirks, and B. Meinterzhage, "Device Modelling," *Proc. IEEE*, Vol. 71, pp. 10–33, 1983.

(4) B. Hoefflinger, "MOS Technologies and Devices for LSI," pp. 399–456 in L. Esaki and G. Soncini (eds.), *Large Scale Integrated Circuits Technology: Stage of the Art and Prospects*, NATO Advanced Study Institute Series, The Hague: Martinus Nijhoff, 1982.

(5) R. S. Muller and T. I. Kamins, *Device Electronics for Integrated Circuits*, 2nd ed., New York: Wiley, 1986.

(6) S. Selberherr, A. Schutz, and H. Potzl, "Two Dimensional MOS Transistor Modelling," pp. 490–581 in P. Antognetti, D. A. Antoniadis, R. W. Dutton, and W. G. Oldham (eds.), *Process and Device Simulation for MOS-VLSI Circuits*, NATO Advanced Study Institute Series, The Hague: Martinus Nijhoff, 1983.

(7) S. M. Sze, *Physics of Semiconductor Devices*, 2nd ed., New York: Wiley, 1981.

(8) C. Turchetti and G. Masetti, "A Charge-Sheet Analysis of Short-Channel Enhancement-Mode MOSFETs," *IEEE J. Solid-State Circuits*, Vol. SC-21, pp. 267–275, 1986.

PROBLEMS

2.1 A MOSFET is fabricated with the following parameters:

$$W = 10\,[\mu\text{m}], \quad L = 6\,[\mu\text{m}], \quad x_{ox} = 800\,[\text{Å}],$$

$$\mu_n = 580\,[\text{cm}^2/\text{V-sec}], \quad V_{T0} = 1.0\,[\text{V}], \quad N_a = 10^{15}\,[\text{cm}^{-3}].$$

The device is subjected to full scaling with $S = 4$. However, the processing limits the oxide thickness to a value $\min(x_{ox}) = 400\,[\text{Å}]$, which requires that the overall device scaling be examined. Neglect second-order effects such as changes in μ_n due to increased doping.

(a) Calculate the original and scaled values of C_{ox} and β for the transistor.

(b) Calculate the original and scaled values of the body bias coefficient $\gamma\,[V^{1/2}]$. Include changes in the substrate doping. Then, assuming that $V'_{T0} = (V_{T0}/S)$ is possible, give a plot of V'_T as a function of $V'_{SB} = (V_{SB}/S)$.

(c) Compare the equations for the scaled device with those that would be obtained if the processing restriction were overcome to allow $x'_{ox} = (x_{ox}/S)$.

(d) Compare the power dissipation for the three cases (i) original, (ii) scaled with min $(x_{ox}) = 400\,[\text{Å}]$, and (iii) complete scaling with no oxide thickness minimum.

(e) As a final case, find the power dissipation if min $(x_{ox}) = 400\,[\text{Å}]$ but where constant voltage scaling is invoked instead.

2.2 Consider a MOSFET that has a gate oxide thickness of $x_{ox} = 300\,[\text{Å}]$. The oxide electric field breakdown value is approximately $\mathscr{E}_{ox,BD} \simeq 5 \times 10^6\,[\text{V/cm}]$. Ignore oxide charge and treat the MOS system as being ideal.

(a) Calculate the maximum oxide voltage V_{ox} that can be applied to the device. Discuss how this limits scaling theory.

(b) Suppose that the gate dimensions are given as $L = W = 2\,[\mu\text{m}]$. Calculate the total number of electrons that is needed to induce breakdown on the gate MOS system. (Note that Gauss's law gives

$$\varepsilon_{ox}\mathscr{E}_{ox} = \rho_s,$$

where $\rho_s\,[\text{C/cm}^2]$ is the surface charge density.)

(c) Suppose that the gate is initially uncharged. A gate current of $I_G = 0.1\,[\mu\text{A}]$ flows. How long will it take to accumulate the critical charge needed to induce breakdown?

2.3 A MOSFET is characterized by the following parameters:

$$x_{ox} = 300\,[\text{Å}], \quad N_a = 8 \times 10^{14}\,[\text{cm}^{-3}], \quad x_j = 0.45\,[\mu\text{m}].$$

The threshold voltage for a "large" MOSFET made in this process is given by $V_{T0} = +0.70\,[\text{V}]$. Assume that $L = 1.4\,[\mu\text{m}]$ and $W = 1.6\,[\mu\text{m}]$.

(a) Calculate the value of x_{dm} in $[\mu\text{m}]$.

(b) Find the short-channel form factor f as predicted by eqn. (2.2-12). Then find $(V_{T0})_{SCE}$ using eqn. (2.2-15).

(c) Find the narrow-width form factor g if the depletion edges are modeled by quarter-circular arcs. Repeat the calculation for a triangular depletion edge with $\xi = 0.75$.

(d) Calculate $(V_{T0})_{NWE}$ for both cases analyzed in part (c).

(e) Estimate $(V_{T0})_{MSE}$ using eqn. (2.2-36). Then compare this with the value found using eqn. (2.3-43). Assume that $\kappa = \xi = 0.75$ for both calculations.

2.4 Apply scaling theory to the equations developed in Section 1.5 for the complete GCA MOSFET equations. Examine the scaling of the depletion widths and the $I - V$ equations. What types of restrictions apply to scaling theory in this case?

2.5 Consider the subthreshold current expression given in eqn. (2.3-12). For $V_{DS} \geq 3(kT/q)$, the drain current is found to be approximately independent of V_{DS} so that

$$I_{sub} \simeq I_1 e^{qAV_{GS}/kT}.$$

Suppose that I_{sub} is measured to be $0.5\,[\mu A]$ when a voltage of $V_{GS} = -0.1\,[V]$ is applied to the MOSFET, which is at $27\,[°C]$.

(a) Assume that $A = 1/2$. Find I_1, and then plot I_{sub} vs. V_{GS} for V_{GS} in the range $[-0.5, 0]\,[V]$. Use a semilog plot and assume $T = 27\,[°C]$.

(b) Repeat the calculation in (a) for $T = 60\,[°C]$.

(c) Now suppose that A is readjusted to a value of $A = 1/2.5$. Find the new value of I_1 for this case, and compute the percent change in I_1 from the case in (a).

(d) Discuss the expected behavior of I_{sub} if the channel length L is reduced.

2.6 The treatment of scaling theory in this chapter assumed that all three dimensions of the MOSFET could be reduced by a single scaling factor S. Suppose instead that the surface dimensions are reduced by S_1 (for y and z) while the vertical dimensions are reduced by $S_2 \neq S_1$ (in the x direction).

Examine the behavior of the Poisson equation (2.1-4) for this case. (The *anisotropic* case here does not simplify too much.)

CHAPTER 3

DC Characteristics of MOS Inverters

This chapter begins the discussion of digital circuits in a MOS-based technology by analyzing the simple inverter. Various configurations will be examined to gain an understanding of the DC switching characteristics. As will be seen in later chapters, inverter concepts can be extended to more complicated logic circuits.

Analytical and computer-based techniques for circuit analysis and design are introduced in the discussion. Studying the algorithms and examples will allow a deeper understanding of the circuit performance. Programming is highly recommended, as it greatly enhances the material. In addition, the approach to inverter design will form the basis for higher-level logic configurations in later chapters.

3.1 Basic Digital Concepts

A generalized nMOS inverter is shown in Fig. 3.1. It consists of a *driver* nMOS transistor (MD) and a *load* device, which is as yet unspecified. Although this is a simple-looking circuit, the complexity of the analysis depends on the load device characteristics. It will be shown later that the inverter can be used as a basis for designing more advanced logic implementations in MOS technology; in fact, the inverter analysis will commonly form the first step for the design of a random logic circuit. Owing to the importance of this circuit, the discussions presented here will be relatively detailed.

An *ideal* inverter is shown in Fig. 3.2(a), which gives the standard logic symbol with its truth table. This symbol is generally used without a direct relationship to the voltages in digital circuits; however, such a connection is easily made. Suppose that the circuitry has a minimum voltage of $0\,[V]$ (or ground) and a maximum voltage of V_{DD}. Then, using a *positive logic* convention,

$$\text{Boolean } 1 = V_{DD}, \qquad \text{Boolean } 0 = 0\,[V] \tag{3.1-1}$$

defines the connection between logic variables and voltages.

The assumed DC voltage transfer characteristic (the *VTC* or the *S-curve*) of the ideal inverter is shown in Fig. 3.2(b) and is simply a plot of the output voltage V_{out} as a function of the input voltage V_{in}. This illustrates that the output switches from V_{DD} to $0\,[V]$ at an input voltage of $(V_{DD}/2)$. For $V_{in} < V_{DD}/2$, the output is high, while $V_{in} > V_{DD}/2$ gives a low output voltage. The inverter characteristics thus become obvious from the VTC structure.

A final point concerning the ideal inverter is obtained in connection with Fig. 3.2(c). With $V_{in}(t)$ taken as a step transition at time t_0, the ideal inverter gives an output voltage $V_{out}(t)$, which falls from V_{DD} to $0\,[V]$ at the same time

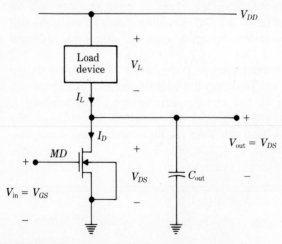

FIGURE 3.1   Basic nMOS inverter circuit.

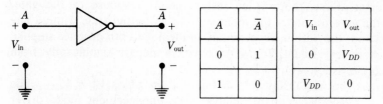

Logic symbol with relation to voltages   Logic and voltage truth tables

(a)

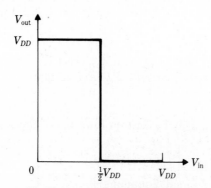

The ideal inverter voltage transfer
characteristic (VTC)

(b)

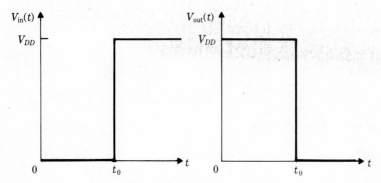

Transient response of the ideal inverter

(c)

FIGURE 3.2   Ideal inverter properties.

t_0 that defines the input transition. This illustrates two transient properties of the ideal inverter. First, there is an instantaneous response of the gate, implying that the gate translates the input state to an output state without any time delay. Second, the output waveform of an ideal gate can produce steplike voltage characteristics. Realistic nMOS inverters depart significantly from these ideal characteristics.

Now consider the nMOS inverter in Fig. 3.1, in which $V_{in} = V_{GS}$, while $V_{out} = V_{DS}$. For purposes of circuit analysis, the enhancement mode driver MOSFET is described by the simplified device equations; these are summarized in Table 3.1 for reference in this chapter. The output node is assumed to be connected to the input of another nMOS gate. In the lowest order of approximation, this node sees a lumped equivalent capacitance C_{out}. The details of finding C_{out} are somewhat complicated and are deferred until the next chapter.

The VTC for the inverter describes $V_{out}(V_{in})$ for DC voltages. Since the load capacitor C_{out} cannot pass DC current, KCL gives that

$$I_L = I_D(V_{in}, V_{out}). \tag{3.1-2}$$

The output voltage $V_{out} = V_{DS}$ may be written in terms of the load voltage V_L

TABLE 3.1   Summary of nMOS Circuit Equations

Circuit Equation Summary
n-Channel MOSFET

Cutoff $(V_{GS} < V_T)$

$I_D \approx 0$

Active Operation $(V_{GS} > V_T)$

Saturation voltage: $V_{DS,sat} = V_{GS} - V_T$

Nonsaturated current: $V_{DS} < V_{DS,sat}$

$$I_D = \frac{\beta}{2}[2(V_{GS} - V_T)V_{DS} - V_{DS}^2]$$

Saturated current: $V_{DS} \geq V_{DS,sat}$

$$I_D = \frac{\beta}{2}(V_{GS} - V_T)^2(1 + \lambda V_{DS})$$

Threshold Voltage

$$V_T = V_{T0} + \gamma(\sqrt{V_{SB} + 2|\phi_F|} - \sqrt{2|\phi_F|})$$

Device Transconductance

$$\beta = k'\left(\frac{W}{L}\right), \qquad k' = \mu_n C_{ox}$$

---

as

$$V_{\text{out}} = V_{DD} - V_L(I_L). \tag{3.1-3}$$

If the functional dependence $V_L(I_L = I_D)$ is known for the load device, then this will serve as a load line equation that can be directly plotted onto the driver characteristics $I_D(V_{DS} = V_{\text{out}})$. Varying $V_{\text{in}} = V_{GS}$ from $0\,[\text{V}]$ to V_{DD} will produce the usual nMOS curves. The VTC can be obtained graphically by plotting V_{out} for each point where the load line intersects the $V_{\text{in}} = V_{GS}$ device curve. Although the graphical load line technique is useful for obtaining a qualitative feeling for the behavior of the different types of load devices, it is not capable of providing an accurate basis for either the analysis or the design of the circuit. Rather, a direct mathematical approach must be developed.

The analytic derivation of the VTC requires a knowledge of the load I_L-V_L relationship for direct substitution into eqn. (3.1-2). To illustrate the process of determining the VTC, suppose that V_{in} is (slowly) varied from $0\,[\text{V}]$ to V_{DD}. When $V_{\text{in}} < V_T$, the driver is in the cutoff mode with $I_D = 0$. This results in most of the voltage being dropped across the drain-source electrodes of the driver, so that $V_{\text{out}} = V_{DS}$ is close to V_{DD}. This particular output voltage is denoted by V_{OH}, which is the *output high* level of the inverter. As V_{in} is increased to values greater than V_T, drain current starts to flow. Initially, $V_{DS} = V_{\text{out}} > (V_{GS} - V_T)$, so that the device is in saturation. Hence,

$$\frac{\beta_D}{2}(V_{\text{in}} - V_T)^2(1 + \lambda V_{\text{out}}) = I_L(V_L), \tag{3.1-4}$$

with $V_L = (V_{DD} - V_{\text{out}})$ providing the required VTC equation. In this equation, λ is the channel length modulation factor, which will usually be taken as $\lambda = 0$ for simplicity. Also, the subscript D is used to denote the driver MOSFET quantities. This remains valid until V_{out} drops to $V_{\text{out}} = (V_{GS} - V_T)$, which defines the border between the saturation and nonsaturation regions of operation. As V_{out} drops below $(V_{GS} - V_T)$, the driver enters the non-saturated operational mode, so that the VTC equation is given by

$$\frac{\beta_D}{2}[2(V_{\text{in}} - V_T)V_{\text{out}} - V_{\text{out}}^2] = I_L(V_L). \tag{3.1-5}$$

Now, as V_{in} starts to approach the power supply level V_{DD}, the output voltage reaches the value of $V_{\text{out}} = V_{OL}$, which stands for the *output low* value of the inverter. The general shape of the nMOS inverter VTC is illustrated in Fig. 3.3.

The nMOS VTC also shows 3 other important *critical voltages* that are commonly used in characterizing the performance of the inverter. These are V_{IL}, V_{IH}, and V_{th}. Consider first the *input low* voltage V_{IL}. This is defined by the point where

$$\frac{dV_{\text{out}}}{dV_{\text{in}}} = -1, \tag{3.1-6}$$

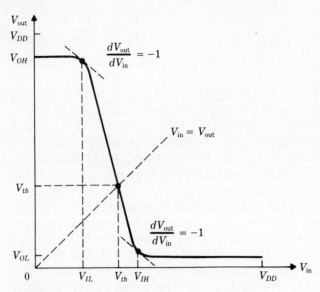

FIGURE 3.3   Generalized VTC of an nMOS inverter.

i.e., where the slope is at -45 degrees. Physically, V_{IL} represents the maximum value that a logic zero input may have and still guarantee a logic 1 (high voltage) at the output. Similarly, V_{IH} is the *input high* voltage and represents the minimum logic 1 input voltage that will guarantee a logic 0 (low voltage) at the output. The actual values of V_{IL} and V_{IH} depend on the specifics of the nMOS inverter circuitry, as does the shape of the transition curve from V_{IL} to V_{IH}.

It is convenient to introduce the concept of a voltage *noise margin,* usually denoted by either NM or VNM. In a binary circuit, there are two voltage noise margins, one for the high states and one for the low states. These are defined by

$$NM_H = V_{OH} - V_{IH}, \qquad NM_L = V_{IL} - V_{OL} \tag{3.1-7}$$

and must be positive numbers for a working digital circuit. The interpretation of the noise margins can be seen in Fig. 3.4, which illustrates the problem of defining high (logic 1) and low (logic 0) voltage states. The basic fact to keep in mind is that voltages are not quantized in nature; a system that has power supply connections V_{DD} and $0\,[V]$ (ground) can exhibit any voltage state between these two limits. This is readily seen by a quick examination of the VTC in Fig. 3.3.

To overcome this particular characteristic, a digital circuit must make use of a range of voltages for the definition of both logic 1 and logic 0 states. The input voltage ranges are

Logic 1: V_{IH} to V_{DD},

Logic 0: 0 to V_{IL}, $\tag{3.1-8a}$

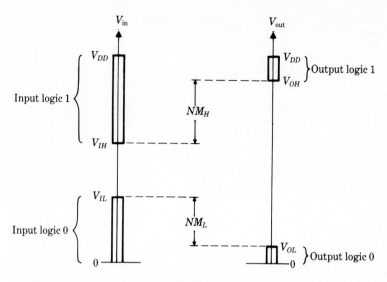

FIGURE 3.4 Noise margin definitions.

while the output ranges are given as

Logic 1: V_{OH} to V_{DD},

Logic 0: 0 to V_{OL}.

(3.1-8b)

The noise margins thus give an indication of how well the circuit operates with regard to "redefining" the two voltage logic levels.

Although the noise margins have been defined in terms of the input and output voltages of an isolated inverter, their name and physical meaning arise from a different situation. Consider two identical gates cascaded as in Fig. 3.5. The output of stage 1 ideally has voltages V_{OL} and V_{OH}. However, when a switching event occurs in the first stage, the properties of the transmission network between the two gates become important. The actual voltage that

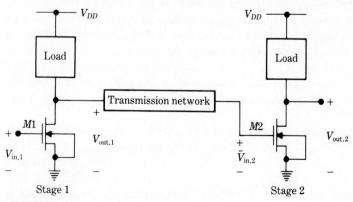

FIGURE 3.5 Two cascaded inverter stages.

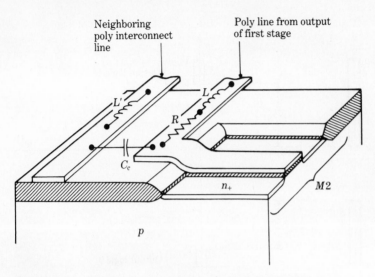

Neighboring
poly interconnect
line

Poly line from output
of first stage

FIGURE 3.6 Model for transmission network problem.

reaches the second stage will be different from that which was sent by the first stage because of losses and line coupling that exist in a realistic chip environment. In general, this says that $V_{in,2} \neq V_{out,1}$.

An example of this type of problem can be seen in Fig. 3.6. The input polysilicon gate to the second stage is used as an interconnect line between the two gates. In addition, the drawing shows another poly interconnect line that has no direct electrical contact with the poly gate line but is in close proximity to it. The basic features of the transmission system problem can be illustrated using the lumped equivalent parasitic elements R, L, L', and C_c introduced to model the system. (Although this model forms only the lowest order of approximation for the analysis, it is sufficient for the present discussion.)

The origins of the parasitic elements are easily understood. First, the polysilicon line will exhibit a sheet resistance R_s, which is typically on the order of about 20 ohms/square; this is modeled as the parasitic resistance R. Next, consider the coupling capacitor C_c, which is inserted between the two poly lines. This is included to account for the fact that the lines will be coupled to one another by electric fields when a voltage difference exists between them. Finally, any line that carries current will exhibit an inductance. The two elements L and L' account for the inductances of the poly gate line and the neighboring line, respectively; these elements can then magnetically couple the two lines together.

To extract the important properties of this system, suppose that $V_{out,1}(t < 0) = V_{OL}$. In this static state, no currents flow, so ideally $V_{in,2} = V_{OL}$; this would indeed be true for an isolated system. However, suppose that a pulse is propagating on the neighboring poly interconnect line, as modeled in Fig. 3.7(a). The simplified model now includes the gate capacitance of $M2$, which is

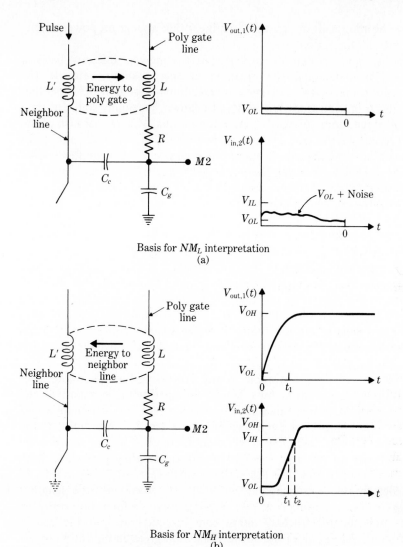

Basis for NM_L interpretation
(a)

Basis for NM_H interpretation
(b)

FIGURE 3.7 Simplified AC circuit model for noise margins.

denoted by C_g. When the pulse passes by the $M2$ input region, electric and magnetic field coupling exists by means of C_c and L, L', respectively. The unwanted voltage induced onto the poly gate line is known as *noise*, thus giving rise to the terminology *noise margin*. (A more correct name for this phenomena would be *electromagnetic interference*, or EMI.) Owing to the energy transfer to the poly gate line, $V_{in,2} > V_{OL}$. As long as $V_{in,2} < V_{IL}$, there are no problems since $M2$ will still interpret this as a logic 0 state. The importance of a reasonable value of NM_L then becomes clear. This type of noise can also be the result of "glitches" (spikes) that appear on power supply lines during normal operation.

The noise margin for high levels NM_H has a somewhat different basis in this example. To understand the meaning of this quantity, refer to the simplified circuit in Fig. 3.7(b). At time $t = 0$, the first stage is switched so that $V_{\text{out},1}$ makes the transition from V_{OL} to V_{OH} as shown, becoming stable at time t_1. The second stage input voltage $V_{\text{in},2}(t)$ will not equal $V_{\text{out},1}(t)$ because of the impedance present in the poly gate line through L, R, and $(C_c + C_g)$. In addition, magnetic coupling via L and L' may transfer some of the energy away from the gate line, reducing the voltage further. Even if this coupling is ignored, it is seen that $V_{\text{in},2}$ will be divided between the RL impedance, which is in series with the capacitive $(C_c + C_g)$ term. Thus, at time t_1, $V_{\text{in},2}(t)$ may not have even reached V_{IH} yet; this possibility is shown in the voltage plot. Eventually, the transients decay away and the V_{OH} level is reached. However, this limits the switching speed of the system, since stage 1 cannot be switched before stage 2 has time to interpret the previous logic state.

This example illustrates typical chip environment situations that must be accounted for in the design of an integrated circuit. The solutions to these types of problems are aided by adhering to a set of "design" or "ground" rules that specify items such as the minimum allowable spacing between poly lines. Each specific fabrication processing line will possess a distinct set of design rules, although it is possible to formulate what are known as "portable design rules" that are, in principle, adaptable to any processing line. (Portable rules have the advantage that chip layouts can be designed without regard to any fabrication specifications; their main disadvantage is that they can many times create layouts that could have been packed into a smaller chip area.) If the design rules are followed, coupling considerations tend to be minimized since spacings are chosen to be large enough to reduce coupling parasitics. With regards to the above example, it is noted that a small value of R is desirable. In terms of a design rule specification, this yields a "maximum line length"; minimization of line resistance is one driving force behind the study of *polysilicides* (polysilicon/refractory metal combinations) for possible use in integrated circuits. At any rate, the above discussion gives an indication of the importance of noise margins in digital IC design.

Returning now to the nMOS VTC in Fig. 3.3, the remaining critical voltage point to be discussed in the VTC is V_{th}, which is termed the *inverter threshold* voltage. This is sometimes denoted by V_M, for *midpoint* voltage, and must not be confused with the device threshold voltage V_T for a MOSFET. As shown in the VTC, V_{th} is defined as the intersection of the VTC curve with the *unity gain* line in which $V_{\text{out}} = V_{\text{in}}$. This voltage serves as a reference for the dividing point in gate transitions and is particularly useful for gauging the effects of input variations. The *voltage noise sensitivities* NS or VNS are defined in a manner analogous to the noise margins with regards to V_{in} being at either V_{OL} or V_{OH}:

$$\text{NS}_L = V_{th} - V_{OL}, \qquad \text{NS}_H = V_{OH} - V_{th}. \tag{3.1-10}$$

For an ideal inverter, $NS_H = NS_L = V_{DD}/2$. The physical interpretation of a noise sensitivity is the amount of voltage necessary at the input to cause a definite output transition.

To complete the discussion of the VTC characteristics, two other voltage differences are useful to introduce. First, the input *transition width* TW is defined as

$$TW = V_{IH} - V_{IL} \tag{3.1-11}$$

and is used to give a measure of the separation between input logic 0 and logic 1 states. Finally, the *logic swing* V_ℓ is defined at the output as

$$V_\ell = V_{OH} - V_{OL}. \tag{3.1-12}$$

The logic swing is simply the output voltage separation between high and low states. The logic swing may be combined with the noise sensitivities to define the circuit *noise immunity* levels,

$$NI_H = \frac{NS_H}{V_\ell}, \qquad NI_L = \frac{NS_L}{V_\ell}, \tag{3.1-13}$$

which are a convenient measure of the ability of the circuit to reject noise.

The generalized nMOS VTC discussion above allows an evaluation of the DC properties of the circuit. The transient analysis, on the other hand, deals explicitly with V_{in} and V_{out} as functions of time. For simplicity, $V_{in}(t)$ will be taken as having step transition properties as illustrated in the voltage plot of Fig. 3.8; note that the input levels are specified by V_{OL} and V_{OH}, which implies a cascaded network. The output waveform $V_{out}(t)$ contains the basic timing parameters that describe the overall transient properties of the inverter.

With a step input voltage, there are 4 fundamental times used in describing the transient response; t_{HL}, t_{LH}, t_{PHL} and t_{PLH}. The first, t_{HL}, is the output *high-to-low* time and gives the time interval necessary for V_{out} to change from the "90% point" to the "10% point," respectively denoted as V_1 and V_0 in the timing diagram. Similarly, t_{LH} is the *low-to-high* time for the output to swing from V_0 to V_1. The *propagation delay times* t_{PHL} (high-to-low) and t_{PLH} (low-to-high) are defined from the input pulse edge to the 50% output points where $V_{out} = \frac{1}{2}(V_{OL} + V_{OH}) = V_{1/2}$, as shown in the voltage plot. The importance of these times is obvious: they establish the maximum operating frequency of the circuit.

The problem of determining t_{HL} is illustrated in Fig. 3.9(a). V_{in} is initially at V_{OL}, so the voltage across C_{out} is $V_{out} = V_{OH}$. When the input is switched at time t_1, the driver is turned on, allowing I_D to flow to ground. The physical basis of t_{HL} is simply the discharging of C_{out} through the driver such that

$$I_D = -C_{out}\frac{dV_{out}}{dt} + I_L, \tag{3.1-14}$$

where the negative sign on the capacitor current indicates a discharge. With

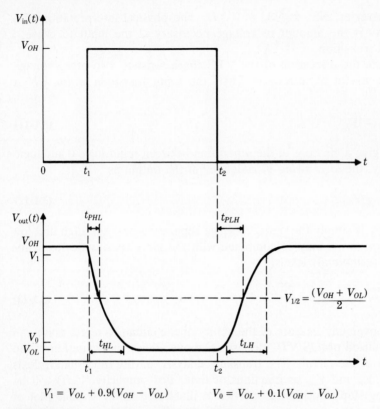

$$V_1 = V_{OL} + 0.9(V_{OH} - V_{OL}) \qquad V_0 = V_{OL} + 0.1(V_{OH} - V_{OL})$$

FIGURE 3.8   Transient response time definitions.

$I_D(V_{out})$ and $I_L(V_{out})$ known, this equation can be integrated for t_{HL}, which is sometimes denoted t_{dis} for *discharge time*.

The low-to-high time, t_{LH}, is determined in a similar fashion. The capacitor initially has a low voltage across it. When the inverter is switched, the driver goes into cutoff, so $I_D = 0$. C_{out} charges through the load device to V_{OH}, as illustrated in Fig. 3.9(b). This charging is governed by

$$I_L(V_{out}) = C_{out}\frac{dV_{out}}{dt}, \tag{3.1-15}$$

so t_{LH} will be a function of the load characteristics. The time interval found by integrating this equation is sometimes denoted as t_{ch} for *charging time*.

The propagation delays t_{PHL} and t_{PLH} are found in a similar manner with different limits of integration. These will be dealt with in detail in Chapter 4, which presents the transient inverter analysis.

This concludes the discussion of the general nMOS inverter characteristics. To make these concepts more meaningful, specific loads will now be analyzed for the inverter.

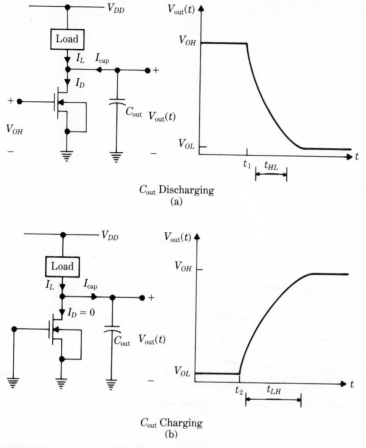

C_{out} Discharging
(a)

C_{out} Charging
(b)

FIGURE 3.9   Physical basis for t_{HL} and t_{LH} transition times.

3.2   Linear Resistors as Load Elements

The simplest load device that can be used in an nMOS inverter is a linear resistor R_L as shown in Fig. 3.10(a). This configuration is straightforward to analyze and illustrates the most important VTC characteristics. However, it is seldom found in practice because of the large chip area consumed by diffused IC resistor patterns. In addition, processing variations do not allow for precise control of resistances, which gives rise to significant variations in the values obtained from batch to batch. This is quite serious, since the discussion below will demonstrate that the transfer properties of the circuit are closely related to the value of R_L.

The discussion of the resistively loaded inverter will be presented in two parts. The initial development will be centered around an analysis of the VTC

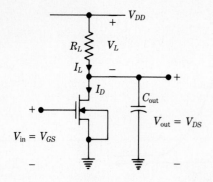

Circuit configuration
(a)

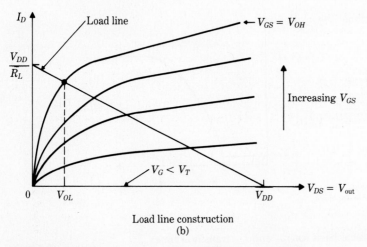

Load line construction
(b)

FIGURE 3.10 Linear resistor as an nMOS inverter load.

for a given R_L and driver (W/L). The critical points will be found in terms of these characterization parameters. Once the basic VTC properties are known, it is possible to introduce the design considerations for the inverter. The next two sections, which deal with using MOSFETs as load, also follow this sectioning scheme. This allows one to comprehend the design of a circuit from an understanding of the VTC properties.

VTC ANALYSIS

The VTC for the resistively loaded inverter is constructed by using the techniques in Section 3.1. First, note that $V_L = I_L R_L$, which gives the output voltage as

$$V_{\text{out}} = V_{DD} - I_L R_L. \tag{3.2-1}$$

Since $I_L = I_D$ and $V_{\text{out}} = V_{DS}$, the load line presented to the driver is simply a straight line that crosses the I_D vs. V_{DS} axes of the device characteristics at V_{DD}/R_L and V_{DD}, respectively. The load line is shown in Fig. 3.10(b), which illustrates how the VTC may be obtained graphically.

To obtain the VTC mathematically, the driver current $I_D(V_{in} = V_{GS})$ must be set equal to the load current through the resistor. I_L is given by Ohm's law as

$$I_L = \frac{V_{DD} - V_{out}}{R_L}.$$ (3.2-2)

To generate the VTC, start with $V_{in} = 0$ ($<V_T$); since the driver is in cut-off, $I_D = 0$ so $V_L = 0$ [V]. With zero voltage across R_L, this establishes $V_{out} = V_{OH} = V_{DD}$. As V_{in} is increased to a value above the threshold voltage of the driver ($V_{in} > V_T$), the device starts conducting current between its drain-source electrodes. The output voltage is initially high, so $V_{out} = V_{DS} > (V_{GS} - V_T)$. The driver is saturated, so the VTC equation is

$$V_{out} = V_{DD} - \frac{\beta_D}{2} R_L (V_{in} - V_T)^2.$$ (3.2-3)

This equation gives the initial portion of the VTC shown in Fig. 3.11.

When V_{out} drops to the value ($V_{in} - V_T$), the device is on the border between being saturated and nonsaturated. Increasing V_{in} lowers the output

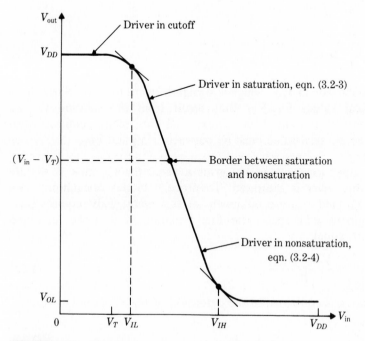

FIGURE 3.11 VTC for a resistor load nMOS inverter.

voltage to $V_{\text{out}} = V_{DS} < (V_{GS} - V_T)$, so the device enters the nonsaturation region where

$$V_{\text{out}} = V_{DD} - \frac{\beta_D}{2} R_L [2(V_{\text{in}} - V_T)V_{\text{out}} - V_{\text{out}}^2]. \tag{3.2-4}$$

This equation provides the remaining portion of the VTC characteristic.

With the VTC form now established, consider the determination of the critical points that dictate the DC properties of the inverter.

V_{OH} It has already been shown that the high value of the output is $V_{OH} = V_{DD}$. This is accurate only to the extent that leakage currents can be ignored.

V_{OL} This quantity is extremely important since it determines NM_L of the circuit. As will be seen in the design discussion, considerable effort will be directed toward setting the actual value of V_{OL}.

To find V_{OL}, note that $V_{\text{out}} = V_{OL}$ implies that $V_{\text{out}} < (V_{\text{in}} - V_T)$, so that the driver is nonsaturated (see eqn. 3.2-4). To ensure that the output is indeed at V_{OL}, take the input voltage to be at $V_{\text{in}} = V_{OH}$. Equating the driver and load currents gives

$$\frac{V_{DD} - V_{OL}}{R_L} = \frac{\beta_D}{2} [2(V_{OH} - V_T)V_{OL} - V_{OL}^2], \tag{3.2-5}$$

or, rearranging,

$$V_{OL}^2 - 2\left(\frac{1}{\beta_D R_L} + V_{DD} - V_T\right)V_{OL} + \frac{2V_{DD}}{\beta_D R_L} = 0. \tag{3.2-6}$$

This is simply a quadratic in V_{OL}, which is straightforward to solve. Since there are two numerical values for V_{OL} that result from the solution to this quadratic, the correct one must be chosen; all this really requires is to remember that the physical value must lie between $0\,[\text{V}]$ and V_{DD}.

V_{IL} The input low voltage V_{IL} gives a value for V_{out} that is slightly below V_{OH}, so the driver is saturated. To proceed in this calculation, note that both $V_{\text{in}} = V_{IL}$ and V_{out} are unknowns at this point. This requires solving two simultaneous equations. The first equation is that the load and driver currents are equal:

$$\frac{\beta_D}{2}(V_{\text{in}} - V_T)^2 = \frac{V_{DD} - V_{\text{out}}}{R_L}. \tag{3.2-7}$$

The second equation is contained in the definition of V_{IL} as the point where

$$\frac{dV_{\text{out}}}{dV_{\text{in}}} = -1. \tag{3.2-8}$$

To compute the derivative condition, the functional form of eqn. (3.2-7) is written as $I_D(V_{in}) = I_L(V_{out})$. Taking differentials of both sides then gives

$$\frac{dI_D}{dV_{in}} dV_{in} = \frac{dI_L}{dV_{out}} dV_{out}, \tag{3.2-9}$$

so

$$\frac{dV_{out}}{dV_{in}} = \frac{(dI_D/dV_{in})}{(dI_L/dV_{out})}. \tag{3.2-10}$$

The derivatives are easily computed using the expressions for $I_D(V_{in})$ and $I_L(V_{out})$ in eqn. (3.2-7). The result is

$$\frac{dV_{out}}{dV_{in}} = -\beta_D R_L(V_{in} - V_T) = -1. \tag{3.2-11}$$

Now, letting $V_{in} = V_{IL}$, the input low voltage is seen to be

$$V_{IL} = V_T + \frac{1}{\beta_D R_L}. \tag{3.2-12}$$

This equation says that V_{IL} is simply given by the driver threshold voltage V_T plus a term whose value is set by β and R_L, which describe the driver transistor and the load resistor, respectively. This is what would be expected from a purely qualitative analysis, since V_T is required to start conduction, with the actual amount of current flowing set by the device properties. This output voltage at $V_{in} = V_{IL}$ is obtained from eqn. (3.2-7) as

$$V_{out}(V_{IL}) = V_{DD} - \frac{1}{2\beta_D R_L}. \tag{3.2-13}$$

This establishes the current at this point as $I_D = I_L = (1/2\beta R_L)$, as can be verified by direct substitution into either side of the current equation.

These results could have been obtained by using a simpler technique that uses $V_{out} = V_{DD} - I_D R_L$ to give

$$\frac{dV_{out}}{dV_{in}} = -R_L \frac{dI_D}{dV_{in}} \tag{3.2-14}$$

since V_{DD} = constant; performing the differentiation then yields the same condition for V_{IL}. The more general approach was used because it will be valid for generalized equations in which $I_D(V_{in}, V_{out}) = I_L(V_{in}, V_{out})$, i.e., when there is a functional dependence of at least one of the currents on both V_{in} and V_{out}.

V_{IH} When $V_{in} = V_{IH}$, $V_{out} < (V_{GS} - V_T)$, so the driver is operating in the nonsaturated region. Equating the driver and load currents thus gives

$$\frac{\beta_D}{2}[2(V_{IH} - V_T)V_{out} - V_{out}^2] = \frac{1}{R_L}(V_{DD} - V_{out}). \tag{3.2-15}$$

Although V_{out} is expected to be slightly greater than V_{OL}, it is an unknown at this point. The calculation of V_{IH} requires the simultaneous solution of eqn. (3.2-15) with the derivative condition $(dV_{out}/dV_{in}) = -1$. For this case, $I_D(V_{in}, V_{out}) = I_L(V_{out})$. Taking differentials gives

$$\frac{\partial I_D}{\partial V_{in}} dV_{in} + \frac{\partial I_D}{\partial V_{out}} dV_{out} = \frac{dI_L}{dV_{out}} dV_{out}, \tag{3.2-16}$$

or

$$\frac{dV_{out}}{dV_{in}} = \frac{\partial I_D / \partial V_{in}}{(dI_L/dV_{out}) - (\partial I_D/\partial V_{out})} = -1. \tag{3.2-17}$$

Computing the derivatives,

$$\frac{\beta_D V_{out}}{\dfrac{1}{R_L} + \beta_D(V_{in} - V_T - V_{out})} = 1, \tag{3.2-18}$$

so that rearranging and setting $V_{in} = V_{IH}$ finally results in

$$V_{out} = \frac{1}{2}(V_{IH} - V_T) + \frac{1}{2\beta_D R_L}, \tag{3.2-19}$$

which is the second equation required for finding the two unknowns V_{IH} and V_{out}. Substituting this expression for V_{out} into eqn. (3.2-15) and regrouping terms yield the form

$$(V_{IH} - V_T)^2 + \frac{2}{\beta_D R_L}(V_{IH} - V_T) - \left(\frac{8V_{DD}}{3\beta_D R_L} - \frac{1}{\beta_D^2 R_L^2}\right) = 0, \tag{3.2-20}$$

which is a quadratic for the quantity $(V_{IH} - V_T)$. The value of V_{IH} is thus found by using the known quadratic solution, again remembering that the proper physical root must be chosen. $V_{out}(V_{IH})$ can be computed from (eqn. 3.2-19), and the current at this point can be found from eqn. (3.2-15).

V_{th} The inverter threshold voltage V_{th} is at the VTC point where $V_{in} = V_{out}$. Since $V_{DS} = V_{GS}$, $V_{DS} > (V_{GS} - V_T)$, so the driver is saturated. The current equation is

$$\frac{\beta_D}{2}(V_{th} - V_T)^2 = \frac{V_{DD} - V_{th}}{R_L}, \tag{3.2-21}$$

yielding the quadratic equation

$$V_{th}^2 - 2\left(V_T - \frac{1}{\beta_D R_L}\right)V_{th} + \left(V_T^2 - \frac{2V_{DD}}{\beta_D R_L}\right) = 0. \tag{3.2-22}$$

This may be directly solved for V_{th} and concludes the VTC critical points.

DESIGN

The design of the resistively loaded inverter is particularly useful for illustrating a simple model that describes the operation of nMOS inverters with arbitrary load devices. In addition, the design criteria will show why this circuit is not practical for random logic implementations and thus give motivation for the study of *active loads* in which a MOSFET is used as a load.

The design approach presented here relies on an analysis of the two output voltages V_{OL} and V_{OH} as providing the design variable information. The input voltages V_{IL} and V_{IH} are treated as secondary in importance, with their values established by the output design. With this in mind, consider first $V_{OH} = (V_{DD} - V_L)$ with $V_{in} < V_T$. As discussed in the VTC analysis, the relation $V_L = I_L R_L$ gives $V_{OH} = V_{DD}$ since $I_D = 0$. A useful viewpoint for nMOS inverters can be introduced by modeling this situation as the series resistive voltage divider in Fig. 3.12. The driver in the cutoff mode is modeled as a resistor of value $R_{off} = V_{OH}/I_D$. This circuit gives the output as

$$V_{OH} = \frac{R_{off}}{R_{off} + R_L} V_{DD}, \tag{3.2-23}$$

so as $R_{off} \to \infty$ (corresponding to an ideal driver with $I_D \to 0$), $V_{OH} \to V_{DD}$. In a realistic MOSFET, some leakage current will flow, giving a finite value for R_{off}. Depending on the value of V_{GS}, this current can range from picoamps (or lower) to the microamp level (for subthreshold leakage). The only design constraint seen at this point is that $R_L \ll R_{off}$ must be satisfied to set V_{OH} close to V_{DD}.

The real problem arises in the design of the circuit for a desired value of V_{OL}. In a well-designed inverter, V_{OL} should be less than about 5% of the power supply voltage V_{DD} to ensure a good value for NM_L; for $V_{DD} = 5\,[V]$, $V_{OL} = 0.25\,[V]$ is a reasonable value to be used for a design target. However, as shown in the following analysis, there exist certain tradeoffs that must be examined before a design set is chosen.

To study V_{OL}, let $V_{in} = V_{OH}$ to ensure that the output is in the desired low state. Equating the nonsaturated driver current to the load current gives, after

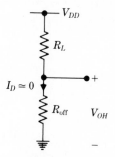

FIGURE 3.12 V_{OH} resistor model.

simple rearrangement,

$$R_L\left(\frac{W}{L}\right)_D = \frac{2(V_{DD} - V_{OL})}{k'[2(V_{OH} - V_T)V_{OL} - V_{OL}^2]},$$
(3.2-24)

where $\beta = k'(W/L)$ has been used. This equation may be viewed as a basic design constraint. It sets the value of the product $R_L(W/L)$ required for a given choice of V_{OL}. As an example, suppose that $V_{DD} = V_{OH} = 5$ [V], $V_T = 1$ [V], $k' = 20$ [μA/V^2], and a value of $V_{OL} = 0.25$ [V] is chosen. The equation requires that $R_L(W/L) \geq 2.45 \times 10^5$ [Ω], which is quite large. Note at this point that once the $R_L(W/L)$ product is set by the choice of V_{OL}, V_{IL} and V_{IH} are automatically established by eqns. (3.2-12) and (3.2-20), respectively. Thus the noise margins are designed into the circuit by this simple calculation.

The tradeoffs between choosing R_L and (W/L) can be seen by noting that the nonsaturated drain current can be used to define the driver *on resistance*

$$R_{on} = \frac{V_{OL}}{I_D} \simeq \frac{1}{k'\left(\dfrac{W}{L}\right)\left[(V_{OH} - V_T) - \dfrac{1}{2}V_{OL}\right]}.$$
(3.2-25)

The resulting resistive model of this situation (see Fig. 3.13) shows that

$$V_{OL} = \frac{R_{on}}{R_{on} + R_L} V_{DD}.$$
(3.2-26)

As R_{on} decreases, corresponding to (W/L) increasing, V_{OL} moves towards the ideal value of 0 [V]. However, a large (W/L) value implies that the driver will require an excessively large chip area, reducing the circuit integration capabilities. Suppose that a reasonable upper limit on (W/L) is specified to be 5. For the process described in the preceding example, $(W/L) = 5$ would give $R_{on} = 2.5$ [kΩ], so a load resistance of $R_L \simeq 50$ [kΩ] is required for the choice $V_{OL} = 0.25$ [V]. If (W/L) is reduced to 2.5 to conserve driver area, then R_L must be increased to about 100 [kΩ]. Such large values of R_L tend to make this

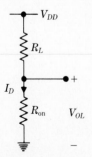

FIGURE 3.13 V_{OL} resistor model.

inverter scheme impractical from chip area conservation principles. It should also be noted here that the choices of R_L and (W/L) play important roles in the transient response of the circuit, since they limit the current that is available to charge and discharge the output capacitance. This will be discussed in detail in Chapter 4, where the individual constraints on both R_L and (W/L) will be presented in terms of transition times.

This simple concept of resistor modeling for V_{OL} illustrates the basic *ratio property* of the static nMOS inverters discussed in this chapter. In these circuits, V_{OL} will be set by a ratio of driver-to-load parameters; this will also establish the VTC characteristics. In the case of a resistive load, the ratio is $k'(W/L)/G_L$, with $G_L = 1/R_L$ the conductance of the load. This has rather striking consequences, as the equations for the VTC input critical points V_{IL} and V_{IH} both contain this ratio and also determine the shape of the transition between these two voltages. Moreover, these ratios will contain information about the chip geometry for the final layout of the circuit. Owing to this observation, the use of driver-to-load ratios is particularly useful when discussing surface geometries.

EXAMPLE 3.2-1

A resistive-load inverter is designed with

$$R_L = 10\,[\text{k}\Omega], \quad (W/L) = 25, \quad k' = 20\,[\mu\text{A/V}^2],$$

$$V_{T0} = +1\,[\text{V}], \quad V_{DD} = +5\,[\text{V}], \quad \gamma = 0.4\,[\text{V}^{1/2}], \quad 2\,|\phi_F| = 0.6\,[\text{V}].$$

Calculate the values of V_{OH}, V_{OL}, V_{IL}, and V_{IH}.

Solution

V_{OH}: Assuming that the leakage is negligible,

$$V_{OH} \simeq V_{DD} = 5\,[\text{V}] \leftarrow,$$

V_{OL}: For this circuit,

$$\beta_D R_L = (25)(20 \times 10^{-6})(10 \times 10^3) = 5\,[\text{V}^{-1}].$$

Equation (3.2-6) is then

$$V_{OL}^2 - (8.4)V_{OL} + 2 = 0,$$

which has solutions

$$V_{OL} \simeq 4.20 \pm 3.95,$$

Thus, the physical solution (in range 0 to V_{DD}) is

$$V_{OL} \simeq 0.25\,[\text{V}] \leftarrow.$$

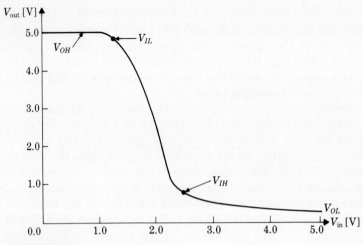

FIGURE E3.1

V_{IL}: Using eqn. (3.2-12),

$$V_{IL} = 1 + \frac{1}{5}, \quad \text{so}$$

$$V_{IL} = 1.2\,[\text{V}] \leftarrow .$$

V_{IH}: Now use eqn. (3.2-20) to write

$$(V_{IH} - V_T)^2 + 0.4(V_{IH} - V_T) - 2.627 = 0,$$

which has a physical solution of

$$V_{IH} - V_T \simeq 1.43\,[\text{V}].$$

Thus,

$$V_{IH} \simeq 2.43\,[\text{V}] \leftarrow .$$

The noise margins are easily seen to be

$$\text{NM}_H \simeq 5 - 2.43 = 2.57\,[\text{V}], \qquad \text{NM}_L \simeq 1.2 - 0.25 = 0.95\,[\text{V}],$$

The SPICE-simulated VTC for this circuit is shown in Fig. E3.1.

```
RESISTOR LOAD INVERTER
VDD 3 0 DC 5VOLTS
MDRIVER 2 1 0 0 EMODE L=5U W=125U
RL 3 2 10KOHM
VS 1 0 DC
.MODEL EMODE NMOS VTO=1 GAMMA=0.4 KP=2.0E-5
.DC VS 0 5 0.1
.PLOT DC V(2)
.END
```

3.3   Enhancement-Mode MOSFETs as Loads

The resistively loaded inverter analyzed in the last section consumes too large an area for practical use in integrated logic circuits. MOSFETs, in contrast, can be made quite small. This section examines the use of an enhancement-mode (E-mode) MOSFET as an *active load* device. The load transistor will be made at the same time the driver is fabricated, so the value of $k' = \mu_n C_{ox}$ is the same for both. To avoid confusion, the parameters that describe the driver MOSFET will be subscripted by D: V_{DSD}, V_{GSD}, V_{TD}, $(W/L)_D$, etc. Similarly, the load MOSFET quantities will carry the subscript L: V_{DSL}, V_{GSL}, V_{TL}, $(W/L)_L$, and so on.

The ability to apply bias to the load MOSFET gate electrode allows the device to be operated in either the saturated or non-saturated region. The driver, of course, is still controlled by $V_{in} = V_{GSD}$. The two different possibilities for the operational characteristics of the load device give two distinct types of inverters.

One point to remember throughout this section is that both the driver and load MOSFETs are fabricated in a common p-type substrate. Hence, the bulk electrodes for both devices are electrically the same. If the driver bulk is grounded, then the load MOSFET body is also at 0 [V]. Consequently, body bias variations will be manifest in V_{TL} being a function of V_{out}. This dependence will become apparent in the analysis and will be seen to set some fundamental limitations on the inverter performance.

3.3.1   Saturated E-Mode MOSFET Loads

This configuration is drawn in Fig. 3.14(a). Both the gate and drain of the load MOSFET are connected to the power supply rail V_{DD}. The common supply voltage gives $V_{GSL} = V_{DSL}$, so $V_{DSL} > (V_{GSL} - V_{TL})$ is automatically satisfied. This forces the load MOSFET to operate in either saturation or cutoff. The saturated load current is given by

$$I_L = \frac{k'}{2}\left(\frac{W}{L}\right)_L (V_{GSL} - V_{TL})^2. \tag{3.3-1}$$

This can be used to obtain the load line equation by noting that $V_{GSL} = (V_{DD} - V_{out})$. Since $I_D = I_L$ and $V_{out} = V_{DSD}$,

$$I_D = \frac{k'}{2}\left(\frac{W}{L}\right)_L [V_{DD} - V_{DSD} - V_{TL}(V_{DSD})]^2 \tag{3.3-2}$$

describes the load line presented to the driver. The load threshold voltage has been explicitly written as a function of $V_{DSD} = V_{out}$ because $V_{SBL} = V_{out}$. The load threshold voltage is thus of the form

$$V_{TL} = V_{TOL} + \gamma(\sqrt{V_{out} + 2|\phi_F|} - \sqrt{2|\phi_F|}), \tag{3.3-3}$$

where V_{TOL} is the $V_{SBL} = 0$ value of the load threshold voltage. It is assumed

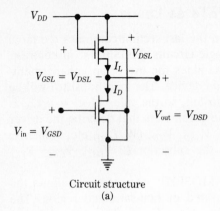

Circuit structure
(a)

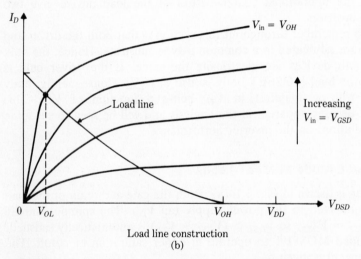

Load line construction
(b)

FIGURE 3.14   Saturated enhancement-load nMOS inverter.

that $V_{TOL} = V_{TOD} = V_{TO}$. The driver threshold voltage is always given by $V_{TD} = V_{TO}$ since $V_{SBD} = 0$. (An external body bias voltage V_{BB} may be applied to both devices. This would give threshold voltages greater than V_{TO}.)

Now the load line equation (3.3-2) is of parabolic form with a minimum at $V_{DSD} = [V_{DD} - V_{TL}(V_{DSD})]$. This is also the point where $I_D = I_L = 0$, corresponding to the driver being in cutoff with $V_{in} < V_{TD}$. The load line is plotted on the driver characteristics in Fig. 3.14(b). One point becomes immediately obvious: the highest output voltage is $V_{out} = V_{OH} = [V_{DD} - V_{TL}(V_{OH})]$. This arises from the use of V_{DD} as the gate bias on the load MOSFET. Since the load is biased into saturation (although it is not conducting here), the device will have an inversion layer formed under its gate oxide. This requires a gate-source voltage of $V_{TL}(V_{OH})$, giving rise to a *threshold voltage loss* (from V_{DD}) at the output.

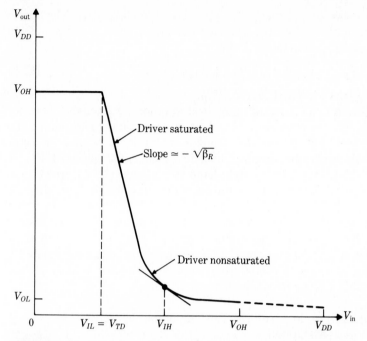

FIGURE 3.15 VTC for saturated enhancement-load inverter.

The VTC in Fig. 3.15 is constructed in the usual manner. With $V_{in} < V_{TD}$, the driver is in cutoff, so $V_{out} = V_{OH}$. When $V_{in} = V_{IL}$, the curve exhibits a discontinuity, which will be discussed in the analysis. As V_{in} increases above V_{TD}, the driver is saturated, so

$$\frac{\beta_D}{2}(V_{in} - V_{TD})^2 = \frac{\beta_L}{2}[V_{DD} - V_{out} - V_{TL}(V_{out})]^2 \qquad \text{(3.3-4)}$$

describes the VTC until $V_{out} = (V_{in} - V_{TD})$. When V_{in} increases further so that $V_{out} < (V_{in} - V_{TD})$, the driver goes into the nonsaturated region, where

$$\frac{\beta_D}{2}[2(V_{in} - V_{TD})V_{out} - V_{out}^2] = \frac{\beta_L}{2}[V_{DD} - V_{out} - V_{TL}(V_{out})]^2. \qquad \text{(3.3-5)}$$

This gives the remaining portion of the transfer curve.

ANALYSIS
The critical VTC points for the E-mode MOSFET loaded configuration are found using the techniques established in the previous sections.

V_{OH}   With the threshold voltage loss discussed above, $V_{OH} = [V_{DD} - V_{TL}(V_{OH})]$, or

$$V_{OH} = V_{DD} - [V_{T0} + \gamma(\sqrt{V_{OH} + 2|\phi_F|} - \sqrt{2|\phi_F|})]. \qquad \text{(3.3-6)}$$

There are two ways to approach the problem of finding V_{OH}. The direct algebraic approach first defines the auxiliary variable $u^2 = V_{OH} + 2|\phi_F|$. Then the equation may be written in the form

$$u^2 + \gamma u - [V_{DD} - V_{T0} + \gamma\sqrt{2|\phi_F|} + 2|\phi_F|] = 0, \tag{3.3-7}$$

which is easily solved for u and, hence, V_{OH}.

The second technique is to use numerical iterations. Although this is not really necessary for the problem at hand, it is convenient to introduce the procedure here so that it will be familiar when it is needed in later developments. To apply the iterative technique, an initial guess of $V_{OH} = (V_{DD} - V_{T0})$ is substituted into the right-hand side of eqn. (3.3-6), which then gives a value of V_{OH}. The process is repeated by using this V_{OH} to generate another value V_{OH}. The cycle is continued until the value used in the right-hand side agrees with the computed V_{OH}, giving the correct consistent answer. While the approach may sound somewhat tedious, the calculations usually converge in 2 or 3 iterations since the square-root terms are relatively insensitive to V_{OH} changes. Moreover, the iterative technique is ideal for use on a programmable calculator or microcomputer system.

V_{OL} The output low voltage can be found by setting $V_{\text{in}} = V_{OH}$. Since the driver is nonsaturated,

$$\frac{\beta_D}{2}[2(V_{OH} - V_{TD})V_{OL} - V_{OL}^2] = \frac{\beta_L}{2}[V_{DD} - V_{OL} - V_{TL}(V_{OL})]^2 \tag{3.3-8}$$

is the equation that determines V_{OL}. The algebraic complexity of this problem can be seen by noting the dependence of V_{TL} on V_{OL} through

$$V_{TL} = V_{T0L} + \gamma(\sqrt{V_{OL} + 2|\phi_F|} - \sqrt{2|\phi_F|}). \tag{3.3-9}$$

Since V_{OL} is inside the square root, eqn. (3.3-8) is actually a quartic (fourth-order polynomial) in the unknown voltage V_{OL}. Although solutions to quartic equations are known, they are extremely tedious to use and introduce a high probability for error in the calculations. As such, a simplified technique based on numerical iterations is well worth investigating.

The simplest approach to use is to change the input voltage from V_{OH} to a generalized value of V_{in}. Then, rearranging eqn. (3.3-8) gives the relation

$$V_{\text{in}} = V_{TD} + \frac{1}{2}V_{OL} + \frac{[V_{DD} - V_{OL} - V_{TL}(V_{OL})]^2}{2\beta_R V_{OL}}, \tag{3.3-10}$$

where $\beta_R = (\beta_D/\beta_L)$ has been introduced as the driver-to-load ratio for this circuit. To find V_{OL}, a reasonable first guess (e.g., $V_{OL} = 0.05V_{DD}$) is made. This is used in eqn. (3.3-10) to compute a value of V_{in}, where it is noted that $V_{\text{in}} = V_{OH}$ is required for a solution. Different choices for V_{OL} are used to generate a set of V_{in} values, which are in turn compared with V_{OH} to the desired degree of accuracy. If the problem is being solved on a programmable system, it is easier to choose a starting value of V_{OL} quite low (say, $0.01V_{DD}$), then increment this value until $(V_{in} - V_{OH})$ goes negative; the magnitude of the increment used in the loop will set the accuracy of the result.

V_{IL} The input low voltage for the circuit is given by

$$V_{IL} = V_{TD},$$

(3.3-11)

implying that the VTC transition begins at the point where the driver enters the active saturation region. This value can be justified by relatively simple arguments. First, note that for $V_{in} < V_{TD}$, the driver is in cutoff, so $I_D = 0$. This places the load into a state of cutoff with

$$I_L = \frac{\beta_L}{2}[V_{DD} - V_{OH} - V_{TL}(V_{OH})]^2 = 0.$$

(3.3-12)

When $V_{in} = V_{TD}$, the driver is at the border between saturation and cutoff, with I_D still zero. However, for V_{in} slightly greater than V_{TD}, current flows as described by

$$\frac{\beta_D}{2}(V_{in} - V_{TD})^2 = \frac{\beta_L}{2}[V_{DD} - V_{out} - V_{TL}(V_{out})]^2.$$

(3.3-13)

Hence, $V_{out} < V_{OH}$, and the VTC transition is under way. This behavior can also be seen by an examination of the load line in Fig. 3.14(b), since V_{in} slightly greater than V_{TD} implies that $V_{out} < V_{OH}$.

This result can be justified mathematically by noting that for $V_{in} > V_{TD}$, eqn. (3.3-13) says $I_D(V_{in}) = I_L(V_{out})$. The VTC slope in this region is computed using

$$\frac{dV_{out}}{dV_{in}} = \frac{(dI_D/dV_{in})}{(dI_L/dV_{out})}$$

(3.3-14)

as in eqn. (3.2-10). Differentiating I_D and I_L gives

$$\frac{dV_{out}}{dV_{in}} = -\frac{\beta_D(V_{in} - V_{TD})}{\left(1 + \frac{dV_{TL}}{dV_{out}}\right)\beta_L[V_{DD} - V_{out} - V_{TL}(V_{out})]},$$

(3.3-15)

which is valid as long as the driver is saturated. This may be rewritten by noting that both devices are described by a saturated current equation of the form

$$I = \frac{\beta}{2}(V_{GS} - V_T)^2,$$

(3.3-16)

so

$$\beta(V_{GS} - V_T) = \sqrt{2\beta I}.$$

(3.3-17)

Using this for both I_D and I_L allows eqn. (3.3-15) to be written as

$$\frac{dV_{\text{out}}}{dV_{\text{in}}} = -\frac{\sqrt{2\beta_D I_D}}{\left(1 + \dfrac{dV_{TL}}{dV_{\text{out}}}\right)\sqrt{2\beta_L I_L}}, \tag{3.3-18}$$

or

$$\frac{dV_{\text{out}}}{dV_{\text{in}}} = -\frac{\sqrt{\beta_R}}{(1 + \eta)}, \tag{3.3-19}$$

where $\beta_R = \beta_D/\beta_L$ is the driver-to-load ratio for the circuit, and

$$\eta = \frac{dV_{TL}}{dV_{\text{out}}} = \frac{\gamma}{2\sqrt{V_{\text{out}} + 2|\phi_F|}} \tag{3.3-20}$$

describes the load body bias effects. Assuming $\eta \ll 1$,

$$\frac{dV_{\text{out}}}{dV_{\text{in}}} \simeq -\sqrt{\beta_R}, \tag{3.3-21}$$

i.e., the slope of the VTC is approximately constant when the driver is saturated. Since $(dV_{\text{out}}/dV_{\text{in}}) = 0$ for $V_{\text{in}} < V_{TD}$, $V_{\text{in}} = V_{TD}$ describes the break point in the VTC, justifying the relation $V_{IL} = V_{TD}$. Also note that this predicts a linear falloff of the VTC in the driver saturation region. Obviously, $\beta_R > 1$ is required for an operational inverter.

One apparent problem that should be noted here is that the point where

$$\frac{dV_{\text{out}}}{dV_{\text{in}}} = -1 \tag{3.3-22}$$

cannot be found explicitly. The preceding analysis implies that in a realistic inverter, this must occur around V_{TD}. The inability of the analysis to compute this point is based on the fact that the MOSFET circuit equations are simplified descriptions of the actual device operational modes. In particular, the circuit equations give $I_D = 0$ in cutoff, which ignores leakage currents. If these are included, then the -1 slope point can be computed. However, the small correction does not warrant the tedious mathematics, so $V_{IL} = V_{TD}$ is generally used as a reasonable estimate.

V_{IH} When $V_{\text{in}} = V_{IH}$, the driver is operating in the nonsaturated region with $V_{\text{out}} < (V_{IH} - V_{TD})$, so

$$\frac{\beta_D}{2}[2(V_{\text{in}} - V_{TD})V_{\text{out}} - V_{\text{out}}^2] = \frac{\beta_L}{2}[V_{DD} - V_{\text{out}} - V_{TL}(V_{\text{out}})]^2. \tag{3.3-23}$$

Since $I_D(V_{\text{in}}, V_{\text{out}}) = I_L(V_{\text{out}})$, the VTC derivative condition is the same as that found in eqn. (3.2-16) of the previous section:

$$\frac{dV_{\text{out}}}{dV_{\text{in}}} = \frac{\partial I_D/\partial V_{\text{in}}}{(dI_L/dV_{\text{out}}) - (\partial I_D/\partial V_{\text{out}})} = -1. \tag{3.3-24}$$

Evaluating the derivatives,

$$\frac{\beta_D V_{\text{out}}}{\beta_L(V_{DD} - V_{\text{out}} - V_{TL})\left(1 + \dfrac{dV_{TL}}{dV_{\text{out}}}\right) + \beta_D(V_{\text{in}} - V_{TD} - V_{\text{out}})} = 1, \qquad (3.3\text{-}25)$$

which must be solved simultaneously with the current equation. To obtain a technique for solution, eqn. (3.3-23) is solved for V_{in} as

$$V_{\text{in}} = V_{TD} + \frac{1}{2}V_{\text{out}} + \frac{1}{2\beta_R V_{\text{out}}}[V_{DD} - V_{\text{out}} - V_{TL}(V_{\text{out}})]^2, \qquad (3.3\text{-}26)$$

where $\beta_R = \beta_D/\beta_L$ as before. Similarly, the derivative condition (3.3-25) allows V_{in} to be expressed as

$$V_{\text{in}} = V_{TD} + V_{\text{out}}\left(2 + \frac{1 + \eta}{\beta_R}\right) - \frac{1}{\beta_R}[V_{DD} - V_{TL}(V_{\text{out}})](1 + \eta), \qquad (3.3\text{-}27)$$

with $\eta = \eta(V_{\text{out}})$ defined in eqn. (3.3-20). These two equations must be solved simultaneously for $V_{\text{in}} = V_{IH}$ and V_{out}. To effect this process, the two expressions for V_{in} are equated. Straightforward rearrangement gives

$$V_{\text{out}} = \frac{\sqrt{[V_{DD} - V_{TL}(V_{\text{out}})]^2 + 2\eta(V_{\text{out}})V_{\text{out}}[V_{DD} - V_{TL}(V_{\text{out}})]}}{\sqrt{3\beta_R + 1 + 2\eta(V_{\text{out}})}}, \qquad (3.3\text{-}28)$$

where both V_{TL} and η have been written as explicit functions of V_{out}. This equation may be solved for V_{out} by the technique of *fixed-point* iteration, which is also referred to as *self-iteration*. First, start by making an initial guess for $V_{\text{out}} > V_{OL}$. This is used to compute $V_{TL}(V_{\text{out}})$ and $\eta(V_{\text{out}})$ on the right-hand side of eqn. (3.3-28), yielding a new value for V_{out}. The iteration begins at this point, with the new value of V_{out} from the first calculation used again to generate the right-hand side of the equation. When the two sides agree, V_{out} is known and may be used in either expression for $V_{\text{in}} = V_{IH}$. This technique for computing V_{out} converges quite rapidly.

If $\beta_R \gg 1$, an approximate value for V_{IH} can be found by neglecting (dI_L/dV_{out}) in the derivative (3.3-24). Enforcing the unity slope condition on $(dV_{\text{out}}/dV_{\text{in}})$ gives the approximation $V_{\text{out}} = \frac{1}{2}(V_{\text{in}} - V_{TD})$; when this is substituted into eqn. (3.3-23), a straightforward algebraic reduction allows $V_{\text{in}} = V_{IH}$ to be written in the approximate form

$$V_{IH} \simeq V_{TD} + \frac{2[V_{DD} - V_{TL}(V_{\text{out}})]}{\sqrt{3\beta_R + 1}}. \qquad (3.3\text{-}29)$$

These two equations may also be solved by self-iteration. First, choose a value for $V_{TL} > V_{T0}$ to compute $V_{\text{in}} = V_{IH}$ in eqn. (3.3-29). This is substituted into $V_{\text{out}} = \frac{1}{2}(V_{\text{in}} - V_{TD})$ for finding a V_{out} value. V_{out} is in turn used to compute $V_{TL}(V_{\text{out}})$, which places the algorithm into a closed loop. The calculation will converge quite rapidly because of the relatively small change in V_{TL} with regard to variations in V_{out}; typically, the calculation will require only 2 or 3 iterations. It is, however, important to remember that $\beta_R \gg 1$ is required for an accurate solution using this approach. The generalized equation (3.3-27) requires a little more work but will give greater accuracy.

V_{th} The inverter threshold voltage V_{th} is found at the point where $V_{in} = V_{out}$. This places the driver in saturation, so

$$\frac{\beta_D}{2}(V_{th} - V_{TD})^2 = \frac{\beta_L}{2}[V_{DD} - V_{th} - V_{TL}(V_{th})]^2 \tag{3.3-30}$$

gives the condition for V_{th}. Solving,

$$V_{th} = \frac{V_{DD} + \sqrt{\beta_R}\,V_{TD} - V_{TL}(V_{th})}{1 + \sqrt{\beta_R}}. \tag{3.3-31}$$

Since $V_{TL} = V_{TL}(V_{th})$, the actual value of V_{th} must be found by means of iterations. A convenient starting point is to set $V_{th} = \frac{1}{2}(V_{OH} + V_{OL})$. The resulting calculation is then seen to be self-iterating and converges quite rapidly.

DESIGN

The first point to note in the design of the circuit is the presence of the threshold voltage loss in V_{OH}:

$$V_{OH} = V_{DD} - [V_{T0L} + \gamma(\sqrt{V_{OH} + 2|\phi_F|} - \sqrt{2|\phi_F|})]. \tag{3.3-32}$$

To ensure a reasonable value for NM_H, the power supply voltage must be relatively large. For example, $V_{DD} = 12\,[V]$ was used in many early (pMOS) chip designs.

The ratioed nature of the inverter can be seen by a direct examination of eqn. (3.3-6). Rearranging,

$$\beta_R = \frac{[V_{DD} - V_{OL} - V_{TL}(V_{OL})]^2}{[2(V_{OH} - V_{TD})V_{OL} - V_{OL}^2]}, \tag{3.3-33}$$

which serves as a design specification on β_R for a given choice of V_{OL}. To understand the consequences of this equation, recall that $\sqrt{\beta_R}$ is the VTC slope in the driver saturation region. In addition, β_R is the primary design variable in the equations that are used to compute V_{IH}. Thus, the specification of a desired V_{OL} voltage sets the minimum β_R. This in turn establishes many of the DC inverter characteristics.

As an example of the design process, suppose that $V_{DD} = 12\,[V]$, while the fabrication sets $V_{T0} = +1.5\,[V]$, $\gamma = 0.4\,[V^{1/2}]$, and $2|\phi_F| = 0.62\,[V]$. Since $V_{IL} = V_{T0} = 1.5\,[V]$, a design value of $V_{OL} = 0.5\,[V]$ is chosen to give $NM_L = 1\,[V]$. Iterating eqn. (3.3-32) with these parameters gives $V_{OH} \simeq 9.6\,[V]$, while $V_{TL}(V_{OL} = 0.5) \simeq 1.61\,[V]$. Equation (3.3-33) then specifies a driver-load ratio of $\beta_R \simeq 12.56$. Integer values are preferred for use in layout, so $\beta_R = 13$ gives the design value. If $k' = \mu_n C_{\text{ox}}$ is the same for both devices, this requires $(W/L)_D = 13(W/L)_L$.

Consider now the resistor model for V_{OL} illustrated in Fig. 3.13 of the previous section. This can be applied to the present case of a saturated E-mode MOSFET load by defining a *large-signal drain-source conductance* G_{DSL} by

$$G_{DSL} = \frac{I_L}{V_{DSL}} . \tag{3.3-34}$$

Since $V_{DSL} = V_{GSL}$ for the saturated load, $I_L = (\beta/2)(V_{GSL} - V_{TL})^2$ may be substituted to give

$$G_{DSL} = \frac{\frac{k'_L}{2}\left(\frac{W}{L}\right)_L [V_{DD} - V_{OL} - V_{TL}(V_{OL})]^2}{(V_{DD} - V_{OL})} , \tag{3.3-35}$$

where V_{out} has been set to V_{OL} and body bias variations (in addition to channel-length modulation) have been neglected. Using the *on conductance* $G_{\text{on}} = 1/R_{\text{on}}$, the resistor analogy gives

$$V_{OL} = \frac{V_{DD}}{\left(1 + \dfrac{G_{\text{on}}}{G_{DSL}}\right)} , \tag{3.3-36}$$

where $G_{\text{on}}/G_{DSL} \propto \beta_R = \beta_D/\beta_L$. This simple equation demonstrates that increasing β_R will lower V_{OL}. Another useful viewpoint is that (W/L) is proportional to the drain-source conductance of a MOSFET.

The design criteria also illustrates the problems introduced by using the saturated MOSFET as a load device. First, the threshold voltage loss requires the use of larger power supply voltages to ensure the integrity of NM_H values. One way to overcome this problem is to employ a MOSFET load that operates in the nonsaturated region. This case is analyzed in the next section and is indeed an alternative arrangement. However, the biasing for a nonsaturated load transistor requires the use of another power supply, which introduces additional complications. If a single power supply circuit is desired, a *dynamic* technique known as *bootstrapping* can be used to raise V_{OH} to V_{DD}. Bootstrapping is analyzed in Chapter 8.

The most significant problem in using the saturated enhancement MOSFET load is the requirement that β_R must be large to attain a sharp VTC transition. In the preceding design example, $\beta_R = 13$ was required, implying that the driver will be quite large when implemented into the chip layout. Although this is a significant savings when compared with the area consumed by the resistively loaded inverter, it is possible to reduce the real estate (chip area) even further by more advanced static and dynamic circuit techniques.

EXAMPLE 3.3-1

An inverter uses a saturated E−mode MOSFET as a load device. The circuit is characterized with the following parameters:

$$V_{T0} = +4\,[\text{V}], \qquad V_{DD} = +12\,[\text{V}], \qquad \gamma = 0.4\,[\text{V}^{1/2}],$$

$$\beta_R = 10, \qquad k' = 20\,[\mu\text{A/V}^2], \qquad 2|\phi_F| = 0.6\,[\text{V}].$$

Compute the critical VTC voltages.

Solution

(a) V_{OH}: Using eqn. (3.3-6) gives

$$V_{OH} = 8 - 0.4(\sqrt{V_{OH} + 0.6} - \sqrt{0.6}).$$

To solve this using iterations, define

$$V_X = 8 - 0.4(\sqrt{V_{OH} + 0.6} - \sqrt{0.6}).$$

The procedure is to guess a value for V_{OH}, then compute V_X. If $V_X = V_{OH}$, then a solution has been found. If not, use V_X as the next guess for V_{OH}.

A straightforward approach to providing the results of the iterations is to use a table like the following:

V_{OH}	V_X
10	7.01
7.01	7.21
7.21	7.19
7.19	7.19 ← $V_X = V_{OH} \simeq 7.19\,[\text{V}]$ is the solution.

A BASIC program that implements this algorithm is as follows:

```
10   READ VDD,VTO,GAMMA,FERMI
20   DATA 12,4,0.4,0.6
30   VOH=10:REM INITIAL GUESS
40   VX=VDD-(VTO+GAMMA*(SQR(VOH+FERMI)-SQR(FERMI)))
50   IF ABS(VOH-VX)<.005 THEN 100
60   VOH=VX:GOTO 40
100  PRINT "VOH=";VOH
```

```
RUN
VOH=7.19223
```

(b) V_{OL}: To compute V_{OL}, use eqn. (3.3-10) to obtain

$$V_{in} = 4 + \frac{1}{2}V_{OL} + \frac{1}{20V_{OL}}[12 - V_{OL} - 4 - 0.4(\sqrt{V_{OL} + 0.6} - \sqrt{0.6})]^2.$$

The simplest approach is to guess values for V_{OL}, then compute the resulting value of V_{in}. When $V_{in} = V_{OH} = 7.19$ [V], a solution has been found.

This type of calculation is summarized by the following chart:

V_{OL}	V_{in}
0.5	9.71
0.6	8.71
0.7	8.00
0.8	7.49
0.85	7.29
0.875	7.197 $\leftarrow$ $V_{in} \simeq V_{OH}$, so $V_{OL} \simeq 0.875$ [V].

A BASIC program for this calculation is shown below. Care must be taken in using this approach because the technique is trial and error, and is not self-iterating.

```
10   READ VDD,VTO,GAMMA,FERMI,BR,VOH
20   DATA 12,4,0.4,0.6,10,7.19
130   VOL=.1 : REM INITIAL GUESS
140   VTL=VTO+GAMMA * (SQR(VOL+FERMI)-SQR(FERMI))
150   VIN=VTO+.5 * VOL+(1/(2 * BR * VOL)) * (VDD-VOL-VTL)^2
160   IF ABS(VIN-VOH) < .01 THEN 200
170   VOL=VOL+.005:GOTO 140
200 PRINT "VOL=";VOL

RUN
VOL=.8749993
```

(c) V_{IL}: From eqn. (3.3-11),

$$V_{IL} \simeq V_{TD} = +4 \text{ [V]}.$$

(d) V_{IH}: First we use eqn. (3.3-28) to find V_{out} when $V_{in} = V_{IH}$. With the given values, this is

$$V_{out} = \frac{8 - 0.4(\sqrt{V_{out} + 0.6} - \sqrt{0.6})}{[31 + 0.4/(2\sqrt{V_{out} + 0.6})]^{1/2}} \, ,$$

where ηV_{out} has been ignored for simplicity. Iterating gives a value of $V_{out} \simeq 1.385$ [V]. The procedure is to guess a value of V_{out}, compute the right-hand side, then compare it with the original guess.

Since V_{out} is known, eqn. (3.3-26) may be used to write

$$V_{IH} \simeq 4 + \frac{1}{2}(1.385) + \frac{1}{(20)(1.385)}(12 - 1.385 - 4.25)^2,$$

where $V_{TL}(1.385) \simeq 4.25$ [V] has been used. The answer is then

$$V_{IH} \simeq 6.16 \text{ [V]}.$$

A BASIC program for this iteration is provided below. This includes terms neglected for the hand calculation.

```
10 READ VDD,VTO,GAMMA,FERMI,BR,VOH
20 DATA 12,4,0.4,0.6,10,7.19
230 VOUT=2:REM INITIAL GUESS
240 VTL=VTO+GAMMA * (SQR(VOUT+FERMI)-SQR(FERMI))
250 ETA=GAMMA/(2 * SQR(VOUT+FERMI))
260 VY=SQR((VDD-VTL)^2+2 * ETA * VOUT * (VDD-VTL))/SQR(3 *
BR+2 * ETA+1)
270 IF ABS(VY-VOUT) < .01 THEN 300
280 VOUT=VY:GOTO 240
300 VIH=VTO+.5 * VOUT+(1/(2 * BR * VOUT)) * (VDD-VOUT-
VTL)^2
310 PRINT "VIH=";VIH

RUN
VIH=6.122183
```

(e) V_{th}: The inverter threshold voltage is computed using eqn. (3.3-31):

$$V_{th} = \frac{20.65 - 0.4(\sqrt{V_{th} + 0.6} - \sqrt{0.6})}{1 + \sqrt{10}},$$

which self-iterates to a value $V_{th} \simeq 4.812\,[\text{V}]$.

The calculation is verified by the following BASIC program.

```
10  READ VDD,VTO,GAMMA,FERMI,BR,VOH
20  DATA 12,4,0.4,0.6,10,7.19
430 VTH=2:REM INITIAL GUESS
440 VTL=VTO+GAMMA * (SQR(VTH+FERMI)-SQR(FERMI))
450 VZ=(VDD+SQR(BR) * VTO-VTL)/(1+SQR(BR))
460 IF ABS(VZ-VTH) < .01 THEN 500
470 VTH=VZ: GOTO 440
500 PRINT "VTH="; VTH

RUN
VTH=4.810475
```

The results of a SPICE simulation on this circuit are shown in Fig. E3.2.

```
SATURATED EMODE INVERTER
VDD 3 0 DC 12VOLTS
MDRIVER 2 1 0 0 EMODE L=5U W=25U
MLOAD 3 3 2 0 EMODE L=10U W=5U
VS 1 0 DC
```

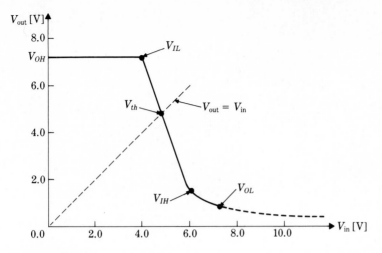

FIGURE E3.2

```
.MODEL EMODE NMOS VTO=4 GAMMA=0.4 KP=2.0E-5
.DC VS 0 12 0.1
.PLOT DC V(2)
.END
```

3.3.2 Nonsaturated Enhancement MOSFET Loads

The threshold voltage loss problem encountered in the previous section can be eliminated using purely static techniques by biasing the E-mode transistor load into the nonsaturation region. This is accomplished by providing another power supply voltage V_{GG} to the load device gate, giving the circuit configuration in Fig. 3.16(a). The requirement on V_{GG} is that

$$V_{GG} > V_{DD} + V_{TL}(V_{DD}). \qquad (3.3\text{-}37)$$

Since $V_{out} = (V_{DD} - V_{DSL})$, this extra bias will ensure that $V_{OH} = V_{DD}$.

The DC operation of the circuit is understood by writing $V_{DSL} = (V_{DD} - V_{out})$ and $V_{GSL} = (V_{GG} - V_{out})$. The nonsaturated load current is then given by

$$I_L = \frac{\beta_L}{2}[2(V_{GG} - V_{out} - V_{TL})(V_{DD} - V_{out}) - (V_{DD} - V_{out})^2]. \qquad (3.3\text{-}38)$$

To obtain the load line equation, set $I_D = I_L$ and $V_{out} = V_{DSD}$. After rearranging,

$$I_D = \frac{\beta_L}{2}(2V_{GG} - V_{DD} - 2V_{TL} - V_{DSD})(V_{DD} - V_{DSD}), \qquad (3.3\text{-}39)$$

which is plotted in Fig. 3.16(b). Note that this form of the load line equation

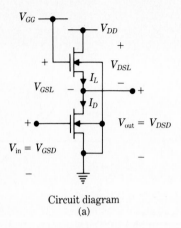

Circuit diagram
(a)

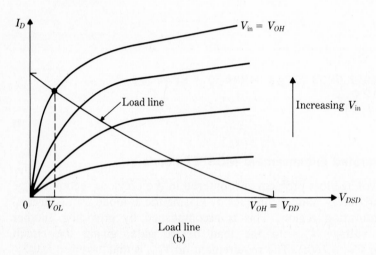

Load line
(b)

FIGURE 3.16 Nonsaturated enhancement-load nMOS inverter.

explicitly shows the zero at $V_{DD} = V_{DSD}$; thus, if eqn. (3.3-37) is satisfied, $V_{OH} = V_{DD}$ for the circuit. Another point that should be observed is the fact that $I_D = I_L$ is *linear* in V_{GG}. If V_{GG} is much greater than V_{DD}, using $V_{DSD,\max} = V_{DD}$ gives

$$I_D \simeq \beta_L V_{GG}(V_{DD} - V_{DSD}), \qquad \textbf{(3.3-40)}$$

i.e., a linear load line. Of course, practical voltage supply levels will not allow for this in a realistic situation. However, this does illustrate the important point that the VTC is a direct function of the load gate voltage V_{GG}.

A typical VTC for the nonsaturated enhancement load is drawn in Fig. 3.17. For $V_{in} < V_{TD}$, $V_{out} = V_{OH} = V_{DD}$. When V_{in} is increased to a value

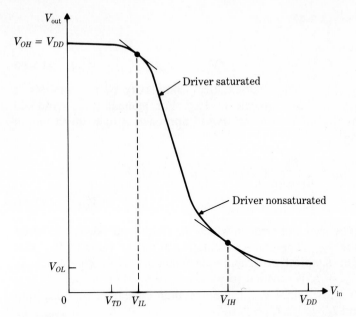

FIGURE 3.17 Voltage transfer characteristics for a nonsaturated enhancement-load nMOS inverter.

greater than V_{TD}, the driver is saturated, giving

$$\frac{\beta_D}{2}(V_{in} - V_{TD})^2 = \frac{\beta_L}{2}[2(V_{GG} - V_{TL} - V_{out})(V_{DD} - V_{out}) - (V_{DD} - V_{out})^2].$$

(3.3-41)

This describes the VTC until $V_{in} = (V_{out} + V_{TD})$, which is the borderline between saturated and nonsaturated driver operation. As V_{in} is increased beyond this voltage, the driver goes into the nonsaturated region, so

$$\frac{\beta_D}{2}[2(V_{in} - V_{TD})V_{out} - V_{out}^2] = \frac{\beta_L}{2}[2(V_{GG} - V_{TL} - V_{out})(V_{DD} - V_{out})$$

$$- (V_{DD} - V_{out})^2] \quad \textbf{(3.3-42)}$$

describes the remainder of the DC curve. These equations demonstrate that the VTC shape is strongly dependent on the specific value of V_{GG}.

ANALYSIS

V_{OH}   As long as eqn. (3.3-37) is satisfied, $V_{OH} = V_{DD}$. This is, of course, one reason for introducing the circuit in the first place.

V_{OL} $V_{out} = V_{OL}$ is computed by setting $V_{in} = V_{OH}$. The driver is nonsaturated, so eqn. (3.3-42) gives

$$\beta_D[2(V_{OH} - V_{TD})V_{OL} - V_{OL}^2] = \beta_L[2(V_{GG} - V_{TL} - V_{OL})$$
$$- (V_{DD} - V_{OL})](V_{DD} - V_{OL}), \quad \text{(3.3-43)}$$

in which $V_{TL} = V_{TL}(V_{OL})$. The mathematical complexity of the analysis is immediately obvious: this is a quartic in V_{OL}. For manual (i.e., handheld calculator) solutions, the most straightforward approach is to write the input voltage as V_{in}, giving

$$V_{in} - V_{TD} = \frac{1}{2}V_{OL} + \frac{\beta_L}{2V_{OL}\beta_D}[2(V_{GG} - V_{TL} - V_{OL})$$
$$- (V_{DD} - V_{OL})](V_{DD} - V_{OL}). \quad \text{(3.3-44)}$$

This can be iterated by trial and error in which the right-hand side is computed for various choices of V_{OL}. The correct solution will be the value of V_{OL} that gives $V_{in} = V_{OH}$. The simplest programmable technique is to iterate V_{OL} in eqn. (3.3-44) until $I_D = I_L$ is satisfied to within a specified error margin. It is easiest to start with a low initial V_{OL} value and then increment within a loop that compares I_D and I_L.

V_{IL} The input low voltage occurs when the driver is saturated, so

$$\frac{\beta_D}{2}(V_{in} - V_{TD})^2 = \frac{\beta_L}{2}[2(V_{GG} - V_{TL} - V_{out})(V_{DD} - V_{out})$$
$$- (V_{DD} - V_{out})^2]. \quad \text{(3.3-45)}$$

This must be solved simultaneously with the VTC slope condition $(dV_{out}/dV_{in}) = -1$. Since $I_D(V_{in}) = I_L(V_{out})$,

$$\frac{dV_{out}}{dV_{in}} = \frac{(dI_D/dV_{in})}{(dI_L/dV_{out})} = -1 \quad \text{(3.3-46)}$$

is valid. Differentiating the currents then gives

$$\beta_D(V_{in} - V_{TD}) = \beta_L[(V_{GG} - V_{TL} - V_{out}) + \eta(V_{DD} - V_{out})], \quad \text{(3.3-47)}$$

which provides the second equation required to solve for the 2 unknowns $V_{in} = V_{IH}$ and V_{out}.

To obtain a tractable technique for solution, note that $V_{in} = V_{IL}$ implies that $\eta(V_{DD} - V_{out})$ will be small. Neglecting this term in the derivative condition gives

$$(V_{GG} - V_{TL} - V_{out}) \simeq \beta_R(V_{in} - V_{TD}) \quad \text{(3.3-48)}$$

in addition to

$$(V_{DD} - V_{out}) \simeq \beta_R(V_{in} - V_{TD}) + (V_{DD} - V_{GG} + V_{TL}), \quad \text{(3.3-49)}$$

which may be substituted into the current equation (3.3-45). After some algebra,

$$V_{in} \simeq V_{TD} + \frac{V_{GG} - V_{DD} - V_{TL}(V_{out})}{\sqrt{\beta_R(\beta_R - 1)}}, \qquad (3.3\text{-}50)$$

which is compared with

$$V_{in} \simeq V_{TD} + \frac{1}{\beta_R}(V_{GG} - V_{TL} - V_{out}) \qquad (3.3\text{-}51)$$

as obtained directly from eqn. (3.3-45). Eliminating $(V_{in} - V_{TD})$ and rearranging then yield

$$V_{out} \simeq V_{GG} - V_{TL}(V_{out}) - \frac{\beta_R}{\sqrt{\beta_R(\beta_R - 1)}}[V_{GG} - V_{DD} - V_{TL}(V_{out})].$$

$$(3.3\text{-}52)$$

Note that if $\beta_R \gg 1$, this gives $V_{out} = V_{DD}$ as expected. This equation for V_{out} may be solved by using self-iteration, where $V_{out} = 0.95V_{DD}$ might be a reasonable first guess. Once V_{out} is known, $V_{in} = V_{IL}$ can be found using $V_{TL}(V_{out})$ in eqn. (3.3-50).

V_{IH} With $V_{in} = V_{IH}$, the driver is nonsaturated with the circuit current equation

$$\beta_D[2(V_{in} - V_{TD})V_{out} - V_{out}^2]$$
$$= \beta_L[2(V_{GG} - V_{TL} - V_{out})(V_{DD} - V_{out}) - (V_{DD} - V_{out})^2]. \quad (3.3\text{-}53)$$

Noting $I_D(V_{in}, V_{out}) = I_L(V_{out})$, the VTC slope around V_{IH} is obtained using eqn. (3.2-15):

$$\frac{dV_{out}}{dV_{in}} = \frac{\partial I_D/\partial V_{in}}{(dI_L/dV_{out}) - (\partial I_D/\partial V_{out})}. \qquad (3.3\text{-}54)$$

The unity slope point occurs when

$$\frac{\beta_D V_{out}}{\beta_L[(V_{GG} - V_{TL} - V_{out}) + \eta(V_{DD} - V_{out})] + \beta_D(V_{in} - V_{TD} - V_{out})} = 1,$$

$$(3.3\text{-}55)$$

which must be solved simultaneously with eqn. (3.3-53). The simplest way to obtain a solution to this problem is to assume that $\eta(V_{DD} - V_{out}) \ll (V_{GG} - V_{TL} - V_{out})$ in the derivative condition. Then

$$V_{in} - V_{TD} \simeq \frac{2\beta_R + 1}{\beta_R}V_{out} - \frac{1}{\beta_R}(V_{GG} - V_{TL}), \qquad (3.3\text{-}56)$$

while the current equation (3.3-52) gives

$$V_{in} - V_{TD} = \frac{1}{2}V_{out} + \frac{1}{2\beta_R V_{out}}[2(V_{GG} - V_{TL} - V_{out})(V_{DD} - V_{out})$$

$$- (V_{DD} - V_{out})^2]. \quad \text{(3.3-57)}$$

Equating these two expressions for $(V_{in} - V_{TD})$ and simplifying yield

$$V_{out} \simeq \sqrt{\frac{2[V_{GG} - V_{TL}(V_{out})]V_{DD} - V_{DD}^2}{3\beta_R + 1}}, \quad \text{(3.3-58)}$$

which can be solved for V_{out} by using the technique of self iterations. Once V_{out} is known, $V_{in} = V_{IH}$ can be computed directly from eqn. (3.3-56).

V_{th}   With $V_{in} = V_{out} = V_{th}$, the driver is saturated so that the inverter threshold voltage occurs when

$$\frac{\beta_D}{2}(V_{th} - V_{TD})^2 = \frac{\beta_L}{2}[2(V_{GG} - V_{TL} - V_{th})(V_{DD} - V_{th}) - (V_{DD} - V_{th})^2],$$

$$\text{(3.3-59)}$$

where $V_{TL} = V_{TL}(V_{th})$. Since this is a fourth-order polynomial in V_{th}, it is quite difficult to solve analytically. The most straightforward numerical technique is to simply guess reasonable values for V_{th} and then iterate until the equation is satisfied.

DESIGN

There are actually two design parameters in this circuit, $\beta_R = \beta_D/\beta_L$ and V_{GG}. If an *a priori* choice for V_{GG} is made (for example, the available power supply voltages for the chip are already established), then β_R is determined directly from the design specification on V_{OL}. This can be seen by using eqn. (3.3-43) to write

$$\beta_R = \frac{\{2[V_{GG} - V_{TL}(V_{OL}) - V_{OL}] - (V_{DD} - V_{OL})\}(V_{DD} - V_{OL})}{2(V_{OH} - V_{TD})V_{OL} - V_{OL}^2}. \quad \text{(3.3-60)}$$

The VTC is structured by this value for β_R, as can be verified by the equations describing the critical points of the curve. To compare this with the saturated MOSFET load inverter, take

$$V_{DD} = 12\,[\text{V}], \quad V_{T0} = 1.5\,[\text{V}], \quad \gamma = 0.4\,[\text{V}^{1/2}],$$
$$2|\phi_F| = 0.62\,[\text{V}], \quad V_{OL} = 0.5\,[\text{V}],$$

and suppose that $V_{GG} = 16\,[\text{V}]$. Noting that $V_{OH} = V_{DD} = 12\,[\text{V}]$, this circuit requires a (rounded integer) value of $\beta_R = 19$. The saturated enhancement load, on the other hand, needed $\beta_R = 13$, which implies a smaller chip area but also gave a reduced NM_H since $V_{OH} < V_{DD}$.

When β_R and V_{GG} are both used as design variables, it is convenient to introduce a *biasing parameter*

$$m = \frac{V_{DD}}{2[V_{GG} - V_{TL}(V_{out})] - V_{DD}},$$ (3.3-61)

where $m = m(V_{out})$ due to body bias effects in V_{TL}. The maximum value of m occurs when $(V_{GG} - V_{TL}) = V_{DD}$, giving $m = 1$. As $V_{GG} \to \infty$, $m \to 0$, indicating that the biasing parameter is restricted to the interval $(0, 1]$. In terms of m, the load current assumes the form

$$I_L = \frac{\beta_L V_{DD}^2}{2m} \left(1 - \frac{V_{out}}{V_{DD}}\right)\left(1 - m\frac{V_{out}}{V_{DD}}\right).$$ (3.3-62)

For a specified design value of V_{OL},

$$\left(1 - \frac{V_{OL}}{V_{DD}}\right)\left(1 - m\frac{V_{OL}}{V_{DD}}\right) = m\beta_R\left[2\left(\frac{V_{OH} - V_{TL}(V_{OL})}{V_{DD}}\right) - \left(\frac{V_{OL}}{V_{DD}}\right)\right]\frac{V_{OL}}{V_{DD}}$$ (3.3-63)

determines the simultaneous requirements on $m = m(V_{OL})$ and β_R. The simplest approach to using this equation in the design process is to compute β_R values for various choices of m. It is noted that a resistive analogy can again be constructed, where now the load equivalent resistor is a function of both V_{GG} and $(W/L)_L$. As with the two previous cases, the absolute values of $(W/L)_D$ and $(W/L)_L$ are dependent on the desired transient response (see Chapter 4).

3.4 Depletion-Mode MOSFET Loads

The static nMOS inverters analyzed thus far have used only enhancement-mode MOSFETs in which $V_T > 0$. In terms of wafer processing considerations, this is a distinct advantage since only a single type of transistor is required. However, it was demonstrated that acceptable VTC characteristics required relatively large driver-load ratios, indicating expensive chip real estate costs. The ratioed nature of static circuits limits their use in high-density integration schemes.

The most efficient (static) nMOS process that allows for good noise margins and sharp VTC transitions without consuming excessive chip area employs both enhancement-mode $(V_T > 0)$ and depletion-mode $(V_T < 0)$ MOSFETs. This type of circuit requires more complicated processing than one that uses only E-mode devices, since additional steps must be added to form the depletion-mode (D-mode) transistors. Basically, a donor ion implant with the associated lithography allows for threshold voltage control of the depletion-mode structures. With the development of ion implantation as an everyday production line technique, however, this does not cause any great problems. The improved circuit integration possibilities more than compensate

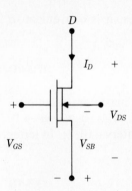

FIGURE 3.18 Depletion-mode MOSFET circuit symbol.

for the extra time and expense involved in performing the extra processing steps.

The circuit symbol for the D-mode n-channel MOSFET is shown in Fig. 3.18, where the implanted n-type channel is suggested by the addition of an extra line to the E-mode schematic designation. The circuit equations for this device are taken to be identical in form to those that describe an E-mode MOSFET. The primary difference between the two is that the $V_{SB} = 0$ threshold voltage V_{T0} is *negative* for a depletion-mode device.

The basic inverter circuit that employs both E-mode and D-mode MOSFETs is shown in Fig. 3.19(a). Note that the depletion-type load has its gate and source electrode connected. This gives $V_{GSL} = 0$, but since $V_{TL} < 0$ for the D-mode transistor, $V_{GSL} = 0 > V_{TL}$ is always satisfied. Consequently, a conducting channel always exists between V_{DD} and the output node through the depletion load. This will give an output high voltage of $V_{OH} = V_{DD}$, as discussed below. The drain-source voltage across the load is $V_{DSL} = (V_{DD} - V_{out})$. Since the D-mode saturation voltage is $V_{DSL,\text{sat}} = (V_{GSL} - V_{TL}) = |V_{TL}|$,

$$V_{DD} - V_{\text{out}} = |V_{TL}| \tag{3.4-1}$$

constitutes the border between saturated and nonsaturated load device operation. Note that

$$V_{TL}(V_{\text{out}}) = V_{T0L} + \gamma_L(\sqrt{V_{\text{out}} + 2\,|\phi_{F,L}|} - \sqrt{2\,|\phi_{F,L}|}), \tag{3.4-2}$$

with $V_{T0L} < 0$ implies that $|V_{TL}|$ decreases as V_{out} increases (since V_{TL} becomes less negative). It will be assumed that $V_{TL} < 0$ is satisfied, i.e., that the body bias term always has a magnitude less than $|V_{T0L}|$. Equation (3.4-1) demonstrates that the depletion load MOSFET will operate in both the saturated and nonsaturated regions, depending on the value of V_{out}.

When V_{out} is small such that $(V_{DD} - V_{\text{out}}) > |V_{TL}(V_{\text{out}})|$, the load is saturated, with

$$I_L = \frac{\beta_L}{2}[V_{GSL} - V_{TL}(V_{\text{out}})]^2 = \frac{\beta_L}{2}[-V_{TL}(V_{\text{out}})]^2. \tag{3.4-3}$$

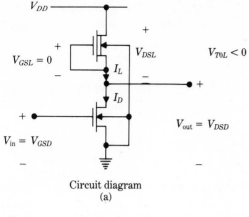

Circuit diagram
(a)

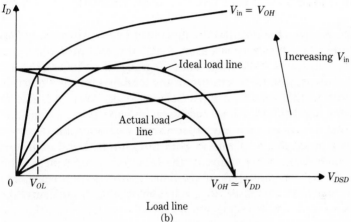

Load line
(b)

FIGURE 3.19 Depletion-mode MOSFET load.

This is particularly interesting when extrapolated to the case of an *ideal* depletion load transistor in which V_{TL} = constant. If body bias effects could be ignored, then eqn. (3.4-3) implies that the saturated D-mode transistor would act as an ideal current source with I_L = constant. Although body bias cannot be neglected in the real world, this remains a useful approximation for the qualitative analysis of the circuit performance. As V_{out} is increased, the load enters the non-saturated region where $(V_{DD} - V_{out}) < |V_{TL}|$. The load current is then described by

$$I_L = \frac{\beta_L}{2}[2\,|V_{TL}(V_{out})|\,(V_{DD} - V_{out}) - (V_{DD} - V_{out})^2], \qquad \textbf{(3.4-4)}$$

which is valid for increasing V_{out} values.

The load line for the depletion MOSFET is constructed by setting $I_L = I_D$ and $V_{out} = V_{DSD}$. For V_{DSL} small, the load is saturated; with the body bias effects contained in $V_{TL}(V_{out})$, eqn. (3.4-3) predicts a decreasing load current as illustrated in Fig. 3.19(b). The load device enters the nonsaturated region when

$$V_{DSD} = V_{DD} - |V_{TL}(V_{DSD})|. \tag{3.4-5}$$

For $V_{out} = V_{DSD}$ greater than this voltage, the load line is obtained from eqn. (3.4-4). This may be rewritten as

$$I_D = \frac{\beta_D}{2} [2|V_{TL}(V_{DSD})| - V_{DD} + V_{DSD}](V_{DD} - V_{DSD}), \tag{3.4-6}$$

giving a zero at $V_{DSD} = V_{DD}$; hence, $V_{OH} = V_{DD}$ for this circuit. This should be compared to the inverters in the previous section that used enhancement-mode MOSFETs as loads. There it was shown that the presence of threshold voltage losses requires using an additional power supply voltage V_{GG} to achieve $V_{OH} = V_{DD}$.

The plot in Fig. 3.19(b) also includes the ideal load line in which body bias variations are ignored, i.e., the load acts as an ideal current source when it is saturated. This demonstrates that body bias degrades the inverter VTC transition. Note, however, that the curvature of the load line for the depletion MOSFET is opposite to that found for enhancement loads. This gives rise to the fact that the depletion MOSFET load configuration will exhibit a transition from V_{OH} to V_{OL} that is sharper than any other nMOS inverter studied.

Since the depletion load can operate in both the saturated and nonsaturated regions, the mathematical derivation of the VTC for this circuit is a bit more involved than that encountered in previous cases. In particular, the presence of two threshold voltages $V_{TD} > 0$ and $V_{TL} < 0$ indicates the existence of two saturation-point voltages (one for the load and one for the driver). The development presented here will assume that

$$V_{TD} < |V_{TL}| < V_{DD} \tag{3.4-7}$$

since this is usually the case for a realistic circuit. The exact relationship between V_{TD} and V_{TL} will be kept arbitrary.

The generation of the VTC initially proceeds in the usual manner. With $V_{in} < V_{TD}$, the driver is in cutoff. Since $V_{GSL} = 0$, the depletion load provides a conducting path from the output node to the power supply rail, so $V_{out} = V_{OH} \simeq V_{DD}$. Now, when V_{in} is increased to a value slightly greater than V_{TD}, the driver enters the saturation region, while $(V_{DD} - V_{out}) < |V_{TL}|$ implies that the load is nonsaturated. Hence,

$$\frac{\beta_D}{2}(V_{in} - V_{TD})^2 = \frac{\beta_L}{2}[2|V_{TL}(V_{out})|(V_{DD} - V_{out}) - (V_{DD} - V_{out})^2] \tag{3.4-8}$$

describes the initial falloff of the VTC from $V_{out} = V_{OH}$.

As V_{in} is increased further, V_{out} decreases according to this equation until either the driver or the load changes operational regions. Which device changes characteristics first depends on the value of V_{out} with regard to the saturation-point constraint for each transistor. Consider V_{out} as it decreases from V_{OH}. If

$$V_{out} < V_{DD} - |V_{TL}| \tag{3.4-9}$$

is satisfied first, then the load will enter the saturation region while the driver is still saturated. On the other hand, if

$$V_{out} < (V_{in} - V_{TD}) \tag{3.4-10}$$

is true before eqn. (3.4-9), then the driver will enter the nonsaturated region while the load is still nonsaturated. To determine which occurs first in a given circuit, first write

$$V_{out} = V_{DD} - |V_{TL}(V_{out})| \tag{3.4-11}$$

as the load borderline voltage; for V_{out} greater than this value, the load is nonsaturated, while V_{out} less than this value indicates a saturated depletion transistor. With $V_{TL}(V_{out})$ given in eqn. (3.4-2), this constitutes a self-iterating equation that can be solved for the V_{out} level that changes the load characteristics. The input voltage required to obtain this particular output level is found by substituting eqn. (3.4-11) into the current equation (3.4-8). Simple rearrangement gives

$$(V_{in} - V_{TD}) = \frac{|V_{TL}|}{\sqrt{\beta_R}}, \tag{3.4-12}$$

where $\beta_R = \beta_D/\beta_L$ as usual and V_{TL} is evaluated at the V_{out} found by solving eqn. (3.4-11). Using the conditions in eqns. (3.4-9) and (3.4-10), it is seen that if V_{out} evaluated from eqn. (3.4-11) is *greater than* $(V_{in} - V_{TD})$ from eqn. (3.4-12), then the load transistor will be the first to change operation regions. On the other hand, if V_{out} is *less than* the value of $(V_{in} - V_{TD})$ from eqn. (3.4-10), the driver will go into the nonsaturated region before the load changes its characteristics. The driver-load ratio β_R plays a major role in determining which event occurs first. To keep the discussion general, both cases will be considered here.

First, suppose that V_{out} from eqn. (3.4-11) is greater than $(V_{in} - V_{TD})$ from eqn. (3.4-12). This implies that the load will change from a nonsaturated to a saturated condition before the driver characteristics are altered. When this occurs, both devices are saturated, so that the VTC for this portion of the curve is described by

$$\frac{\beta_D}{2}(V_{in} - V_{TD})^2 = \frac{\beta_L}{2}[-V_{TL}(V_{out})]^2. \tag{3.4-13}$$

This predicts a very rapid falloff of the VTC. To understand the properties of this equation, note that an ideal load in which V_{TL} = constant would give a steplike VTC transition. The VTC remains quite sharp even when body bias effects are included. Now, as V_{out} continues to decrease (corresponding to larger input voltages), the condition $V_{\text{out}} < (V_{\text{in}} - V_{TD})$ will eventually be satisfied. The driver then enters the nonsaturated mode of operation, so that the remaining portion of the VTC is obtained from

$$\frac{\beta_D}{2}[2(V_{\text{in}} - V_{TD})V_{\text{out}} - V_{\text{out}}^2] = \frac{\beta_L}{2}[-V_{TL}(V_{\text{out}})]^2. \tag{3.4-14}$$

The other possibility for the central portion of the VTC is that $V_{\text{out}} < (V_{\text{in}} - V_{TD})$ is satisfied before eqn. (3.4-9). For this case, the driver is the first device to change characteristics, going from a saturated to a nonsaturated mode. The VTC is obtained from

$$\frac{\beta_D}{2}[2(V_{\text{in}} - V_{TD})V_{\text{out}} - V_{\text{out}}^2] = \frac{\beta_L}{2}[2|V_{TL}(V_{\text{out}})|(V_{DD} - V_{\text{out}})$$
$$- (V_{DD} - V_{\text{out}})^2]. \tag{3.4-15}$$

This remains valid until V_{out} drops to a level in which eqn. (3.4-9) is satisfied. Further decreases in V_{out} take the load into saturation, so that the VTC for large input voltages is given by

$$\frac{\beta_D}{2}[2(V_{\text{in}} - V_{TD})V_{\text{out}} - V_{\text{out}}^2] = \frac{\beta_L}{2}[-V_{TL}(V_{\text{out}})]^2, \tag{3.4-16}$$

which is identical to eqn. (3.4-14). This analysis demonstrates that only the central portion of the VTC (which determines the sharpness of the VTC curve) is sensitive to changes of load and driver operational regions. Some examples of depletion load VTC curves are shown in Fig. 3.20.

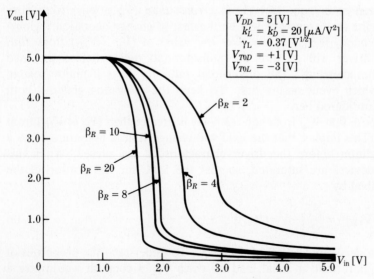

FIGURE 3.20   VTCs for depletion-mode loaded inverter.

ANALYSIS

V_{OH} The output high voltage of this circuit is usually taken to be $V_{OH} = V_{DD}$, as discussed previously. This is, of course, an approximation since it ignores the resistance of the load device. The nature of the approximation may be understood by recalling that

$$V_{OH} = V_{DD} - V_{DSL} \mid_{V_{in}=0}. \qquad (3.4\text{-}17)$$

To compute the drain-source voltage across the load transistor, note that the magnitude of the load current I_L is restricted to the driver leakage level. The conductance of the nonsaturated load MOSFET is

$$G_{DSL} \equiv \frac{I_L}{V_{DSL}} = \frac{\beta_L}{2}[2\,|V_{TL}(V_{OH})| - (V_{DD} - V_{OH})], \qquad (3.4\text{-}18)$$

so that approximating $V_{DSL} = I_L/G_{DSL}$ gives

$$V_{OH} = V_{DD} - \frac{I_L}{\dfrac{k'_L}{2}\left(\dfrac{W}{L}\right)_L [2\,|V_{TL}(V_{OH})| - (V_{DD} - V_{OH})]}. \qquad (3.4\text{-}19)$$

This can be used for self-iterations to find a more realistic value of V_{OH} for the circuit (assuming that the driver leakage current is known). So long as this current is small (below the microampere range), most current fabrication processes will yield device parameters such that $V_{OH} = V_{DD}$ is a very good approximation. However, if leakage currents become large, there will be a noticeable lowering of V_{OH} from the power supply level. One way to counteract this is to increase the area of the load, i.e., use a larger $(W/L)_L$ value; this has the detrimental effect of increasing the chip area.

V_{OL} V_{OL} is computed by setting $V_{in} = V_{OH}$ and $V_{out} = V_{OL}$ in eqn. (3.4-16):

$$\beta_R[2(V_{OH} - V_{TD})V_{OL} - V_{OL}^2] = |V_{TL}(V_{OL})|^2. \qquad (3.4\text{-}20)$$

In the strictest sense, this forms a quartic in V_{OL}. However, a technique for solution may be obtained by first rearranging the equation to read

$$V_{OL}^2 - 2(V_{OH} - V_{TD})V_{OL} + \frac{1}{\beta_R}|V_{TL}(V_{OL})|^2 = 0. \qquad (3.4\text{-}21)$$

Ignoring for the moment the body bias effects in V_{TL}, this can be viewed as quadratic in V_{OL} with solutions

$$V_{OL} = (V_{OH} - V_{TD}) - \sqrt{(V_{OH} - V_{TD})^2 - \frac{1}{\beta_R}|V_{TL}(V_{OL})|^2}. \qquad (3.4\text{-}22)$$

The actual value of V_{OL} can be found very quickly by the self-iterative technique. A reasonable first guess for V_{OL} is made, which allows calculation of $V_{TL}(V_{OL})$ and, therefore, the right side of the equation. The new value of V_{OL} is substituted into the right-hand side of the same equation until the input V_{OL} is consistent with that found from using eqn. (3.4-22). Since body bias variations are small when $V_{\text{out}} = V_{OL}$, the convergence is quite rapid, often requiring only 2 or 3 iterations.

V_{IL} When $V_{\text{in}} = V_{IL}$, the driver is saturated and the load is operating in the nonsaturated region. At this point, eqn. (3.4-8) gives

$$\beta_D(V_{\text{in}} - V_{TD})^2 = \beta_L[2\,|V_{TL}(V_{\text{out}})|\,(V_{DD} - V_{\text{out}}) - (V_{DD} - V_{\text{out}})^2], \quad \textbf{(3.4-23)}$$

which must be solved simultaneously with the VTC slope condition $(dV_{\text{out}}/dV_{\text{in}}) = -1$. Since the currents have the functional dependence $I_D(V_{\text{in}}) = I_L(V_{\text{out}})$,

$$\frac{dV_{\text{out}}}{dV_{\text{in}}} = \frac{(dI_D/dV_{\text{in}})}{(dI_L/dV_{\text{out}})} = -1 \qquad\qquad \textbf{(3.4-24)}$$

is valid. Computing the derivatives gives the second equation

$$\beta_D(V_{\text{in}} - V_{TD}) = \beta_L[|V_{TL}(V_{\text{out}})| - (1 + \eta)(V_{DD} - V_{\text{out}})] \qquad \textbf{(3.4-25)}$$

for the two unknowns $V_{\text{in}} = V_{IL}$ and V_{out}.

To find a value for V_{IL}, assume that $\eta \ll 1$. Then eqn. (3.4-25) gives the approximation

$$(V_{DD} - V_{\text{out}}) \simeq [|V_{TL}(V_{\text{out}})| - \beta_R(V_{\text{in}} - V_{TD})], \qquad \textbf{(3.4-26)}$$

which may be substituted into the right side of eqn. (3.4-23). Performing the algebra yields

$$(V_{\text{in}} - V_{TD}) \simeq \frac{|V_{TL}(V_{\text{out}})|}{\sqrt{\beta_R(\beta_R + 1)}}, \qquad\qquad \textbf{(3.4-27)}$$

which allows for the calculation of $V_{\text{in}} = V_{IL}$ once V_{out} at this point is known. To obtain V_{out}, note that eqn. (3.4-26) says that

$$(V_{\text{in}} - V_{TD}) \simeq [|V_{TL}(V_{\text{out}})| - (V_{DD} - V_{\text{out}})]\frac{1}{\beta_R}. \qquad \textbf{(3.4-28)}$$

Equating these two expressions for $(V_{\text{in}} - V_{TD})$ gives

$$V_{\text{out}} \simeq V_{DD} - |V_{TL}(V_{\text{out}})|\left[1 - \frac{\beta_R}{\sqrt{\beta_R(\beta_R + 1)}}\right], \qquad \textbf{(3.4-29)}$$

which forms a self-iterating equation for V_{out}. The approximate value of V_{IL} can thus be computed using relatively straightforward techniques.

V_{IH} When $V_{\text{in}} = V_{IH}$, the circuit is in a state described by

$$\beta_D[2(V_{\text{in}} - V_{TD})V_{\text{out}} - V_{\text{out}}^2] = \beta_L\,|V_{TL}(V_{\text{out}})|^2 \qquad \textbf{(3.4-30)}$$

since the driver is nonsaturated while the load is in saturation. Using $I_D(V_{in}, V_{out}) = I_L(V_{out})$, the required VTC slope condition

$$\frac{dV_{out}}{dV_{in}} = \frac{\partial I_D/\partial V_{in}}{(dI_L/dV_{out}) - (\partial I_D/\partial V_{out})} = -1 \qquad (3.4\text{-}31)$$

is evaluated as

$$\beta_D V_{out} = \beta_D(V_{in} - V_{TD} - V_{out}) - \beta_L \eta(V_{out}) |V_{TL}(V_{out})|. \qquad (3.4\text{-}32)$$

Rearranging,

$$(V_{in} - V_{TD}) = 2V_{out} + \frac{\eta |V_{TL}|}{\beta_R}. \qquad (3.4\text{-}33)$$

This may be substituted into the current eqn. (3.4-30). After some algebra, the resulting equation is

$$V_{out}^2 + \frac{2\eta |V_{TL}|}{3\beta_R} V_{out} - \frac{|V_{TL}|^2}{3\beta_R} = 0, \qquad (3.4\text{-}34)$$

which may be initially viewed as an approximate quadratic in V_{out}. Solving gives a self-iterating equation for V_{out} of the form

$$V_{out} = -\frac{\eta(V_{out}) |V_{TL}(V_{out})|}{3\beta_R} + \frac{|V_{TL}(V_{out})|}{\sqrt{3\beta_R}} \sqrt{\frac{\eta^2}{3\beta_R} + 1}, \qquad (3.4\text{-}35)$$

which is suitable for programming. If hand calculations are used, an estimate for V_{out} can be obtained by taking $\eta = 0$ so

$$V_{out} \simeq \frac{|V_{TL}(V_{out})|}{\sqrt{3\beta_R}}. \qquad (3.4\text{-}36)$$

This is also self-iterating and yields an approximate value for V_{out} with straightforward calculations. Once V_{out} is known, $V_{in} = V_{IH}$ can be computed from eqn. (3.4-33) as

$$V_{IH} = V_{TD} + 2V_{out} + \frac{\eta(V_{out}) |V_{TL}(V_{out})|}{\beta_R}. \qquad (3.4\text{-}37)$$

V_{th} The inverter threshold voltage occurs at the intersection of the VTC and the unity-gain line, so $V_{th} = V_{in} = V_{out}$ implies that the driver is saturated. Since V_{th} is between V_{IL} and V_{IH}, the load may be in either operating region depending on the device characteristics.

If the load is saturated, V_{th} is found from

$$\beta_D(V_{th} - V_{TD})^2 = \beta_L |V_{TL}(V_{th})|^2, \qquad (3.4\text{-}38)$$

or

$$V_{th} = \frac{|V_{TL}(V_{th})|}{\sqrt{\beta_R}} + V_{TD}, \qquad (3.4\text{-}39)$$

which can be self-iterated. On the other hand, if the load is nonsaturated, then

$$\beta_D(V_{th} - V_{TD})^2 = \beta_L[2\,|V_{TL}(V_{th})|\,(V_{DD} - V_{th}) - (V_{DD} - V_{th})^2]. \qquad \text{(3.4-40)}$$

The simplest approach in this case is to find V_{th} by trial and error.

DESIGN

The design of the depletion load nMOS inverter proceeds in a relatively straightforward manner. Interest is directed toward obtaining desired output levels for V_{OH} and V_{OL} by means of two design variables, β_R and V_{T0L}. It should be noted that the zero-bias load threshold voltage V_{T0L} is set by the depletion-mode ion implant during the fabrication process; as such, it is not available for use as a circuit design variable. Including V_{T0L} in the design set will keep the discussion in general terms while simultaneously illustrating the effects V_{TL} has on the VTC characteristics.

The output high voltage was given in eqn. (3.4-19) as

$$V_{OH} = V_{DD} - \cfrac{I_L}{\dfrac{k'_L}{2}\left(\dfrac{W}{L}\right)_L\,[2\,|V_{TL}(V_{OH})| - (V_{DD} - V_{OH})]}, \qquad \text{(3.4-41)}$$

where I_L is constrained to be the driver MOSFET leakage current. The design requirements are extracted from the second term by recalling that the ideal output high voltage is $V_{OH} = V_{DD}$. To approach this ideal level, $(W/L)_L$ can be made large at the expense of consuming a larger chip area. Also note that

$$V_{TL}(V_{OH}) = V_{T0L} + \gamma_L(\sqrt{V_{OH} + 2\,|\phi_{F,L}|} - \sqrt{2\,|\phi_{F,L}|}). \qquad \text{(3.4-42)}$$

Since a large V_{OH} is desired, the body bias term will tend to offset the fact that V_{T0L} is negative, which in turn reduces $|V_{TL}(V_{OH})|$. This consideration is one factor that leads to choosing relatively high values for $|V_{T0L}|$.

The most crucial design constraints for the circuit arise when setting the desired V_{OL} value. Rearranging eqn. (3.4-20) gives

$$\beta_R = \frac{|V_{TL}(V_{OL})|^2}{2(V_{OH} - V_{TD})V_{OL} - V_{OL}^2}, \qquad \text{(3.4-43)}$$

where the driver-load ratio is

$$\beta_R = \frac{\beta_D}{\beta_L} = \frac{k'_D\left(\dfrac{W}{L}\right)_D}{k'_L\left(\dfrac{W}{L}\right)_L}. \qquad \text{(3.4-44)}$$

β_R has been explicitly written in terms of the process transconductance parameters k' to illustrate a subtle point. Since the inverter uses both enhancement-mode and depletion-mode MOSFETs, k'_D may not be equal to k'_L. This may be understood in a simplistic manner by recalling from basic semiconductor physics that the electron mobility μ_n is a function of the doping density. Owing to the fact that the E-mode threshold voltage adjustment acceptor ion implant will not necessarily be at the same level as the D-mode donor implant (reflecting, among other things, the differences in V_{T0D} and V_{T0L}), the mobilities in k' will be different for the driver and the load. Although many current processes give $k'_D \simeq k'_L$ for a certain range of device geometries, this point should be checked before the reduction for $\beta_R = (W/L)_D/(W/L)_L$ is made.

Consider now the basic design information contained in eqn. (3.4-43). This allows for the determination of the β_R value required to set the output low voltage V_{OL}. The expression also demonstrates the dependence of β_R on the load threshold voltage through the factor $|V_{TL}(V_{OL})|$. As an example of the use of this equation, suppose that $V_{DD} = +5\,[\text{V}] = V_{OH}$, $V_{TD} = +1\,[\text{V}]$, $\gamma_L = 0.40\,[\text{V}^{1/2}]$, and $2|\phi_{F,L}| = 0.60\,[\text{V}]$ are chosen as representative nominal parameters. The equation then gives the set of curves illustrated in Fig. 3.21 where each curve corresponds to a different value of V_{T0L}.

To understand the design constraints, suppose that $V_{T0L} = -4\,[\text{V}]$. As a first calculation, assume that $V_{OL} = 0.20\,[\text{V}]$ (4% V_{DD}) is used as a design specification; then eqn. (3.4-43) gives a minimum required value of $\beta_R = 10$. Using eqns. (3.4-27) and (3.4-29), the resulting circuit will have an input low voltage of $V_{IL} \simeq 1.32\,[\text{V}]$, giving a logic 0 voltage noise margin of $NM_L = (V_{IL} - V_{OL}) \simeq 1.12\,[\text{V}]$.

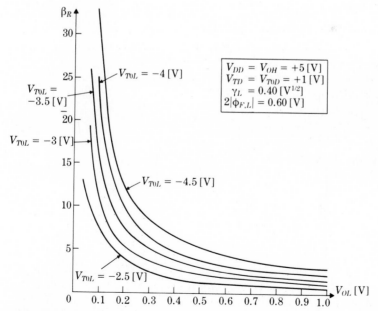

FIGURE 3.21 The driver-load ratio $\beta_R(V_{OL})$ for a depletion-load nMOS inverter.

Next, suppose that the circuit is designed for $V_{OL} = 0.50$ [V] (10% V_{DD}); the minimum driver-load ratio is then $\beta_R = 4$. The circuit design with this value of β_R will have $V_{IL} \simeq 1.76$ [V], so $NM_L \simeq 1.26$ [V]. This gives the interesting result that an inverter with smaller area can actually provide a better noise margin for logic 0 states. However, to place this example in proper perspective, the logic 1 (high) noise margins $NM_H = (V_{OH} - V_{IH})$ should be examined. The values of V_{IH} for the two configurations can be computed by eqns. (3.4-36) and (3.4-37). Assuming that $V_{OH} = V_{DD} = 5$ [V], the calculations give $NM_H \simeq 2.53$ [V] when $\beta_R = 10$ ($V_{OL} = 0.20$ [V]) and $NM_H \simeq 1.67$ [V] when $\beta_R = 4$ ($V_{OL} = 0.50$ [V]). Thus, the inverter with smaller area exhibits a lower NM_H value. This result could have been anticipated qualitatively by noting the differences in the V_{IL} values for the two circuits. In general, increasing β_R will improve the logic 1 noise margin NM_H while simultaneously reducing the logic 0 noise margin NM_L.

The terminology "noise margin" may appear to imply that $NM_L > NM_H$ is the optimum design criterion. However, recall from Section 3.1 that the logic 1 noise margin NM_H accounts for transmission losses between stages. In a realistic design setting, the inverter output must be connected to the input of other gates (i.e., the fan out is FO > 1). This gives a different design constraint on β_R.

Consider the 3-gate inverter arrangement illustrated in Fig. 3.22. Since the input to gate 1 is grounded, it has an output voltage of $V_{OH,1} = V_{DD}$. This logic 1 state must be transmitted to the inputs of both gate 2 and gate 3. Gate 2 is shown as having a direct connection to the output of gate 1. In a chip

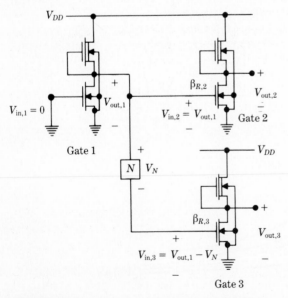

FIGURE 3.22 Model for calculation of $\beta_R(V_{in})$.

environment, this implies that the interconnect line used as a transmission network between the two stages has a negligible effect on the voltage levels. Consequently, $V_{in,2} = V_{DD}$, so that the driver-load ratio $\beta_{R,2}$ for gate 2 may be computed using eqn. (3.4-43) as described above to set the desired value of $V_{OL,2}$.

Now note that gate 3 is connected to gate 1 through a transmission network N, which exhibits a voltage drop V_N. Gate 3 thus has a maximum input voltage of $V_{in,3} = (V_{DD} - V_N)$. Although this particular situation is not representative of a realistic static nMOS DC analysis, a similar problem does arise in the clocked logic circuits discussed in Chapter 8. The design of gate 3 requires finding the driver-load ratio $\beta_{R,3}$ needed to set the desired value of $V_{OL,3}$. This is accomplished by replacing V_{OH} by $V_{in,3}$ in eqn. (3.4-43). In general, the basic design equation becomes

$$\beta_R(V_{in}) = \frac{|V_{TL}(V_{OL})|^2}{2(V_{in} - V_{TD})V_{OL} - V_{OL}^2},$$
(3.4-45)

which may be used to compute β_R for arbitrary input voltages.

To illustrate a calculation, suppose that

$$V_{DD} = V_{OH,1} = 5\,[\text{V}], \quad V_{TD} = +1\,[\text{V}], \quad \gamma_L = 0.4\,[\text{V}^{1/2}],$$
$$2\,|\phi_{F,L}| = 0.6\,[\text{V}], \quad V_N = +1\,[\text{V}], \quad V_{T0L} = -4\,[\text{V}],$$

and the circuit specifications require $V_{OL,2} = V_{OL,3} = 0.5\,[\text{V}]$. From the previous example, $\beta_{R,2} = 4$ is sufficient for gate 2. However, $V_{in,3} = (V_{DD} - V_N) = +4[\text{V}]$, so eqn. (3.4-45) gives $\beta_{R,3} = 5.5$ as the minimum driver-load ratio for gate 3. This illustrates the fact that larger β_R values are required to compensate for reduced logic 1 input voltages. The values for V_{DD}, V_{TD}, γ_L, and $2\,|\phi_{F,L}|$ in this example are used to generate the plots of $\beta_R(V_{in})$ in Fig. 3.23. Each graph represents a specific choice for V_{T0L}; the set of curves on each graph is generated by varying V_{OL} in eqn. (3.4-45).

Since the depletion load inverter forms the basis of many logic gates analyzed in later discussions, it is useful to examine the design problem in the context of a simplified resistive network. Recall from Section 3.2 that the driver on resistance is given by

$$R_{on} \equiv \frac{V_{OL}}{I_D} = \frac{1}{\dfrac{k_D'}{2}\left(\dfrac{W}{L}\right)_D [2(V_{OH} - V_{TD}) - V_{OL}]} = \frac{1}{G_{on}}$$
(3.4-46)

and is used in series with a load resistor to form a simple voltage divider. With the load in saturation, the equivalent drain-source resistance presented by the D-mode MOSFET is

$$R_{DSL} = \frac{1}{G_{DSL}} = \frac{V_{DD} - V_{OL}}{\dfrac{k_L'}{2}\left(\dfrac{W}{L}\right)_L |V_{TL}(V_{OL})|^2}.$$
(3.4-47)

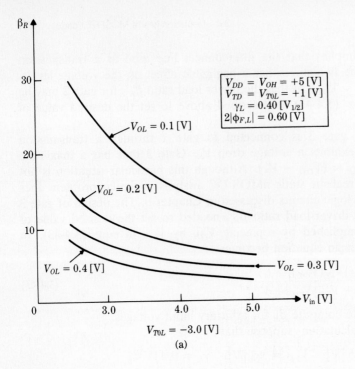

$$
\begin{aligned}
V_{DD} &= V_{OH} = +5\,[\mathrm{V}] \\
V_{TD} &= V_{T0L} = +1\,[\mathrm{V}] \\
\gamma_L &= 0.40\,[\mathrm{V}_{1/2}] \\
2|\phi_{F,L}| &= 0.60\,[\mathrm{V}]
\end{aligned}
$$

$V_{T0L} = -3.0\,[\mathrm{V}]$

(a)

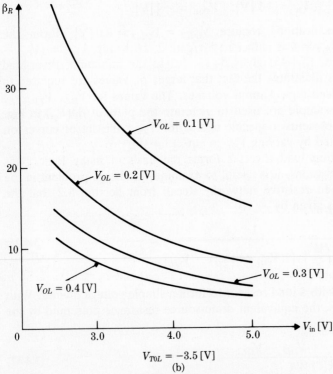

$V_{T0L} = -3.5\,[\mathrm{V}]$

(b)

FIGURE 3.23 $\beta_R(V_{\mathrm{in}})$ for various V_{OL} choices.

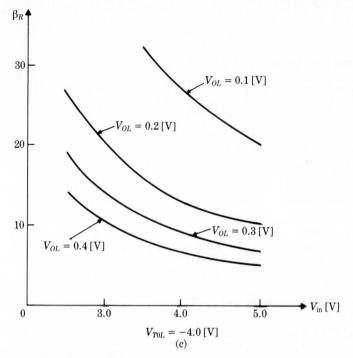

$V_{T0L} = -4.0 \,[\text{V}]$
(c)

FIGURE 3.23   (*contd.*)

Thus,

$$V_{OL} = \frac{R_{\text{on}}}{R_{\text{on}} + R_{DSL}} V_{DD} \tag{3.4-48}$$

implies that increasing β_R will lower the value of V_{OL}. Again note that the device conductances are proportional to their (W/L) ratios. This simple observation will become very useful when random logic gates are discussed.

EXAMPLE 3.4-1

A depletion load inverter has threshold voltages of $V_{T0D} = +1\,[\text{V}]$ and $V_{T0L} = -3.5\,[\text{V}]$. The power supply is $V_{DD} = +5\,[\text{V}]$. Compute the value of V_{OL} if $\beta_R = 6$. Assume that $\gamma = 0.4\,[\text{V}^{1/2}]$ and $2\,|\phi_F| = 0.6\,[\text{V}]$.

Solution
Using eqn. (3.4-22) with the given values yields

$$V_{OL} = 4 - \sqrt{16 - \frac{1}{6}|V_{TL}(V_{OL})|^2},$$

where

$$V_{TL}(V_{OL}) = -3.5 + 0.4(\sqrt{V_{OL} + 0.6} - \sqrt{0.6}).$$

To solve this by iterations, define

$$V_X = 4 - \sqrt{16 - \frac{1}{6}|V_{TL}(V_{OL})|^2}.$$

Then guess a value for V_{OL}. If $V_X = V_{OL}$, the solution has been found. Otherwise, use V_X for the next guess of V_{OL}.

This type of iteration can be illustrated by using a table:

V_{OL}	$|V_{TL}(V_{OL})|$	V_X
1	3.304	0.234
0.234	3.444	0.255
0.255	3.440	0.255 ← $V_{OL} = V_X \approx 0.255$ [V]

A BASIC program that implements this algorithm is as follows:

```
10 READ VDD,VTD,VTOL,GAMMA,FERMI,BETAR
20 DATA 5,1,-3.5,0.4,0.6,6
30 VOL=1:REM INITIAL GUESS
40 VTL=VTOL+GAMMA*(SQR(VOL+FERMI)-SQR(FERMI))
50 VX=(VDD-VTD)-SQR((VDD-VTD)^2-(1/BETAR)*(ABS(VTL)^2))
60 IF ABS(VX-VOL) < .0005 THEN 100
70 VOL=VX:GOTO 40
100 PRINT "VOL=";VOL


RUN
VOL=.2546237
```

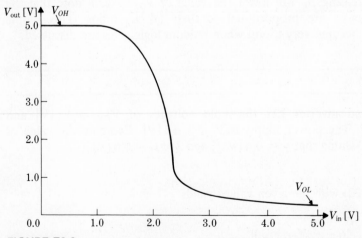

FIGURE E3.3

The SPICE-generated VTC is shown in Fig. E3.3.

```
DEPLETION LOAD INVERTER
VDD 3 0 DC 5VOLTS
MDRIVER 2 1 0 0 EMODE L=5U W=15U
MLOAD 3 2 2 0 DMODE L=10U W=5U
VS 1 0 DC
.MODEL EMODE NMOS VTO=1 GAMMA=0.40 KP=2.5E-5
.MODEL DMODE NMOS VTO=-3.5 GAMMA=0.40 KP=2.5E-5
.DC VS 0 5 0.1
.PLOT DC V(2)
.END
```
■

EXAMPLE 3.4-2

A depletion load inverter is characterized by $V_{DD} = 5$ [V], $V_{TOD} = +1$ [V], and $V_{TOL} = -3.5$ [V]. The driver-load ratio is $\beta_R = 5$, and it is assumed that $\gamma = 0.4$ [V$^{1/2}$] and $2|\phi_F| = 0.6$ [V]. Compute V_{IL} and V_{IH} for the circuit.

Solution
(a) V_{IL}: When $V_{in} = V_{IL}$, the output voltage is given by eqn. (3.4-29):

$$V_{out} = 5 - 0.087 |V_{TL}(V_{out})|,$$

where

$$V_{TL} = -3.5 + 0.4(\sqrt{V_{out} + 0.6} - \sqrt{0.6}).$$

To solve this, self-iteration is used. The results are summarized in the following table.

V_{out}	$|V_{TL}(V_{out})|$	$5 - 0.087\,|V_{TL}|$
4	2.95	4.74
4.74	2.89	4.75
4.75	2.88	4.75 ← $V_{out} \approx 4.75$ [V]

This value of V_{out} is now used in eqn. (3.4-27) to compute V_{IL}:

$$V_{IL} = 1 + \frac{(2.88)}{\sqrt{(5)(6)}} \approx 1.53 \text{ [V]}.$$

The following BASIC program solves this algorithm directly.

```
10  READ VDD,VTOL,VTD,GAMMA,FERMI,BR
20  DATA 5,-3.5,1,0.4,0.6,5
30  VOUT=4: REM INITIAL GUESS FOR VOUT
40  VTL=VTOL+GAMMA * (SQR(VOUT+FERMI)-SQR(FERMI))
50  VX=VDD-ABS(VTL) * (1-BR/(SQR(BR * (BR+1))))
```

```
 60  IF ABS(VOUT-VX) < .001 THEN 80
 70  VOUT=VX: GOTO 40
 80  VIL=VTD+ABS(VTL)/(SQR(BR*(BR+1)))
100  PRINT "VIL=";VIL
```

```
RUN
VIL=1.526682
```

(b) V_{IH}: This also requires that V_{out} be computed first. Assuming that η is neglible for simplicity, eqn. (3.4-36) gives

$$V_{out} = \frac{|V_{TL}(V_{out})|}{\sqrt{15}},$$

where

$$V_{TL} = -3.5 + 0.4(\sqrt{V_{out} + 0.6} - \sqrt{0.6}).$$

To find V_{out}, guess a value and then compute the right-hand side. Self-iterating gives the following results:

V_{out}	$\dfrac{|V_{TL}(V_{out})|}{\sqrt{15}}$
1	0.85
0.85	0.86
0.86	0.86 $\leftarrow$ $V_{out} \simeq 0.86\,[\text{V}]$

This value of V_{out} is then used in eqn. (3.4-37) to compute V_{IH}:

$$V_{IH} \simeq 1 + 2(0.86) = 2.72\,[\text{V}].$$

The following BASIC program also assumes η is negligible for the calculation. However, the η terms are easily added if desired.

```
 10  READ VDD,VTOL,VTD,GAMMA,FERMI,BR
 20  DATA 5,-3.5,1,0.4,0.6,5
130  VOUT=1: REM INITIAL GUESS
140  VTL=VTOL+GAMMA*(SQR(VOUT+FERMI)-SQR(FERMI))
150  VY=ABS(VTL)/SQR(3*BR)
160  IF ABS(VOUT-VY) < .001 THEN 180
170  VOUT=VY: GOTO 140
180  VIH=VTD+2*VOUT
200  PRINT "VIH=";VIH
```

```
RUN
VIH=2.7184
```

A SPICE input file is shown below, and output VTC for this circuit is given in Fig. E3.4.

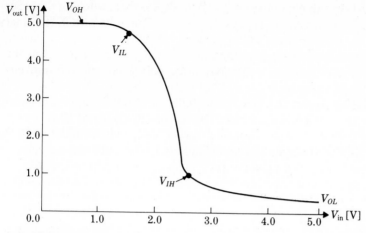

FIGURE E3.4

```
DEPLETION LOAD INVERTER
VDD 3 0 DC 5VOLTS
MDRIVER 2 1 0 0 EMODE L=5U W=12.5U
MLOAD 3 2 2 0 DMODE L=10U W=5U
VS 1 0 DC
.MODEL EMODE NMOS VTO=1 GAMMA=0.40 KP=2.5E-5
.MODEL DMODE NMOS VTO=-3.5 GAMMA=0.40 KP=2.5E-5
.DC VS 0 5 0.1
.PLOT DC V(2)
.END
```

3.5 The CMOS Inverter

All of the circuits presented so far are fabricated using an nMOS technology in which n-channel MOSFETs are the only active devices. In this section, p-channel (pMOS) transistors are introduced for the purpose of constructing a CMOS (complementary MOS) inverter. A CMOS structure employs both an n-channel and a p-channel MOSFET in a manner that allows the pMOS device to complement the nMOS characteristics.

Historically, pMOS was the first LSI (large-scale integration) process available, primarily because p-channel enhancement-mode MOSFETs with $V_{Tp} < 0$ are relatively simple to fabricate. The disadvantage of using pMOS as a basis for digital circuit implementations is that p-channel MOSFETs conduct drain-source current using positively charged holes. Since the hole mobility μ_p is always less than the electron mobility μ_n, pMOS transistors intrinsically exhibit lower process transconductance parameters $k_p' = \mu_p C_{ox}$ than those found in equivalent n-channel devices. While this does not significantly affect the (DC) voltage transfer characteristics, it leads to a slower transient response for pMOS circuits. In the next chapter it will be shown that switching times for an inverter have an approximate $(1/\mu)$ dependence. Thus, the pMOS transient response will be slower than an equivalent nMOS circuit by a factor $\mu_n/\mu_p \simeq 2.5$. This was, in fact, a strong motivational factor which led to the development of nMOS E-mode technology using poly gates and ion implants for threshold voltage adjustments.

As mentioned above, CMOS uses both n-channel and p-channel MOSFETs in a complementary manner. The major advantage of CMOS over nMOS is that CMOS circuits dissipate power only when undergoing a switching event; if a CMOS gate has stable input voltages, only leakage current levels are required from the power supply. This is to be compared with an nMOS-based design in which current flows from V_{DD} to ground whenever the driver is on. In particular, an nMOS inverter consumes power in the static situation where $V_{out} = V_{OL}$.

The power required to operate a digital integrated circuit is important for a number of reasons. First, the power per unit area dissipated by the circuit must be kept below a certain maximum value, which is determined by both the semiconductor properties and the ability of the IC package (which houses the chip) to dissipate heat. If this maximum surface power density is exceeded, the circuit will not operate correctly; in the worst case situation, the circuit will be destroyed by the Joule heating effects. Another consideration is a bit more pragmatic. If the chips require large current levels, then the cost of the system power supply will increase. This makes them less attractive for use in large system designs.

The average current levels required for CMOS circuits are small enough to allow for the use of batteries as a power supply. This fact has been exploited in the development of completely portable microcomputer systems. There are other advantages to using a CMOS-based design. As will be shown in the analysis, the CMOS inverter is capable of producing high and low output levels of $V_{OH} = V_{DD}$ and $V_{OL} = 0\,[V]$, respectively. The VTC will exhibit a sharp transition, and it is possible to design a CMOS circuit that operates over a wide range of power supply voltages V_{DD}. Because of these attractive features, CMOS usage has grown at a remarkable rate.

The above discussion may make it seem that CMOS is an ideal technology. However, disadvantages must also be mentioned to form the complete picture. First, the processing for CMOS is more complex than for nMOS since p-channel devices must be incorporated into the fabrication line. The difficulty encountered here may be seen by noting that nMOS transistors are made in a p-type background, while a pMOS device requires an n-type background. If a p-type wafer is used for the CMOS circuit, extra processing steps must be added to create n-tub areas where the pMOS transistors can be formed. In addition, an extra ion implant step must be added to adjust the threshold voltage of the p-channel device.

The basic CMOS process must also include steps to prevent a condition known as *latch-up* in the circuits. Latch-up can occur in a CMOS configuration because of the presence of parasitic bipolar junction transistors (BJTs) in the chip integration structure. In the simplest model, two BJTs can collectively act as a silicon-controlled rectifier (SCR). When a CMOS circuit undergoes a latch-up, the SCR is switched on, drawing large currents and locking the logic state of the gate. The large current levels experienced in latch-up can be sufficient to destroy the chip. The only way to release a circuit from a latched condition is to remove the power supply voltage. *Guard rings,* which are small regions of *n* and *p* material surrounding the transistors, can be added to prevent latch-up from taking place. However, the additional processing steps increase the complexity and costs involved in producing CMOS chips.

Another drawback of CMOS technology is that the circuits generally require more transistors than equivalent gate design in nMOS. Although this will not be apparent in the CMOS inverter circuit (which uses only 2 MOSFETs), the device count will rise significantly with the use of random logic circuits (discussed in Chapter 6). Depending on the real estate costs, this may present a problem to the chip designer. It also serves as a limitation on the integration density possible with standard CMOS.

CMOS development has progressed rapidly in recent years. In general, state-of-the-art CMOS is technologically superior to existing nMOS processes, which sometimes obscures the relative problem areas introduced by going to CMOS in the first place. It is important to keep the overall properties of the technology in proper perspective. After all, engineering is concerned with solving problems by considering all relevant parameters.

3.5.1 The *p*-Channel MOSFET

The device structure and circuit symbol for a *p*-channel enhancement-mode MOSFET are shown in Fig. 3.24. The circuit equations for the pMOS transistor can be derived directly from the device physics by using the techniques established in Chapter 1. However, as an alternative approach, the discussion here will obtain the pMOS characteristics by directly modifying the *n*-channel MOSFET equations. This will serve the dual purpose of finding the pMOS circuit performance and simultaneously illustrating the complementary nature of the two devices.

Consider first the pMOS threshold voltage V_{T0p}, which is valid for $V_{BSp} = 0$. By definition, $V_{T0p} < 0$ for the enhancement-mode *p*-channel transistor. This has the same general form as V_{T0n}:

$$V_{T0p} = \Phi_{GS} - 2\phi_{Fn} - \frac{1}{C_{\text{ox}}}(Q_{ss} + Q_{\text{ox}}) - \frac{1}{C_{\text{ox}}} Q_{B0n}. \tag{3.5-1}$$

The differences between this equation and the expression for V_{T0n} are as follows. First, $\Phi_{GS} = (\Phi_G - \Phi_S)$ must now account for the fact that the substrate potential Φ_S is in the *n*-type body region with doping N_d. Second, the

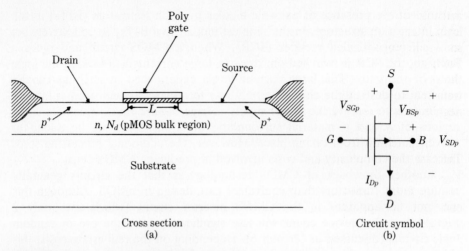

Cross section
(a)

Circuit symbol
(b)

FIGURE 3.24 *p*-channel MOSFET (pMOS).

bulk Fermi potential is changed to

$$\phi_{Fn} = \frac{kT}{q} \ln\left(\frac{N_d}{n_i}\right). \tag{3.5-2}$$

Finally, the bulk depletion charge density is given by

$$Q_{B0n} = \sqrt{2q\varepsilon_{Si}N_d(2\phi_{Fn})}, \tag{3.5-3}$$

which is positive corresponding to ionized donors. The final working value of V_{T0p} will be set by an ion implant step. Assuming a total ion dose of D_I, this adds (approximately) a term $\pm qD_I/C_{ox}$ to the equation, where the plus sign is used for an acceptor implant (which can partially compensate for N_d) and the minus sign refers to a donor ion implantation. When body bias V_{BSp} is applied to the device,

$$V_{Tp} = V_{T0p} - \gamma_p(\sqrt{V_{BSp} + 2\phi_{Fn}} - \sqrt{2\phi_{Fn}}), \tag{3.5-4}$$

where

$$\gamma_p = \frac{\sqrt{2qN_d\varepsilon_{Si}}}{C_{ox}} \tag{3.5-5}$$

is the body bias parameter for the *p*-channel structure.

The operational modes of the pMOS transistor can now be deduced by referring to Fig. 3.24, where $V_{BSp} = 0$ will be assumed. The polarities shown for the positive voltages V_{SGp} and V_{SDp} are those required for active operation. However, to make analogies with the nMOS transistor, the relations

$$V_{SGp} = -V_{GSp}, \qquad V_{SDp} = -V_{DSp}, \tag{3.5-6}$$

with V_{GSp} and V_{DSp} both negative, will be used. Now if the device is to exhibit active operation in which drain current I_{Dp} flows, a p-type (hole) inversion layer must be formed underneath the gate oxide to connect the p^+ drain and source regions. The creation of this channel is controlled by $V_{SGp} = -V_{GSp}$. For small gate-source voltages such that $V_{Tp} < V_{GSp}$, the negative potential on the gate will not be sufficient to induce the formation of a channel. Hence, the transistor is in cutoff with $I_{Dp} = 0$. Note that since both V_{Tp} and V_{GSp} are negative, the cutoff requirement may be rewritten using absolute values as

$$|V_{GSp}| < |V_{Tp}|. \tag{3.5-7}$$

This is to be compared with an nMOS device where cutoff occurs when $V_{GSn} < V_{Tn}$.

To obtain active device operation, the source-gate voltage must be large enough to ensure that $|V_{Tp}| < V_{SGp}$ is satisfied. This is equivalent to requiring

$$|V_{GSp}| > |V_{Tp}|, \tag{3.5-8}$$

which is similar in form to the nMOS criterion that $V_{GSn} > V_{Tn}$. As with the n-channel transistor, the pMOS device has two active regions of operation depending on the value of $V_{SDp} = -V_{DSp}$. If

$$|V_{DSp}| < |V_{GSp} - V_{Tp}|, \tag{3.5-9}$$

then the pMOS transistor is nonsaturated. To find I_{Dp} (which is defined positive flowing out of the drain) for this case, recall that the nMOS nonsaturated current is

$$I_{Dn} = \frac{\beta_n}{2}[2(V_{GSn} - V_{Tn})V_{DSn} - V_{DSn}^2]. \tag{3.5-10}$$

Since the pMOS polarities are exactly opposite to the nMOS voltages, the relations in eqn. (3.5-6) may be used to write

$$I_{Dp} = \frac{\beta_p}{2}[2(V_{SGp} + V_{Tp})V_{SDp} - V_{SDp}^2] \tag{3.5-11}$$

as the nonsaturated p-channel MOSFET current. In this equation, $\beta_p = k_p'(W/L)_p$, where $k_p' = \mu_p C_{ox}$. The transistor is saturated when

$$|V_{DSp}| > |V_{GSp} - V_{Tp}|. \tag{3.5-12}$$

By direct analogy with the nMOS transistor, the saturated pMOS current is then given by

$$I_{Dp} = \frac{\beta_p}{2}(V_{SGp} + V_{Tp})^2 \tag{3.5-13}$$

where channel-length modulation has been ignored. The complementary nature of the nMOS and pMOS transistors thus becomes clear.

As a prelude to the CMOS analysis, consider the pMOS inverter in Fig. 3.25, which uses a resistor R_L as a load. This circuit is the exact complement of

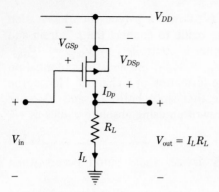

FIGURE 3.25 Basic pMOS inverter using a load resistor.

the nMOS case. Applying KVL gives

$$V_{in} - V_{GSp} = V_{DD}, \qquad V_{out} = V_{DD} + V_{DSp} = I_{Dp}R_L, \tag{3.5-14}$$

where both V_{GSp} and V_{DSp} are negative. The operation of the inverter is as follows. If $V_{in} = V_{DD}$, then $|V_{GSp}| = 0 < |V_{Tp}|$, so the transistor is in cutoff with $I_{Dp} = 0$. Hence, $V_{out} = I_{Dp}R_L = 0\,[\text{V}]$. On the other hand, if $V_{in} = 0\,[\text{V}]$, then $|V_{GSp}| = V_{DD} > |V_{Tp}|$, and the transistor is nonsaturated. The output voltage for this case is $V_{out} = I_{Dp}R_L = V_{DD}$, assuming that $|V_{DSp}|$ is small. This short discussion thus demonstrates the complementary relation between nMOS and pMOS inverter structures.

3.5.2 CMOS Inverter Characteristics

The basic CMOS inverter circuit is shown in Fig. 3.26. The circuit topology is such that the gates of the two transistors are connected to form the input with

$$V_{in} = V_{GSn} = V_{DD} - V_{SGp} \tag{3.5-15}$$

while the two drain electrodes are tied together, giving the output voltage of

$$V_{out} = V_{DSn} = V_{DD} - V_{SDp}. \tag{3.5-16}$$

The complementary nature of the circuit can be seen from eqn. (3.5-15): as $V_{in} = V_{GSn}$ increases, V_{GSp} decreases, and vice versa. Since V_{in} controls the gate bias of both MOSFETs, the concepts of "driver" and "load" have no meaning. Rather, the nMOS and pMOS trsnsistors act in a complementary fashion to set the inverter characteristics. Note that there are no body bias effects in the circuit since $V_{SBn} = 0 = V_{BSp}$.

To construct the VTC for the CMOS inverter, start with input voltages $V_{in} < V_{Tn}$. Since the nMOS transistor is in cutoff, $I_{Dn} = I_{Dp} = 0$. However, the p-channel MOSFET has an inversion layer established since V_{SGn} is large. The output voltage is thus

$$V_{out} = V_{OH} = V_{DD} - V_{SDp} \simeq V_{DD}, \tag{3.5-17}$$

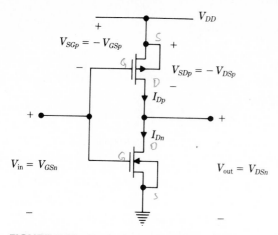

FIGURE 3.26   Basic CMOS inverter circuit.

where $V_{SDp} = 0$ is valid since a conducting channel exists between the output node and V_{DD} through the pMOS transistor.

When V_{in} is increased to a level slightly greater than V_{Tn}, the nMOS transistor enters the saturation region, while eqn. (3.5-16) implies that the pMOS device is nonsaturated. This portion of the VTC is described by

$$\frac{\beta_n}{2}(V_{in} - V_{Tn})^2 = \frac{\beta_p}{2}[2(V_{DD} - V_{in} - |V_{Tp}|)(V_{DD} - V_{out})$$

$$- (V_{DD} - V_{out})^2], \quad \textbf{(3.5-18)}$$

where eqns. (3.5-15) and (3.5-16) have been used with eqn. (3.5-11) in writing I_{Dp}. As V_{in} continues to increase, V_{out} falls as described by eqn. (3.5-18) until

$$(V_{DD} - V_{out}) > (V_{DD} - V_{in} - |V_{Tp}|) \qquad \textbf{(3.5-19)}$$

is satisfied. When this condition is true, both devices are saturated, so the central portion of the VTC is obtained from

$$\frac{\beta_n}{2}(V_{in} - V_{Tn})^2 = \frac{\beta_p}{2}(V_{DD} - V_{in} - |V_{Tp}|)^2. \qquad \textbf{(3.5-20)}$$

This predicts a vertical falloff of the VTC since channel-length modulation has been ignored; even with channel-length modulation effects, the slope is very steep. When V_{out} falls to a level where

$$V_{out} < (V_{in} - V_{Tn}) \qquad \textbf{(3.5-21)}$$

is satisfied, the nMOS transistor goes into a nonsaturated state, giving the VTC equation

$$\frac{\beta_n}{2}[2(V_{in} - V_{Tn})V_{out} - V_{out}^2] = \frac{\beta_p}{2}(V_{DD} - V_{in} - |V_{Tp}|)^2. \qquad \textbf{(3.5-22)}$$

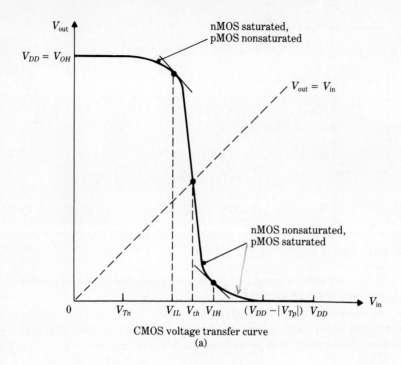

CMOS voltage transfer curve
(a)

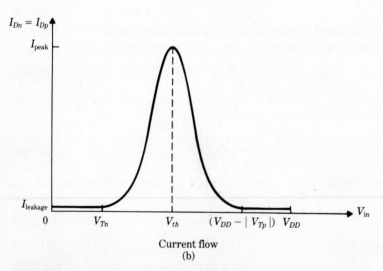

Current flow
(b)

FIGURE 3.27   CMOS inverter characteristics.

Finally, when V_{in} is increased to the point where

$$(V_{DD} - V_{in}) < |V_{Tp}| \tag{3.5-23}$$

holds, the pMOS transistor goes into cutoff, giving $I_{Dn} = I_{Dp} = 0$. The n-channel MOSFET still has an inversion layer established, so

$$V_{out} = V_{OL} = V_{DSn} \simeq 0\,[\text{V}]. \tag{3.5-24}$$

A typical VTC as obtained from these equations is shown in Fig. 3.27(a).

The low power requirements of the CMOS circuit can be seen in the plot of I_D vs. V_{in} drawn in Fig. 3.27(b). So long as $V_{in} < V_{Tn}$ or $(V_{DD} - V_{in}) < |V_{Tp}|$, no current flows between V_{DD} and ground. Since these input voltages correspond respectively to logic 0 and logic 1 states, a CMOS circuit only admits leakage current when the inputs are stable. Significant current flow is present only when a switching event takes place. This further illustrates the complementary roles of the two MOSFETs in the circuit.

ANALYSIS

V_{OH} With $V_{in} < V_{Tn}$, the nMOS transistor is in cutoff while the pMOS transistor has an inversion layer established. Hence,

$$V_{OH} \simeq V_{DD}, \tag{3.5-25}$$

as discussed previously.

V_{OL} When $(V_{DD} - V_{in}) < |V_{Tp}|$, the p-channel transistor is in cutoff while the nMOS transistor provides a conducting channel to ground. Thus,

$$V_{OL} \simeq 0\,[\text{V}] \tag{3.5-26}$$

is a good approximation.

V_{IL} The input low voltage occurs when the nMOS transistor is saturated while the pMOS device is nonsaturated. Equating currents gives

$$\frac{\beta_n}{2}(V_{IL} - V_{Tn})^2 = \frac{\beta_p}{2}[2(V_{DD} - V_{IL} - |V_{Tp}|)(V_{DD} - V_{out})$$

$$- (V_{DD} - V_{out})^2], \tag{3.5-27}$$

which is one of the required equations for the two unknowns $V_{in} = V_{IL}$ and V_{out}. Since the functional dependence of the currents is $I_{Dn}(V_{in}) = I_{Dp}(V_{in}, V_{out})$, the unity-slope point where $(dV_{out}/dV_{in}) = -1$ is determined by

$$\frac{dV_{out}}{dV_{in}} = \frac{(dI_{Dn}/dV_{in}) - (\partial I_{Dp}/\partial V_{in})}{\partial I_{Dp}/\partial V_{out}} = -1. \tag{3.5-28}$$

Evaluating the derivatives provides the second equation

$$V_{IL}\left(1 + \frac{\beta_n}{\beta_p}\right) = 2V_{\text{out}} + \frac{\beta_n}{\beta_p} V_{Tn} - V_{DD} - |V_{Tp}|, \tag{3.5-29}$$

which must be solved simultaneously with eqn. (3.5-27). Owing to the absence of body bias variations in the CMOS inverter, combining these two equations will yield a quadratic for V_{IL}. Instead of deriving the general quadratic equation (which is somewhat messy), it is usually simpler to reduce both eqn. (3.5-27) and (3.5-29) by substituting the given numerical values for the parameters. Calculating V_{IL} then reduces to straightforward algebra.

V_{IH} When $V_{\text{in}} = V_{IH}$, the nMOS transistor is nonsaturated while the pMOS device is saturated. The current equation at this point is

$$\frac{\beta_n}{2}[2(V_{IH} - V_{Tn})V_{\text{out}} - V_{\text{out}}^2] = \frac{\beta_p}{2}(V_{DD} - V_{IH} - |V_{Tp}|)^2. \tag{3.5-30}$$

Noting that $I_{Dn}(V_{\text{in}}, V_{\text{out}}) = I_{Dp}(V_{\text{in}})$, the VTC derivative is evaluated by

$$\frac{dV_{\text{out}}}{dV_{\text{in}}} = \frac{(dI_{Dp}/dV_{\text{in}}) - (\partial I_{Dn}/\partial V_{\text{in}})}{\partial I_{Dn}/\partial V_{\text{out}}} = -1, \tag{3.5-31}$$

which gives

$$V_{IH}\left(1 + \frac{\beta_p}{\beta_n}\right) = 2V_{\text{out}} + V_{Tn} + \frac{\beta_p}{\beta_n}(V_{DD} - |V_{Tp}|). \tag{3.5-32}$$

Equations (3.5-30) and (3.5-32) combine to form a quadratic for V_{IH}, which can be solved by standard techniques.

V_{th} The inverter threshold $V_{th} = V_{\text{in}} = V_{\text{out}}$ occurs when both transistors are saturated such that

$$\frac{\beta_n}{2}(V_{th} - V_{Tn})^2 = \frac{\beta_p}{2}(V_{DD} - V_{th} - |V_{Tp}|)^2. \tag{3.5-33}$$

Thus,

$$V_{th} = \frac{V_{Tn} + \sqrt{\beta_p/\beta_n}(V_{DD} - |V_{Tp}|)}{(1 + \sqrt{\beta_p/\beta_n})} \tag{3.5-34}$$

gives the intersection of the VTC with the unity-gain line $V_{\text{in}} = V_{\text{out}}$.

DESIGN
The complementary pMOS/nMOS configuration that serves as a basis for the CMOS inverter automatically gives output logic voltages of $V_{OH} = V_{DD}$ and $V_{OL} = 0$ [V]. Consequently, unless leakage currents are a problem, the circuit design can be accomplished without reference to these quantities. This

situation is quite different from nMOS, where the fundamental design constraints arise in setting V_{OL}.

One approach to CMOS design is concerned with setting the inverter threshold voltage V_{th} as given in eqn. (3.5-34). The geometrical design parameters of the circuit are contained in β_p and β_n through

$$\frac{\beta_p}{\beta_n} = \frac{\mu_p \left(\dfrac{W}{L}\right)_p}{\mu_n \left(\dfrac{W}{L}\right)_n}, \tag{3.5-35}$$

where it is assumed that C_{ox} is the same for both devices (i.e., that the gate oxides are grown at the same time). The value of this ratio required to establish a given inverter threshold voltage is obtained from

$$\sqrt{\frac{\beta_n}{\beta_p}} = \frac{(V_{DD} - V_{th} - |V_{Tp}|)}{(V_{th} - V_{Tn})}. \tag{3.5-36}$$

Now recall that an *ideal* inverter has $V_{th} = V_{DD}/2$. Using this value in eqn. (3.5-36) gives the requirement that

$$\sqrt{\frac{\beta_n}{\beta_p}} = \frac{\left(\dfrac{1}{2}V_{DD} - |V_{Tp}|\right)}{\left(\dfrac{1}{2}V_{DD} - V_{Tn}\right)}. \tag{3.5-37}$$

To exploit the possibility of obtaining a VTC that is close to the ideal shape, CMOS processing lines may set $V_{Tn} = |V_{Tp}|$. Then $\beta_p = \beta_n$, so the device aspect ratios are related by

$$\frac{\left(\dfrac{W}{L}\right)_p}{\left(\dfrac{W}{L}\right)_n} = \frac{\mu_n}{\mu_p}. \tag{3.5-38}$$

Since $\mu_n/\mu_p \simeq 2.5$, a minimum-area CMOS inverter will have $(W/L)_n \simeq 1$ and $(W/L)_p \simeq 2.5$.

To understand the consequences of choosing $V_{th} = V_{DD}/2$ and $V_{Tn} = |V_{Tp}|$, note that the resulting VTC for this case is completely symmetric about $V_{DD}/2$. The input low voltage V_{IL} can be found by reducing eqn. (3.5-29) to the form

$$V_{out} = V_{IL} + \frac{1}{2}V_{DD}, \tag{3.5-39}$$

which is valid for this case. Substituting into eqn. (3.5-27) and rearranging results in the linear equation,

$$V_{IL} = \frac{1}{4}\left(V_{Tn} + \frac{3}{2}V_{DD}\right). \tag{3.5-40}$$

The input high level V_{IH} is found in a similar manner:

$$V_{IH} = \frac{1}{4}\left(\frac{5}{2}V_{DD} - V_{Tn}\right). \tag{3.5-41}$$

Summing these two expressions gives the useful relation

$$V_{IL} + V_{IH} = V_{DD}. \tag{3.5-42}$$

Although the specific values of V_{IL} and V_{IH} depend on the power supply voltage, eqn. (3.5-42) demonstrates that the *shape* of the CMOS VTC is invariant with respect to the choice for V_{DD}. Thus, CMOS circuits can be operated over a wide range of power supply values. The absolute minimum power supply voltage is $V_{DD} = (V_{Tn} + |V_{Tp}|) = 2V_{Tn}$; however, since this gives a zero input transition width $(V_{IL} = V_{IH})$, $V_{DD} = 3V_{Tn}$ is a more practical minimum value. The maximum V_{DD} level is set by the drain-source breakdown voltages of the transistors. As an example, the CMOS 4000 series SSI (small-scale integration) circuits have a recommended V_{DD} range of 3 [V] to 15 [V].

As a final comment on this circuit design, the noise margins are calculated by

$$NM_H = NM_L = V_{IL} = \frac{1}{4}\left(V_{Tn} + \frac{3}{2}V_{DD}\right). \tag{3.5-43}$$

Thus, increasing V_{DD} improves the noise immunity but also increases the power consumption of the circuit.

EXAMPLE 3.5-1

A symmetric CMOS inverter has $V_{T0n} = 0.8\,[\text{V}] = |V_{T0p}|$ and $\beta_n = \beta_p$. Compute V_{IL} and V_{IH} for $V_{DD} = 5\,[\text{V}]$ and $V_{DD} = 10\,[\text{V}]$.

Solution
(a) $V_{DD} = 5\,[\text{V}]$: The parameters substituted into eqns. (3.5-40) and (3.5-41) give

$$V_{IL} = \frac{1}{4}\left(0.8 + \frac{15}{2}\right) = 2.075\,[\text{V}] \leftarrow$$

$$V_{IH} = \frac{1}{4}\left(\frac{25}{2} - 0.8\right) = 2.925\,[\text{V}] \leftarrow .$$

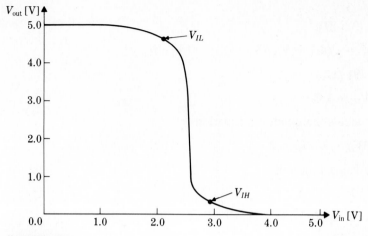

FIGURE E3.5

(b) $V_{DD} = 10\,[\text{V}]$: For this case,

$$V_{IL} = \frac{1}{4}\left(0.8 + \frac{30}{2}\right) = 3.95\,[\text{V}] \leftarrow$$

$$V_{IH} = \frac{1}{4}\left(\frac{50}{2} - 0.8\right) = 6.05\,[\text{V}] \leftarrow .$$

Note that $V_{IL} + V_{IH} = V_{DD}$ for this design.

The SPICE-generated VTC for the case $V_{DD} = 5\,[\text{V}]$ is shown in Fig. E3.5.

```
SYMMETRIC CMOS INVERTER
VDD 3 0 DC 5VOLTS
MNT 2 1 0 0 MN L=5U W=5U
MPT 3 1 2 3 MP L=5U W=12.5U
VS 1 0 DC
.MODEL MN NMOS VTO=1 GAMMA=0.37 KP=2.5E-5
.MODEL MP PMOS VTO=-1 GAMMA=0.4 KP=1.0E-5
.DC VS 0 5 0.1
.PLOT DC V(2)
.END
```
■

EXAMPLE 3.5-2

A nonsymmetric CMOS inverter is designed with $\beta_n = 2.5\beta_p$, and the processing gives $V_{T0n} = +1\,[\text{V}]$ and $V_{T0p} = -0.9\,[\text{V}]$. Compute V_{IL}, V_{IH}, and V_{th} if $V_{DD} = +5\,[\text{V}]$.

Solution

V_{IL}: From eqn. (3.5-27),

$$2.5(V_{IL} - 1)^2 = [2(4.1 - V_{IL})(5 - V_{out}) - (5 - V_{out})^2],$$

while eqn. (3.5-29) gives

$$V_{out} = 1.75V_{IL} + 1.7.$$

Eliminating V_{out} yields the quadratic equation

$$V_{IL}^2 + 2.133V_{IL} - 6.628 = 0,$$

which has a physical solution of

$$V_{IL} = 1.72\,[\text{V}] \leftarrow .$$

This occurs at a value of $V_{out} = 4.71\,[\text{V}]$.

V_{IH}: Now, eqn. (3.5-30) gives

$$2.5[2(V_{IH} - 1)V_{out} - V_{out}^2] = [16.81 - 8.2V_{IH} + V_{IH}^2],$$

which must be solved with eqn. (3.5-32):

$$V_{out} = 0.7V_{IH} - 1.32.$$

A quadratic equation can be obtained. However, this can also be solved using an iterative computer solution as follows:

```
10 VIH=2.2:REM INITIAL GUESS
20 VOUT=.7 * VIH-1.32
30 F1=2.5 * (2 * (VIH-1) * VOUT-VOUT^2)
40 F2=16.81-8.2 * VIH+VIH^2
50 IF ABS (F1-F2) < .005 THEN 100
60 VIH=VIH+.001:GOTO 20
100 PRINT "VIH=";VIH

RUN
VIH=2.47798        ← VIH ≈ 2.48 [V]
```

V_{th}: Use eqn. (3.5-34) to calculate

$$V_{th} = \frac{1 + \sqrt{0.4}\,(5 - 0.9)}{(1 + \sqrt{0.4})}$$

$$\simeq 2.2\,[\text{V}].$$

The VTC generated by a SPICE simulation is shown in Fig. E3.6.

```
NON-SYMMETRIC CMOS INVERTER
VDD 3 0 DC 5VOLTS
MNF 2 1 0 0 MN L=5U W=5U
```

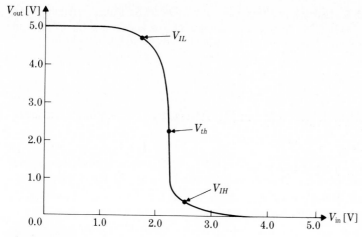

FIGURE E3.6

```
MPF 2 1 3 3 MP L=5U W=5U
VS 1 0 DC
.MODEL MN NMOS VTO=1 GAMMA=0.4 KP=2.5E-5
.MODEL MP PMOS VTO=-0.9 GAMMA=0.4 KP=1.0E-5
.DC VS 0 5 0.1
.PLOT DC V(2)
.END
```

■

REFERENCES

The following books emphasize digital MOS circuits and systems

(1) W. N. Carr and J. P. Mize, *MOS/LSI Design and Application,* New York: McGraw-Hill, 1972.

(2) M. I. Elmasry, "Digital MOS Integrated Circuits: A Tutorial," pp. 4-27 in M. I. Elmasry (ed.), *Digital MOS Integrated Circuits,* New York: IEEE Press (Wiley), 1981.

(3) L. A. Glasser and D. W. Dobberpuhl, *The Design and Analysis of VLSI Circuits,* Reading, MA: Addison-Wesley, 1985.

(4) D. A. Hodges and H. G. Jackson, *Analysis and Design of Digital Integrated Circuits,* New York: McGraw-Hill, 1983.

(5) J. Mavor, M. A. Jack, and P. B. Denyer, *Introduction to MOS LSI Design,* London: Addison-Wesley, 1983.

(6) C. A. Mead and L. Conway, *Introduction to VLSI Systems,* Reading, MA: Addison-Wesley, 1980.

(7) W. M. Penny and L. Lau (eds.), *MOS Integrated Circuits,* New York: Van Nostrand Reinhold, 1972.

(8) N. Weste and K. Eshraghian, *Principles of CMOS VLSI Design*, Reading, MA: Addison-Wesley, 1985. (The Appendix, "Computing Noise Margins for CMOS and NMOS Inverters" by A. Trivedi and N. S. Vasanthavada should be noted.)

PROBLEMS

3.1 The 3-terminal device shown in Fig. P3.1(a) has $I - V$ characteristics of

$$I_a = 0, \qquad I_c = \alpha V_{AC} + \theta V_{BC},$$

where $\alpha = 2\,[\mu A/V]$, $\theta = 14\,[\mu A/V]$. The device is used as a load device in an nMOS inverter, as illustrated in Fig. P3.1(b). Assume that the driver is described by $k' = 25\,[\mu A/V^2]$, $(W/L) = 2$, $\gamma = 0.40\,[V^{1/2}]$, $2\,|\phi_F| = 0.60\,[V]$, and $V_{DD} = +5\,[V]$, $V_{TO} = +1.0\,[V]$.

Calculate V_{OH}, V_{OL}, V_{IL}, and V_{IH} for the circuit. Assume $V_{in} = 0$ for the V_{OH} calculation.

3.2 A resistor of value $50\,[k\Omega]$ is used as an inverter load. The driver characteristics are given by

$$V_{TO} = +4\,[V], \quad k' = 25\,[\mu A/V^2], \quad \gamma = 0.4\,[V^{1/2}],$$

$$2\,|\phi_F| = 0.6\,[V], \quad V_{DD} = 12\,[V].$$

(a) Compute the value of (W/L) for the driver needed to give $V_{OL} = 0.5\,[V]$. Repeat the calculation for $V_{OL} = 0.75\,[V]$.

(b) Assume that the inverter is designed using your (W/L) value for $V_{OL} = 0.75\,[V]$ in part (a). Calculate V_{IL}, V_{IH}, and the noise margins for the circuit.

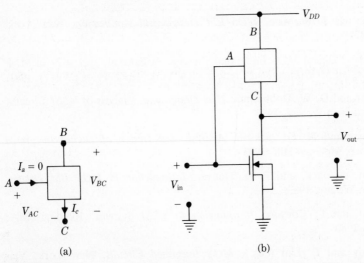

(a)   (b)

FIGURE P3.1

*(c) Run a SPICE simulation to produce the VTC for the circuit if $V_{OL} = 0.75$ [V].

3.3 An inverter is built using a saturated E-mode MOSFET as a load with $\beta_R = 8$. The processing is characterized by circuit parameters

$$V_{DD} = 5\,[\text{V}], \quad k' = 25\,[\mu\text{A}/\text{V}^2], \quad V_{T0} = +1.0\,[\text{V}],$$
$$\gamma = 0.39\,[\text{V}^{1/2}], \quad 2\,|\phi_F| = 0.6\,[\text{V}].$$

(a) Calculate the value of V_{OH}.
(b) Calculate V_{OL}.
(c) Calculate V_{IH}.
(d) Calculate V_{IL}.
(e) Calculate V_{th}.
*(f) Prepare a SPICE simulation file to extract the VTC.

3.4 Design a saturated enhancement-load inverter that gives $V_{OL} = 0.20$ [V] using the parameters in Problem 3.3. Assume that the minimum linewidth constraint is 4 [μm], and determine $(W/L)_L$ and $(W/L)_D$ that will implement your DC design.

* Simulate your circuit using SPICE and extract the critical VTC points from the output data.

3.5 An inverter is made using a nonsaturated load MOSFET. The base process is described by

$$V_{DD} = 10\,[\text{V}], \quad V_{T0} = +3\,[\text{V}], \quad 2\,|\phi_F| = 0.60\,[\text{V}],$$
$$\gamma = 0.53\,[\text{V}^{1/2}], \quad k' = 20\,[\mu\text{A}/\text{V}^2].$$

(a) Calculate the smallest permissible value of V_{GG} needed to set $V_{OH} = V_{DD}$.
(b) Assume that $V_{GG} = 15$ has been set. Find the value of β_R needed to design the circuit for $V_{OL} = 0.5$ [V]. Then round the β_R value to the nearest integer value that will guarantee $V_{OL} \le 0.5$ [V] and use this for the remainder of the problem.
(c) Calculate the minimum and maximum values of the bias parameter for your design (look at the differences induced by variations in V_{out}).
(d) Calculate V_{IL} for the circuit. Use programming techniques if available.
(e) Find V_{IH} using the simplified equation (3.3-58).
*(f) Simulate the circuit using SPICE and compare results. Comment on any significant deviations from the analytic calculations.

3.6 A depletion-mode MOSFET is made with $k' = 25\,[\mu\text{A}/\text{V}^2]$, $W = 4\,[\mu\text{m}]$, $L = 8\,[\mu\text{m}]$, and $V_{T0} = -3.5$ [V] and is operated such that V_{DS} ranges from 0 [V] to $V_{DD} = +5$ [V]. Ignore body bias in this problem.
(a) Plot I_D vs. V_{DS} for $V_{GS} = 0$ [V].
(b) Plot I_D vs. V_{DS} for $V_{GS} = -1$ [V] and $V_{GS} = +1$ [V] on the same axis.

(c) Suppose that an enhancement-mode MOSFET is made that has the same β value but is designed with $V_{T0} = +1 \, [V]$. Calculate the value of V_{GS} that must be applied to this device to obtain the same current flow as the $V_{GS} = 0$ case for the depletion-mode transistor. Assume that $V_{DS} = +3 \, [V]$ for both devices.

3.7 A depletion-load nMOS process has parameters

$$\text{E-mode:} \qquad V_{T0} = 0.9 \, [V],$$

$$\text{D-mode:} \qquad V_{T0} = -3.2 [V]$$

$$\text{Both devices:} \qquad k' = 25 \, [\mu A/V^2], \quad \gamma = 0.37 \, [V^{1/2}],$$

$$2 \, |\phi_F| = 0.58 \, [V].$$

The circuits operate at a power supply level of $V_{DD} = 5 \, [V]$.

(a) Calculate the value of β_R needed for an inverter that has $V_{OL} = 0.3 \, [V]$.

(b) Suppose that the smallest linewidth permitted by the fabrication process gives $\min(W) = 4 \, [\mu m] = \min(L)$. Find the values of W and L for the driver and the load that minimize the chip area. Use the nearest integer β_R that gives $V_{OL} \leq 0.3 \, [V]$.

(c) Rework the problem if a bulk bias of $V_{BB} = 3 \, [V]$ is applied (negative to the substrate).

3.8 A depletion-load inverter is described by the following parameters:

$$k'_L = 25 \, [\mu A/V^2], \quad (W/L)_L = 0.4, \quad V_{T0L} = -3.0 \, [V],$$

$$k'_D = 20 \, [\mu A/V^2], \quad (W/L)_D = 4.0, \quad V_{T0D} = +0.9 \, [V].$$

Assume that, for both devices, $\gamma = 0.4 \, [V^{1/2}]$, $2 \, |\phi_F| = 0.6 \, [V]$, $\mu_n = 580 \, [cm^2/V\text{-sec}]$. The power supply is given as $V_{DD} = +5 \, [V]$.

(a) Calculate the value of $V_{OL} \, [V]$.

(b) Calculate $V_{IH} \, [V]$ and $V_{IL} \, [V]$.

(c) Suppose that the input voltage is set at $V_{in} = 0 \, [V]$. Leakage current through the reverse-biased pn junction is measured, giving a load current of $I_L = 1 \, [\mu A]$.

Starting with the equation for $I_L(V_{DSL})$, find a self-iterating equation for V_{DSL}. Then solve the equation to find the actual value of V_{OH} for the circuit.

Repeat the calculation for a value of $I_L = 0.5 \, [\mu A]$ leakage current.

3.9 A depletion-load inverter is designed in a technology where

$$k'_L = 30 \, [\mu A/V^2], \quad V_{T0L} = -3.0 \, [V], \quad \gamma_L = 0.36 \, [V^{1/2}],$$

$$k'_D = 24 \, [\mu A/V^2], \quad V_{T0D} = +1.0 \, [V], \quad \gamma_D = 0.38 \, [V^{1/2}].$$

The value $2 \, |\phi_F| = 0.58 \, [V]$ is valid for both devices. The driver is designed with $W = 14\,[\mu\text{m}]$, $L = 3.5\,[\mu\text{m}]$, while the load has a geometry of $W = 4.0\,[\mu\text{m}]$, $L = 7.6\,[\mu\text{m}]$. The power supply is set at $V_{DD} = 5\,[\text{V}]$.

(a) Calculate β_R for the design.

(b) Find V_{OL}, V_{IL}, and V_{IH} for the design.

(c) Suppose that processing variations give $k'_L = 29\,[\mu\text{A/V}^2]$ for a wafer lot. Calculate the new value of V_{OL} for this batch.

*(d) Simulate the original circuit using SPICE. Perform a .DC analysis to obtain the VTC, and find the maximum value of I_L that flows.

3.10 Two inverters are cascaded as shown in Fig. P3.2. The device parameters are given as

$$\text{E-mode:} \quad V_{T0} = +1\,[\text{V}], \quad k' = 25\,[\mu\text{A/V}^2], \quad \gamma = 0.37\,[\text{V}^{1/2}],$$
$$2\,|\phi_F| = 0.58\,[\text{V}],$$

$$\text{D-mode:} \quad V_{T0} = -3.5\,[\text{V}], \quad k' = 30\,[\mu\text{A/V}^2], \quad \gamma = 0.37\,[\text{V}^{1/2}],$$
$$2\,|\phi_F| = 0.58\,[\text{V}].$$

The first stage is designed with $\beta_R = 8$, while the second stage has $\beta_R = 10$.

(a) Calculate the value of V_{OH} from stage 1 (i.e., max (V_X)).

(b) Assume that max (V_{in}) into the first stage is equal to V_{OH} from part (a). Find the value of V_{OL} from stage 1 (i.e., min (V_X)).

(c) Compute V_{OL} of stage 2 (= min (V_{out})) when the inverters are cascaded as shown.

(d) Redesign the second stage to give a value of $V_{OL} = \text{min}\,(V_{\text{out}}) = 0.50\,[\text{V}]$.

(e) The processing sets the minimum value of both L and W at $4\,[\mu\text{m}]$.

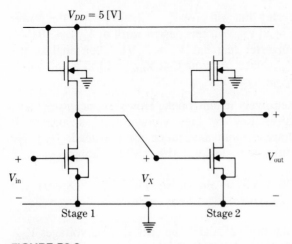

$V_{DD} = 5\,[\text{V}]$

V_{in}

V_X

V_{out}

Stage 1   Stage 2

FIGURE P3.2

Find the driver and load dimensions for the second stage with the original $\beta_R = 10$ if a minimum-area circuit is used. Calculate the value of the 2nd-stage driver gate capacitance $C_g = C_{ox}(WL)_D$ in units of [fF]. (There are different ways to design this circuit and still meet the specifications.)

3.11 A CMOS process is described by the base parameters

$$V_{TOn} = +0.8\,[\text{V}], \quad k_n' = 40\,[\mu\text{A/V}^2],$$

$$V_{TOp} = -0.8\,[\text{V}], \quad k_p' = 16\,[\mu\text{A/V}^2].$$

(a) A symmetric inverter is built with $\beta_n = \beta_p$. Calculate the values of V_{IL} and V_{IH} for power supply voltages of $V_{DD} = 5\,[\text{V}]$, $10\,[\text{V}]$, and $15\,[\text{V}]$.
(b) Calculate the current flow through the circuit when $V_{in} = V_{th}$ for all the power supply values.

3.12 A nonsymmetric CMOS inverter is fabricated in the process described by the base parameter set in Problem 3.11. The device aspect ratios are chosen to be $(W/L)_n = 1.4$ and $(W/L)_p = 4$. The power supply voltage is $V_{DD} = 5\,[\text{V}]$.
(a) Calculate V_{IL}.
(b) Calculate V_{IH}.
(c) Calculate V_{th}.
*(d) Verify the results with a SPICE simulation. Assume a minimum linewidth of $4\,[\mu\text{m}]$ (if desired).

3.13 A CMOS process gives process values of

$$k_n' = 40\,[\mu\text{A/V}^2], \quad V_{TOn} = +0.8\,[\text{V}],$$

$$k_p' = 16\,[\mu\text{A/V}^2], \quad V_{TOp} = -0.9\,[\text{V}].$$

(a) Design a CMOS inverter that will give $V_{th} = 2.5\,[\text{V}]$ when the power supply is set at $V_{DD} = 5\,[\text{V}]$. Will this design work at $V_{DD} = 10\,[\text{V}]$?
(b) Instead, design an inverter that has $V_{th} = 3\,[\text{V}]$. Then analyze the circuit for V_{IL} and V_{IH} values. Assume that $V_{DD} = 5\,[\text{V}]$ is still valid.
*(c) Simulate both circuits using SPICE.

3.14 Positive logic is used exclusively in this book. However, negative logic, where a logic 0 is represented by high voltages and a logic 1 is represented by low voltages, could have been used instead. Negative logic is sometimes useful when dealing with pMOS designs or with subcircuits in CMOS.

Consider the pMOS inverter shown in Fig. 3.25. Discuss the operation of this circuit if negative logic were employed.

*3.15 Write a computer program that calculates the critical VTC voltages V_{OL}, V_{OH}, V_{IL}, V_{IH}, and V_{th} for a depletion-load inverter. The input variables should be β_R or $(W/L)_D$, $(W/L)_L$. Assume circuit parameters of

$$V_{DD} = 5\,[\text{V}], \quad V_{TOD} = +0.9\,[\text{V}], \quad V_{TOL} = -3.3\,[\text{V}],$$

$$\gamma = 0.37\,[\text{V}^{1/2}], \quad 2\,|\phi_F| = 0.6\,[\text{V}], \quad k'_D = 25\,[\mu\text{A}/\text{V}^2] = k'_L$$

unless another set is available. Provide plotting of $V_{\text{out}}(V_{\text{in}})$ if possible.

*3.16 Write a computer program that calculates the critical VTC voltages V_{IL}, V_{IH}, and V_{th} for a CMOS inverter with an input of $(W/L)_n$ and $(W/L)_p$. Use V_{DD} as a variable, but specify the circuit parameters

$$V_{TOn} = 0.8\,[\text{V}], \quad k'_n = 40\,[\mu\text{A}/\text{V}^2], \quad V_{TOp} = -0.7\,[\text{V}],$$

$$k'_p = 15\,[\mu\text{A}/\text{V}^2]$$

unless another set is convenient. Include an output plot of $V_{\text{out}}(V_{\text{in}})$ if permitted by the computer system.

CHAPTER 4

Switching of MOS Inverters

The transient switching characteristics of digital circuits set the fundamental limitations on system speed. This chapter deals with the various time intervals that are used to characterize digital switching events. It will be shown that the design directly determines the transient response of the circuit to input excitations. This completes the design cycle introduced in Chapter 3 to the point where the interplay between DC and transient properties is obvious.

Emphasis is placed on understanding the circuit operation. Important parameters are extracted and analyzed, and approximations are introduced whenever reasonable. SPICE is used to provide answers when the analysis becomes intractable. Programming practice is again encouraged while reading the material.

4.1 The Output High-to-Low Time

The first nMOS transient time interval that will be computed is the output high-to-low time t_{HL}, also referred to as the *fall time* t_f. Figure 4.1(a) illustrates the problem. At time t_1, the input voltage undergoes a step transition from $V_{in} = V_{OL}$ ($t < t_1$) to $V_{in} = V_{OH}$ ($t > t_1$). The output responds by falling from $V_{out} = V_{OH}$ at t_1 to $V_{out} = V_{OL}$ at time t_2. The time required for the output to change from a stable logic 1 state to a stable logic 0 state is denoted by t_{HL}.

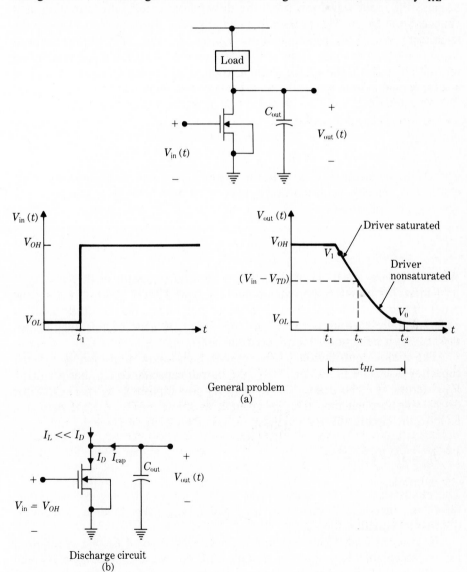

General problem
(a)

Discharge circuit
(b)

FIGURE 4.1 Calculation of the fall time.

The total fall time of the inverter is actually the sum of two intervals and may be expressed as

$$(t_2 - t_1) = \tau + t_{\text{dis}}. \tag{4.1-1}$$

t_{dis} is the *discharge* time and will be discussed in detail. τ is referred to as the channel *transit time* and is the average time required for an electron to traverse the driver MOSFET channel from source to drain. This enters the calculation because the input waveform takes the driver from cutoff into saturation as it changes from V_{OL} to V_{OH} at time t_1. For $t < t_1$, no inversion layer exists in the transistor. When V_{in} is switched to V_{OH} at t_1, the gate voltage induces formation of the channel. Conduction through the MOSFET is not established until electrons from the source make it to the drain, which requires a time τ; τ is a fundamental limit that establishes the absolute minimum value of $(t_2 - t_1)$.

An estimate of τ can be obtained in a simplistic manner by writing

$$\tau = \frac{L}{v}, \tag{4.1-2}$$

where v is the electron velocity in the channel. For small drain-source electric fields, the velocity is proportional to the electric field by $v = \mu_n \mathscr{E}$. Since $\mathscr{E} \simeq V_{DS}/L$, this gives

$$\tau \simeq \frac{L^2}{\mu_n V_{DS}}. \tag{4.1-3}$$

When $\mathscr{E}$ is large ($\geq 10^4$ [V/cm]), velocity saturation occurs so that $v = v_s \simeq 10^7$ [cm/sec] is a better approximation. In a typical MOS process, τ is on the order of picoseconds [ps]. The time interval $(t_2 - t_1)$, on the other hand, is experimentally measured to be on the order of nanoseconds [ns]. Thus, $\tau \ll t_{\text{dis}}$, so it need not be considered further.

The major contribution to the output high-to-low transition is t_{dis}, the capacitor discharge time. For $t < t_1$, the output capacitance C_{out} has a voltage V_{OH} across it. The transition to V_{OL} is accomplished by the capacitor discharging through the driver to ground, as shown in Fig. 4.1(b). Applying KCL to the circuit, the voltage decay is described by $I_D = (I_L + I_{\text{cap}})$, or

$$I_D = I_L - C_{\text{out}} \frac{dV_{\text{out}}}{dt}. \tag{4.1-4}$$

The minus sign arises because I_{cap} is defined positive leaving the capacitor. Given the functional forms of $I_D(V_{\text{out}})$ and $I_L(V_{\text{out}})$, this yields a (nonlinear) differential equation for $V_{\text{out}}(t)$.

Now recall from Chapter 3 that a driver-load ratio of $\beta_R > 1$ is required for an acceptable V_{OL} value in a static nMOS inverter. Since $I_D \propto \beta_D$ and $I_L \propto \beta_L$, $I_D > I_L$ in a practical design. To obtain a differential equation that can be solved by direct analytic techniques, this observation will be taken one step further by assuming that $I_D \gg I_L$. Then

$$I_D(V_{\text{out}}) \simeq -C_{\text{out}} \frac{dV_{\text{out}}}{dt} \tag{4.1-5}$$

will allow for the calculation of $(t_2 - t_1) = t_{\text{dis}}$. In the strictest sense, $t_{HL} \neq (t_2 - t_1)$ since t_{HL} is defined as the time required for the output voltage to fall from $V_1 = V_{OL} + 0.9(V_{OH} - V_{OL})$ to $V_0 = V_{OL} + 0.1(V_{OH} - V_{OL})$, which define the 90% and 10% points, respectively (see Fig. 3.8). However, the present discussion will take $t_{HL} = (t_2 - t_1) = t_{\text{dis}}$, so the time interval computed here will actually represent the full swing from V_{OH} to V_{OL}. While this modification is not really necessary, it simplifies the final expressions.

To compute $t_{HL} = (t_2 - t_1)$, eqn. (4.1-5) is rearranged and integrated to read

$$t_{HL} \equiv \int_{t_1}^{t_2} dt \simeq -C_{\text{out}} \int_{V_{\text{out}}(t_1)=V_{OH}}^{V_{\text{out}}(t_2)=V_{OL}} \frac{dV_{\text{out}}}{I_D(V_{\text{out}})}. \tag{4.1-6}$$

Since $V_{\text{out}}(t)$ is decreasing, the driver changes from a saturated to a nonsaturated state when $V_{\text{out}} = (V_{\text{in}} - V_{TD})$. This occurs at time t_x, as shown in Fig. 4.1(a). Since I_D is different for the two operational regions, eqn. (4.1-6) is decomposed as

$$t_{HL} = (t_2 - t_x) + (t_x - t_1), \tag{4.1-7}$$

where

$$(t_x - t_1) = -C_{\text{out}} \int_{V_{OH}}^{V_{OH}-V_{TD}} \frac{dV_{\text{out}}}{\dfrac{\beta_D}{2}(V_{OH} - V_T)^2} \tag{4.1-8}$$

describes the initial fall from V_{OH} to $(V_{OH} - V_{TD})$ when the driver is saturated, and

$$(t_2 - t_x) = -C_{\text{out}} \int_{V_{OH}-V_{TD}}^{V_{OL}} \frac{dV_{\text{out}}}{\dfrac{\beta_D}{2}[2(V_{OH} - V_{TD})V_{\text{out}} - V_{\text{out}}^2]} \tag{4.1-9}$$

gives the final decay to V_{OL} with the MOSFET exhibiting nonsaturated characteristics.

The first integral in eqn. (4.1-8) is trivial since I_D is not a function of V_{out} (ignoring channel-length modulation). Thus,

$$(t_x - t_1) = \frac{2C_{\text{out}}V_{TD}}{\beta_D(V_{OH} - V_{TD})^2}. \tag{4.1-10}$$

To perform the integration in eqn. (4.1-9), note the indefinite integral

$$\int \frac{dx}{x(a + bx^n)} = \frac{1}{an} \ln\left(\frac{x^n}{a + bx^n}\right) \tag{4.1-11}$$

as found in any standard table. Using $n = 1$, $a = 2(V_{OH} - V_{TD})$ and $b = -1$, this gives directly that

$$(t_2 - t_x) = \frac{C_{\text{out}}}{\beta_D(V_{OH} - V_{TD})} \ln\left[\frac{2(V_{OH} - V_{TD})}{V_{OL}} - 1\right]. \tag{4.1-12}$$

Combining the results of these two integrations gives

$$t_{HL} = \tau_D\left\{\frac{2V_{TD}}{(V_{OH} - V_{TD})} + \ln\left[\frac{2(V_{OH} - V_{TD})}{V_{OL}} - 1\right]\right\} \tag{4.1-13}$$

as the output fall time. In this expression,

$$\tau_D \equiv \frac{C_{\text{out}}}{\beta_D(V_{OH} - V_{TD})} \tag{4.1-14}$$

is the time constant that characterizes the discharging event. To understand this terminology, note that $[\beta_D(V_{OH} - V_{TD})]^{-1}$ has units of ohms; thus τ_D may be viewed as a type of RC time constant for the discharge of the capacitor through the resistance of the driver. Although this interpretation is not strictly correct (since the MOSFET is at best a nonlinear resistor), it remains a useful concept.

Now to maximize the switching frequency of the inverter, t_{HL} must be made as small as possible. Assuming that the voltages are set, this requirement reduces to minimizing τ_D as a primary design objective. Since τ_D is proportional to C_{out}, decreasing the capacitance seen by the output node will improve the transient response of the circuit. As discussed later, the value of C_{out} is determined by many interrelated factors. Included in this set are considerations such as the fan-out of the circuit, the layout geometry, and the doping densities in the devices.

The expression for t_{HL} in eqn. (4.1-13) illustrates the dependence of the transient response on the DC inverter design. Recall from Chapter 3 that increasing the driver-load ratio $\beta_R = \beta_D/\beta_L$ decreases the output low voltage V_{OL}. Because of the presence of β_D and V_{OL} in the expression for t_{HL}, the DC design choice for β_R will set the fall time of the inverter. To extract the functional dependence of t_{HL} on τ_D, first note that increasing β_R will reduce the discharge time constant τ_D. This correlates with the physical picture, since a larger β_D gives the driver greater current flow capabilities, which in turn allows for a quicker discharge of C_{out}. As an alternative viewpoint, increasing β_R reduces the value of the equivalent driver drain-source resistance.

The time interval t_{HL} also has an indirect dependence on β_R through V_{OL}. As β_R increases, the magnitude of the log term in eqn. (4.1-13) increases. However, this growth is relatively slow compared with the $1/\beta_R$ decrease experienced by τ_D. Consequently, when these two terms are multiplied together, it is seen that increasing β_R gives smaller fall times t_{HL}. This behavior is illustrated in Fig. 4.2, which shows t_{HL}/C_{out} [ns/pF] as a function of both V_{OL} and β_R for a depletion load inverter using the circuit process parameters listed in the drawing.

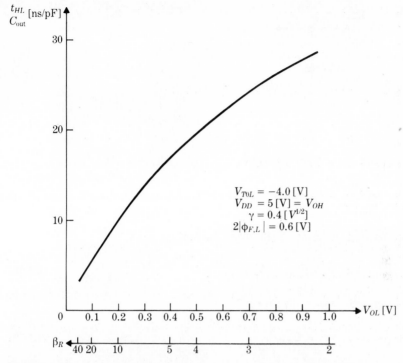

FIGURE 4.2 t_{HL}/C_{out} as a function of V_{OL} and β_R for a depletion load.

Although the above discussion may seem to imply that a large value of β_R is desirable for fast switching times, it must be remembered that the analysis has only dealt with a single isolated gate. In a realistic digital system, β_R values should be kept as small as possible. One reason for this requirement is to minimize the real estate consumed by the circuit, as discussed in the last chapter. However, another important consideration is that large driver-load ratios may actually slow down the overall *system* response. To understand this remark, note that the *input* gate capacitance of the driver MOSFET is approximately $C_g = C_{\text{ox}}(WL)_D$. C_g is proportional to the driver gate area, so that implementing a large β_R design by increasing $(W/L)_D$ gives a large value for this capacitance. The problem arises because C_g contributes to the capacitance seen by the output node of the *previous stage*. Thus, using a large β_R will slow down the switching speed of the gate that feeds the inverter, which may in turn degrade the system response as a whole. This problem will be analyzed in greater detail in Section 4.7.

Consider now the problem illustrated in Fig. 4.3(a), where a parasitic interconnect line resistance has been modeled by including R_{line} between the output inverter node and C_{out}. To find t_{HL} for this circuit, note that KVL gives

$$V_{\text{out}}(t) = V_R(t) + V_{DS}(t). \qquad \text{(4.1-15)}$$

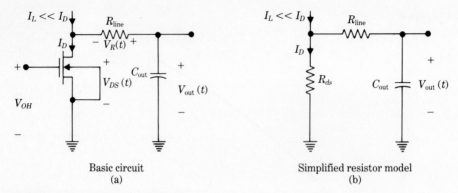

FIGURE 4.3   Parasitic interconnect resistance in the discharge problem.

This can be written in terms of the drain current as

$$-\frac{1}{C_{\text{out}}} \int I_D(t)\, dt = I_D R_{\text{line}} + V_{DS}. \tag{4.1-16}$$

Differentiating with respect to time,

$$-I_D = R_{\text{line}} C_{\text{out}} \frac{dI_D}{dt} + C_{\text{out}} \frac{dV_{DS}}{dt}, \tag{4.1-17}$$

or, since $I_D = I_D[V_{DS}(t)]$,

$$I_D(V_{DS}) = -\left[C_{\text{out}} + R_{\text{line}} C_{\text{out}} \frac{dI_D}{dV_{DS}} \right] \frac{dV_{DS}}{dt}. \tag{4.1-18}$$

Comparing this with eqn. (4.1-5) shows that the interconnect resistance introduces another term into the time integration. Since the mathematics becomes somewhat messy, it is simpler (and also instructive) to approximate the effects of R_{line} on t_{HL} by using the model in Fig. 4.3(b).

When $R_{\text{line}} = 0$, the discharge time constant is $\tau_D = R_{ds} C_{\text{out}}$, where

$$R_{ds} = \frac{1}{\beta_D(V_{OH} - V_{TD})} \tag{4.1-19}$$

represents the driver drain-source resistance in the model. Including R_{line} increases the total resistance to $R = R_{ds} + R_{\text{line}}$. Consequently, τ_D may be approximated by

$$\tau_D \simeq \left[\frac{1}{\beta_D(V_{OH} - V_{TD})} + R_{\text{line}} \right] C_{\text{out}}. \tag{4.1-20}$$

The fall time is then found by using this in eqn. (4.1-13) for t_{HL}. Although this is only a first-order approximation, it does illustrate the fact that interconnect resistance degrades the switching response of an nMOS digital circuit.

EXAMPLE 4.1-1

An nMOS inverter has a driver described by $(W/L)_D = 4$, $k' = 25\,[\mu A/V^2]$, and $V_{T0D} = 1\,[V]$. The circuit operates at $V_{DD} = 5\,[V]$ and is designed to have $V_{OL} = 0.25\,[V]$.

Calculate t_{HL} if $C_{out} = 100\,[fF]$.

Solution
Using eqn. (4.1-14), the driver time constant is

$$\tau_D = \frac{(100 \times 10^{-15})}{(25 \times 10^{-6})(4)(5 - 1)} = 0.25\,[ns].$$

Then eqn (4.1-13) gives

$$t_{HL} = (0.25)\left[\frac{2}{4} + \ln\left(\frac{8}{0.25} - 1\right)\right],$$

so

$$t_{HL} \simeq 0.98\,[ns] \leftarrow. \qquad\blacksquare$$

4.2 nMOS Rise Times

This section deals with calculating the output low-to-high time t_{LH} for various nMOS loads. The general problem is illustrated in Fig. 4.4. For times $t < t_a$, $V_{in} = V_{OH}$, so the output voltage across the capacitor C_{out} is V_{OL}. At $t = t_a$, the input makes a step transition downward to $V_{in} = V_{OL}$. This places the driver in cutoff, giving $I_D = 0$. The load current I_L is diverted to the capacitor, allowing C_{out} to charge up toward V_{OH}; it reaches this level at time t_b. In the literature, t_{LH} is also called the rise time (t_r) or the charging time (t_{ch}) for obvious reasons.

The charging current is described by

$$C_{out}\frac{dV_{out}}{dt} \simeq I_L(V_{out}). \tag{4.2-1}$$

Noting that t_{LH} is defined as the time required for the output to charge from V_0 (the 10% point) at time t_0 to V_1 (the 90% point) at time t_1, this gives

$$t_{LH} \equiv \int_{t_1}^{t_2} dt = C_{out}\int_{V_0}^{V_1}\frac{dV_{out}}{I_L(V_{out})}. \tag{4.2-2}$$

Since I_L as a function of $V_L = (V_{DD} - V_{out})$ is dependent on the specific load device used in the circuit, a separate calculation will be performed for each possibility. One result of the different load currents is that nonsymmetrical nMOS switching (where $t_{HL} \neq t_{LH}$) characterizes the circuits.

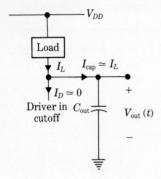

General charging circuit
(a)

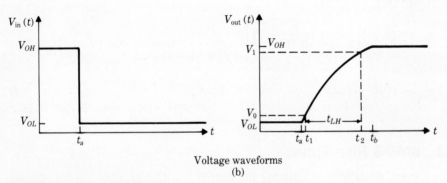

Voltage waveforms
(b)

FIGURE 4.4 Rise time circuit.

RESISTIVE nMOS LOAD

The simplest case is that where a linear load resistor R_L is used as the nMOS inverter load. This is shown in Fig. 4.5 and is simply the problem of charging a capacitor through a resistor. Since $V_R = (V_{DD} - V_{out})$, Ohm's law gives

$$t_{LH} = R_L C_{out} \int_{V_0}^{V_1} \frac{dV_{out}}{(V_{DD} - V_{out})}. \tag{4.2-3}$$

Integrating,

$$t_{LH} = R_L C_{out} \ln\left(\frac{V_{DD} - V_0}{V_{DD} - V_1}\right), \tag{4.2-4}$$

so $\tau_L = R_L C_{out}$ is the time constant for the charging event.

The tradeoffs between the DC design and the resulting transient response can be seen from this expression. Recall that increasing R_L decreases V_{OL} and also provides for a sharper VTC transition. This, however, will give longer charge times since both τ_L and $(V_{DD} - V_0)$ will increase. The 90% point voltage V_1 must be used as the upper limit in eqn. (4.2-3) since integrating to

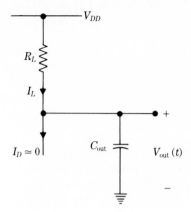

FIGURE 4.5 Charging circuit for a resistive load.

the DC high level $V_{OH} = V_{DD}$ would give an infinite rise time. As a final comment, low C_{out} values are required to reduce the rise time interval.

SATURATED ENHANCEMENT LOAD

When a saturated E-mode MOSFET is used as a load, the charging current is given by

$$I_L = \frac{\beta_L}{2}(V_{DD} - V_{out} - V_{TL})^2. \tag{4.2-5}$$

The circuit in Fig. 4.6 shows that body bias effects are present since V_{TL} is a function of V_{out}. The derivative dV_{TL}/dV_{out} is positive, implying that V_{TL} increases as V_{out} increases toward V_1. This in turn reduces the load current available to charge C_{out}. Although body bias can be included in computing t_{LH}, it is much simpler to approximate the rise time by assuming V_{TL} = constant for

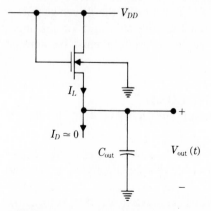

FIGURE 4.6 Charging circuit for a saturated enhancement load.

the integration. Then eqn. (4.2-2) becomes

$$t_{LH} = \frac{2C_{\text{out}}}{\beta_L} \int_{V_0}^{V_1} \frac{dV_{\text{out}}}{(V_{DD} - V_{\text{out}} - V_{TL})^2},$$ (4.2-6)

which integrates directly to

$$t_{LH} = \frac{2C_{\text{out}}}{\beta_L} \left[\frac{1}{V_{DD} - V_{TL} - V_1} - \frac{1}{V_{DD} - V_{TL} - V_0} \right].$$ (4.2-7)

Body bias effects can be included in an approximate manner by evaluating V_{TL} in this equation at the midpoint where $V_{\text{out}} = \frac{1}{2}(V_0 + V_1)$. The worst-case value for t_{LH} can be found by taking $V_{TL} = V_{TL}(V_{OH})$, as this provides the smallest charging current level. Since $V_{OH} = [V_{DD} - V_{TL}(V_{OH})]$ for the saturated enhancement load, eqn. (4.2-7) reduces to

$$t_{LH,\text{max}} = \frac{2C_{\text{out}}}{\beta_L(V_{OH} - V_0)} \left[\frac{V_{OH} - V_0}{V_{OH} - V_1} - 1 \right].$$ (4.2-8)

This equation is particularly useful for examining the tradeoffs between the DC design and the resulting transient characteristics of the circuit. Recall that the enhancement load nMOS inverter requires a driver-load ratio of $\beta_R = \beta_D/\beta_L \gg 1$ to achieve a small V_{OL} value and a sharp VTC transition. For a design that minimizes both the chip area and the driver input capacitance, β_L should be as small as possible. This, however, leads to longer rise times because t_{LH} is inversely proportional to β_L. In fact, comparing the rise time in eqn. (4.2-8) with the fall time t_{HL} in eqn. (4.1-13) shows that $\beta_D \gg \beta_L$ generally gives $t_{LH} \gg t_{HL}$.

With these observations in mind, it is possible to summarize the complete design process for the circuit. First, the desired value of V_{OL} is used to compute the driver-load ratio β_R. To find the value of $\beta_L = k_L'(W/L)_L$, the load aspect ratio $(W/L)_L$ must be adjusted to give an acceptable charging time t_{LH}. This is the most important transient quantity to consider since it limits the switching frequency of the circuit. Once $(W/L)_L$ is found, the driver geometry is determined by $(W/L)_D = \beta_R(W/L)_L$. If either the chip area or the input capacitance becomes excessively large, then the procedure must be iterated with larger V_{OL} values.

NON-SATURATED ENHANCEMENT LOAD

The charging circuit for this configuration is shown in Fig. 4.7. The current supplied by the load is given by

$$I_L = \frac{\beta_L}{2} [(V_{GG} - V_{TL} - V_{\text{out}})(V_{DD} - V_{\text{out}}) - (V_{DD} - V_{\text{out}})^2],$$ (4.2-9)

which must be substituted into eqn. (4.2-2). The easiest way to perform the

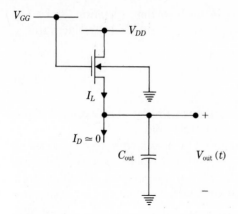

FIGURE 4.7 Charging circuit for a nonsaturated enhancement load.

integration is to first rewrite the current as

$$I_L = \frac{\beta_L V_{DD}^2}{2m} \left(1 - \frac{V_{out}}{V_{DD}}\right)\left(1 - m\frac{V_{out}}{V_{DD}}\right).$$ **(4.2-10)**

In this equation, m is the biasing parameter

$$m = \frac{V_{DD}}{2(V_{GG} - V_{TL}) - V_{DD}}$$ **(4.2-11)**

introduced in Section 3.3.2; recall that m is restricted to the interval $(0, 1]$. The rise time integral thus assumes the form

$$t_{LH} = \frac{2mC_{out}}{\beta_L V_{DD}} \int_{V_0/V_{DD}}^{V_1/V_{DD}} \frac{d(V_{out}/V_{DD})}{\left(1 - \frac{V_{out}}{V_{DD}}\right)\left(1 - m\frac{V_{out}}{V_{DD}}\right)},$$ **(4.2-12)**

where m is taken as a constant for the integration. Using the known integral

$$\int \frac{dx}{(1 - x)(1 - mx)} = \left(\frac{1}{1 - m}\right) \ln\left(\frac{1 - mx}{1 - x}\right)$$ **(4.2-13)**

gives

$$t_{LH} = \frac{2C_{out}}{\beta_L V_{DD}} \left(\frac{m}{1 - m}\right) \ln\left\{\frac{\left(1 - m\frac{V_1}{V_{DD}}\right)\left(1 - \frac{V_0}{V_{DD}}\right)}{\left(1 - m\frac{V_0}{V_{DD}}\right)\left(1 - \frac{V_1}{V_{DD}}\right)}\right\}$$ **(4.2-14)**

for the rise time of the circuit.

To interpret this result, note that $m \to 0$ as $V_{GG} \to \infty$. Equation (4.2-9) shows that using large values for the load bias voltage V_{GG} provides increased charging current levels. This is reflected in the rise time equation, since t_{LH} decreases as $m \to 0$. In the limit where $V_{GG} \to \infty$, eqn. (4.2-14) reduces to t_{LH} for a resistive load with

$$R_L = \frac{2}{\beta_L V_{DD}} \left(\frac{m}{1 - m} \right) \tag{4.2-15}$$

as the equivalent load resistance. This correlates with the observation that the DC load line becomes approximately linear in this limit.

Body bias effects in V_{TL} have been ignored by assuming m to be a constant in the rise time integral. As seen in eqn. (4.2-9), body bias leads to a decrease in the charging current supplied by the nonsaturated load, which in turn increases the rise time interval. Reduced current flow levels can be accounted for in an average manner by computing $V_{TL}(V_{\text{out}})$ at $V_{\text{out}} = \frac{1}{2}(V_0 + V_1)$ and then using this threshold voltage to find m for use in t_{LH}. The most pessimistic rise time values are obtained by taking V_{TL} at $V_{\text{out}} = V_1$ or V_{OH}.

The overall design procedure for this circuit follows the same lines established for the saturated enhancement load, except that now the extra load bias voltage V_{GG} must be included. It should be remembered that the actual value of V_{GG} is generally not a design variable but is set by overall system power supply considerations.

DEPLETION MOSFET LOAD

The final case that will be examined is that where a depletion-mode MOSFET is used as the load. This gives the charging circuit in Fig. 4.8(a). Since $V_{GSL} = 0$ and $V_{DSL} = (V_{DD} - V_{\text{out}})$, the transistor changes its operational characteristics as the capacitor voltage increases. Initially, V_{out} is small,

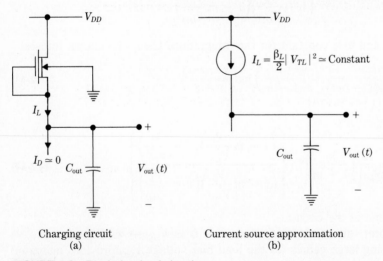

Charging circuit
(a)

Current source approximation
(b)

FIGURE 4.8 Depletion load rise time.

so $(V_{DD} - V_{out}) > |V_{TL}|$; the load MOSFET is thus saturated with

$$I_L = \frac{\beta_L}{2}|V_{TL}(V_{out})|^2. \tag{4.2-16}$$

If body bias effects were not present (i.e., $\gamma_L = 0$), this would describe an ideal current source and reduce the problem to the circuit shown in Fig. 4.8(b). Although this simplification ignores many important factors (including the change in the operational mode of the load device), it is useful for obtaining a first estimate of t_{LH}. To this end, note that the change in the capacitor voltage is $\Delta V = (V_1 - V_0)$. Then

$$t_{LH} \simeq \frac{C_{out}(\Delta V)}{I_L} = \frac{2C_{out}(V_1 - V_0)}{\beta_L|V_{TL}|^2} \tag{4.2-17}$$

gives the rise time of the idealized circuit. The most important point to note here is that increasing $|V_{T0L}|$ will improve the transient response since a larger charging current will be available. Moreover, setting the gate-source voltage to $V_{GSL} = 0$ keeps I_L approximately constant. This indicates that the depletion MOSFET load will provide for faster charging of C_{out} when compared with the other possible load devices.

Returning to the problem at hand, the capacitor will increase V_{out} to the point where $(V_{DD} - V_{out}) < |V_{TL}|$ is satisfied. The depletion load transistor changes to a nonsaturated conducting mode with

$$I_L = \frac{\beta_L}{2}[2|V_{TL}(V_{out})|(V_{DD} - V_{out}) - (V_{DD} - V_{out})^2]. \tag{4.2-18}$$

This describes the charging current for the remaining portion of the rise time interval.

A more accurate expression for t_{LH} is found using eqn. (4.2-2) and noting that the load transistor changes from a saturated to a nonsaturated mode when $V_{out} = (V_{DD} - |V_{TL}|)$. The integral then splits into

$$t_{LH} = C_{out}\int_{V_0}^{V_{DD}-|V_{TL}|} \frac{dV_{out}}{I_{L(sat)}} + C_{out}\int_{V_{DD}-|V_{TL}|}^{V_1} \frac{dV_{out}}{I_{L(nonsat)}}, \tag{4.2-19}$$

where body bias effects will be included in an average manner after the integrations are performed. The first term is easily evaluated as

$$C_{out}\int_{V_0}^{V_{DD}-|V_{TL}|} \frac{dV_{out}}{I_{L(sat)}} = \frac{2C_{out}(V_{DD} - |V_{TL}| - V_0)}{\beta_L|V_{TL}|^2} \tag{4.2-20}$$

since $I_{L(sat)}$ in eqn. (4.2-16) is taken as a constant. To compute the second term, $I_{L(nonsat)}$ from eqn. (4.2-18) is used to write

$$C_{out}\int_{V_{DD}-|V_{TL}|}^{V_1} \frac{dV_{out}}{I_{L(nonsat)}} =$$

$$-\frac{2C_{out}}{\beta_L}\int_{|V_{TL}|}^{V_{DD}-V_1} \frac{d(V_{DD} - V_{out})}{[2|V_{TL}|(V_{DD} - V_{out}) - (V_{DD} - V_{out})^2]}, \tag{4.2-21}$$

where the variable of integration has been changed from V_{out} to $(V_{DD} - V_{out})$. With this form, the indefinite integral in eqn. (4.1-11) may be used with $x = (V_{DD} - V_{out})$, $n = 1$, $a = 2|V_{TL}|$, and $b = -1$. Evaluating at the limits yields

$$C_{out} \int_{V_{DD}-|V_{TL}|}^{V_1} \frac{dV_{out}}{I_{L(nonsat)}} = \frac{C_{out}}{\beta_L |V_{TL}|} \ln \left[\frac{2|V_{TL}| - (V_{DD} - V_1)}{(V_{DD} - V_1)} \right] \tag{4.2-22}$$

as the nonsaturated contribution to the rise time. Combining eqns. (4.2-20) and (4.2-22) gives the complete expression

$$t_{LH} = \frac{C_{out}}{\beta_L |V_{TL}|} \left\{ \frac{2(V_{DD} - |V_{TL}| - V_0)}{|V_{TL}|} + \ln \left[\frac{2|V_{TL}| - (V_{DD} - V_1)}{(V_{DD} - V_1)} \right] \right\} \tag{4.2-23}$$

for the depletion-type nMOS load. Body bias may be included in the final expression by evaluating V_{TL} at the average output voltage $V_{out} = \frac{1}{2}(V_0 + V_1)$. By inspection, increasing $|V_{T0L}|$ will reduce the charging time.

The overall design process for the depletion load nMOS inverter is similar to that described for the previous load types. The design specification for V_{OL} is first used to compute the required value for $\beta_R = \beta_D/\beta_L$. The value of β_L used in the circuit will set t_{LH}, so eqn. (4.2-23) should be checked for various $(W/L)_L$ values. To ensure a minimum area configuration with the smallest inverter input capacitance, the smallest $(W/L)_L$ that satisfies the switching time requirements should be used. The driver aspect ratio $(W/L)_D$ is then determined by this choice. It is again seen that C_{out} should be minimized to minimize the charging interval.

It is possible to define a load charge time constant τ_L for the depletion mode transistor by writing

$$\tau_L = \frac{C_{out}}{\beta_L |V_{TL}|}. \tag{4.2-24}$$

This is particularly useful since it contains the important device parameters β_L and $|V_{TL}|$ and allows for first estimate comparisons of t_{LH} on a circuit-to-circuit basis. Also, the presence of a parasitic interconnect line resistance R_{line} can be included in the rise time calculation by approximating

$$\tau_L \simeq \left(\frac{1}{\beta_L |V_{TL}|} + R_{line} \right) C_{out}, \tag{4.2-25}$$

which is analogous to eqn. (4.1-20) of the previous section for the discharge analysis.

This section has been concerned with computing the various rise time expressions for t_{LH}. To appreciate the overall meaning of the rise and fall times, consider an inverter in which V_{out} switches from V_{OL} (logic 0) to V_{OH}

(logic 1) and then back to V_{OL} again. The minimum time interval required for this response is $(t_{HL} + t_{LH})$. Consequently, the *maximum switching frequency* f_{max} of the circuit is given by

$$f_{max} = \frac{1}{t_{HL} + t_{LH}}. \tag{4.2-26}$$

If the circuit is operated at a switching frequency $f > f_{max}$, then the output voltage will not have sufficient time to stabilize to coherent logic state values. This must be avoided, since it may lead to a false interpretation of the data by the next stage.

EXAMPLE 4.2-1

A depletion load inverter has a load MOSFET with

$$V_{T0L} = -3.5\,[V], \quad \gamma_L = 0.40\,[V^{1/2}], \quad 2\,|\phi_{F,L}| = 0.60\,[V],$$
$$k'_L = 25 \times 10^{-6}\,[A/V^2], \quad (W/L)_L = 0.65.$$

It is used with the driver described in Example 4.1-1 so that $V_{OL} = 0.25\,[V]$. Calculate t_{LH} for $C_{out} = 100\,[fF]$.

Solution
First find the 10% and 90% voltages

$$V_0 = V_{OL} + 0.1V_\ell = 0.725\,[V],$$
$$V_1 = V_{OL} + 0.9V_\ell = 4.525\,[V].$$

To include some body bias effects, calculate V_{TL} at

$$V_{out} = \frac{1}{2}(V_0 + V_1) = 2.625\,[V]$$

so that

$$V_{TL} = -3.5 + 0.4(\sqrt{2.625 + 0.6} - \sqrt{0.6}) \simeq -3.09\,[V].$$

The load time constant is

$$\tau_L = \frac{(100 \times 10^{-15})}{(25 \times 10^{-6})(0.65)(3.09)} \simeq 1.99\,[ns].$$

Then eqn. (4.2-23) gives

$$t_{LH} \simeq (1.99)\left[\frac{2(5 - 3.09 - 0.725)}{3.09} + \ln\left(\frac{2(3.09) - (5 - 4.525)}{(5 - 4.525)}\right)\right],$$

so $t_{LH} \simeq 5.71\,[ns]$.   ■

EXAMPLE 4.2-2

This example uses SPICE to show the difference in transient response using a depletion MOSFET load and a saturated E-mode load. Both circuits were simulated using $C_{out} = 30$ [fF]. The device parameters can be extracted from the SPICE listings.

(a) Depletion load:

```
DMODE INVERTER TRANSIENT ANALYSIS
VDD 3 0 DC 5VOLTS
MDRIVER 2 1 0 0 EMODE L=5U W=20U
MLOAD 3 2 2 0 DMODE L=10U W=5U
VS 1 0 DC PULSE (0 5 2NS IPS 1PS 5NS)
CLOAD 2 0 3.0E-14
.MODEL EMODE NMOS VTO=1 GAMMA=0.37 KP=2.5E-5
.MODEL DMODE NMOS VTO=-3 GAMMA=0.37 KP=2.5E-5
.TRAN 1NS 25NS
.PLOT TRAN V(2)
.END
```

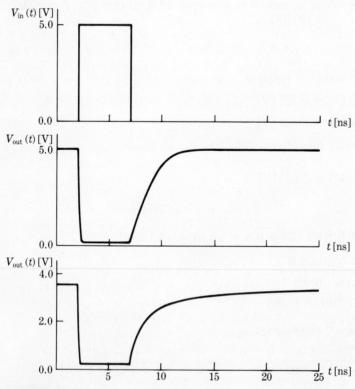

FIGURE E4.1

(b) Saturated E-mode load:

```
SATURATED EMODE INVERTER TRANSIENT ANALYSIS BR=10
VDD 3 0 DC 5VOLTS
MDRIVER 2 1 0 0 EMODE L=5U W=25U
MLOAD 3 3 2 0 EMODE L=10U W=5U
VS 1 0 DC PULSE (0 5 2NS 1PS 1PS 5NS)
CLOAD 2 0 3.0E-14
.MODEL EMODE NMOS VTO=1 GAMMA=0.37 KP=2.5E-5
.TRAN 1NS 25NS
.PLOT TRAN V(2)
.END
```

The plots in Fig. E4.1 show the difference between the rise times t_{LH} for the two circuits. (Note that $\beta_R = 8$ for the depletion load circuit, while $\beta_R = 10$ for the saturated enhancement load.)   ∎

4.3   nMOS Propagation Delay Times

The propagation delay times t_{PHL} and t_{PLH} are important for determining the overall signal delay introduced by the finite time intervals required to respectively discharge and charge the output capacitance of the inverter. As shown in Fig. 4.9, t_{PHL} is defined as the time required for the output voltage to fall from V_{OH} to $V_{1/2} = \frac{1}{2}(V_{OL} + V_{OH})$. Similarly, t_{PLH} is the rise time interval needed to charge C_{out} from V_{OL} to $V_{1/2}$. Once these two quantities are known for a given circuit configuration, the propagation (delay) time is given by

$$t_p = \frac{1}{2}(t_{PHL} + t_{PLH}), \tag{4.3-1}$$

i.e., t_p is the simple average of the two.

The average delay time t_p is commonly used to characterize the *ring oscillator circuit* shown in Fig. 4.10. A ring oscillator is constructed by cascading an odd number of identical inverter stages to induce oscillations, which may be measured at any input or output node. The oscillatory behavior of the circuit is accomplished by the feedback provided in connecting the output of the final stage to the input of the first stage. With the conventions adopted here, the output of a given stage is viewed as initiating a change in logic states as V_{out} crosses the 50% point $V_{1/2}$. A more precise definition of t_p would set this point at the inverter threshold voltage V_{th}. However, since V_{th} varies with the specific design of the inverter, it is easier to use $V_{1/2}$ as a reasonable approximation. At any rate, t_p gives an indication of the time required for the signal to propagate through a single stage in the circuit. Note that the input voltage waveform to a given stage in the oscillator will not have the step characteristics shown in Fig. 4.9(a); rather, the cascaded arrangement

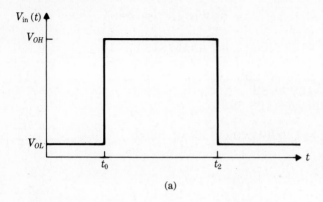

(a)

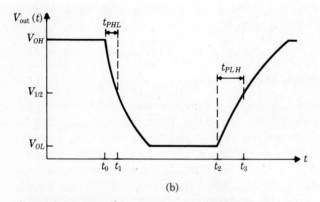

(b)

FIGURE 4.9 Propagation delay time definitions.

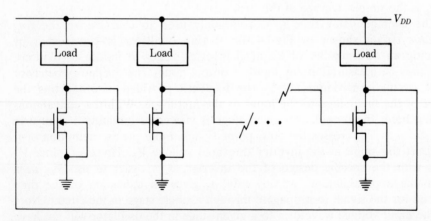

FIGURE 4.10 Ring oscillator with *n* (odd) stages.

will yield an input voltage that is of the form shown for the *output* of a single inverter in Fig. 4.9(b). It is, however, much simpler to use the step input voltage approximation in the analytic study.

The ring oscillator also demonstrates the reasoning behind defining the input VTC critical voltages V_{IL} and V_{IH} as the points where $(dV_{out}/dV_{in}) = -1$. Since V_{out} and V_{in} are equally scaled (both range linearly from $0\,[V]$ to V_{DD}), these slope points allow for a precise borderline between stable input logic levels and the inverter transition region. The operation of the ring oscillator depends on the existence of such critical voltage points. To understand this point, note that if V_{in} is slightly less than V_{IL}, then the input will stabilize into a logic 0 state, while V_{in}, slightly greater than V_{IL} implies that the system will be unstable and tend to fall toward the logic 1 state. This instability sustains oscillations in the circuit.

Consider first the high-to-low propagation time t_{PHL}. By definition, this delay occurs during the capacitor discharge when the driver is in the active region conducting current to ground. The basic analysis in Section 4.1 is therefore valid, so t_{PHL} can be determined by

$$t_{PHL} = -C_{out} \int_{V_{OH}}^{V_{1/2}} \frac{dV_{out}}{I_D(V_{out})}, \tag{4.3-2}$$

which is just eqn. (4.1-6) rewritten with the different upper limit. Assuming that the driver will operate in both the saturated and nonsaturated regions during the discharge, the integral splits into two terms:

$$t_{PHL} = -C_{out} \int_{V_{OH}}^{V_{OH}-V_{TD}} \frac{dV_{out}}{\dfrac{\beta_D}{2}(V_{OH} - V_{TD})^2}$$

$$- C_{out} \int_{V_{OH}-V_{TD}}^{V_{1/2}} \frac{dV_{out}}{\dfrac{\beta_D}{2}[2(V_{OH} - V_{TD})V_{out} - V_{out}^2]}. \tag{4.3-3}$$

The first term describes the discharge during the time interval when the driver is saturated and is identical to the integral found in Section 4.1. The second term arises from the driver going into the nonsaturated region, which assumes that $V_{1/2} < (V_{OH} - V_{TD})$. The indefinite integral given in eqn. (4.1-11) is again valid, where it is noted that the upper limit of integration is different from that encountered when computing t_{HL}. Performing the integrations gives

$$t_{PHL} = \tau_D \left\{ \frac{2V_{TD}}{(V_{OH} - V_{TD})} + \ln\left[\frac{4(V_{OH} - V_{TD})}{(V_{OH} + V_{OL})} - 1 \right] \right\}, \tag{4.3-4}$$

where the driver discharge time constant τ_D is again given by

$$\tau_D = \frac{C_{out}}{\beta_D(V_{OH} - V_{TD})}. \tag{4.3-5}$$

The low-to-high propagation time t_{PLH} is shown in Fig. 4.9(b) and corresponds to the output capacitance charging from V_{OL} to $V_{1/2}$. Since the rate of charging is dependent on the load device characteristics, a separate expression must be found for each of the possible load configurations, as was done for t_{LH} in the previous section. The integrals are basically the same with the generic form

$$t_{PLH} = C_{\text{out}} \int_{V_{OL}}^{V_{1/2}} \frac{dV_{\text{out}}}{I_L(V_{\text{out}})}. \tag{4.3-6}$$

This is simply eqn. (4.2-2) with the new limits of integration.

RESISTOR LOAD
For the case where a resistor R_L is used as a load, the integral is

$$t_{PLH} = R_L C_{\text{out}} \int_{V_{OL}}^{V_{1/2}} \frac{dV_{\text{out}}}{(V_{DD} - V_{\text{out}})}. \tag{4.3-7}$$

Integrating gives

$$t_{PLH} = R_L C_{\text{out}} \ln\left(\frac{V_{DD} - V_{OL}}{V_{DD} - V_{1/2}}\right). \tag{4.3-8}$$

As a lowest-order approximation for this case, suppose that the circuit is designed such that $V_{OL} \ll V_{OH} = V_{DD}$. Then $(V_{DD} - V_{1/2}) = \frac{1}{2}V_{DD}$ and

$$t_{PLH} \simeq R_L C_{\text{out}} \ln(2), \tag{4.3-9}$$

where it is noted that $\ln(2) \simeq 0.693$.

SATURATED ENHANCEMENT LOAD
The load current is now given by

$$I_L(V_{\text{out}}) = \frac{\beta_L}{2}[V_{DD} - V_{\text{out}} - V_{TL}(V_{\text{out}})]^2. \tag{4.3-10}$$

Ignoring body bias effects by assuming V_{TL} to be a constant, the integration in eqn. (4.3-6) yields

$$t_{PLH} = \frac{2C_{\text{out}}}{\beta_L}\left(\frac{1}{V_{DD} - V_{TL} - V_{1/2}} - \frac{1}{V_{DD} - V_{TL} - V_{OL}}\right). \tag{4.3-11}$$

For this case, the average body bias can be included by evaluating $V_{TL}(V_{\text{out}})$ at $\frac{1}{4}(V_{OH} + V_{OL})$, i.e., at the midpoint voltage of the propagation time charge interval.

NONSATURATED ENHANCEMENT LOAD
When the load gate is at a voltage $V_{GG} > [V_{DD} + V_{TL}(V_{DD})]$, the load MOSFET is nonsaturated. In terms of the biasing parameter m defined in eqn.

(4.2-11), the equation for computing t_{PLH} becomes

$$t_{PLH} = \frac{2mC_{\text{out}}}{\beta_L V_{DD}} \int_{V_{OL}/V_{DD}}^{V_{1/2}/V_{DD}} \frac{d(V_{\text{out}}/V_{DD})}{\left(1 - \dfrac{V_{\text{out}}}{V_{DD}}\right)\left(1 - m\dfrac{V_{\text{out}}}{V_{DD}}\right)}. \tag{4.3-12}$$

Using the indefinite integral in eqn. (4.2-13) gives

$$t_{PLH} = \frac{2C_{\text{out}}}{\beta_L V_{DD}} \left(\frac{m}{1-m}\right) \ln\left\{ \frac{\left(1 - m\dfrac{V_{1/2}}{V_{DD}}\right)\left(1 - \dfrac{V_{OL}}{V_{DD}}\right)}{\left(1 - m\dfrac{V_{OL}}{V_{DD}}\right)\left(1 - \dfrac{V_{1/2}}{V_{DD}}\right)} \right\}. \tag{4.3-13}$$

The body bias can be included in an average manner by computing V_{TL} at $V_{\text{out}} = \frac{1}{4}(V_{OH} + V_{OL})$, with the worst-case situation occurring at $V_{\text{out}} = V_{1/2}$.

DEPLETION LOAD

The last case that needs to be considered is that where a depletion MOSFET is used as the load. The high-to-low propagation time is given by

$$t_{PLH} = C_{\text{out}} \int_{V_{OL}}^{V_{DD}-|V_{TL}|} \frac{dV_{\text{out}}}{I_{L(\text{sat})}} + C_{\text{out}} \int_{V_{DD}-|V_{TL}|}^{V_{1/2}} \frac{dV_{\text{out}}}{I_{L(\text{nonsat})}}, \tag{4.3-14}$$

where is has been assumed that $V_{1/2} > (V_{DD} - |V_{TL}|)$; this is generally true, since $|V_{TOL}|$ is usually chosen to be a significant fraction (e.g., greater than 60%) of the power supply V_{DD} level. Except for the change in the limits, the integrals are identical to those encountered in computing t_{LH}. Thus,

$$t_{PLH} = \frac{C_{\text{out}}}{\beta_L |V_{TL}|} \left\{ \frac{2(V_{DD} - |V_{TL}| - V_{OL})}{|V_{TL}|} + \ln\left[\frac{2|V_{TL}| - (V_{DD} - V_{1/2})}{(V_{DD} - V_{1/2})}\right] \right\} \tag{4.3-15}$$

gives the propagation delay for the high-to-low transition when a depletion-mode MOSFET is used as a load.

The results in this section allow for the direct calculation of the propagation delay time t_p experienced by a signal passing through the gate. Minimizing C_{out} gives the shortest delay times. Also, the interplay between the DC design and the resulting transient characteristics is always evident in static nMOS. The effects of parasitic line resistance can be included in t_{PHL} and t_{PLH} by modifying the driver and load time constants τ_D and τ_L as discussed earlier.

EXAMPLE 4.3-1

Calculate the propagation delay time t_p for the inverter circuit described in Example 4.1-1 (driver) and Example 4.2-1 (load).

Solution
First calculate

$$V_{1/2} = V_{OL} + 0.5V_\ell = 2.625\,[\text{V}].$$

Equation (4.3-4) gives the discharge propagation delay as

$$t_{PHL} \simeq (0.25)\left[\frac{2}{4} + \ln\left(\frac{4(5-1)}{5.25} - 1\right)\right] \simeq 0.304\,[\text{ns}].$$

The charging time for logic propagation is computed using eqn. (5.3-15):

$$t_{PLH} \simeq (1.99)\left\{\frac{2(5 - 3.09 - 0.25)}{3.09} + \ln\left[\frac{2(3.09) - (5 - 2.625)}{(5 - 2.625)}\right]\right\}$$

$$\simeq 2.007\,[\text{ns}].$$

Then

$$t_p \simeq \frac{1}{2}(0.304 + 2.007) = 1.16\,[\text{ns}]. \qquad\blacksquare$$

EXAMPLE 4.3-2

The step input voltage is used for analytic simplicity. This example shows the SPICE simulations for (a) a step voltage and (b) $V_{in}(t)$ with finite rise and fall times.

The plots in Figs. E4.2 and E4.3 illustrate that the step input voltage approximation is surprisingly accurate, giving t_{HL} and t_{LH} values on the same order of magnitude.

(a) Step input voltage:

```
DMODE INVERTER TRANSIENT ANALYSIS
VDD 3 0 DC 5VOLTS
MDRIVER 2 1 0 0 EMODE L=5U W=20U
MLOAD 3 2 2 0 DMODE L=10U W=5U
VS 1 0 DC PULSE (0 5 2NS 1PS 1PS 5NS)
CLOAD 2 0 3.0E-14
.MODEL EMODE NMOS VTO=1 GAMMA=0.37 KP=2.5E-5
.MODEL DMODE NMOS VTO=-3 GAMMA=0.37 KP=2.5E-5
.TRAN 1NS 20NS
.PLOT TRAN V(2)
.END
```

(b) Finite rise and fall times:

```
DMODE INVERTER TRANSIENT ANALYSIS
VDD 3 0 DC 5VOLTS
```

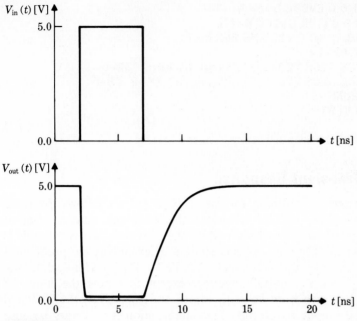

FIGURE E4.2

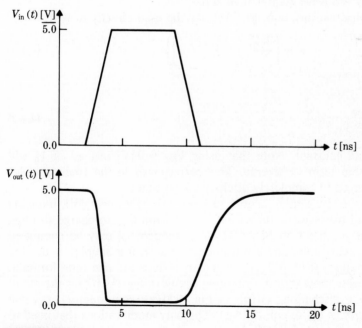

FIGURE E4.3

```
MDRIVER 2 1 0 0 EMODE L=5U W=20U
MLOAD 3 2 2 0 DMODE L=10U W=5U
VS 1 0 DC PULSE ( 0 5 2NS 2NS 2NS 5NS )
CLOAD 2 0 3.0E-14
.MODEL EMODE NMOS VTO=1 GAMMA=0.37 KP=2.5E-5
.MODEL DMODE NMOS VTO=-3 GAMMA=0.37 KP=2.5E-5
.TRAN 1NS 20NS
.PLOT TRAN V(2)
.END                                                            ■
```

4.4 CMOS Transient Response

Calculating the transient response for the CMOS inverter follows the same lines as for the nMOS circuits. Consider the discharge time t_{HL}. Since $V_{in} = V_{OH}$, the nMOS transistor is on while the pMOS device is off; this is shown in Fig. 4.11(a). The time t_{HL} is computed by examining the problem of discharging C_{out} through the n-channel MOSFET. This is identical to the calculation of t_{HL} for the nMOS inverter, except that now the limits of integration must be taken from V_1 to V_0, which are respectively the 90% and 10% points of the output waveforms. This modification is necessary because the CMOS inverter has $V_{OH} = V_{DD}$ and $V_{OL} = 0\,[\text{V}]$, i.e., the logic swing of the inverter is the full power supply range. This is to be compared with nMOS in which $V_\ell < V_{DD}$ always holds because of the ratioed nature of the circuits; in particular, V_{OL} is always greater than zero.

With these observations, eqn. (4.1-13) may be used directly to write

$$t_{HL} = \tau_n \left\{ \frac{2V_{Tn}}{(V_1 - V_{Tn})} + \ln \left[\frac{2(V_1 - V_{Tn})}{V_0} - 1 \right] \right\}, \tag{4.4-1}$$

in which

$$\tau_n = \frac{C_{out}}{\beta_n(V_1 - V_{Tn})} \tag{4.4-2}$$

is the nMOS time constant. Note that using $V_{OL} = 0\,[\text{V}]$ instead of V_0 will cause the logarithm term to diverge. This corresponds to the (theoretically) infinite time required to completely discharge a capacitor.

To compute the rise time t_{LH}, note that now $V_{in} = V_{OL} = 0\,[\text{V}]$, implying that the n-channel transistor is turned off. This allows C_{out} to charge through the pMOS device as shown in Fig. 4.11(b). Although t_{LH} may be computed directly from the generic integral given in eqn. (4.2-2), it is simpler to use the complementary nature of the CMOS circuit to deduce the rise time formula. To effect this process, note that the charging circuit in Fig. 4.11(b) is the direct complement of the discharging circuit in Fig. 4.11(a); consequently, t_{LH} will have the same *form* as t_{HL} in eqn. (4.4-1). The only modifications that need to be made are to replace the nMOS device parameters with pMOS quantities,

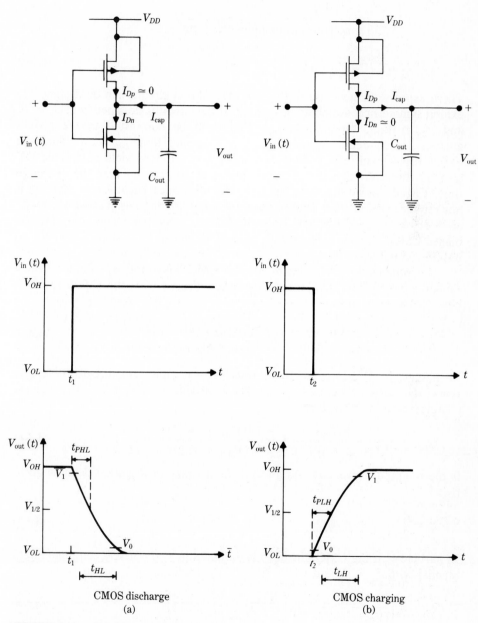

FIGURE 4.11 CMOS transient analysis

i.e., $V_{Tn} \to |V_{Tp}|$ and $\beta_n \to \beta_p$. Thus,

$$t_{LH} = \tau_p \left\{ \frac{2|V_{Tp}|}{(V_1 - |V_{Tp}|)} + \ln \left[\frac{2(V_1 - |V_{Tp}|)}{V_0} - 1 \right] \right\}, \qquad (4.4\text{-}3)$$

where the pMOS time constant assumes the form

$$\tau_p = \frac{C_{\text{out}}}{\beta_p(V_1 - |V_{Tp}|)}. \qquad (4.4\text{-}4)$$

These expressions give rise to the fact that CMOS circuits can be designed to exhibit symmetrical switching properties. If the circuit is designed with $\beta_n = \beta_p$ and $V_{Tn} = |V_{Tp}|$, then $t_{LH} = t_{HL}$. This is to be contrasted to the nMOS analysis, which demonstrated that $\beta_R > 1$ implies that $t_{LH} > t_{HL}$. Since an nMOS driver-load ratio greater than unity is required to yield acceptable DC transfer characteristics, the charging time is always greater than the discharge time. CMOS, however, does not have this drawback because the inverter is not ratioed. (Recall that CMOS does not allow for the designation of "load" and "driver" since the MOSFETs complement each other.) This consideration makes CMOS system timing easier to deal with than equivalent designs in an nMOS technology.

The propagation delay times for the CMOS inverter are computed in a similar manner. First, t_{PHL} may be obtained directly from the nMOS result in eqn. (4.3-4). This gives

$$t_{PHL} = \tau_n \left\{ \frac{2V_{Tn}}{(V_{OH} - V_{Tn})} + \ln \left[\frac{4(V_{OH} - V_{Tn})}{(V_{OH} + V_{OL})} - 1 \right] \right\}. \qquad (4.4\text{-}5)$$

To find t_{PLH}, all that needs to be done is to complement this expression to reflect the fact that the pMOS transistor is providing the path for the capacitor charging current. Therefore,

$$t_{PLH} = \tau_p \left\{ \frac{2|V_{Tp}|}{(V_{OH} - |V_{Tp}|)} + \ln \left[\frac{4(V_{OH} - |V_{Tp}|)}{(V_{OH} + V_{OL})} - 1 \right] \right\} \qquad (4.4\text{-}6)$$

is the required expression. As always, the average propagation delay is given by

$$t_p = \frac{1}{2}(t_{PHL} + t_{PLH}). \qquad (4.4\text{-}7)$$

In particular, if the CMOS inverter is designed in a completely complementary manner with $\beta_n = \beta_p$ and $V_{Tn} = |V_{Tp}| = V_T$, then $t_{PHL} = t_{PLH}$, so

$$t_p = \tau_n \left\{ \frac{2V_T}{(V_{DD} - V_T)} + \ln \left[\frac{4(V_{DD} - V_T)}{V_{DD}} - 1 \right] \right\}, \qquad (4.4\text{-}8)$$

where the output levels of $V_{OH} = V_{DD}$ and $V_{OL} = 0\,[\text{V}]$ have been used in arriving at the final expression.

Some general observations concerning the CMOS transient response can now be made. First, note that the rise and fall times are proportional to C_{out}. Although this general behavior is the same as that encountered in nMOS circuits, more capacitance contributions exist in a CMOS-based system design. To understand this comment, note that the input to a CMOS inverter is connected to the gates of both the nMOS and pMOS transistors in a parallel manner. If two CMOS inverters are cascaded, then the output capacitance seen by the first stage will include the sum of the capacitance contributions from both MOSFETs which form the input to the second stage. In an nMOS system, only the driver gate capacitance is (directly) seen at the input node. Consequently, nMOS designs tend to exhibit faster switching speeds than equivalent CMOS systems, assuming all else equal. The details of the analysis are covered later in this chapter.

Finally, note that the analysis can be extended to include the presence of interconnect line resistance R_{line} by a straightforward approximation. Following the nMOS treatment, the extended rise and fall times can be computed by modifying the nMOS and pMOS time constants to

$$\tau_n \simeq \left[\frac{1}{\beta_n(V_1 - V_{Tn})} + R_{line} \right] C_{out} \qquad (4.4\text{-}9)$$

and

$$\tau_p \simeq \left[\frac{1}{\beta_p(V_1 - |V_{Tp}|)} + R_{line} \right] C_{out}, \qquad (4.4\text{-}10)$$

respectively. These may then be used in the appropriate equations to estimate the increased time intervals.

EXAMPLE 4.4-1

A symmetric CMOS inverter has

$$\beta_n = \beta_p = 25\,[\mu A/V^2], \quad V_{T0n} = 1\,[V] = |V_{T0p}|.$$

The power supply voltage is $V_{DD} = 5\,[V]$, and the output capacitance is estimated to be $C_{out} = 30\,[fF]$. Compute $t_{LH} = t_{HL}$.

Solution
First calculate the time constant τ_n:

$$\tau_n = \frac{(30 \times 10^{-15})}{(25 \times 10^{-6})(3.5)} \simeq 0.343\,[ns],$$

where $V_1 = 0.9V_{DD} = 4.5\,[V]$ has been used. Then, since $V_0 = 0.5\,[V]$, eqn. (4.4-1) gives

$$t_{HL} \simeq 0.343 \left\{ \frac{2(1)}{3.5} + \ln\left[\frac{2(3.5)}{0.5} - 1 \right] \right\},$$

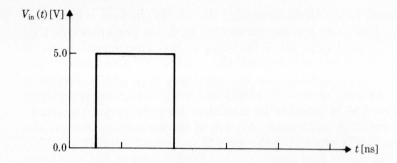

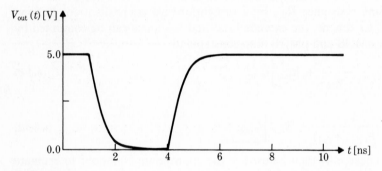

FIGURE E4.4

so

$$t_{HL} = t_{LH} \simeq 1.08 \,[\text{ns}].$$

A SPICE simulation of the circuit is obtained using the following input listing. The resulting transient response is shown in Fig. E4.4.

```
SYMMETRIC CMOS INVERTER
VDD 3 0 DC 5VOLTS
MNF 2 1 0 0 MN L=5U W=5U
MPF 2 1 3 3 MP L=5U W=12.5U
VS 1 0 DC PULSE ( 0 5 1NS 1PS 1PS 3NS )
CLOAD 2 0 3.0E-14
.MODEL MN NMOS VTO=1 GAMMA=0.37 KP=2.5E-5
.MODEL MP PMOS VTO=-1 GAMMA=0.4 KP=1.0E-5
.TRAN 1NS 10NS
.PLOT TRAN V(2)
.END
```

■

EXAMPLE 4.4-2

Calculate the propagation time t_p for the circuit in Example 4.4-1.

Solution
Using eqn. (4.4-5),

$$t_{PHL} \simeq (0.343)\left[\frac{2(1)}{4} + \ln(3.2 - 1)\right] \simeq 0.442\,[\text{ns}],$$

so by symmetry,

$$t_{PHL} = t_{PLH}$$
and
$$t_p \simeq 0.442\,[\text{ns}]. \qquad\blacksquare$$

EXAMPLE 4.4-3

Again consider the circuit described in Ex. 4.4-1, but now assume that $\beta_n = 25\,[\mu\text{A}/\text{V}^2]$ and $\beta_p = 10\,[\mu\text{A}/\text{V}^2]$. Calculate t_{HL} and t_{LH} for this case.

Solution
Since β_n is unchanged, $t_{HL} \simeq 1.08\,[\text{ns}]$ is still valid. Charging C_{out} is now described by the time constant

$$\tau_p = \frac{(30 \times 10^{-15})}{(10 \times 10^{-6})(3.5)} = 2.5\tau_n \simeq 0.857\,[\text{ns}],$$

which gives

$$t_{LH} \simeq 2.69\,[\text{ns}]$$

by substituting into eqn. (4.4-3).

The nonsymmetric switching is explicitly shown in the following SPICE file. Figure E4.5 plots $V_{\text{in}}(t)$ and $V_{\text{out}}(t)$, allowing for determination of t_{HL} and t_{LH}.

```
CMOS INVERTER
VDD 3 0 DC 5VOLTS
MNF 2 1 0 0 MN L=5U W=5U
MPF 3 1 2 3 MP L=5U W=5U
VS 1 0 DC PULSE (0 5 1NS 1PS 1PS 3NS)
CLOAD 2 0 3.0E-14
.MODEL MN NMOS VTO=1 GAMMA=0.37 KP=2.5E-5
.MODEL MP PMOS VTO=-1 GAMMA=0.4 KP=1.0E-5
.TRAN 1NS 10NS
.PLOT TRAN V(2)
.END
```
$\qquad\blacksquare$

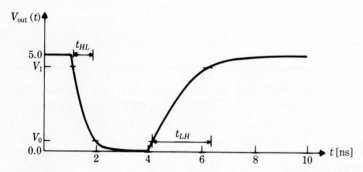

FIGURE E4.5

4.5 The Power-Delay Product

The *power-delay product* (PDP) is used to characterize the overall performance of a digital gate circuit. It is given by

$$PDP = P_{av}t_p, \tag{4.5-1}$$

where P_{av} is the average power dissipated by the gate and t_p is the average propagation delay time. The units of the PDP are [watts][sec] = [J] (joules), so it represents an average energy. Typically, MOS-based digital gates display power-delay products on the order of a few picojoules [pJ]. A PDP can be computed for any switching circuit. It is commonly used to compare the performance of various logic families or processing technologies. A small PDP value is desirable, as this implies both lower power consumption and fast switching speeds.

To compute the power-delay product, the time-averaged power P_{av} consumed by the gate must be known. This in turn requires that the input waveform $V_{in}(t)$ be specified. For simplicity, $V_{in}(t)$ is often taken as a square wave with period T [sec], as illustrated in Fig. 4.12(a); a more realistic input waveform that displays the finite rise and fall times is shown in Fig. 4.12(b).

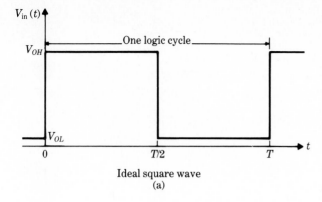

Ideal square wave
(a)

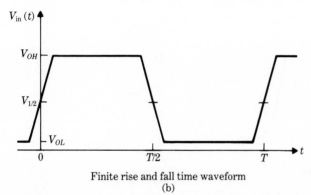

Finite rise and fall time waveform
(b)

FIGURE 4.12 Input voltage waveforms for the power-delay products.

Both conventions will be studied in the discussion here, but it should be remembered that the PDP is dependent on the actual waveform measured at the input node of a gate.

As a first step toward understanding the meaning of the PDP, suppose that an ideal square wave $V_{in}(t)$ (Fig. 4.12a) is applied to the resistively load nMOS inverter shown in Fig. 4.13(a); the output voltage $V_{out}(t)$ then assumes the form drawn in Fig. 4.13(b). To find the power-delay product for this situation, both t_p and P_{av} must be computed. The average propagation delay time t_p may be found by using the techniques in the previous sections of this chapter. However, to illustrate the *general* dependence of the PDP on the circuit quantities, the analysis here will initially use the simplified forms

$$t_{PHL} \simeq \tau_D = R_{on}C_{out}, \qquad t_{PLH} \simeq \tau_L = R_L C_{out}, \qquad (4.5\text{-}2)$$

where R_{on} is the on-resistance of the driver; note that $R_{on} = R_{DS}$. With these approximations, the average propagation delay is

$$t_p \simeq \frac{1}{2}(R_{on} + R_L)C_{out}. \qquad (4.5\text{-}3)$$

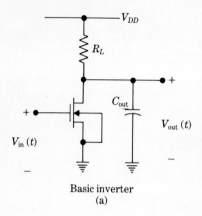

Basic inverter
(a)

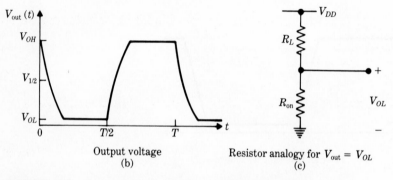

Output voltage
(b)

Resistor analogy for $V_{\text{out}} = V_{OL}$
(c)

FIGURE 4.13   Power-delay product in a resistively loaded inverter.

Consider now the power dissipated by the circuit. This is given by

$$P_{\text{av}} = I_{\text{av}} V_{DD}, \tag{4.5-4}$$

where I_{av} is the average power supply current. To compute I_{av}, the current flow is separated into two contributions: the constant (DC) current flow when the output is stable with $V_{\text{out}} = V_{OL}$ and the transient current that flows during the rise and fall times. The DC current is computed with the aid of the resistor analogy shown in Fig. 4.13(c). Using Ohm's law, the average DC power dissipation during the period T is given by

$$P_{\text{av}} \simeq \frac{V_{DD}^2}{2(R_{\text{on}} + R_L)}, \tag{4.5-5}$$

where the factor of $\frac{1}{2}$ arises because this power is dissipated during (approximately) half of the cycle, i.e., when $V_{\text{out}} = V_{OL}$. Now, combining eqns. (4.5-3) and (4.5-5), the PDP that results from the constant DC current flow only is given by

$$(\text{PDP})_{DC} \simeq \frac{1}{4} C_{\text{out}} V_{DD}^2. \tag{4.5-6}$$

Minimization of the power-delay product can be accomplished by reducing the output capacitance C_{out} or by using a lower power supply voltage. Although this constitutes at best a crude approximation for the PDP, it is often used in the literature. Note that the expression is just one-half of the energy stored in the output capacitor with a voltage V_{DD}; the calculation thus correlates with what is expected qualitatively. The important point to remember is that this contribution to the PDP is proportional to $C_{\text{out}}V_{DD}$.

The total power-delay product for the circuit must also account for the average power consumed by the gate during the rise and fall time intervals. Consider first the charging current supplied by V_{DD} during the rise time t_{LH}. Since the driver is in cutoff, this can be estimated by

$$I_{\text{av}} \simeq C_{\text{out}}\frac{(\Delta V)}{(\Delta t)} = C_{\text{out}}\frac{V_\ell}{t_{LH}}, \tag{4.5-7}$$

with $V_\ell = V_{DD}$ being the logic swing. The resulting PDP contribution due to this current is then

$$(\text{PDP})_{LH} \simeq C_{\text{out}}V_{DD}^2\frac{t_p}{t_{LH}}. \tag{4.5-8}$$

The power supply current used by the inverter during the discharge time t_{HL} is approximated by

$$I_{\text{av}} \simeq \frac{1}{2}(I_{\text{initial}} + I_{\text{final}}), \tag{4.5-9}$$

where

$$I_{\text{initial}} = \frac{1}{R_L}(V_{DD} - V_{OH}), \qquad I_{\text{final}} = \frac{1}{R_L}(V_{DD} - V_{OL}), \tag{4.5-10}$$

respectively, give the current at the beginning and end of the discharging event. Thus, assuming $V_{OL} \ll V_{OH} = V_{DD}$,

$$I_{\text{av}} \simeq \frac{V_{DD}}{2R_L}. \tag{4.5-11}$$

Now, noting that $t_{PHL} \simeq \tau_D$, a first-order estimate for the rise time t_{HL} is

$$t_{HL} \simeq 2\tau_D = 2R_{\text{on}}C_{\text{out}}. \tag{4.5-12}$$

Forming the power-delay product for this time interval gives the term

$$(\text{PDP})_{HL} \simeq C_{\text{out}}V_{DD}^2\frac{R_{\text{on}}}{R_L}\frac{t_p}{t_{HL}}. \tag{4.5-13}$$

The complete expression for the PDP is obtained by summing all contributions:

$$\text{PDP} \simeq C_{\text{out}}V_{DD}^2\left(\frac{1}{4} + \frac{t_p}{t_{LH}} + \frac{R_{\text{on}}}{R_L}\frac{t_p}{t_{HL}}\right). \tag{4.5-14}$$

This can be simplified by noting that $R_{on} \ll R_L$ will be valid in a well-designed inverter. The propagation delay time is then $t_p \simeq (\tau_L/2)$. Using this in conjunction with the approximations $t_{LH} \simeq 2\tau_L$ and $t_{HL} \simeq 2\tau_D$ gives

$$\text{PDP} \simeq \frac{3}{4} C_{out} V_{DD}^2 \qquad (4.5\text{-}15)$$

as the lowest-order approximation for the total PDP. Again note that the power-delay product is proportional to $C_{out} V_{DD}$.

Although the above calculations were greatly simplified, they do show the general dependence of the PDP in a realistic nMOS circuit. Equation (4.5-15) may be used as a first estimate of the actual PDP value; for example, if $C_{out} = 0.5 \, [\text{pF}]$ and $V_{DD} = 5 \, [\text{V}]$, then this equation predicts that PDP $\simeq 9.4 \, [\text{pJ}]$. Since the average energy required per logic cycle is proportional to the output capacitance, minimizing C_{out} is again seen as a primary design consideration.

Now that the general properties of the power-delay product have been extracted, attention will be directed toward obtaining more precise calculational tools for PDP determinations. In particular, the depletion load nMOS inverter and the CMOS inverter power-delay products will be examined in detail. Since the average propagation delay time expressions have already been derived for both circuits, t_p is assumed known at this point. Only the average power dissipation P_{av} remains to be found. With $V_{in}(t)$ as illustrated in Fig. 4.12(b) (which shows explicitly the finite slope during the input transitions), the power supply current as a function of time may be plotted for both circuits as shown in Fig. 4.14. It has been assumed that the logic cycle period T is much greater than $(t_{LH} + t_{HL})$. Figure 4.14(b) demonstrates that an nMOS inverter draws significant current when $V_{out} = V_{OL}$, in addition to the transient current required when the input is switched. This is to be contrasted with the CMOS current waveform in Fig. 4.14(c), which shows significant current flow only during the transition times.

Consider first the nMOS inverter, which employs a D-mode MOSFET as a load. The average DC power dissipated in the circuit during a single T-cycle is

$$(P_{av})_{DC} \simeq \frac{1}{2} I_{max} V_{DD} \qquad (4.5\text{-}16)$$

where I_{max} is the maximum power supply current required when $V_{out} = V_{OL}$; the factor of $\frac{1}{2}$ again arises from the assumption that the inverter is in this state during approximately half the logic cycle. Explicitly, this power can be computed by noting that $I_{max} = I_D(V_{out} = V_{OL})$, so

$$(P_{av})_{DC} \simeq \frac{\beta_D}{4} [2(V_{OH} - V_{TD})V_{OL} - V_{OL}^2]V_{DD}$$

$$\simeq \frac{\beta_L}{4} [V_{TL}(V_{OL})]^2 V_{DD} \qquad (4.5\text{-}17)$$

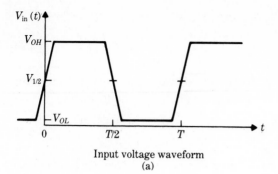

Input voltage waveform
(a)

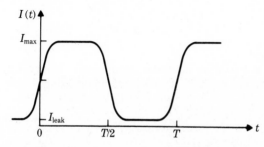

Power supply current for an nMOS inverter
(b)

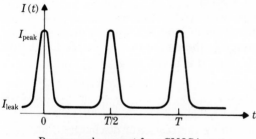

Power supply current for a CMOS inverter
(c)

FIGURE 4.14   Current waveforms for the power-delay product calculations.

since the driver is nonsaturated while the load is saturated in this logic state.

The current supplied to the nMOS inverter during the charging time t_{LH} is computed by

$$(I_{av})_{LH} = \frac{1}{T} \int_0^{t_{LH}} I_L(t)\, dt. \qquad (4.5\text{-}18)$$

Although this can be evaluated directly by substituting the known expressions for the load current, it is simpler to note that the charging event is described by

KCL as

$$I_L(t) = I_D(t) + C_{out}\frac{dV_{out}}{dt}, \tag{4.5-19}$$

i.e., the load current is the sum of the driver and capacitor currents. Substituting this into eqns. (4.5-18) then gives $(I_{av})_{LH}$ in the form

$$(I_{av})_{LH} = \frac{1}{T}\int_0^{t_{LH}} I_D(t)\, dt + \frac{1}{T}C_{out}\int_0^{t_{LH}} \frac{dV_{out}}{dt}\, dt. \tag{4.5-20}$$

The first term may be rewritten by defining the average driver rise time current $I_{D,LH}$ as

$$I_{D,LH} \equiv \frac{1}{t_{LH}}\int_0^{t_{LH}} I_D(t)\, dt. \tag{4.5-21}$$

note that $I_{D,LH}$ is zero if $V_{in}(t)$ is taken as an ideal square wave (since the driver would be in cutoff). The second integral is easily evaluated using

$$\int_0^{t_{LH}} \frac{dV_{out}}{dt}\, dt = \int_{V_0}^{V_1} dV_{out}, \tag{4.5-22}$$

so

$$(I_{av})_{LH} = \frac{1}{T}[I_{D,LH}t_{LH} + C_{out}(V_1 - V_0)] \tag{4.5-23}$$

gives the average power supply current during the charging event.

The power supply current delivered to the circuit during the discharge time t_{HL} is computed in a similar fashion. First it is noted that

$$(I_{av})_{HL} = \frac{1}{T}\int_0^{t_{HL}} I_L(t)\, dt \tag{4.5-24}$$

is the average power supply current during the capacitor discharge. Since

$$I_L(t) = I_D(t) + C_{out}\frac{dV_{out}}{dt}, \tag{4.5-25}$$

$(I_{av})_{HL}$ may be written as

$$(I_{av})_{HL} = \frac{1}{T}[I_{D,HL}t_{HL} - C_{out}(V_1 - V_0)], \tag{4.5-26}$$

where

$$I_{D,HL} \equiv \frac{1}{t_{HL}}\int_0^{t_{HL}} I_D(t)\, dt \tag{4.5-27}$$

represents the average driver current during the fall time.

The total transient current consumed by the inverter during one logic cycle is now seen to be

$$(I_{av})_{transient} = \frac{1}{T}(I_{D,LH}t_{LH} + I_{D,HL}t_{HL}), \tag{4.5-28}$$

where the terms proportional to C_{out} have canceled one another. Since the average power dissipation is $P_{av} = I_{av}V_{DD}$, the total power-delay product resulting from both DC and transient current flow is

$$PDP \simeq \frac{1}{2}I_{max}V_{DD}t_p + (I_{D,LH}t_{LH} + I_{D,HL}t_{HL})V_{DD}\frac{t_p}{T}. \tag{4.5-29}$$

To understand this expression, assume that $I_{D,LH} = I_{D,HL} = I_{D,av}$. The logic switching frequency is $f = 1/T$, while the maximum switching frequency is

$$f_{max} = \frac{1}{t_{HL} + t_{LH}}. \tag{4.5-30}$$

The PDP can therefore be written in the form

$$PDP \simeq \frac{1}{2}I_{max}V_{DD}t_p + I_{D,av}V_{DD}t_p\frac{f}{f_{max}}. \tag{4.5-31}$$

Now recall that the derivation assumed $T \gg (t_{LH} + t_{HL})$, i.e., $f \ll f_{max}$. For this case, the DC term dominates the PDP. However, as the logic switching frequency f is increased, the transient power consumption increases proportionately. Faster switching also gives a decrease in the DC power, since the factor $\frac{1}{2}$ is no longer a good approximation but must be replaced by a smaller fraction. In the limit where $f = f_{max}$, the inverter is *never* in the stable state $V_{out} = V_{OL}$, so

$$PDP \simeq I_{D,av}V_{DD}t_p, \tag{4.5-32}$$

indicating that the power dissipation arises purely from the transient contributions. As a final point, note that $t_p \propto C_{out}$, while the currents are proportional to voltages. Consequently, the expression shows the approximate dependence of

$$PDP \propto C_{out}V_{DD} \times (\text{Voltage}), \tag{4.5-33}$$

which is in the same form found in the simplified analysis.

The power-delay product for the CMOS inverter is computed by using the current waveform in Fig. 4.14(c). Since current flows only during a switching event, the average power supply current required during a single logic cycle T can be written by analogy with eqn. (4.5-28) as

$$I_{av} = \frac{1}{T}[I_{Dn,LH}t_{LH} + I_{Dn,HL}t_{HL}]. \tag{4.5-34}$$

In this equation,

$$I_{Dn,LH} \equiv \frac{1}{t_{LH}} \int_0^{t_{LH}} I_{Dn}(t)\, dt \tag{4.5-35}$$

gives the average current during the rise time, while

$$I_{Dn,HL} \equiv \frac{1}{t_{HL}} \int_0^{t_{HL}} I_{Dn}(t)\, dt \tag{4.5-36}$$

is the average fall time current. For a completely symmetric CMOS inverter, $I_{Dn,LH} = I_{Dn,HL} = I_{Dn,\text{av}}$, so the power-delay product is given by

$$\text{PDP}_{\text{CMOS}} = I_{Dn,\text{av}} V_{DD} t_p \frac{f}{f_{\max}}. \tag{4.5-37}$$

The power-delay products derived above for the nMOS and CMOS inverters may now be used to compare the two configurations. Assuming all else equal, it is seen that the PDP for the CMOS inverter will be smaller than that for an equivalent nMOS system as long as $f \ll f_{\max}$. However, as the switching frequency is increased, the CMOS PDP approaches the nMOS PDP value. Consequently, CMOS circuits tend to lose their low-power characteristics when used in high-speed logic applications.

To place the preceding comparison into proper perspective, it must be remembered that the comments apply only to the PDP values computed for isolated inverters. The overall picture is much more complicated. First, note that a realistic digital system will be quite complex, consisting of, perhaps, thousands of gates. Because of the nature of digital logic steering, it is highly unlikely that all of the gates on the chip will be switched simultaneously. Since a significant fraction of the total gate count will be in stable logic states, the CMOS low-power characteristic is still an important factor in determining the overall system performance, regardless of the switching frequency. Also, recall from Chapter 3 that CMOS logic layouts require larger chip areas than equivalent nMOS designs (discussed in detail in Chapter 6). This in turn implies that the on-chip power dissipation per unit area $[\text{W/cm}^2]$ will generally be smaller in a CMOS layout since the gates cannot achieve the packing densities possible in nMOS. Finally, note that the comparison assumed that the nMOS and CMOS had equal propagation delays. As mentioned previously, CMOS output capacitances are larger than those encountered in equivalent nMOS circuits. Consequently, $(t_p)_{\text{CMOS}} > (t_p)_{\text{nMOS}}$ if the technologies (i.e., minimum linewidths, etc.) are equal. This then implies that high-speed digital CMOS requires a more advanced processing technology to achieve the lower nMOS PDP levels. As discussed in Section 3.5, the nMOS-CMOS tradeoffs should be considered in detail in any system design.

4.6   MOSFET Capacitances

The calculations in this chapter have demonstrated that all digital MOS transient response times are proportional to the output node capacitance. Minimizing C_{out} thus becomes a primary design objective in an integrated high-speed logic circuit. The attainment of this goal is not a trivial matter, since MOS chip capacitances are complicated functions of the fabrication processes and the layout geometries. Moreover, many of the lumped-equivalent capacitor structures that can be introduced to model the problem are nonlinear (voltage-dependent), so C_{out} can be computed only in an average sense. A more accurate analysis of the transient response requires a computer simulation since the overall problem is considered to be totally intractable from the viewpoint of hand calculations.

Although an exact determination of the output node capacitance is not possible, estimates of C_{out} can be obtained from relatively straightforward modeling. This is accomplished by first isolating the intrinsic MOSFET capacitances from those associated with the interconnect lines. The MOSFET capacitance model to be developed in this section will be (approximately) valid for every transistor in the circuit. The value of C_{out} for an arbitrary logic gate can then be found by combining the MOSFET and line capacitances using standard network techniques. Luckily, the most important contributions are in parallel, so C_{out} will simply be the sum of the individual capacitors.

The large-signal MOSFET capacitance model that will be used to compute C_{out} is based on the self-aligned, poly gate LOCOS (for local oxidation of silicon) structure depicted in Fig. 4.15; the details concerning the processing steps required to fabricate this device are covered in Chapter 5. Although the LOCOS MOSFET has been singled out for the analysis (since it is typical of current processing schemes), the model developed here is generally applicable to any MOSFET regardless of the technology base. Figure 4.16(a) shows the

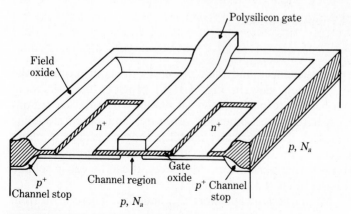

FIGURE 4.15   Basic LOCOS MOSFET structure.

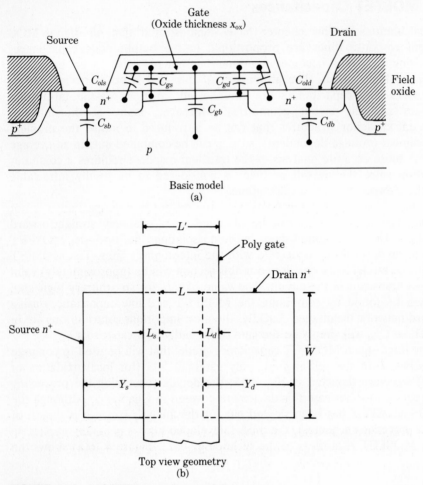

FIGURE 4.16   MOSFET capacitor model.

basic lumped-element capacitances and their physical origins in terms of the device cross section. The top view in Fig. 4.16(b) provides the additional geometric information needed to perform the calculations. The device regions used to define the capacitors are somewhat arbitrary, and other modeling schemes can be formulated. This particular model is chosen because it allows the capacitors to be divided into contributions that may be computed directly from the device and processing parameters. Moreover, it can be easily related to the simple MOS model discussed in Chapter 1.

Consider first the MOS *overlap* capacitors C_{ols} and C_{old}. These arise from noting that the *physical length* L' of the polysilicon gate is given by

$$L' = L_s + L + L_d, \tag{4.6-1}$$

where L is the actual channel length of the MOSFET (i.e., between the drain and source n^+ regions). L_s and L_d give the gate-source and gate-drain overlap distances, respectively. It is necessary to have $L_s > 0$ and $L_d > 0$ for an operational device to ensure that the inversion layer makes contact with both the drain and source n^+ regions, which in turn allows drain-source current flow. Since Fig. 4.16(b) shows that both n^+ regions have width W, the overlap capacitances are given by

$$C_{ols} = C_{ox}WL_s, \qquad C_{old} = C_{ox}WL_d, \tag{4.6-2}$$

where

$$C_{ox} = \frac{\varepsilon_{ox}}{x_{ox}} \tag{4.6-3}$$

is the gate oxide capacitance per unit area; note that both C_{ols} and C_{old} are constants for a given geometry.

The overlap capacitances are parasitic elements that originate from the basic fabrication steps. In the self-aligned process, the polysilicon gate is employed as a mask to define the n^+ drain and source regions; directly after this step, $L_s = L_d = 0$ and $L' = L$. The overlaps occur because the remaining processing steps require heating of the wafer. This gives rise to *lateral diffusion* of the n^+ dopants (i.e., in a direction parallel to the surface of the wafer), so the poly gate overlaps the drain and source regions of the final structure. The symmetry of the MOSFET implies that $L_s = L_d = L_o$, with typical overlap distances being less than a few tenths of a micron.

The above discussion illustrates that the circuit designer does not have control over the overlap distances. Rather, these are fabrication parameters that are defined by the processing steps. With regard to the overlap capacitances, W is the only layout parameter that can be controlled at the circuit design level. This is often implied in a design rule set by defining an overlap capacitance per unit gate width by

$$C_o = C_{ox}L_o. \tag{4.6-4}$$

Then

$$C_{ols} = C_{old} = C_o W \tag{4.6-5}$$

gives the capacitance values in terms of the gate width W.

Now consider the capacitors denoted as C_{gs}, C_{gd}, and C_{gb} in Fig. 4.16(a). These contributions are associated with the basic MOS system of the transistor. The gate-source capacitance C_{gs} is really the gate-to-channel capacitance as seen between the gate and source; similarly, C_{gd} represents the gate-drain capacitance when the channel is acting as a conductor to the drain n^+ region. The voltage-dependent nature of the channel implies that these elements are nonlinear. Since the intrinsic MOS gate capacitance of the channel region is $C_{ox}WL$, these two contributions may be written in the form

$$C_{gs} = C_{ox}WLf_1(V_{GS}; V_{GD}), \qquad C_{gd} = C_{ox}WLf_2(V_{GS}; V_{GD}), \tag{4.6-6}$$

where f_1 and f_2 are functions that describe the nonlinear dependence on the voltages. C_{gb} is the gate-bulk capacitance and consists of the gate capacitance in series with the depletion capacitance established by the p-type space charge region. This is also voltage-dependent and may be expressed as

$$C_{gb} = C_{ox}WLf_3(V_{GS}; V_{GD}; V_{SB}),\qquad(4.6\text{-}7)$$

where f_3 includes body bias effects through V_{SB}. The voltage functions f_1, f_2, and f_3 can be approximated analytically by making some relatively straightforward assumptions about the voltage variation in the channel. However, since interest is currently directed toward obtaining the *average* MOSFET contributions to C_{out}, the details of the derivations will not be presented here.

These capacitors may be understood qualitatively by examining the MOSFET in the three operational regions of cutoff, nonsaturation, and saturation, as depicted in Fig. 4.17. In cutoff ($V_{GS} < V_T$), there is no inversion layer channel; hence, $C_{gs} = C_{gd} = 0$. The gate-bulk capacitance is thus approximated by

$$C_{gb} = C_{ox}WL,\qquad(4.6\text{-}8)$$

which is the (channel) gate capacitance of the transistor.

When the transistor is biased into the active region, $V_{GS} > V_T$ implies that the inversion layer is formed. The presence of the channel tends to *shield* the bulk electrode from the gate since the inversion layer acts as a conductor to the drain and source n^+ regions. This gives $C_{gb} = 0$ as a reasonable approximation in the active region. With the device in a nonsaturated mode, the gate-source and gate-drain capacitors are roughly equal; this can be seen in Fig. 4.17(b),

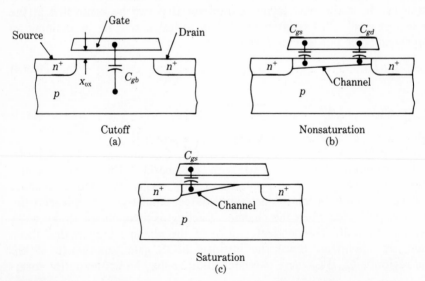

Cutoff
(a)

Nonsaturation
(b)

Saturation
(c)

FIGURE 4.17   MOSFET gate capacitances in the three operational regions.

which shows that the channel extends across the entire transistor. The nonsaturated contributions are thus described by

$$C_{gs} \simeq \frac{1}{2} C_{ox} WL \simeq C_{gd}. \tag{4.6-9}$$

On the other hand, when the MOSFET is saturated, the channel is pinched off and does not contact the drain n^+ region; this is depicted in Fig. 4.17(c). The capacitances for this operational mode are given by

$$C_{gs} = \frac{2}{3} C_{ox} WL, \qquad C_{gd} = 0. \tag{4.6-10}$$

The three capacitors as functions of the gate-source voltage V_{GS} are plotted in Fig. 4.18. Note that the sum $(C_{gs} + C_{gd} + C_{gb})$ has a minimum value of approximately $\frac{1}{3} C_{ox} WL$ (at the cutoff/saturation border) and a maximum value of $C_{ox} WL$ (when the MOSFET either is in cutoff or is nonsaturated).

The mathematical analysis of the problem gives the capacitances as functions of the drain source voltage V_{DS}. The basic equations that describe the capacitances are listed below for reference.

(a) Cutoff:

$$C_{gb} \simeq C_{ox} WL$$
$$C_{gs} \simeq 0$$
$$C_{gd} \simeq 0 \tag{4.6-11}$$

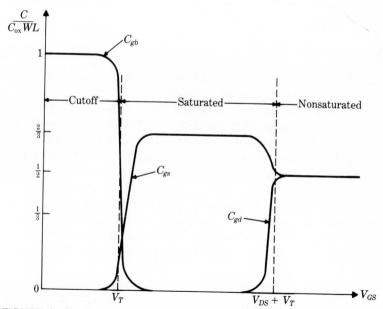

FIGURE 4.18 Gate capacitances as functions of gate-source voltage.

(b) Nonsaturation:

$$C_{gb} \simeq 0$$

$$C_{gs} \simeq \frac{1}{2} C_{ox} WL \left(1 + \frac{V_{DS}}{3V_{DS,\text{sat}}} \right)$$

$$C_{gd} \simeq \frac{1}{2} C_{ox} WL \left(1 - \frac{V_{DS}}{V_{DS,\text{sat}}} \right) \tag{4.6-12}$$

(c) Saturation:

$$C_{gb} \simeq 0$$

$$C_{gs} \simeq \frac{2}{3} C_{ox} WL$$

$$C_{gd} \simeq 0 \tag{4.6-13}$$

In these equations, the saturation voltage is given by $V_{DS,\text{sat}} = (V_{GS} - V_T)$. Note that the general results are similar to those graphed in Fig. 4.18.

It is useful to combine the MOS capacitances with the overlap contributions when estimating circuit values. The total gate capacitance is given by

$$C_G = C_{ox} WL', \tag{4.6-14}$$

where $L' = L + 2L_o$ is the poly gate length. The total gate-source capacitance is defined by

$$C_{GS} = C_{ols} + C_{gs}, \tag{4.6-15}$$

while the total gate-drain capacitance is

$$C_{GD} = C_{old} + C_{gd}. \tag{4.6-16}$$

The appropriate values of these elements when computing C_{out} for a digital gate will be discussed later in the context of examples.

The two remaining capacitors in the model of Fig. 4.16(a) are C_{sb} and C_{db}. These represent the voltage-dependent depletion capacitances that result from the pn junctions at the drain and source regions. The problem of determining these elements is aided by using the expanded drawing in Fig. 4.19. This shows an n^+ well in a p-type bulk region and is representative of either a drain or a source; note that a p^+ region surrounds the n^+ sidewalls. The actual doping profile around the pn junction is generally quite complicated. A step (or abrupt) doping will be assumed for simplicity. The step pn junction analysis is summarized in the Appendix. (It should be noted that the analysis can be extended to the case of a linearly graded pn junction by straightforward modifications.)

Consider a step profile pn junction with respective p and n doping densities of N_a and N_d. The reverse-biased depletion capacitance per unit area of the

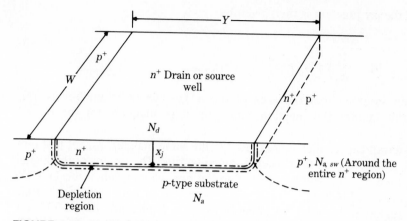

FIGURE 4.19 Expanded view of an n^+ drain or source region for computing depletion capacitances.

junction is given by

$$C = \frac{C_{j0}}{\left(1 + \dfrac{V_r}{\phi_o}\right)^{1/2}}, \tag{4.6-17}$$

where V_r is the *magnitude* of the reverse-bias voltage applied to the junction; ϕ_o is the built-in potential

$$\phi_o = \left(\frac{kT}{q}\right) \ln \left(\frac{N_d N_a}{n_i^2}\right); \tag{4.6-18}$$

and C_{j0} is the zero-bias ($V_r = 0$) capacitance per unit area computed from

$$C_{j0} = \sqrt{\frac{q\varepsilon_{Si}}{2\left(\dfrac{1}{N_a} + \dfrac{1}{N_d}\right)\phi_o}}. \tag{4.6-19}$$

To apply these equations to the problem at hand, first note that the n^+ well is specified as having a donor doping of N_d. The acceptor concentration, on the other hand, depends on whether the pn junction is taken at the bottom of the n^+ well or at the sidewalls, since the two doping levels are different. In general, both contributions are important in determining C_{sb} and C_{db}.

The zero-bias capacitance per unit area C_{j0} at the bottom of the well is obtained by simply using N_a and N_d in eqns. (4.6-18) and (4.6-19), with N_a taken as the p-type bulk substrate doping. The total capacitance for this portion of the pn junction is then given by

$$C_{\text{bottom}} = \frac{C_{j0}WY}{\left(1 + \dfrac{V_r}{\phi_o}\right)^{1/2}}.$$

(4.6-20)

It has been assumed that the area of the n^+/p-type bulk junction is WY; although this ignores the junction curvature, it is adequate for the present analysis.

The sidewall capacitance is computed in the same manner, except that now the p^+ doping of $N_{a,sw}$ must be used in the equations. The origins of the higher sidewall p-type dopings will be clarified in the next chapter when the LOCOS process is discussed in detail. For now it is sufficient to note than an n^+ drain or source region will border on either a LOCOS field oxide or a MOSFET channel (assuming the simplified geometry shown in Fig. 4.15). A border on a field oxide region will see the large $N_{a,sw}$ due to an acceptor field implant step in the processing. This p^+ region is provided to ensure that the semiconductor surface below the field oxide does not invert, i.e., it aids in isolating the device by preventing the formation of unwanted channels. At an n^+-to-channel border, the channel region has a higher doping than that found in the p-type bulk because of the threshold ion implant. This gives increased sidewall doping values there. The field implant and the threshold voltage adjustment implants are not at the same level since they are performed in different processing steps; however, it will be assumed for simplicity that $N_{a,sw}$ is a constant around the entire n^+ region. Typically, $N_{a,sw}$ is about 10 times greater than the bulk doping N_a.

The sidewall capacitance is usually computed by first taking the sidewall capacitance per unit area as

$$C_{j0sw} = \sqrt{\frac{q\varepsilon_{\text{Si}}}{2\left(\dfrac{1}{N_{a,sw}} + \dfrac{1}{N_d}\right)\phi_{osw}}},$$

(4.6-21)

where

$$\phi_{osw} = \left(\frac{kT}{q}\right)\ln\left(\frac{N_d N_{a,sw}}{n_i^2}\right)$$

(4.6-22)

is the sidewall built-in potential. Since the n^+ has a junction depth of x_j, the *sidewall capacitance per unit length* C_{jsw} is taken as

$$C_{jsw} = C_{j0sw}x_j.$$

(4.6-23)

The total sidewall capacitance is then given by

$$C_{sw} = \frac{C_{jsw}\ell}{\left(1 + \dfrac{V_r}{\phi_{osw}}\right)^{1/2}},$$

(4.6-24)

where ℓ is the total sidewall perimeter length. In terms of the configuration shown in Fig. 4.19, $\ell = 2W + 2Y$; this is, however, dependent on the specifics of the layout and will vary from device to device.

The total depletion capacitance of the *pn* junction can now be computed by summing C_{bottom} and C_{sw} as given in eqns. (4.6-20) and (4.6-24). Assuming that $\phi_o = \phi_{osw}$ in the voltage terms, this gives

$$C_d(V_r) = \frac{C_T}{\left(1 + \dfrac{V_r}{\phi_o}\right)^{1/2}}, \tag{4.6-25}$$

where

$$C_T = C_{j0}WY + C_{jsw}\ell \tag{4.6-26}$$

represents the total zero-bias capacitance. The nonlinear nature of C_d is obvious. When applied to a MOSFET, V_r will be either V_{SB} or V_{DB}, depending on whether C_{sb} or C_{db} is being computed. Since the device voltages are changing during a transient switching event (i.e., a charge or discharge), the problem of determining values for C_{sb} and C_{db} for use in C_{out} becomes complicated to the point where the analysis is considered intractable. Interest is instead directed toward obtaining average values for the depletion capacitances, which may then be used in the circuit analysis.

The average (linear) depletion capacitance C_{av} may be defined by

$$C_{\text{av}} = \frac{1}{V_2 - V_1} \int_{V_1}^{V_2} C_d(V_r)\, dV_r, \tag{4.6-27}$$

where is has been assumed that the reverse voltage across the depletion regions changes from $V_r = V_1$ to $V_r = V_2$. Substituting $C_d(V_r)$ from eqn. (4.6-25) and integrating gives

$$C_{\text{av}} = \frac{2\phi_o C_T}{(V_2 - V_1)}\left[\left(1 + \frac{V_2}{\phi_o}\right)^{1/2} - \left(1 + \frac{V_1}{\phi_o}\right)^{1/2}\right]. \tag{4.6-28}$$

This may be cast into a simpler-looking form by defining a dimensionless voltage factor $K(V_1, V_2)$ by

$$K(V_1, V_2) = \frac{C_{\text{av}}}{C_T} = \frac{2\phi_o}{(V_2 - V_1)}\left[\left(1 + \frac{V_2}{\phi_o}\right)^{1/2} - \left(1 + \frac{V_1}{\phi_o}\right)^{1/2}\right] < 1. \tag{4.6-29}$$

Then

$$C_{\text{av}} = K(V_1, V_2)C_T \tag{4.6-30}$$

gives the average depletion capacitance that can be used to model both C_{sb} and C_{db} in the MOSFET model. Note that this requires a knowledge of the DC voltage levels in the circuit.

The general discussion of the MOSFET capacitances is complete. All of the elements in Fig. 4.16 can be approximated by relatively straightforward techniques. The results presented in this section will be required to complete the analysis of the basic transient response in static logic circuits. Moreover, it will be seen later that the existence of the transistor capacitances allows for the design of *dynamic logic circuits* in which charge is stored on various circuit nodes with the aid of an external clocking signal.

EXAMPLE 4.6-1

A MOSFET n-channel process is characterized by the following parameters:

$$N_a = 10^{15}\,[\text{cm}^{-3}], \quad N_{a,sw} = 2 \times 10^{16}\,[\text{cm}^{-3}], \quad N_d = 10^{20}\,[\text{cm}^{-3}],$$

$$x_{ox} = 500\,[\text{Å}], \quad x_j = 0.5\,[\mu\text{m}], \quad L_o = 0.4\,[\mu\text{m}].$$

Compute C_o, C_{j0}, and C_{jsw}.

Solution

C_o: Compute

$$C_{ox} = \frac{(3.9)(8.854 \times 10^{-14})}{0.05 \times 10^{-4}} \simeq 6.91 \times 10^{-8}\,[\text{F/cm}^2],$$

so the overlap capacitance from eqn. (4.6-4) is

$$C_o \simeq (6.91 \times 10^{-8})(0.4 \times 10^{-4}) = 2.76\,[\text{pF/cm}],$$

or, in more convenient units for hand calculations,

$$C_o \simeq 0.276\,[\text{fF}/\mu\text{m}].$$

C_{j0}: The built-in potential of the n^+-substrate junction is

$$\phi_o \simeq (0.026)\ln\left[\frac{(10^{15})(10^{20})}{(1.45 \times 10^{10})^2}\right] \simeq 0.879\,[\text{V}].$$

Equation (4.6-19) then gives

$$C_{j0} = \left[\frac{(1.6 \times 10^{-19})(11.8)(8.854 \times 10^{-14})}{2(10^{-15} + 10^{-20})(0.879)}\right]^{1/2}$$

or

$$C_{j0} \simeq 9.75 \times 10^{-9}\,[\text{F/cm}^2] = 0.0975\,[\text{fF}/\mu\text{m}^2].$$

C_{jsw}: For the sidewall pn junction,

$$\phi_{osw} \simeq (0.026)\ln\left[\frac{(2 \times 10^{16})(10^{20})}{(1.45 \times 10^{10})^2}\right] \simeq 0.957\,[\text{V}].$$

The zero-bias sidewall capacitance/cm² is

$$C_{j0sw} \simeq \left[\frac{(1.6 \times 10^{-19})(11.8)(8.854 \times 10^{-14})}{2(5 \times 10^{-17} + 10^{-20})(0.975)} \right]^{1/2}$$

$$\simeq 4.18 \times 10^{-8} \, [\text{F/cm}^2].$$

Since the junction depth is $x_j = 0.5 \times 10^{-4} \, [\text{cm}]$,

$$C_{jsw} = (4.18 \times 10^{-8})(0.5 \times 10^{-4}) \simeq 2.09 \, [\text{pF/cm}],$$

or

$$C_{jsw} \simeq 0.209 \, [\text{fF}/\mu\text{m}].$$

The numbers calculated in this example will be used in later problems as typical values. ■

EXAMPLE 4.6-2

Consider the MOSFET shown in Fig. E4.6. Using the results of Example 4.6-1, calculate C_{ols} and C_G. Then find C_{av} of an n^+ region if the reverse voltage on the *pn* junction ranges from 0.3 [V] to 5 [V].

Solution
C_{ols}: This is simply

$$C_{ols} = C_{old} = C_o W \simeq (0.276)(6) \simeq 1.66 \, [\text{fF}].$$

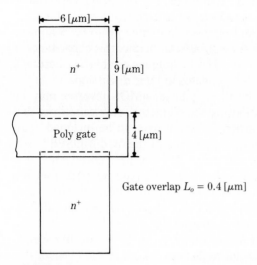

FIGURE E4.6

C_G: Use

$$C_G = C_{ox}WL' = (6.91 \times 10^{-8})(6 \times 10^{-4})(4 \times 10^{-4})\,[\text{F}],$$

or

$$C_G \simeq 16.58\,[\text{fF}].$$

C_{av}: Assuming that $\phi_o \simeq \phi_{osw} = 0.879\,[\text{V}]$ for simplicity,

$$K(0.3, 5) \simeq \frac{2(0.879)}{4.7}\left[\left(1 + \frac{5}{0.879}\right)^{1/2} - \left(1 + \frac{0.3}{0.879}\right)^{1/2}\right]$$

$$\simeq 0.534.$$

Next calculate

$$C_T = (0.095)(6)(9.4) + (0.209)(12 + 18.8),$$

where the overlap L_o length has been included in the area and perimeter calculation. This gives

$$C_T \simeq 11.8\,[\text{fF}],$$

so

$$C_{av} \simeq (0.534)(11.8) \simeq 6.30\,[\text{fF}]. \qquad\blacksquare$$

4.7 Inverter Output Capacitance

Once the device geometries have been established, the MOSFET capacitance model developed in the last section may be used to estimate a value for C_{out}. Knowledge of C_{out} allows the transient response time intervals to be computed and also completes the design cycle. As will be seen below, this capacitance is quite sensitive to the overall chip layout. The techniques established here will be used to evaluate more complicated logic gates in later discussions.

The most important contributions to C_{out} in an nMOS inverter may be modeled as shown in Fig. 4.20. In choosing the elements, only capacitors that undergo a voltage change during a switching event need to be included. This eliminates, for example, the drain-bulk capacitance of the load MOSFET $M2$. Also, the gate-source capacitance of the driver $M1$ can be ignored since it is assumed to be charged by the input voltage source. The n^+-bulk junction capacitances C_{db1} and C_{sb2} are taken to be the zero-bias values calculated from eqn. (4.6-26). Consequently, a factor of $K(V_{OL}, V_{OH})$ must be included to obtain the average values that are used to compute C_{out}.

A quick inspection of the circuit shows that the capacitors are in parallel. Thus, a first-order estimate of the output node capacitance of the inverter is

$$C_{out} \simeq C_{GD1} + C_{GD2} + K(V_{OL}, V_{OH})[C_{db1} + C_{sb2}] + C_{line} + C_{G3}. \qquad \textbf{(4.7-1)}$$

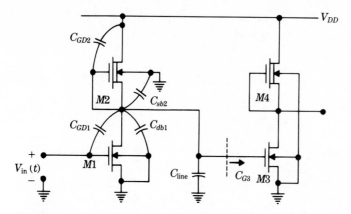

FIGURE 4.20 Approximation used for C_{out} in cascaded nMOS inverters.

Given the values of V_{OL} and V_{OH} as found from the DC analysis, this equation allows for a numerical estimate of C_{out}.

C_{line} is the interconnect line capacitance that enters the problem as a parasitic element from the layout. A simplified example of this contribution is illustrated in Fig. 4.21. This represents the output line as a metal interconnect run that feeds the output of the next stage. In the strictest sense, the analysis of the interconnect properties should be based on transmission line theory in which the structure is treated in a distributed manner. However, a lumped-element approximation is generally sufficient for an initial circuit calculation, and its use greatly simplifies the problem. The line capacitance is estimated by

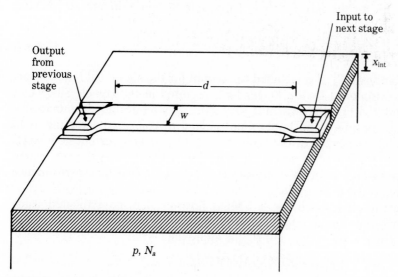

FIGURE 4.21 Simplified interconnect scheme for line capacitance.

the parallel-plate formula

$$C_{\text{line}} \simeq C_{\text{int}} A_{\text{line}}, \tag{4.7-2}$$

where

$$C_{\text{int}} = \frac{\varepsilon_{\text{ox}}}{x_{\text{int}}} [\text{F/cm}^2] \tag{4.7-3}$$

represents the capacitance per unit area formed between the line and the substrate, x_{int} is the oxide thickness under the line, and $A_{\text{line}} = wd$ is the total area of the interconnect. Consequently, short interconnects should be used to reduce the overall value of C_{out}. This is particularly critical since C_{line} can dominate in the output capacitance expression.

Another parasitic that should be mentioned is the line resistance R_{line}. Assuming that the interconnect has a sheet resistance of $R_s[\Omega]$, this lumped-equivalent element may be approximated by means of

$$R_{\text{line}} = nR_s \, [\Omega], \tag{4.7-4}$$

where $n = (d/w)$ is the number of squares with area w^2 as seen in the direction of current flow. The effects of R_{line} on the circuit transient response has already been discussed in previous sections of this chapter.

The last term, C_{G3}, in C_{out} is the input capacitance of the next stage; this is illustrated in the original circuit of Fig. 4.20. The value of C_{G3} is found using eqn. (4.6-14) and is dependent on the dimensions of MOSFET $M3$. Note that this situation represents a fan-out (FO) of 1. An important point that arises here is that C_{out} increases with the FO of the stage. Figure 4.22 provides an example of this for the case FO = 3. To include the extra drivers, C_{G3} in eqn. (4.7-1) is replaced by

$$C_{G3} \rightarrow C_{G3} + C_{G4} + C_{G5} + (\Delta C_{\text{line}}). \tag{4.7-5}$$

The contribution (ΔC_{line}) is included to account for the additional interconnect capacitance introduced by the longer lines required in the layout. Since C_{out} increases with the FO, the switching properties may be degraded to unacceptable levels if the FO is too large. One method to compensate for this is to increase the (W/L) ratios of the transistors so that greater current flow levels are possible. However, this approach simultaneously increases the real estate consumption of the circuit. A compromise between the switching performance and chip area is usually required.

The output capacitance for a CMOS inverter may be estimated using similar techniques. The important contributions are shown in Fig. 4.23. Since all of the elements are in parallel, C_{out} is approximated by

$$C_{\text{out}} \simeq C_{GDn} + C_{GDp} + K(V_{OL}, V_{OH})(C_{dbp} + C_{dbn}) + C_{\text{line}} + C_G. \tag{4.7-6}$$

The term C_G represents the input capacitance seen looking into the next stage

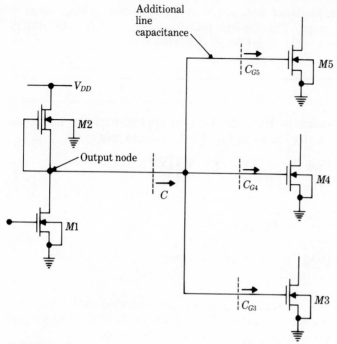

FIGURE 4.22   Capacitance calculation for FO = 3.

and is given by

$$C_G = C_{Gn} + C_{Gp} \tag{4.7-7}$$

since both MOSFETs contribute gate capacitance that must be charged or discharged during a switching event. This leads to the situation where the input capacitance of a CMOS stage is generally larger than that of an equivalent-function nMOS circuit. CMOS geometries must be carefully designed to

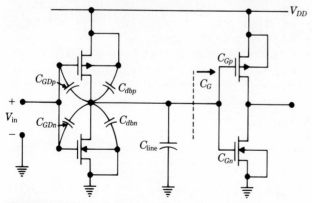

FIGURE 4.23   Approximation used for C_{out} in cascaded CMOS inverters.

ensure that the input capacitance does not present an excessively large value of C_G to the preceding stage. The fan-out problems discussed for the nMOS inverter are also important for high CMOS switching rates.

EXAMPLE 4.7-1

Use the capacitance values in Example 4.6-1 to approximate the maximum value of C_{out} for the inverter shown in Fig. E4.7. Assume that

$$V_{TOD} = +1\,[V], \quad V_{TOL} = -3.5\,[V], \quad \gamma = 0.4\,[V^{1/2}],$$
$$2\,|\phi_F| = 0.6\,[V], \quad V_{DD} = 5\,[V],$$

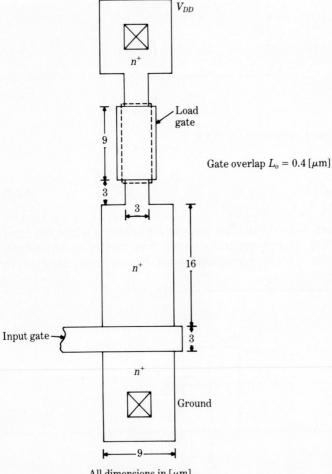

All dimensions in [μm]

FIGURE E4.7

and use a built-in potential of $\phi_o \simeq 0.918$ for both the n^+-substrate and n^+-sidewall K calculation.

Solution

It is easiest to calculate the various contributions and then sum the results.

(a) $(C_{GD})_{\text{driver}}$: The overlap capacitance is

$$C_{old} = (0.276)(9) \simeq 2.484 \text{ [fF]},$$

while the maximum value of C_{gd} is (note the overlap of $L_o = 0.4 \, [\mu\text{m}]$)

$$C_{gd,\max} = \frac{1}{2} C_{\text{ox}} WL = \frac{1}{2}(0.691)(2.2)(9) \simeq 6.841 \text{ [fF]}.$$

The maximum value of C_{GD} from the driver is then

$$(C_{GD})_{\text{driver}} \simeq 2.484 + 6.841 = 9.325 \text{ [fF]}.$$

(b) $(C_{GD})_{\text{load}}$: Using the same approach as in (a),

$$(C_{GD})_{\text{load}} = (0.276)(3) + \frac{1}{2}(0.691)(8.2)(3) = 9.325 \text{ [fF]}$$

is the largest value of the load gate-drain capacitance.

(c) $K(V_{OH}, V_{OL})(C_{dbD} + C_{sbL})$: First, compute the zero-bias capacitance. For the geometry shown,

$$\begin{aligned} C_{dbD} + C_{sbL} &= C_{j0}(\text{Area}) + C_{jsw}(\text{Perimeter}) \\ &= (0.0975)[(16.4)(9) + (3)(3.4)] \\ &\quad + (0.209)2(16.4 + 9 + 3.4) \\ &\simeq 27.424 \text{ [fF]}. \end{aligned}$$

It is not necessary to split up driver and load contributions.

To compute $K(V_{OH}, V_{OL})$, V_{OL} must be found. The geometry gives

$$\beta_R = \frac{(9/2.2)}{(3/8.2)} \simeq 11.182,$$

so V_{OL} is found by iterating

$$V_{OL} = 4 - \sqrt{16 - \frac{1}{11.182} |-3.5 + 0.4(\sqrt{V_{OL} + 0.6} - \sqrt{0.6})|^2},$$

which gives $V_{OL} \simeq 0.137 \text{ [V]}$. Since $V_{OH} \simeq V_{DD} = 5 \text{ [V]}$,

$$\begin{aligned} K(V_{OH}, V_{OL}) &\simeq \frac{2(0.918)}{(5 - 0.137)} \left[\left(1 + \frac{5}{0.918}\right)^{1/2} - \left(1 + \frac{0.137}{0.918}\right)^{1/2} \right] \\ &\simeq 0.554 \end{aligned}$$

Thus,

$$K(V_{OH}, V_{OL})(C_{dbD} + C_{sbL}) \simeq (0.554)(27.424) \simeq 15.193 \, [\text{fF}].$$

The total approximate capacitance is thus

$$C_{\text{out}} \simeq 9.325 + 9.325 + 15.193,$$

or

$$C_{\text{out}} \simeq 33.84 \, [\text{fF}].$$

The line capacitance C_{line} and capacitance seen looking into the next stage must also be found for a complete analysis.   ∎

EXAMPLE 4.7-2

To examine the accuracy of the hand-calculated results in Example 4.7-1, the circuit was simulated (a) using SPICE for the case $C_{\text{out}} = 31.3 \, [\text{fF}]$ a constant and (b) using SPICE modeling of the depletion and gate capacitances.

The input files and outputs are as follows. The hand calculations tended to overestimate the values of t_{HL} and t_{LH} in this case (see Fig. E4.8). However, the results are the same order of magnitude.

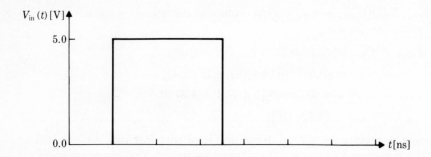

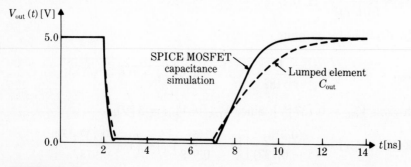

FIGURE E4.8

(a) Constant C_{out}:

```
DMODE INVERTER TRANSIENT ANALYSIS
VDD 3 0 DC 5VOLTS
MDRIVER 2 1 0 0 EMODE L=2.2U W=9U
MLOAD 3 2 2 0 DMODE L=8.2U W=3U
VS 1 0 DC PULSE (0 5 2NS 1PS 1PS 5NS)
CLOAD 2 0 3.136E-14
.MODEL EMODE NMOS VTO=1 GAMMA=0.4 KP=2.5E-5
.MODEL DMODE NMOS VTO=-3.5 GAMMA=0.4 KP=2.5E-5
.TRAN 1NS 15NS
.PLOT TRAN V(2)
.END
```

(b) SPICE-simulated capacitances:

```
DMODE INVERTER TRANSIENT ANALYSIS
VDD 3 0 DC 5VOLTS
MDRIVER 2 1 0 0 EMODE L=2.2U W=9U AD=147.6P PD=47.8U
MLOAD 3 2 2 0 DMODE L=8.2U W=3U AS=10.2P PS=9.8U
VS 1 0 DC PULSE (0 5 2NS 1PS 1PS 5NS)
.MODEL  EMODE  NMOS  VTO=1  GAMMA=0.4  KP=2.5E-5  CGSO=276P
+ CGDO=276P MJ=0.5 MJSW=0.5 CJ=97.5U CJSW=209P
.MODEL  DMODE  NMOS  VTO=-3.5  GAMMA=0.4  KP=2.5E-5  CGSO=276P
+ CGDO=276P MJ=0.5 MJSW=0.5 CJ=97.5U CJSW=209P
.TRAN 1NS 15NS
.PLOT TRAN V(2)
.END                                                        ∎
```

EXAMPLE 4.7-3

This example illustrates the total design cycle, which includes both the DC and transient specifications. Suppose that the nMOS process specifies

$$V_{DD} = 5\,[\text{V}], \quad V_{T0D} = 1\,[\text{V}], \quad V_{T0L} = -3.5\,[\text{V}], \quad \gamma = 0.4\,[\text{V}^{1/2}],$$
$$2\,|\phi_F| = 0.6\,[\text{V}]$$

and that the capacitance levels in the previous examples are still valid. It is desired to design an inverter that cascades into an identical circuit with $V_{OL} < 0.25\,[\text{V}]$ and $t_{LH} < 10\,[\text{ns}]$.

Using eqn. (3.4-43) for $V_{OL} = 0.25$ gives a minimum $\beta_R = 6.11$. This is used to create the layout shown in Fig. E4.9. Since the lateral doping is specified as $L_o = 0.4\,[\mu\text{m}]$,

$$(W/L)_L = \frac{4}{7.2} \quad \text{and} \quad (W/L)_D = \frac{12}{3.2},$$

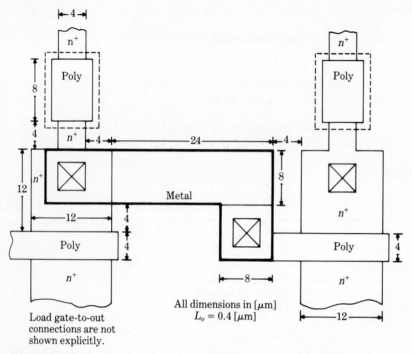

FIGURE E4.9

so $\beta_R = 6.75$. The actual DC voltage V_{OL} is then obtained by iterating eqn. (3.4-12) in the form

$$V_{OL} = 4 - \sqrt{16 - \frac{1}{6.75}|-3.5 + 0.4(\sqrt{V_{OL} + 0.6} - \sqrt{0.6})|^2},$$

which gives $V_{OL} = 0.226\,[\mathrm{V}]$. This is within the design specification.

The crucial transient performance depends on the capacitance C_{out}. To estimate this, break it into parts as follows:

(a) C_{int}: the interconnect capacitance consists of metal and poly over the field oxide. Assume that these are given by

$$C_{m\text{-}f} = 0.0345\,[\mathrm{fF}/\mu\mathrm{m}^2]\quad \text{(metal to field)}$$

and

$$C_{p\text{-}f} = 0.0576\,[\mathrm{fF}/\mu\mathrm{m}^2]\quad \text{(poly to field)}.$$

(The origin of these numbers will be discussed in Section 5.4.)

The interconnect capacitance is then

$$C_{\mathrm{int}} = 0.0345(24 \times 8) + 0.0576[8^2 + (4)(8)] \approx 12.154\,[\mathrm{fF}].$$

This can be added to the input gate capacitance of the next stage, which is given by

$$C_G = 0.691(12)(4) \simeq 33.168\,[\text{fF}],$$

to give a total MOS capacitance of

$$C \simeq 45.322\,[\text{fF}].$$

(b) $C_{\text{depletion}}$: Next examine the n^+ depletion capacitance at the output of the first stage. The zero-bias value is

$$(C_{dbD} + C_{sbL}) = (0.0975)(12^2 + 4^2)$$
$$+ (0.209)[2(12) + 2(12.4) + 2(4.4)]$$
$$= 27.638\,[\text{fF}],$$

where the overlap region has been included in the perimeter calculation. To calculate $K(V_{OH}, V_{OL})$, assume that $\phi_o = 0.879\,[\text{V}]$ for both the bottom and sidewall. This gives $K = 0.539$, so

$$C_{\text{av}} = 0.539(27.638) = 14.9\,[\text{fF}]$$

is the average value.

(c) $C_{GDD} + C_{GSL}$: The maximum values of these are found from

$$(C_{GDD} + C_{GDL}) = \frac{0.691}{2}[(8)(4) + (12)(4)] = 27.64\,[\text{fF}].$$

The total value of C_{out} from these approximations is

$$C_{\text{out}} = 45.32 + 14.9 + 27.64 \simeq 87.86\,[\text{fF}].$$

To check to see if this meets the specification on t_{LH}, note that $V_1 = 4.52\,[\text{V}]$, $V_0 = 0.70\,[\text{V}]$, and evaluate the load threshold voltage at $V = 2.61\,[\text{V}]$, which gives $V_{TOL}(2.61) = -3.09\,[\text{V}]$. Then, assuming $k'_L = 25 \times 10^{-6}\,[\text{A/V}^2]$,

$$t_{LH} \simeq \frac{87.86 \times 10^{-15}}{(4/7.2)(25 \times 10^{-6})(3.09)} \left\{ \frac{2(5 - 3.09 - 0.7)}{3.09} \right.$$
$$\left. + \ln\left[\frac{2(3.09) - 0.48}{0.48} \right] \right\}$$

directly from eqn. (4.2-23). Evaluating,

$$t_{LH} \simeq 6.67\,[\text{ns}] < 10\,[\text{ns}],$$

so the layout meets the specification. If this were not satisfied, the layout and circuit design would have to be redone with a larger $(W/L)_L$.   ∎

4.8 Scaled Inverter Performance†

The transient analysis discussed in this chapter can be used to approximate the switching performance of a scaled circuit. Assume that the device dimensions are scaled by $S > 1$ such that

$$(\text{Length})' = \frac{(\text{Length})}{S}. \tag{4.8-1}$$

Primed quantities refer to the scaled device. Scaling theory is applied to the layout in general, so the length reduction applies to all geometries in the chip.

   Consider first the case of full scaling. The general behavior of the scaled circuit can be established by examining a typical transient time interval. Equation (4.1-13) gives

$$t_{HL} = \tau_D \left\{ \frac{2V_{TD}}{(V_{OH} - V_{TD})} + \ln \left[\frac{2(V_{OH} - V_{TD})}{V_{OL}} - 1 \right] \right\} \tag{4.8-2}$$

for the nMOS high-to-low time. Full scaling requires voltage reduction by $V' = (V/S)$. Since voltage ratios remain unchanged, the terms inside the curly bracket are constant under the scaling transformation.

   The important scaling properties are contained in the driver time constant

$$\tau_D = \frac{C_{\text{out}}}{\beta_D(V_{OH} - V_{TD})}. \tag{4.8-3}$$

The voltages scale according to

$$(V_{OH} - V_{TD})' = \frac{(V_{OH} - V_{TD})}{S}, \tag{4.8-4}$$

while the upward scaling of oxide capacitance by $C'_{\text{ox}} = SC_{\text{ox}}$ shows that

$$\beta'_D = S\beta_D. \tag{4.8-5}$$

   C_{out} consists of oxide and depletion capacitances. Invoking the linear upward scaling of oxide capacitances gives

$$(C')_{\text{oxide}} = C'_{\text{ox}}(\text{Area})' = \frac{(C)_{\text{oxide}}}{S}, \tag{4.8-6}$$

which holds for both gate and interconnect (field) capacitances. Depletion contributions do not scale linearly but tend to increase as S^m, where m is the grading constant ($\frac{1}{2}$ for step profile junctions) due to increased doping levels. As a low-order approximation, assume the simpler form

$$(C')_{\text{junction}} \simeq \frac{(C)_{\text{junction}}}{S}, \tag{4.8-7}$$

† This section is based on the material in Section 2.1. It may be skipped without loss of continuity.

which underestimates the capacitance reduction. Then

$$C'_{out} \simeq \frac{C_{out}}{S} \tag{4.8-8}$$

represents the overall approximate behavior of C_{out}.

Using these results in eqn. (4.8-3) gives the scaled time constant as

$$\tau'_D \simeq \frac{\tau_D}{S}, \tag{4.8-9}$$

which shows a decrease in discharging time. This behavior is also found for charging times t_{LH}, which are governed by load time constants $\tau'_L = (\tau_L/S)$. The maximum switching frequency is

$$f'_{max} = \frac{1}{t'_{HL} + t'_{LH}} \simeq Sf_{max}, \tag{4.8-10}$$

representing an upward scaling.

Constant voltage scaling gives different results. The driver time constant is

$$\tau'_D \simeq \frac{\tau_D}{S^2}, \tag{4.8-11}$$

which leads to

$$f'_{max} \simeq S^2 f_{max}. \tag{4.8-12}$$

This agrees with physical reasoning since the current flow levels are increased by a factor of S, which in turn gives faster charging and discharging.

Although these results imply simple scaling of the transient response times, it must be remembered that scaling theory constitutes only a first-order theory. The real analysis is much more complicated and often requires a change in the basic device physics.

REFERENCES

(1) P. E. Allen and E. Sanchez-Sinencio, *Switched Capacitor Circuits*, New York: Van Nostrand Reinhold, 1984.

(2) M. Annaratone, *Digital CMOS Circuit Design*, Hingham, MA: Kluwer Academic Publishers, 1986.

(3) A. Barna, *VHSIC*, New York: Wiley, 1981.

(4) M. I. Elmasry (ed.), *Digital MOS Integrated Circuits*, New York: IEEE Press (Wiley), 1981.

(5) A. B. Glaser and G. E. Subak-Sharpe, *Integrated Circuit Engineering*, Reading, MA: Addison-Wesley, 1979.

(6) L. A. Glasser and D. W. Doberpuhl, *The Design and Analysis of VLSI Circuits,* Reading, MA: Addison-Wesley, 1985.

(7) D. A. Hodges and H. G. Jackson, *Analysis and Design of Digital Integrated Circuits,* New York: McGraw-Hill, 1983.

(8) E. T. Lewis, "Design and Performance of '1.25-μm' CMOS for Digital Circuit Applications," *Proc. IEEE,* vol. 73, pp. 419–432, 1985.

(9) J. Mavor, M. A. Jack, and P. B. Denyer, *Introduction to MOS LSI Design,* London: Addison-Wesley, 1983.

(10) O. J. McCarthy, *MOS Device and Circuit Design,* Belfast: Wiley, 1982.

(11) S. Muroga, *VLSI System Design,* New York: Wiley, 1982.

(12) W. M. Penney and L. Lau (eds.), *MOS Integrated Circuits,* New York: Van Nostrand Reinhold, 1972.

(13) P. Subramaniam, "Modeling MOS VLSI Circuits for Transient Analysis," *IEEE J. Solid-State Circuits,* vol. SC-21, pp. 276–285, 1986.

(14) Y. Tsividis and P. Antognetti (eds.), *Design of MOS VLSI Circuits for Telecommunications,* Englewood Cliffs, NJ: Prentice-Hall, 1985.

(15) Y. Tsividis, *Operation and Modeling of the MOS Transistor,* New York: McGraw-Hill, 1987.

PROBLEMS

4.1 A depletion load nMOS inverter is designed with $(W/L)_D = 4$ and $(W/L)_L = 0.4$ in a process where

$$k'_D = 26\,[\mu\text{A/V}^2], \quad V_{TOD} = +0.8\,[\text{V}], \quad k'_L = 32\,[\mu\text{A/V}^2],$$
$$V_{TOL} = -3.5\,[\text{V}], \quad \gamma_L = 0.26\,[\text{V}^{1/2}], \quad 2\,|\phi_F| = 0.57\,[\text{V}].$$

The power supply is $5\,[\text{V}]$ and the output load capacitance is estimated to be $95\,[\text{fF}]$.

(a) Calculate V_{OL} and then find the values of V_0 and V_1.

(b) Find the discharge time $t_{HL}\,[\text{ns}]$.

(c) Find the charging time interval t_{LH}. Then calculate $f_{\max}$ for this circuit.

(d) Suppose that the load is redesigned to a value $(W/L)_L = 0.5$ in an effort to decrease the charge time t_{LH}. Calculate the new value of $t_{LH}\,[\text{ns}]$ and compare it with the original value found in (c). Discuss the tradeoffs involved in this redesign.

*(e) Simulate the original circuit by performing a SPICE transient analysis

4.2 An inverter that uses a saturated enhancement-mode transistor as a load is characterized by

$$k' = 26 \, [\mu\text{A}/\text{V}^2], \quad V_{T0} = +0.8 \, [\text{V}], \quad V_{DD} = +5 \, [\text{V}],$$
$$\gamma = 0.39 \, [\text{V}^{1/2}], \quad 2 \, |\phi_F| = 0.58 \, [\text{V}].$$

The driver has $(W/L)_D = 24$, and the load is specified by $(W/L)_L = 2$. Assume that the output capacitance is approximated as $C_{\text{out}} = 225 \, [\text{fF}]$.

(a) Perform the DC analysis to find V_{OL}, V_0, and V_1.
(b) Calculate the fall time $t_{HL} \, [\text{ns}]$.
(c) Calculate the rise time $t_{LH} \, [\text{ns}]$.
(d) Find the value of $(W/L)_L$ needed to decrease the charging time by 20%. What are the consequences of this design change?
*(e) Perform a transient analysis on the original circuit using SPICE. Then change the aspect ratios to $(W/L)_D = 18$, $(W/L)_L = 3$, and simulate the new circuit. Compare results in terms of circuit performance.

4.3 Two inverters are made in an nMOS process described by

$$V_{DD} = 5 \, [\text{V}], \quad V_{T0} = +1 \, [\text{V}] \text{ (E-mode)}, \quad V_{T0} = -3.5 \, [\text{V}] \text{ (D-mode)},$$
$$\gamma = 0.4 \, [\text{V}^{1/2}], \quad 2 \, |\phi_F| = 0.6 \, [\text{V}], \quad k' = 25 \, [\mu\text{A}/\text{V}^2] \text{ (all MOSFETs)}.$$

One inverter uses a load resistor R_L for a load, while the other employs a depletion load MOSFET with $(W/L)_L = 0.5$. Both are designed to have $V_{OH} = V_{DD}$ and $V_{OL} = 0.25 \, [\text{V}]$ and to drive a capacitance of $C_{\text{out}} = 100 \, [\text{fF}]$.

(a) Calculate the charge time t_{LH} for the depletion-load circuit.
(b) Find the value of load resistance R_L that must be used to achieve the same t_{LH} value in the resistively loaded inverter.
(c) Calculate the driver aspect ratio for each circuit. Comment on the differences in $(W/L)_D$ between the two design choices.

Include body bias effects (where appropriate) by averaging.

4.4 An nMOS inverter uses a depletion-mode MOSFET as a load device. The circuit has

$$V_{DD} = 5 \, [\text{V}], \quad V_{T0D} = +0.8 \, [\text{V}], \quad V_{T0L} = -3.3 \, [\text{V}],$$
$$\gamma = 0.37 \, [\text{V}^{1/2}], \quad 2 \, |\phi_F| = 0.58 \, [\text{V}], \quad k' = 28 \, [\mu\text{A}/\text{V}^2].$$

Assume $C_{\text{out}} = 175 \, [\text{fF}]$.

(a) Calculate t_{HL} and t_{LH} in units of [ns].
(b) Find the propagation delay t_p.
(c) Calculate the PDP for a 50% duty cycle at $f \ll f_{\text{max}}$.

4.5 A *super buffer* circuit is shown in Fig. P4.1. The first stage is just a standard depletion load inverter. The second stage looks like an inverter circuit, except that (i) the driver MOSFET gate is connected to the first

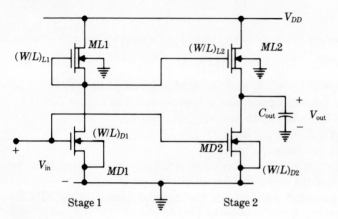

FIGURE P4.1

stage gate (V_{in}) and (ii) the gate of the load MOSFET is driven by the output of the first stage. This circuit increases the current flow to C_{out} during t_{LH}, thus improving the switching speed.

Assume that the circuit is designed with

$$\beta_R = \frac{k_D'(W/L)_{D1}}{k_L'(W/L)_{L1}} = \frac{k_D'(W/L)_{D2}}{k_L'(W/L)_{L2}}.$$

(a) Suppose that the capacitor is initially uncharged, with $V_{out}(0) = 0\,[\text{V}]$. The input voltage V_{in} is switched to $V_{in} = 0\,[\text{V}]$ at time $t = 0$. Determine the conducting state (saturated or nonsaturated) of the second-stage load MOSFET *ML2*. Then find an expression for the charging current into C_{out}.

(b) Integrate the expression in part (a) to find t_{LH} for this circuit. Discuss the results when compared with a standard inverter.

(c) Now suppose that the input is switched to $V_{in} = V_{DD}$. Calculate the DC output low-voltage V_{OL} from the super buffer for this case.

(d) Design a 2-stage super buffer that does *not* invert the input signal. (*Hint*: Look at the connections to the second stage to understand why voltage inversion occurs.)

4.6 The inverter shown in Fig. P4.2 is built in a process described by the following parameters:

$$N_a = 10^{15}\,[\text{cm}^{-3}], \quad k' = 25\,[\mu\text{A/V}^2]\ (\text{both transistors}),$$
$$\gamma = 0.47\,[\text{V}^{1/2}],$$
$$C_{j0} = 1.10 \times 10^{-8}\,[\text{F/cm}^2], \quad C_{jsw} = 6.50\,[\text{pF/cm}],$$
$$\phi_o = 0.717\,[\text{V}], \quad \phi_{osw} = 0.975\,[\text{V}], \quad L_o = 0.3\,[\mu\text{m}]\ (\text{not shown}).$$

Assume that $V_{TOD} = +0.8\,[\text{V}]$, $V_{TOL} = -3.3\,[\text{V}]$, and $V_{DD} = 5\,[\text{V}]$.

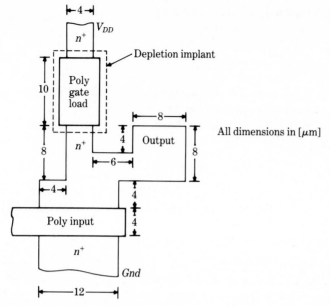

All dimensions in [μm]

Overlap regions are not shown explicitly

FIGURE P4.2

(a) Calculate the value of $(C_{dbD} + C_{sbL})$ in units of [pF]. (These are the zero-bias values.)

(b) Find the maximum value of C_{GD} for the driver MOSFET.

(c) Find the maximum value of C_{GD} for the load MOSFET.

(d) Calculate V_{OL} from the layout geometry assuming that max $(V_{in}) = V_{DD}$. Then find the value of $K(V_{OH}, V_{OL})$.

(e) Use your results to find C_{out}.

(f) Calculate t_{HL}, t_{LH}, and f_{max} for the circuit.

*(g) Perform a SPICE simulation on the circuit using (i) the lumped element C_{out} found above and (ii) full SPICE simulation of the capacitances.

*4.7 Design a depletion load inverter with $V_{OL} \leq 0.25$ [V] using the process described in Problem 4.6. Implement a simplified layout plan and then calculate the value of C_{out} due to the inverter.

Examine the transient response by using a SPICE simulation. Extract the important time intervals from your data. Compare the SPICE results with those obtained in a hand calculation.

4.8 The inversion charge in a MOSFET was obtained in Chapter 1 as

$$Q_I(y) = -C_{ox}[V_{GS} - V_T - V(y)] \, [\text{C/cm}^2],$$

where $V(y)$ is the channel voltage approximated by

$$V(y) \simeq (V_{GS} - V_T)\left(1 - \sqrt{1 - \frac{y}{L}}\right).$$

(see Problem 1.8).

(a) Calculate the total charge Q_c [C] in the channel with dimensions W and L. (This requires an integration step.)

(b) The channel transit time τ can be defined by the alternate form

$$\tau = \frac{|Q_c|}{I_{D,\text{sat}}}.$$

Find τ using Q_c from part (a). Calculate τ for a device with $L = 4$ [μm], $\mu_n = 580$ [cm^2/V-sec], and $(V_{GS} - V_T)$ values of 1 [V], 2 [V], and 4 [V].

(c) Calculate the gate-source capacitance using

$$C_{GS} = \left|\frac{\partial Q_c}{\partial V_{GS}}\right|,$$

and thus verify eqn. (4.6-13).

4.9 A CMOS inverter is designed with $\beta_n = \beta_p = 35$ [μA/V^2] in a process where $V_{T0n} = +0.9$ [V] and $V_{T0p} = -0.8$ [V]. The output capacitance is approximated as $C_{\text{out}} = 125$ [fF]. Assume $V_{DD} = 5$ [V].

(a) Compute t_{HL} and t_{LH} for the circuit.

(b) Find the propagation time t_p.

(c) Estimate the PDP if the circuit is driven with a square wave (50% duty cycle) at a frequency $f = 0.1 f_{\text{max}}$. Use eqn. (4.5-1) and estimate I_{av} by taking the simple average of the minimum and maximum current flow values.

4.10 The layout for a CMOS inverter is shown in Fig. P4.3. This uses a p-type substrate for the nMOS transistors and an n-tub region for the pMOS transistors. The oxide capacitance is $C_{\text{ox}} = 6.91 \times 10^{-8}$ [F/cm^2] for both MOSFET types. The depletion capacitance levels are given as follows:

nMOS: (n^+ to p-substrate)

$$C_{j0} = 0.0975 \text{ [fF/}\mu\text{m}^2], \qquad \phi_o = 0.879 \text{ [V]},$$
$$C_{jsw} = 0.107 \text{ [fF/}\mu\text{m]}, \qquad \phi_{osw} = 0.921 \text{ [V]}.$$

pMOS: (p^+ to n-tub)

$$C_{j0} = 0.0298 \text{ [fF/}\mu\text{m}^2], \qquad \phi_o = 0.939 \text{ [V]},$$
$$C_{jsw} = 0.362 \text{ [fF/}\mu\text{m]}, \qquad \phi_{osw} = 0.985 \text{ [V]}.$$

All dimensions in the drawing are in units of microns [μm]. Assume that

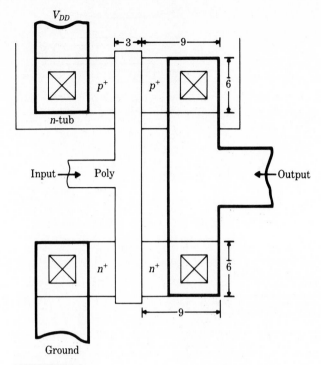

FIGURE P4.3

the overlap distance is $L_o = 0.3\,[\mu\text{m}]$; this is not shown explicitly in the figure but should be included in the calculations. Use $V_{DD} = 5\,[\text{V}]$.

(a) Calculate the maximum values of C_{GDp} and C_{GDn} in [fF].

(b) Find the zero-bias values of C_{dbp} and C_{dbn}. Keep the sidewall and bottom contributions separate in the calculations.

(c) Calculate $K(V_{OH}, V_{OL})$ for the inverter, and thus find the total C_{out} due to the MOSFET layout (i.e., not including the line interconnect or C_G values).

(d) Calculate t_{HL} and t_{LH} for the circuit if the transistor characteristics are as follows:

nMOS: $V_{T0n} = +0.8\,[\text{V}], \quad k_n' = 40\,[\mu\text{A}/\text{V}^2],$

$\gamma = 0.26\,[\text{V}^{1/2}], \quad 2\,|\phi_{Fp}| = 0.58\,[\text{V}],$

pMOS: $V_{T0p} = -0.8\,[\text{V}], \quad k_p' = 16\,[\mu\text{A}/\text{V}^2],$

$\gamma = 0.84\,[\text{V}^{1/2}], \quad 2\phi_{Fn} = 0.70\,[\text{V}].$

Include C_{out} as computed in (c).

*(e) Perform a SPICE simulation of the circuit. First include the lumped-element C_{out} value as an input parameter. Then rewrite the

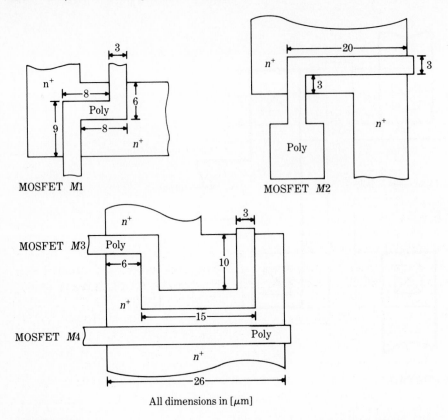

MOSFET $M1$

MOSFET $M2$

MOSFET $M3$

MOSFET $M4$

All dimensions in [μm]

FIGURE P4.4

input file in a manner that allows SPICE to calculate the depletion and gate capacitances directly. Compare results. (*Note*: Interconnect capacitances are added to this circuit in Problem 5.11.)

4.11 MOSFETs with nonrectangular gate geometries are often used to reduce the real estate consumption and attain more compact layouts. Gate serpentine patterns are quite common and consist of gate regions that contain 90° bends. Examples are shown in Fig. P4.4.

The problem of determining effective aspect ratio values is quite complicated and generally requires a computer simulation. However, simplified rules of thumb can be developed to assist in initial design and layout. The simplest rule is to take L as the channel length from n^+ to n^+ as set by a "straight" gate section, while W is determined by measuring the centerline of the gate pattern.

Determine the value of $(W/L)_{\text{eff}}$ for each of the MOSFETs shown. Also find the gate capacitance assuming that $C_{\text{ox}} = 0.691$ [fF/μm^2]. Take the lateral overhang distance as $L_o = 0.3$ [μm].

4.12 The simplified treatment of line resistances R_{line} resulted in the ap-

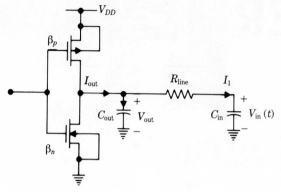

FIGURE P4.5

proximation that the time constants are increased. However, this ignores the fact that C_{out} is made up of contributions that originate from different regions of the chip layout. Figure P4.5 shows the case where C_{out} represents the local output capacitance of a CMOS inverter, while C_{in} is the input capacitance to the next stage. R_{line} is included between the two to model the line resistance. The voltages across the two capacitors are different because of the voltage drop across R_{line}.

(a) Use KCL at the inverter output node to write I_{out} in terms of V_{in}, V_{out}, and the capacitances. Discuss how the equation is modified in the limit $R_{line} \rightarrow 0$.

(b) Write I_1 in terms of R_{line} and the voltages. Then find I_1 in terms of C_{in}. Use the results to find a differential equation for $V_{in}(t)$ with $V_{out}(t)$ acting as a source.

(c) Consider a charging event with $V_{in}(0) = 0$ [V], and suppose that

$$V_{out}(t) = V_{DD} \frac{t}{t_1} \qquad (0 \le t \le t_1)$$

models the inverter output voltage. Find $V_{in}(t)$ by solving the differential equation in (b) subject to the proper initial condition.

(d) A discharge can be modeled by the linear ramp

$$V_{out}(t) = V_{DD} \left(1 - \frac{t}{t_2} \right) \qquad (0 \le t \le t_2).$$

Find $V_{in}(t)$ for this case by assuming an initial condition $V_{in}(0) = V_{DD}$.

(e) Examine the effect of splitting the capacitance by comparing $V_{in}(t)$ with $V_{out}(t)$ for both cases.

4.13 Examine the scaling of a CMOS inverter. Analyze both full scaling and constant-voltage scaling as applied to the basic circuit parameters.

CHAPTER  5

MOS Integrated Circuit Fabrication

A VLSI processing facility constitutes the major financial commitment for integrated circuit production. Vast amounts of money and engineering time are invested in an effort to advance and refine the state of the art. Higher throughput, circuit densities, and yields are all important to the economic stability of the industry.

Circuit design is structured by the processing. The interplay between chip fabrication and circuit design cannot be separated. Rather, they merge to become a single discipline. Process engineers must understand circuit design, and circuit designers must have a knowledge of fabrication if they are to be effective.

The fundamentals of silicon chip fabrication are discussed in this chapter, with an emphasis on MOS-based processes, as suggested by the chapter title. The level is introductory and is not meant to replace a detailed course in the subject. Instead, the objective is to bring out important points relevant to circuit design.

Persons with experience in the subject matter may choose to simply glance through most of the chapter in a first reading. The mathematics can generally be omitted without loss of continuity in later chapters. Note, however, that some of the drawings are worth attention.

5.1 Overview of Basic Silicon Processing Steps

This section will center around a brief examination of the basic wafer processing steps used to fabricate silicon integrated circuits. The discussion is intended only as an overview. A reader desiring a more in-depth treatment should consult one of the texts referenced at the end of the chapter.

5.1.1 Thermal Oxidation

Thermal oxidation is the process used to create a "native" silicon dioxide (SiO_2) layer on a silicon surface. The SiO_2 grown by this process is used in different ways. For example, it serves as the insulator in the MOS system, is needed to perform patterning during the lithography, and allows for the technique of LOCOS (local oxidation of silicon) device isolation.

A silicon dioxide layer may be grown by passing an oxygen-rich gas over a silicon surface. Pure oxygen can be used as described by the reaction

$$Si + O_2 \rightarrow SiO_2.$$

Steam is also used to grow oxide layers. The reaction is

$$Si + 2H_2O \rightarrow SiO_2 + 2H_2,$$

where it is seen that free hydrogen is left as a by-product. Both reactions use heat as a catalyst, thus giving rise to the term "thermal" oxidation. Oxide growth using O_2 is called a "dry" process to distinguish it from "wet" oxidation in steam.

The kinetics of the oxide growth process can be understood by referring to Fig. 5.1. Figure 5.1(a) shows that the oxygen can react directly with the surface

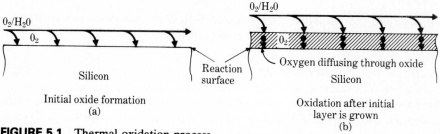

O_2/H_2O

O_2

Silicon

Reaction surface

Oxygen diffusing through oxide

Silicon

Initial oxide formation
(a)

Oxidation after initial
layer is grown
(b)

FIGURE 5.1 Thermal oxidation process.

silicon atoms to form SiO_2. This gives a high initial oxidation rate. After an oxide layer forms, the oxygen must *diffuse* through the existing SiO_2 (as shown in Fig. 5.1b) before it can react with the silicon atoms. Consequently, the oxide growth rate decreases from its initial value. An analysis of the reactant fluxes using a simple steady-state model for the growth kinetics shows that the oxide thickness x_{ox} as a function of time t is approximated by the quadratic equation

$$x_{ox}^2(t) + Ax_{ox}(t) = B(t + t_0), \tag{5.1-1}$$

where A and B are quantities characterizing the growth variables, such as temperature and pressure, and t_0 is included to account for any existing oxide at time $t = 0$. Denoting this initial oxide thickness by $x_{ox}(0)$, t_0 is computed from

$$t_0 = \frac{1}{B}[x_{ox}^2(0) + Ax_{ox}(0)]. \tag{5.1-2}$$

The solution to the oxide growth equation is

$$x_{ox}(t) = \frac{A}{2}\left[\sqrt{1 + \frac{4B}{A^2}(t + t_0)} - 1 \right], \tag{5.1-3}$$

which allows for the growth rate (dx_{ox}/dt) to be calculated.

To study the oxidation process, assume that $x_{ox}(0) = 0$ so that $t_0 = 0$ for simplicity. Then eqn. (5.1-3) reduces to

$$x_{ox}(t) = \frac{A}{2}\left[\sqrt{1 + \frac{4Bt}{A^2}} - 1 \right]. \tag{5.1-4}$$

For small values of t corresponding to the initial oxidation, the square-root term may be expanded in a Taylor series to give

$$x_{ox}(t) \simeq \frac{B}{A}t. \tag{5.1-5}$$

Since the oxide thickness is proportional to the oxidation time t, this is called the *linear growth region* with (B/A) the *linear rate constant*. As the oxidation time is increased, $(4Bt/A^2)$ dominates the square-root term, so for large values of t,

$$x_{ox}(t) \simeq \sqrt{Bt}. \tag{5.1-6}$$

This describes the *parabolic growth region* with B the *parabolic rate constant*.

It is possible to express the rate constants in the form

$$\begin{aligned} B &= C_1 e^{-E_1/kT}, \\ (B/A) &= C_2 e^{-E_2/kT}, \end{aligned} \tag{5.1-7}$$

where C_1 and C_2 contain the information on the kinetic growth variables, while E_1 and E_2 are (approximately) activation energies for the process. Typical values for (111) and (100) silicon surfaces are given in Table 5.1 for both dry

TABLE 5.1 Oxide Growth Rate Constants*

	Parabolic Rate Constant: $B = C_1 e^{-E_1/kT}$ Linear Rate Constant: $(B/A) = C_2 e^{-E_2/kT}$	
Source	(111) Silicon	(100) Silicon
Dry O_2	$C_1 = 7.72 \times 10^2 \, [\mu m^2/hr]$ $C_2 = 6.23 \times 10^6 \, [\mu m/hr] \qquad C_2 = 3.71 \times 10^6 \, [\mu m/hr]$ $E_1 = 1.23 \, [eV]$ $E_2 = 1.96 \, [eV]$	
Steam (@ 640 torr.)	$C_1 = 3.86 \times 10^2 \, [\mu m^2/hr]$ $C_2 = 1.63 \times 10^8 \, [\mu m/hr] \qquad C_2 = 0.97 \times 10^8 \, [\mu m/hr]$ $E_1 = 0.78 \, [eV]$ $E_2 = 2.05 \, [eV]$	

* From S. K. Ghandhi, *VLSI Fabrication Principles,* New York: Wiley, © 1983. Reprinted with permission from John Wiley & Sons, Inc.

and wet oxide growth rates. At a given temperature T, steam oxidation is much faster than that obtained by dry O_2. This faster growth rate is offset by the fact that the presence of excess H_2O found in the steam-grown oxide leads to inferior SiO_2 properties. Critical SiO_2 layers such as those required for MOSFET gate oxides are usually grown in dry oxygen. Examples of oxide thickness as a function of growth time are shown in Fig. 5.2. Increased pressure enhances the oxide growth rate, so *high-pressure* oxidation is commonly used when large x_{ox} values are needed in the chip design. Heavy doping in the silicon also increases the growth rate of an oxide layer.

Thermal oxidation uses silicon atoms at the Si-SiO_2 interface to create molecules of silicon dioxide. The location of the silicon surface will thus be different after the oxide growth is completed. This is illustrated in Fig. 5.3. The thickness x_{Si} of the silicon lost to the oxide growth can be related to the oxide thickness x_{ox} using simple arguments. First, recall that $N_{Si} = 5 \times 10^{22}$ [Si atoms/cm^3] are in the crystal. The silicon dioxide has approximately $N_{SiO_2} = 2.3 \times 10^{22}$ [molecules/cm^3]. Since it takes one atom of silicon to create one molecule of SiO_2,

$$N_{Si} x_{Si} = N_{SiO_2} x_{ox} \qquad (5.1\text{-}8)$$

expresses the proper balance. Thus,

$$x_{Si} = \frac{N_{SiO_2}}{N_{Si}} x_{ox} \approx 0.46 x_{ox} \qquad (5.1\text{-}9)$$

may be used to calculate the amount of silicon consumed in the oxidation process. This will become particularly important when the LOCOS isolation technique is discussed.

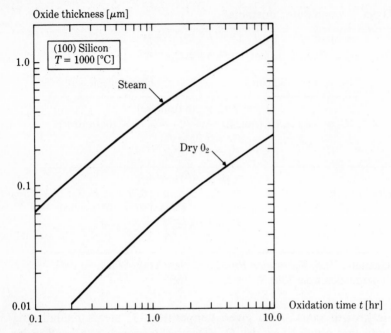

FIGURE 5.2   Example of thermal oxide growth.

It has been found that stray ionic charges, most notably alkali ions such as Na^+, are trapped in thermally grown oxide layers. This leads to problems in MOS threshold voltage variations through the flatband voltage V_{FB}. The amount of ionic charge can be reduced by performing the oxidation in a chlorinated atmosphere, such as that obtained by adding HCl or chlorine gas. The addition of chlorine to the recipe allows for trapping of Na^+ ions into neutral subsystems. This reduces charge at the Si-SiO_2 interface to levels where acceptable MOSFET performance can be attained. Phosphosilicate glass (PSG) may also be used as a method to attract mobile ions to a region of the oxide where they can be neutralized or removed.

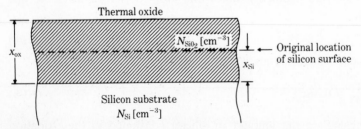

FIGURE 5.3   Geometry for calculation of silicon consumed in the thermal oxidation process.

EXAMPLE 5.1-1

An oxide layer is grown on (111) silicon using steam. The oxidation is performed at 1000 [°C] and lasts for 1 hour. Calculate x_{ox}.

Solution
First find the growth rate constants. For (111) silicon in steam,

$$B = 3.86 \times 10^2 e^{-0.78/(8.62 \times 10^{-5})(1273)} \simeq 0.316 \, [\mu m^2/hr],$$

while

$$(B/A) = 1.63 \times 10^8 e^{-2.05/(8.62 \times 10^{-5})(1273)} \simeq 1.244 \, [\mu m/hr].$$

Dividing,

$$A = \frac{B}{(B/A)} \simeq 0.252 \, [\mu m].$$

Using eqn. (5.1-4) then gives

$$x_{ox} \simeq \frac{(0.252)}{2} \left[\sqrt{1 + \frac{4(0.316)(1)}{(0.252)^2}} - 1 \right],$$

so $x_{ox} \simeq 4500 \, [\text{Å}]$. ∎

5.1.2 Impurity Diffusion

The diffusion of impurity atoms into silicon crystal has been used since the beginnings of the semiconductor industry. This doping technique is based on the fact that dopant atoms gain thermal kinetic energy at elevated temperatures. When this motion is combined with a gradient in the doping density $N(x)$, a net flux of particles $F \, [cm^2\text{-sec}]^{-1}$ is created such that the particles tend to move away from regions of high concentration. Mathematically this motion is described by *Fick's law* in the (one-dimensional) form

$$F = -D \frac{\partial N(x)}{\partial x}, \tag{5.1-10}$$

where $D \, [cm^2/\text{sec}]$ is the *diffusion coefficient*. D depends on the temperature, the dopant species, and the concentration of dopants in the diffusion path. Diffusions are typically performed in the temperature range 900–1100 [°C].

The general properties of diffusive motion are obtained by writing a balance equation for particle densities. Consider the situation shown in Fig. 5.4 where a concentration gradient causes particles to diffuse from region 1 to region 2. Since the number of particles lost from region 1 is (dN/dt), the flow

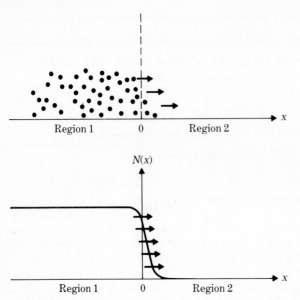

FIGURE 5.4 Particle flow established by a concentration gradient.

is described by

$$\frac{\partial N(x, t)}{\partial t} = -\frac{\partial F(x)}{\partial x}. \tag{5.1-11}$$

Classical diffusion theory assumes that D is not a function of position x. Thus, substituting the flux from eqn. (5.1-10) gives the *diffusion equation*

$$\frac{\partial N(x, t)}{\partial t} = D\frac{\partial^2 N(x, t)}{\partial x^2}. \tag{5.1-12}$$

This partial differential equation can be solved subject to appropriate boundary conditions to determine the impurity profile diffused into the silicon. The diffusion coefficient D in this equation is interpreted as an average value and varies with dopant species and diffusion temperature T [°K]. It may be computed using the formula

$$D = D_0 e^{-E_{av}/kT}, \tag{5.1-13}$$

in which E_{av} is an average activation energy corresponding to the dopant "jumping" from site to site. Approximate values of D_0 and E_{av} are provided in Table 5.2 for B, P, and As.

Dopant profiles are controlled by the manner in which the dopants diffuse into the silicon. Two types of profiles are of particular interest. The first is obtained using a *constant surface concentration* (or *predeposit*) process in which the surface dopant density $N_S = N(0)$ is maintained at a value N_0 [cm^{-3}]. N_0 is

TABLE 5.2 Average Diffusion Coefficients for Common Silicon Dopants

$D = D_0 e^{-E_{av}/kT}$ [cm^2/sec]		
Dopant	D_0 [cm^2/sec]	E_{av} [eV]
Boron	5.853	3.62
Phosphorus	5.853	3.62
Arsenic	0.091	3.37

known as the *solid solubility limit* and represents the maximum dopant density that can be "dissolved" in the silicon. N_0 depends on the dopant species and temperature as illustrated by the graph of Fig. 5.5. This type of diffusion problem is shown in Fig. 5.6(a). Assuming that the diffusion cycle lasts a time t_p and is carried out at a temperature T_p, the final impurity profile is found to be

$$N(x) = N_0 \operatorname{erfc}\left(\frac{x}{2\sqrt{D_p t_p}}\right),$$

(5.1-14)

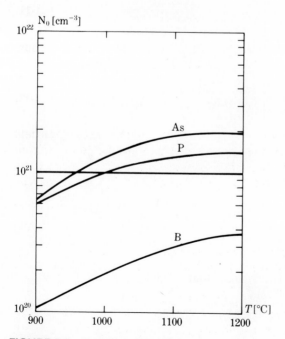

FIGURE 5.5 Solid solubility limits for boron, arsenic, and phosphorus.

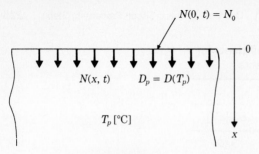

$N(0, t) = N_0$

$N(x, t)$ $\quad$ $D_p = D(T_p)$

$T_p \, [°C]$

Basic diffusion problem
(a)

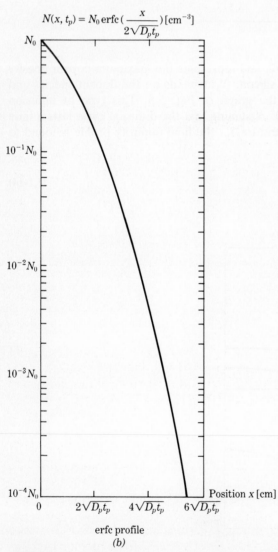

$$N(x, t_p) = N_0 \, \text{erfc} \left(\frac{x}{2\sqrt{D_p t_p}} \right) [\text{cm}^{-3}]$$

erfc profile
(b)

FIGURE 5.6 Constant surface concentration diffusion.

where $D_p = D(T_p)$. The *complementary error function* erfc (y) is given by

$$\text{erfc}\,(y) = 1 - \text{erf}\,(y), \tag{5.1-15}$$

where erf (y) is the *error function* defined by the integral representation

$$\text{erf}\,(y) = \frac{2}{\sqrt{\pi}} \int_0^y e^{-u^2}\, du. \tag{5.1-16}$$

A plot of the erfc profile is shown in Fig. 5.6(b). The total number of dopant atoms Q introduced into the wafer during this step is

$$Q = A \int_0^\infty N(x)\, dx = 2N_0 A \sqrt{\frac{D_p t_p}{\pi}}, \tag{5.1-17}$$

with A being the surface area. This type of diffusion is used for shallow doping profiles and also as a first step in deep diffusions.

The second type of diffusion problem is one where the dopants have been introduced into the wafer by a predeposit step and are then subjected to a higher-temperature *drive-in* diffusion at temperature T_d for time t_d. This is shown in Fig. 5.7(a) and is also known as a *limited source* diffusion. Assuming that Q atoms have been left by the predeposit, solving eqn. (5.1-12) gives the Gaussian profile

$$N(x) = \frac{Q}{A\sqrt{\pi D_d t_d}}\, e^{-x^2/4D_d t_d}, \tag{5.1-18}$$

with $D_d = D(T_d)$. Using Q from eqn. (5.1-17) finally yields the result of this *2-step* process as

$$N(x) = \frac{2N_0}{\pi} \sqrt{\frac{D_p t_p}{D_d t_d}}\, e^{-x^2/4D_d t_d}. \tag{5.1-19}$$

This profile is illustrated in Fig. 5.7(b) and is valid as long as $(D_d t_d) \gg (D_p t_p)$ is satisfied. Usually the wafer will be subjected to additional heat treatment steps, which will cause the dopants to diffuse deeper into the silicon. This may be accounted for by replacing the factor $(D_d t_d)$ in eqn. (5.1-19) by an *effective* value

$$\overline{Dt} = D_d t_d + D_1 t_1 + D_2 t_2 + \cdots, \tag{5.1-20}$$

where each term $D_j t_j$ in the sum represents the effect of a separate heating step in the wafer processing.

The doping profiles in eqns. (5.1-14) and (5.1-19) are reasonable approximations as long as the doping levels are low enough to ignore the dependence of D on the carrier densities. High-concentration diffusions used in certain VLSI processes do not result in erfc or Gaussian profiles, and the analysis must be modified to describe these situations. The most general

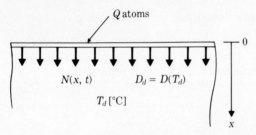

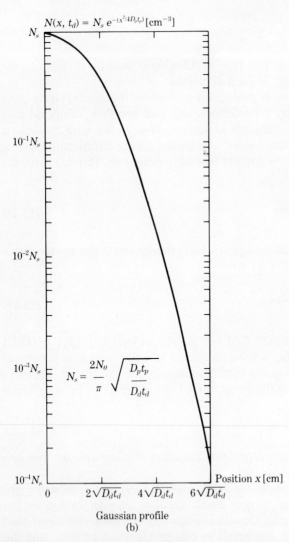

Basic diffusion process
(a)

$$N(x, t_d) = N_s \, e^{-(x^2/4D_p t_p)} \, [\text{cm}^{-3}]$$

$$N_s = \frac{2N_0}{\pi} \sqrt{\frac{D_p t_p}{D_d t_d}}$$

Gaussian profile
(b)

FIGURE 5.7 Limited source diffusion.

TABLE 5.3 Concentration-Dependent Diffusion
Coefficient Quantities*

$$D = h\left[D_i^o + D_i^-\left(\frac{n}{n_i}\right) + D_i^=\left(\frac{n}{n_i}\right)^2 + D_i^+\left(\frac{p}{n_i}\right)\right]$$

$$D_i^x = D_0^x e^{-E_a^x/kT}$$

Dopant	Boron	Phosphorus	Arsenic
D_0^o	0.091	3.85	0.38
E_a^o	3.36	3.66	3.58
D_0^-	–	4.44	22.9
E_a^-	–	4.0	4.1
$D_0^=$	–	44.2	–
$E_a^=$	–	4.37	–
D_0^+	166.3	–	–
E_a^+	4.08	–	–

D_0^x in [cm²/sec], E_a^x in [eV]
* From R. A. Colclaser, *Microelectronics Processing and Device Design*, New York: Wiley, © 1980. Reprinted with permission from John Wiley & Sons, Inc.

situation is described by writing the diffusion coefficient in the form

$$D = h\left[D_i^o + D_i^-\left(\frac{n}{n_i}\right) + D_i^=\left(\frac{n}{n_i}\right)^2 + D_i^+\left(\frac{p}{n_i}\right)\right], \tag{5.1-21}$$

where h is called the *field enhancement factor*. It accounts for *field-aided* diffusion under the influence of a built-in electric field that is created by the separation of electrons from their host dopant atoms. h has a value between 1 and 2. n and p are the carrier densities, while n_i is the intrinsic concentration at the diffusion temperature. The quantity D_i is the intrinsic diffusion coefficient for the x *charge-state vacancy*. Intrinsic values are computed from

$$D_i^x = D_0^x e^{-E_a^x/kT}, \tag{5.1-22}$$

where E_a is the activation energy for the state. Table 5.3 provides values of D_0 and E_a for B, P, and As. Note that only certain terms in eqn. (5.1-21) contribute to D for a given dopant.

Boron is a p-type (acceptor) dopant that exhibits neutral and positive charge-state vacancy interactions. The diffusion coefficient has the form

$$D \simeq D_i^o + D_i^+\left(\frac{p}{n_i}\right), \tag{5.1-23}$$

where the field enhancement factor has been set to unity. It is convenient to define an average intrinsic diffusion coefficient by

$$D_i = D_i^o + D_i^+. \tag{5.1-24}$$

A constant surface concentration boron diffusion is then found to be described by the profile

$$N(x, t_p) \simeq N_0(1 - Y^{2/3}), \tag{5.1-25}$$

where

$$Y = \left(\frac{n_i x^2}{6N_0 D_i t_p}\right)^{3/2}. \tag{5.1-26}$$

This type of diffusion result is shown in Fig. 5.8 where it is compared to the

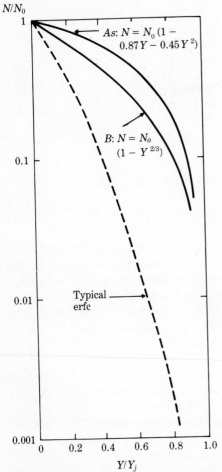

FIGURE 5.8 High-density constant surface concentration diffusion profile comparison.

erfc distribution. The diffusion edge is much sharper than that predicted by the classical theory. When this type of boron diffusion is performed into an n-type background, a pn junction is formed. The sharpness of the curve allows the location x_j of the junction to be found by setting $N(x_j, t_p) = 0$, i.e., $Y = 1$. Thus,

$$x_j \simeq 2.45\sqrt{D_i t_p}\,\sqrt{\frac{N_0}{n_i}} \tag{5.1-27}$$

allows for an estimate of the diffusion depth.

High-density arsenic (n-type) diffusions are characterized by

$$D = h\left[D_i^o + D_i^-\left(\frac{n}{n_{ie}}\right)\right], \tag{5.1-28}$$

showing interactions with neutral and single-negative charge-state vacancies. The *effective intrinsic carrier density* n_{ie} is introduced to account for *bandgap narrowing* from the high doping levels. The effective density is computed from

$$n_{ie}^2 = n_i^2 e^{\Delta E_g/kT} > n_i^2 \tag{5.1-29}$$

with ΔE_g the reduction in the energy gap. Figure 5.9 shows n_i and n_{ie} over the standard range of diffusion temperatures.

It is found that the field enhancement factor for As is approximately

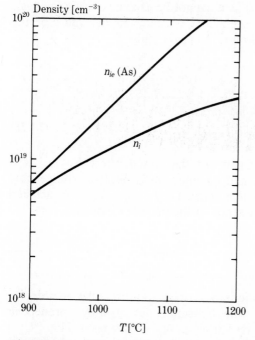

FIGURE 5.9 n_i and n_{ie} as functions of temperature.

$h = 2$. Using this with the assumption that $(n/n_{ie}) \gg 1$ reduces the diffusion coefficient to

$$D \simeq 2D_i^- \left(\frac{n}{n_{ie}} \right). \tag{5.1-30}$$

A constant surface concentration diffusion lasting a time t_p results in a doping profile that is approximated by

$$N(x, t_p) = N_0(1 - 0.87Y - 0.45Y^2), \tag{5.1-31}$$

with Y the dimensionless variable

$$Y = \sqrt{\frac{n_{ie}x^2}{8N_0 D_i^- t_p}}. \tag{5.1-32}$$

This type of distribution is also plotted in Fig. 5.8. When As is diffused into a lightly doped p-type sample, the pn junction depth may be calculated by setting $N(x_j, t_p) = 0$. Solving the quadratic results in

$$x_j \simeq 1.616\sqrt{2D_i^- t_p \left(\frac{N_0}{n_{ie}} \right)}, \tag{5.1-33}$$

which provides a good estimate of the diffusion depth.

The diffusive behavior of high-density phosphorus diffusions is quite complicated and is termed *anomalous* as it cannot be explained using the basic theory. The form of the diffusion coefficient depends on the value of the electron concentration with respect to a reference density

$$n_e \simeq 4.65 \times 10^{21} e^{-0.39/kT}. \tag{5.1-34}$$

In regions near the surface where $n > n_e$,

$$D = h\left[D_i^o + D_i^= \left(\frac{n}{n_{ie}} \right)^2 \right] \tag{5.1-35}$$

is valid. However, in the deeper regions of the crystal where $n < n_e$, it is found that phosphorus has neutral and complex single-negative charge-state interactions. This changes D to what is termed the "tail" value. A detailed discussion of this topic is beyond the scope of the present treatment.

EXAMPLE 5.1-2

A 2-step phosphorus diffusion is made into a p-type wafer that has an acceptor doping density of $N_a = 10^{15}$ [cm^{-3}]. The diffusion is described by a predeposit for 30 minutes at 900 [°C] followed by a drive-in for 60 minutes at 1100 [°C].

Calculate the depth x_j of the pn junction formed by this process.

Solution

First read $N_0 = 6 \times 10^{20}$ [cm^{-3}] for the solid solubility. Next compute the diffusion coefficients at the two temperatures:

$$D_p = 5.853e^{-3.62/(8.62 \times 10^{-5})(1173)} \simeq 1.66 \times 10^{-15} \text{ [cm}^2/\text{sec]},$$

$$D_d = 5.853e^{-3.62/(8.62 \times 10^{-5})(1373)} \simeq 3.15 \times 10^{-13} \text{ [cm}^2/\text{sec]}.$$

Then use eqn. (5.1-19) to find the doping profile:

$$N(x) = \frac{2(6 \times 10^{20})}{\pi} \sqrt{\frac{(1.66 \times 10^{-15})(30)}{(3.15 \times 10^{-13})(60)}} e^{-x^2/4(3.15 \times 10^{-13})(60)(60)}.$$

This gives

$$N(x) \simeq 1.97 \times 10^{19} e^{-x^2/0.454} \text{ [cm}^{-3}],$$

where x is in units of [μm].

To calculate the junction depth, just look for the point where the donor doping exactly equals the p-type acceptor level N_a, i.e.,

$$N(x_j) = N_a.$$

Solving,

$$x_j \simeq \sqrt{0.454 \ln \left(\frac{1.97 \times 10^{19}}{10^{15}} \right)} \simeq 2.12 \, [\mu\text{m}]. \qquad \blacksquare$$

EXAMPLE 5.1-3

A high-density arsenic diffusion is performed at 1000 [°C] for 30 minutes into a lightly doped p-type wafer. Calculate the junction depth x_j.

Solution

First read $N_0 \simeq 6.5 \times 10^{20}$ [cm^{-3}] and $n_{ie} \simeq 2 \times 10^{19}$ [cm^{-3}] from the charts. Next use the data in Table 5.3 to calculate

$$D_i^- = 22.9e^{-4.1/(8.62 \times 10^{-5})(1273)} \simeq 1.36 \times 10^{-15} \text{ [cm}^2/\text{sec]}.$$

Equation (5.1-33) then gives

$$x_j \simeq 1.616 \sqrt{2(1.36 \times 10^{-15})(30)(60) \frac{6.5 \times 10^{20}}{2 \times 10^{19}}},$$

or $x_j \simeq 0.204$ [μm]. $\qquad \blacksquare$

5.1.3 Ion Implantation

Doping by ion implantation has already been mentioned in Section 1.1 for setting MOSFET threshold voltages. An ion implanter is an apparatus that first

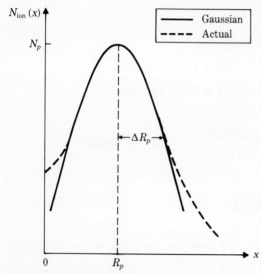

FIGURE 5.10 Gaussian approximation for ion implant profile.

ionizes a dopant source and then accelerates the ions to high energies. The ions form a beam that is focused to scan over the wafer. The dopant ions are literally smashed into the substrate. This results in a doping profile that is approximated to first order by the Gaussian

$$N_{\text{ion}}(x) = \frac{D_I}{\sqrt{2\pi}(\Delta R_p)} e^{-(1/2)[(x-R_p)/\Delta R_p]^2}. \qquad (5.1\text{-}36)$$

The projected range R_p represents the average penetration depth of the implanted ions. R_p increases with increasing energies and is also a function of the dopant mass. The straggle (ΔR_p) is the standard deviation associated with the implanted depth and depends on both the incident ion energy and the ion mass relative to the silicon atomic mass. The general shape of the Gaussian approximation is illustrated in Fig. 5.10, where it is compared to a more realistic implant profile.

The number of ions implanted into a wafer is specified by using the dose D_I [cm^{-2}] defined as

$$D_I = \int_{-\infty}^{+\infty} N_{\text{ion}}(x)\, dx = N_p(\Delta R_p)\sqrt{2\pi}, \qquad (5.1\text{-}37)$$

where N_p represents the peak concentration (at $x = R_p$) in units of [cm^{-3}]. The dose is a carefully controlled quantity and can be measured using the *current integrator* technique shown in Fig. 5.11. This approach to *dosimetry* measures the current supplied to the wafer to offset the charge imbalance created by the ion beam. To understand the system, let I represent the beam

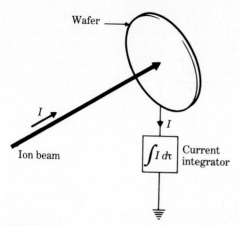

FIGURE 5.11 Current integrator dosimetry arrangement.

current so that

$$\Delta Q_o = \int_0^{t_I} I(\tau)\,d\tau \tag{5.1-38}$$

is the total charge delivered to the wafer during an implant time t_I. The charge on an individual ion can be written as $(\pm sq)$, where $s = 1$ or 2 depending on whether the ions are singly or doubly charged. The total number of ions implanted into the wafer is thus given by $(\Delta Q_o/sq)$. Denoting the implanted area by A yields the dose expression

$$D_I = \frac{1}{sqA} \int_0^{t_I} I(\tau)\,d\tau. \tag{5.1-39}$$

This shows that the implant dose can be computed directly from knowledge of the beam current. Dose measurements typically exhibit less than 1% error, which serves to make ion implantation one of the most carefully controlled processes available in silicon wafer fabrication technology.

Ion implant profiles result from the statistical nature of the *ion-stopping process*. An incident ion is assumed to have a mass M_I and energy E_I, as shown in Fig. 5.12. The ion loses energy to the crystal by collisions with electrons and silicon nuclei. The maximum energy E_T transferred to a *target* mass M_T (and hence lost by the ion) is found to be

$$E_T = 4\frac{M_I M_T}{(M_I + M_T)^2} E_I, \tag{5.1-40}$$

where M_T is either the electron mass m or the nuclear mass M_{Si}, depending on the event. Coulombic forces also contribute to the ion energy loss. The overall effects are contained in the *stopping power* S [eV/cm] such that

$$S = S_n(E_I) + S_e(E_I) \tag{5.1-41}$$

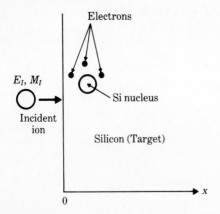

FIGURE 5.12   Basic ion-stopping problem.

gives the nuclear and electronic contributions. The ion energy loss is then described by

$$-\frac{dE_I}{dx} = N_{\text{Si}}[S_n(E_I) + S_e(E_I)]. \tag{5.1-42}$$

This may be used to find the implanted distance R for an ion by integrating in the form

$$R = \int_0^R dx = -\int_{E_I}^0 \frac{dE_I}{N_{\text{Si}}[S_n(E_I) + S_e(E_I)]}. \tag{5.1-43}$$

The limits on the energy integral correspond to having an incident ion energy E_I at $x = 0$ and to an ion at rest $(E = 0)$ at $x = R$. The projected range R_p is the average value of R for all of the ions. The straggle ΔR_p represents the statistical variations in the ion penetration distances.

The preceding brief description of the ion-stopping process illustrates two important problems introduced by the implant. First, the collisions with the targets will tend to knock Si atoms out of their equilibrium lattice sites, which causes cracks and defects in the crystal. Second, the ions will be randomly distributed in the lattice and cannot act as substitutional impurities until the implant is *activated*. This requires that the dopants find positions corresponding to normal lattice sites in the crystal. To solve these two problems, the wafer is subjected to an *annealing* step that simultaneously repairs the damage and activates the implant.

Thermal annealing consists of heating the wafer for a short time to allow for diffusion of both the dopants and the silicon atoms. Assuming that this is performed at a temperature T_A for a time t_A, the thermal anneal has the effect of changing the straggle to

$$(\Delta R_p) \rightarrow \sqrt{(\Delta R_p)^2 + 2D_A t_A}, \tag{5.1-44}$$

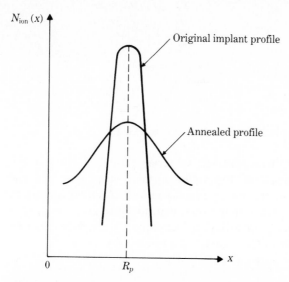

FIGURE 5.13   Effects of implant thermal annealing.

where $D_A = D(T_A)$ for the implanted species. A typical anneal cycle might be 30 minutes at 1000 [°C]. Using this in eqn. (5.1-36) gives the final implant form

$$N_{ion}(x) = \frac{D_I}{\sqrt{2\pi}\sqrt{(\Delta R_p)^2 + 2D_A t_A}} \exp\left[-\frac{1}{2}\left(\frac{x - R_p}{\sqrt{(\Delta R_p)^2 + 2D_A t_A}}\right)^2\right]. \qquad \textbf{(5.1-45)}$$

As shown by the plots in Fig. 5.13, the thermal anneal causes the dopants to diffuse away from the projected range point R_p. This spreads the distribution and reduces the peak concentration. A more accurate analysis would account for reflections and out-diffusions at the silicon surface.

Rapid anneal techniques have recently become of interest. This approach involves heating the wafer to high temperatures for very short time periods. An example of a rapid anneal technique is *laser annealing,* which employs a pulsed high-energy laser beam focused to scan the wafer. This induces localized melting and recrystallization. The primary advantage of this approach is that the annealed profile is very close to the original implant distribution, i.e., the spread described by eqn. (5.1-44) is not found for this type of annealing. The main disadvantage is that the *wafer throughput* (the number of wafers processed per hour) is greatly reduced.

5.1.4   Chemical Vapor Deposition

Chemical vapor deposition (CVD) is used to obtain layers of various materials in silicon wafer fabrication. As implied by its name, CVD produces layers by using chemical reactions that deposit the desired material onto the substrate. The reaction kinetics involved in this type of process depend on the temperature, the pressure, and the reactant species.

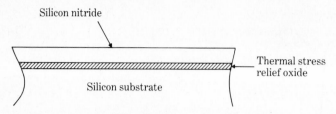

FIGURE 5.14 Nitride with stress-relief oxide to prevent cracking.

Silicon nitride (Si_3N_4) is an important material produced by a CVD process. Nitride is used to *passivate* the wafer surface, since it is a dense dielectric that forms an effective barrier against alkali ions. Another important property of Si_3N_4 is that it does not oxidize very well; this property is exploited to provide the basis for the LOCOS isolation technique. A common reaction for producing nitride layers uses silane (SiH_4) in ammonia by means of

$$3SiH_4 + 4NH_3 \rightarrow Si_3N_4 + 12H_2.$$

The reaction temperature for this process is on the order of 700 [°C]. One problem that arises is that nitride layers exhibit very high tensile stress, which can lead to cracks when the nitride is deposited directly onto a silicon surface. To avoid this situation, a thin oxide layer is usually grown on a silicon surface prior to the nitride deposition, as shown in Fig. 5.14. This *stress relief oxide* tends to relieve some of the surface tension and allows for reliable coverage by the nitride.

CVD oxide layers are required whenever a silicon surface is not available for a thermal oxide growth. One of the more common reactions used for this purpose is the pyrolysis of silane in oxygen as described by

$$SiH_4 + 2O_2 \rightarrow SiO_2 + 2H_2O \quad (600\text{–}1000 \text{ [°C]}),$$
or
$$SiH_4 + O_2 \rightarrow SiO_2 + 2H_2 \quad (300\text{–}500 \text{ [°C]}).$$

The lower-temperature reaction produces hydrogen as a by-product and results in higher-quality films than those obtained using the high-temperature water-producing process. Phosphosilicate glass (PSG) can be deposited by simultaneous pyrolysis of silane and phosphine (PH_3). Phosphine reacts with oxygen according to

$$2PH_3 + 4O_2 \rightarrow P_2O_5 + 3H_2O.$$

The P_2O_5 is then incorporated into the silica during the deposition process. PSG has a lower tensile stress than pure silica, which tends to make PSG more reliable when many thermal cycles are required in the process flow.

Polycrystalline silicon (poly) may be deposited using the direct pyrolysis of silane

$$SiH_4 \rightarrow Si + 2H_2.$$

In a low-pressure system, this occurs at a reaction temperature of about 600 [°C]. Typical deposition rates are on the order of 200 [Å/min]. The polycrystalline nature of the silicon deposited in this manner is a strong function of reaction temperature. For temperatures less than about 575 [°C], the poly layer is found to be amorphous. In contrast, layers grown at temperatures greater than about 625 [°C] tend to exhibit localized columnar crystal patterns. The orientation and diameter of these columns are dependent on the specifics of the reaction conditions. Poly may be doped during the deposition process by the addition of diborane (B_2H_6), arsine (AsH_3), or phosphine to the system. In MOS technology it is common to deposit intrinsic poly and then selectively dope regions of the layer during the normal process flow.

5.1.5 Metallization and Silicides

Metal layers are used to provide low-resistance interconnects. Aluminum (and its alloys) is the most commonly used metal for this application, with a room temperature resistivity of about 2.7 [mΩ-cm]. Gold is also useful for certain applications. Metallic layers of this type are obtained by first melting and then evaporating the metal in a closed vacuum system. The solid samples are heated using standard resistance wiring, electron beams, or rf-induction heating techniques. Placing the wafers into the evaporant flux allows for a complete and (almost) uniform coverage of the surface. Condensation of the metal onto the wafer is attained by keeping the temperature of the wafer lower than that of the evaporant.

Refractory (high-temperature) metals are finding increased use in VLSI structures. The main advantage of using a refractory metal is that low-resistance patterned interconnect levels can be formed at any time in the chip process flow. The high melting point associated with these metals keeps the pattern intact even if the wafer is subjected to additional heat treatments. (Al has a low melting point—660 [°C]—and can be used only after all the high-temperature cycles are completed.) This property allows for multiple-level interconnects to be created, which aids in the layout of complex chips. Refractory metals of current interest include titanium (Ti), platinum (Pt), molybdenum (Mo), and tungsten (W). These may be evaporated using electron-beam heating or may be *sputtered* onto the wafer. In sputtering, an inert gas such as Ar is ionized and used to bombard an electrode that is coated with the desired substance. The collisions knock the atoms off the electrode, creating a particle flux directed toward the wafer.

Silicides are compounds formed using silicon and a refractory metal. These have been introduced for the same reason as the elemental refractory metals, i.e., to create high-conductivity paths that can withstand high-temperature processes without melting. Some of the better-studied silicides are $TiSi_2$, $PtSi_2$, $TaSi$, and WSi_2. In this group, titanium disilicide ($TiSi_2$) has shown the lowest resistivity, with about $\rho = 2.5$ [mΩ-cm] for cosputtered samples.

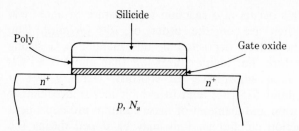

FIGURE 5.15 A polycide gate MOSFET.

Polycides are simply combinations of polysilicon and silicides. A typical polycide consists of a lower level of poly with an upper coating of silicide. These are particularly useful because they exhibit the best features of both polysilicon and silicides, including good stability, adherence, and coverage (from the poly) and a high conductivity path (from the silicide). A MOSFET fabricated with a polycide gate is shown in Fig. 5.15. It should be noted that the threshold voltage of this device is dependent on the specifics of the layering used for the gate. If a thick poly layer (>1500 [Å]) is used, then the gate threshold voltage will be the same as if only poly were used. However, if the poly layer is thin, a shift in the gate-substrate work function Φ_{GS} will occur because of the silicide. This must be accounted for in the threshold voltage adjustment implant.

5.2 Lithography

An integrated circuit may be viewed as a set of patterned layers. Each layer is a material with specific electrical properties. For example, n^+ regions, silicon dioxide, and aluminum form layers of doped semiconductor, insulator, and metal, respectively. In general, a layer must be patterned before the next layer of material is applied. A device such as a MOSFET is a subset of the patterned chip layers. Device characteristics result from the 3-dimensional structure created by both the layering order and the geometric pattern on each layer.

The process used to transfer a pattern to a layer of an integrated circuit is called *lithography*. Since each layer has its own distinct patterning requirements, the lithographic sequence must be repeated for every layer used to build the chip. The main requirements for a lithographic process are (a) a *mask* or *reticle* that is a "master copy" of the desired pattern; (b) an *exposure* or *printing* system that allows the transfer of the pattern to the wafer; and (c) a *resist* material that coats the wafer and is itself patterned by the printing system as the first step in creating the patterned layer.

5.2.1 Basic Pattern Transfer

The easiest way to illustrate the important aspects of a lithographic process is through a specific example. The one chosen here is basic *optical* lithography

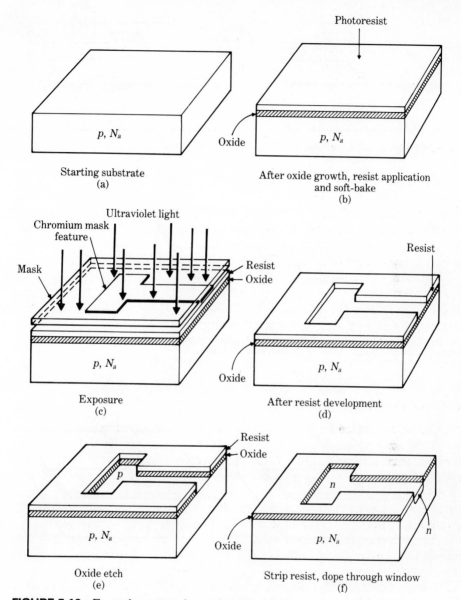

FIGURE 5.16   Example process for optical lithography.

with *contact/proximity* printing. The process will be discussed in conjunction with the wafer fabrication steps in Fig. 5.16. This set of drawings will show how a patterned n^+ region is created in a p-type substrate such that the n^+ *well* is accessed by a metal interconnect layer.

The starting point for the example is the section of p-type wafer with constant doping N_a shown in Fig. 5.16(a). The wafer is first subjected to a

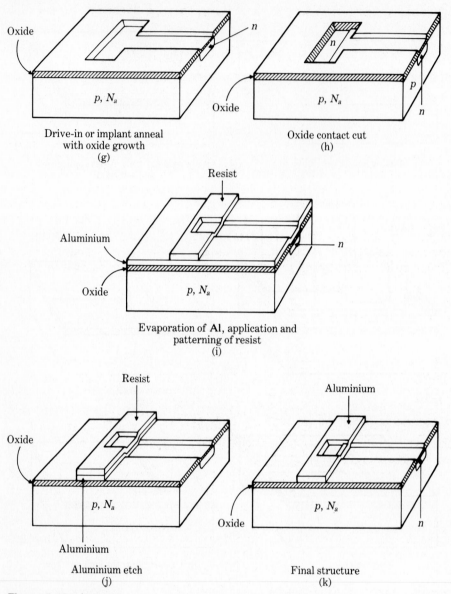

Drive-in or implant anneal
with oxide growth
(g)

Oxide contact cut
(h)

Evaporation of **Al**, application and
patterning of resist
(i)

Aluminium etch
(j)

Final structure
(k)

Figure 5.16 *(contd.)*

thermal oxide growth and is then coated with a layer of *photoresist*. At this point, the wafer has the layering of Fig. 5.16(b). Photoresist is a light-sensitive organic polymer (generally a type of plastic) that is initially in a liquid state when it is applied to the wafer. The wafer is spun to allow for an even coating of resist, and it is then subjected to a *soft-bake* heating step that solidifies the resist enough for handling. Photoresist is available in two types, negative and positive. Negative resists react to light by absorbing photons, which creates cross-linking of the polymers when the resist is developed. Regions of negative photoresist that are exposed to light become hardened after the developing process. Positive photoresist works in the opposite manner, with absorbed photons contributing toward increasing the solubility of the resist base *resin*. Exposing positive resist to light creates soluble regions that are dissolved away in the developing solution. The example here will assume that negative photoresist is used.

Figure 5.16(c) shows the actual exposure step in the first lithographic cycle. The pattern for the n^+ geometry is shown by the opaque regions on the mask. The mask itself is high-quality quartz glass on which a 1-to-1 (or 1 ×) copy of the pattern has been made. The most common substance used to create opaque mask regions for optical lithography is chromium. Ultraviolet (UV) light is used to expose the photoresist through the mask as shown. Photon absorption occurs in the photoresist underneath the clear mask regions. In contact printing, the mask actually touches the surface of the resist layer. This results in the best pattern transfer resolution but tends to wear out the mask rather quickly. Proximity printing solves this problem by maintaining a small spacing (on the order of 20–40 $[\mu m]$) between the mask and wafer surface. Although the mask does not degrade as rapidly, the resolution of the image is reduced from that obtained by contact printing due to diffraction.

After exposure, the resist is developed to obtain the patterning as in Fig. 5.16(d). Regions of negative resist that were exposed to the UV light through the mask have become hardened. This allows for the silicon dioxide to be selectively etched into the pattern, illustrated by Fig. 5.16(e). The silica layer may be etched using a wet chemical bath of buffered hydrofluoric (HF) acid. Alternatively, a *dry etch* process based on a gaseous plasma may be used. It is important to note that this step has opened a patterned oxide window down to the bare p-type silicon substrate.

The next step in the processing is to strip the photoresist off the wafer. This is required because resist generally has a low melting temperature, which prohibits leaving it on during high-temperature cycles. The pattern provided by the oxide window is then used to selectively dope the wafer with donors to form the patterned n-type well in the p-type substrate. Either diffusion or implantation may be used since the oxide acts as a mask to keep the dopants out of the silicon regardless of the doping technique employed. The oxide itself becomes doped during this process but is assumed to be thick enough to keep the dopants from reaching the silicon surface. The resulting n-type well that is created in this step is shown in Fig. 5.16(f).

After the initial doping is completed, the wafer is heated for either a drive-in (for a diffusion) or an anneal (with an ion implant). The atmosphere is assumed to be nitrogen and oxygen, so an oxide layer is grown over the entire surface. This is drawn in Fig. 5.16(g). The oxide is introduced at this point as an insulator between the semiconductor and the metal layer.

Access to the n^+ well is obtained by another lithography step, which results in the oxide contact window shown in Fig. 5.16(h). Aluminum is then evaporated over the entire surface of the wafer, and the desired metal pattern is formed on the resist layer with the next lithography sequence. The wafer at this point in the processing appears as illustrated by Fig. 5.16(i). Etching the aluminum leaves the patterned layers shown in Fig. 5.16(j). The last step in this example is to strip the resist, which leaves the final structure shown in Fig. 5.16(k).

This example demonstrates that the masking process is repeated for every layer that needs patterning. In particular, masks were used to pattern the n^+ doping, the oxide contact cut, and the metal geometry. Owing to the importance of obtaining patterned layers that are accurate images of the mask, good lithography is crucial for making integrated circuits. Many variations in the lithography sequence are possible. Although space prohibits a detailed discussion of the subject, some of the possibilities are discussed.

The first to be examined is an exposure technique known as *projection printing*. In this approach, the mask is kept some distance away from the wafer, and the pattern transfer is effected by optically projecting the image through a lens system. One example of this type of printing system is the *wafer stepper* shown in Fig. 5.17(a). The layer patterning is contained on the reticle, which is typically a 5× or 10× enlargement of a single die area. Pattern transfer is accomplished by optically reducing the image to the correct dimensions while simultaneously focusing it onto a specified region of the wafer. After the portion of the wafer is exposed, the system "steps" the image

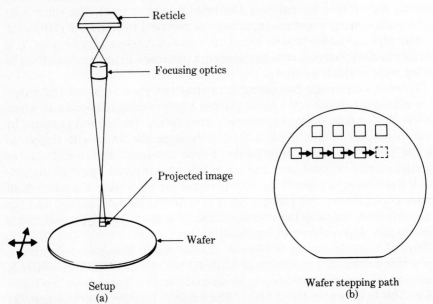

Setup
(a)

Wafer stepping path
(b)

FIGURE 5.17   Wafer stepper projection printing system.

to the next die site, as schematically illustrated by Fig. 5.17(b). This stepping process is continued until the wafer is completely exposed.

The direct wafer stepping technique is useful because it allows for wafer surface variations due to warping. This is particularly important as wafer diameters increase. The use of a 5× or 10× reticle pattern allows for larger tolerances in the pattern definition, making it more accurate than that possible with a 1× system. Of course, diffraction effects compensate for some of this because of the nature of the projection system. Another advantage of this technique is that the pattern transfer is not greatly affected by dust particles because the rectile is shielded with a sheath of clear nitrocellulose (or similar material) called a *pellicle*. The system is arranged so that any dust accumulation is restricted to the pellicle surface. Image control is maintained by ensuring that the distance between the reticle and the pellicle is sufficient to keep the dust particles out of the focusing range of the lens system. Consequently, the dust particles do not image onto the wafer. The main disadvantage of the wafer stepper is that the wafer throughput is low compared with that of a single-exposure process.

Negative photoresists are based on light-sensitive organic polymeric resins to which photosensitizing agents have been added to aid in the photon absorption process. Typical negative-working resins are polyvinyl cinnamate and cyclized polyisoprene. The resin itself must have the basic properties that it can adhere to the desired surface and can be developed into a hardened film that is resistant to the basic layer etchants. The specific sensitizer used to create a working negative resist depends on the resin system since the two must be compatible. Negative photoresist can be made very sensitive to light, allowing for rapid exposure. This makes negative photoresists useful for high wafer throughput. However, they have the undesirable property of swelling during the development process because the developing agent is absorbed by the polymer. Owing to the distortion introduced by the swelling, negative photoresists tend to be limited to minimum linewidths on the order of 2 [μm].

Positive photoresists do not absorb the developing solution and are therefore the prime candidates for fine-line optical VLSI lithography. The polymeric chemistry associated with positive photoresists is quite different from that described above for negative-working resists. Resins used in positive photoresists are not generally sensitive to light but are chosen for their basic characteristics, such as solubility and adherence. The exposure properties of positive resists arise from the reaction of the added light sensitizers with the base resin chains. Positive photoresists tend to require longer exposure times than those needed for negative resists. Consequently, improved resolution capabilities are gained at the expense of reduced wafer throughput.

X-ray lithography is similar to the optical UV technique described above except that X-rays are used as the exposure source. This approach is of interest because the shorter wavelengths of X-rays allow for finer resolution in the pattern transfer process while still maintaining high wafer throughput rates.

The most difficult aspect of X-ray lithography is mask making. The requirements are somewhat different from those used in optical lithography in that the opaque regions of an X-ray mask are made to absorb (rather than reflect) the radiation. Gold is typically used as an absorber for this type of mask. The walls for the gold must be maintained perpendicular to the mask surface to ensure good resolution.

Electron-beam lithography offers the highest-resolution pattern formation by directly writing onto the wafer. The electron beam is columnated to form the desired *spot size,* and the beam is then used to scan the wafer surface, as shown in Fig. 5.18(a). The pattern information is contained in the e-beam scan control system, which selectively turns the beam on and off depending on the wafer coordinates of the beam as it scans the surface. Since beam diameters as small as 0.01 [μm] are possible, direct e-beam writing allows for extremely fine-line patterning. The major drawback of this technique is that the wafer throughput is low because of the time required to scan the wafer at a rate slow enough to expose the resist. The throughput can be increased by enlarging the spot size, but this obviously reduces the resolution. A typical production line system providing a minimum linewidth of 2 [μm] might use an e-beam as large as 0.5 [μm]. Other problems are that the circular beam shape makes it difficult to form right-angle (90°) turns in the pattern and that electron backscattering, illustrated in Fig. 5.18(b), tends to widen the actual exposure track.

5.2.2 Lateral Doping Effects

Doping through an oxide window results in a doped region that is larger than the window itself. This occurs because of *lateral doping* effects in which there is a "spreading" of the dopants outward from the window edge in directions parallel to the silicon surface. Both diffused and implanted patterns exhibit lateral spreading. The MOSFET overlap capacitances C_{ols} and C_{old} discussed in Section 4.6, are direct consequences of this phenomenon.

Lateral diffusion is illustrated in Fig. 5.19(a). It is seen that the oxide window serves as a mask for the patterning but that dopants at the edge of the oxide window diffuse outward. This results in the diffused well geometry shown in Fig. 5.19(b). The amount of lateral spreading is usually characterized by the lateral junction distance y_j. This can be quite large, with $(y_j/x_j) = 0.8$ a typical order of magnitude for a deep 2-step Gaussian profile. The actual value of y_j is dependent on the background doping N_{BC} of the substrate and the specifics of the diffusion cycle.

Lateral doping in ion implantation initially results from ion scattering outside the oxide "window plane" during the ion-stopping process. This is shown in Fig. 5.20(a). Thermal annealing of the implant will induce further spreading because of lateral diffusive motion during the heat treatment. The final width of the implant is thus larger than the oxide window, as depicted in Fig. 5.20(b). Lateral doping from an ion implant is generally much less than

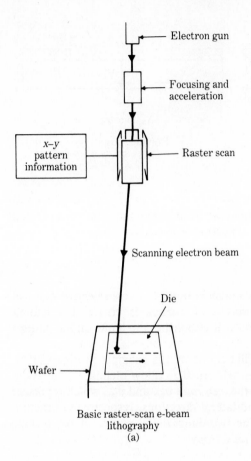

Basic raster-scan e-beam
lithography
(a)

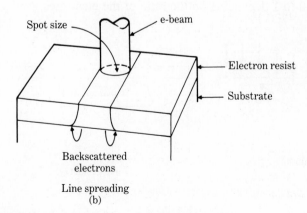

Line spreading
(b)

FIGURE 5.18 Electron-beam lithography.

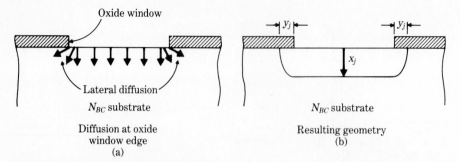

FIGURE 5.19 Lateral diffusion effects.

that for an impurity diffusion. Consequently, ion implantation is better suited for the fine-line geometries needed in VLSI chip designs.

5.2.3 Etching

Pattern transfer uses a resist layer as a etchant mask to allow selective removal of the underlying material. The process of etching is critical to creating fine-line features on integrated circuits as it establishes the limits on the pattern resolution.

The basic problem in etching is illustrated by the example in Fig. 5.21(a), where a mask (resist) layer has been patterned to allow etching of an oxide. The etch process is characterized by the *etch rates* v_{ver} and v_{lat}, which represent the rate of etching in the vertical and lateral directions, respectively. Practical units for etch rates are [Å/min]. The two-dimensional nature of the etching process is specified by the *degree of anisotropy*

$$A = 1 - \frac{v_{\text{lat}}}{v_{\text{ver}}}, \qquad (5.2\text{-}1)$$

where A varies from 0 to 1 depending on the ratio of the etch rates.

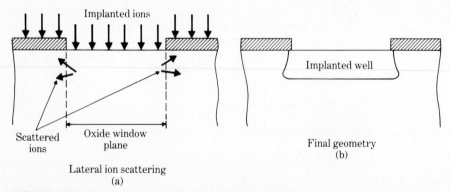

FIGURE 5.20 Lateral doping effects in an ion implantation.

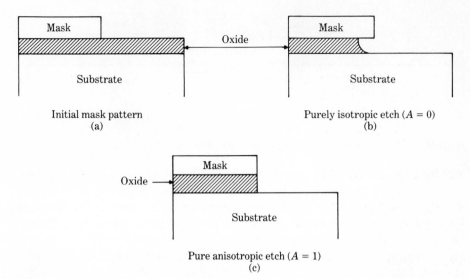

FIGURE 5.21 Basic etching problem.

Purely isotropic etching occurs when $v_{lat} = v_{ver}$, i.e., $A = 0$. This results in the structure shown in Fig. 5.21(b), where the oxide etch undercuts the resist mask. A *totally anisotropic* etch is defined by $v_{lat} = 0$, giving $A = 1$. In this case, only vertical etching is present so that the oxide wall is perpendicular to the substrate surface as shown in Fig. 5.21(c). Anisotropic etching thus results in the ideal structure.

Figure 5.22 shows the general case where a film of thickness h is etched with an amount ℓ lost in the lateral undercut. This is described by

$$A = 1 - \frac{\ell}{h} \qquad (5.2\text{-}2)$$

since the amount of etching is just the etch rate multiplied by the etch time t_e. The drawing also gives a more realistic case in which the masking material itself has undergone some etching. This leaves a slight tilt in the mask edge.

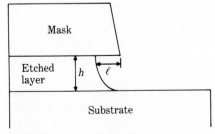

FIGURE 5.22 Geometry for a partially anisotropic etch.

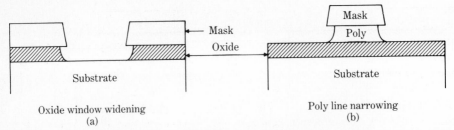

Oxide window widening
(a)

Poly line narrowing
(b)

FIGURE 5.23   Window and line effects in a partially anisotropic etch.

The difference in etched contours for the mask and etched layer arises from the fact that two materials exhibit different etch rates for a given process.

The importance of achieving a value of A close to unity is illustrated in Fig. 5.23. In Fig. 5.23(a) it is seen that an isotropic etch component will open an oxide window that is larger than the resist pattern. If the oxide is then used as a dopant mask, the curved oxide walls will lead to increased lateral doping spreads. Figure 5.23(b) shows the opposite problem, where a polysilicon line feature is defined by a resist pattern. The isotropic portion of the poly etch from v_{lat} gives a linewidth smaller than the resist pattern as shown by the cantilever resist structure. This decreases the cross-sectional area of the current flow, which increases the sheet resistance of the line. In addition, the undercutting requires that the mask pattern be larger than the final chip feature.

Etching techniques are generally classified as being either *wet* or *dry*. *Wet etching*, as implied by its name, employs aqueous solutions of acids (or caustics) to etch the desired layers. The wafers are immersed directly in the solution, which allows the chemical reaction to take place over the exposed surface regions. Wet etches tend to be isotropic when used on amorphous or polycrystalline material. This then limits the use of wet etching to lithographic features larger than about 3 [μm].

Silicon dioxide may be wet-etched in a buffered hydrofluoric acid solution containing HF, NH_4F, and H_2O. The HF readily attacks the SiO_2 layer but does not etch bare silicon. This property of *selectivity* makes it useful for etching oxide windows down to a silicon substrate since the vertical etch rate goes to zero when the substrate is reached. The HF solution is buffered to slow down the etch rate to a controllable level.

Phosphosilicate glass can be etched using a solution of HF and nitric acid (HNO_3) in water. Increasing the ratio of HNO_3 to HF yields a solution that will etch polysilicon. Silicon nitride (Si_3N_4) is wet-etched in heated phosphoric acid (H_3PO_4). Solutions for etching aluminum tend to be based on H_3PO_4 with HNO_3 and CH_3COOH (acetic acid) in water.

Dry etching systems are based on the properties of a weakly ionized gas consisting of electrons, ions, and free radicals. This collection of particles constitutes a *plasma,* so this approach is generically termed *plasma-assisted* etching. The plasma is usually produced by an rf glow discharge in which electrons are accelerated by high electric fields and induce ionization by inelastic collisions with gas molecules. Plasma-assisted etching can be divided into two categories, *plasma etching* and *reactive ion etching* (RIE), depending on the mechanisms involved. Although a realistic dry etch arrangement will use both types of etching, one will tend to dominate the overall process.

Plasma etching is based on the reaction of the surface material with a neutral species in the plasma. This type of etch produces by-products of the reaction that are carried away from the wafer by controlling the gas flow. Since the etching mechanism requires the neutral species in the reaction, the etching characteristics of the system tend to degrade with increased wafer area. This effect is termed *loading* and requires an adjustment of the system parameters according to the number of wafers being processed. Straight plasma etches are isotropic because of the dependence of the surface reaction. In practice, however, the presence of directed ions in the plasma can enhance v_{ver} to give a more anisotropic profile.

Reactive ion etching uses a directed ion flux that etches the material by both momentum transfer (i.e., sputtering or *ion milling*) and a reactive etch mechanism. RIE systems are generally designed so that the ions impact the wafer vertically to ensure that $v_{\text{ver}} \gg v_{\text{lat}}$. This directionality allows for a minimization of the lateral etching, so that highly anisotropic etch contours are possible. RIE is thus a good choice for fine-line VLSI structures.

The gases used to create the plasma determine both the etch rates and profiles for a given layer. Polysilicon and silicon compounds (SiO_2, Si_3N_4, PSG, etc.) can be etched using carbon tetrafluoride (CF_4). The basic plasma forming reaction is

$$CF_4 + e^- \rightarrow CF_3^+ + F^* + 2e^-,$$

which shows how an electron (e^-) impacts with a CF_4 molecule to produce the CF_3^+ ion and the F^* free radical. The F^* radical is the active species for reactively etching silicon by forming SiF_4, which is stable and can be removed from the wafer surface. A similar situation holds for silicon nitride etching. Silicon dioxide relies on the CF_3^+ ion for intermediate reduction to SiO, which then reacts to form SiF_4. It is found that addition of oxygen (and other gases) can greatly affect the etching characteristics of the plasma for certain materials.

The refractory metals Mo, W, Ti, and Ta can be etched using a CF_4 plasma. The silicides formed with these metals also exhibit this property. Aluminum, on the other hand, cannot be etched with CF_4. Rather, a chlorine-containing gas such as carbon tetrachloride (CCl_4) is used to induce a reaction that forms $AlCl_3$ as a stable by-product. Polysilicon can also be etched using a chlorine-based plasma.

An important consideration in dry etching centers around the selectivity of the plasma. As mentioned above, selectivity refers to the ability to etch only certain materials in a given process. This is particularly crucial in etching multilayer patterns since each layer reacts differently to the etch mechanisms. In wet etching, selectivity is attained by using acids that attack only a single

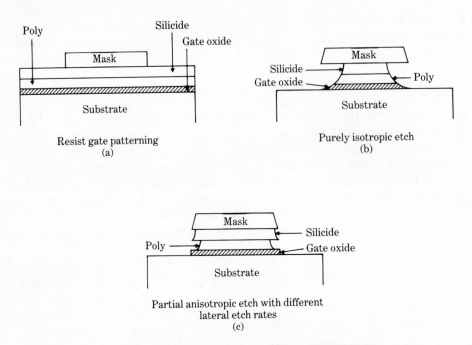

FIGURE 5.24 Etch selectivity problem for a polycide MOSFET gate.

layer of material at a time. This can be partially achieved in a purely reactive dry etch by varying the gas composition to preferentially etch a given layer at a faster rate. Note, however, that a plasma made with CF_4 can potentially etch any silicon-containing layer, so that high selectivity is difficult to obtain. Another problem in dry etching is that ion bombardment of the wafer sputters off material at a rate that is relatively insensitive to the composition. Consequently, there are no built-in mechanisms to stop the process after the layer is etched. This must be controlled externally by monitoring the etching time to prevent over-etching the wafer.

An interesting example of the multilayer dry etching problem is provided by the polycide-gate MOSFET in Fig. 5.15 of the previous section. Patterning of the gate requires etching through the silicide, the polysilicon, and the gate oxide. Figure 5.24(a) shows the initial layering with the gate etch mask. If a purely isotropic plasma etch is used, the gate will have the structure shown in Fig. 5.24(b). Although this gives a smooth transition at the layer interfaces, there is significant undercutting of the silicide underneath the mask. This results in increased sheet resistances, which act against the basic reason for introducing the silicide layer in the first place. Figure 5.24(c) shows the structure obtained using an RIE in which the lateral etch rate for the poly layer is greater than that for the silicide. This produces an undesirable silicide cantilever in the MOSFET. Problems of this type illustrate the importance of characterizing the etching process used in the wafer fabrication.

5.3   Isolation Techniques

Isolation in an integrated circuit deals with the need to electrically isolate neighboring devices on the chip. MOSFETs are fabricated on planar portions of the silicon known as *active areas*. Active areas are surrounded by field regions, which are included for the isolation. Interconnects are patterned conducting layers for "wiring" the devices together according to the circuit design. Isolation sets the characteristics of the field regions to prevent unwanted conduction paths between active areas. This, of course, is required to ensure that the circuit operates properly.

A MOS technology is partially self-isolating in that the MOSFETs have *pn* junctions that are usually reverse-biased. For example, an *n*-channel MOSFET consists of n^+ drain and source wells in a *p*-type substrate. Since only leakage current flows across the reverse biased junctions, some isolation is automatically provided when the transistors are placed into the chip environment.

The main complicating factor in MOS isolation is that unwanted inversion layers may form under the interconnect lines that are patterned in the field regions. If the field silicon surface becomes inverted, a conduction path may be established between MOSFETs in adjacent active areas. Isolation techniques are usually directed toward dealing with this problem.

The basic requirements for an isolation scheme are that (a) the technique should eliminate the possibility of forming unwanted field inversion layers and (b) the initial active area regions must be flat for fabricating the MOSFETs. Another desirable feature is that the overall chip process yield planar surface topologies after each major step. There are different approaches used for isolation in MOS integrated circuits. Three are presented below.

5.3.1   Etched Field Oxide Isolation

This basic MOS isolation approach uses the sequence shown in Fig. 5.25 to define and isolate active areas. The starting point is a lightly doped *p*-type

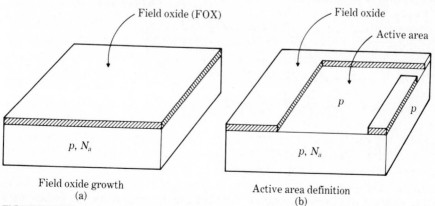

Field oxide growth
(a)

Active area definition
(b)

FIGURE 5.25   Etched field oxide isolation.

wafer that serves as the substrate. The first step is to thermally grow a thick *field oxide* (FOX) over the entire surface of the wafer, as illustrated by Fig. 5.25(a). Active areas are defined by a masking step that selectively etches the field oxide down to the bare silicon. The patterning for the present example is shown in Fig. 5.25(b). The field is defined by regions of the wafer where the FOX remains after this etch step. Active area patterns are set by the surface geometry requirements for the MOSFET to be placed in each location. This process also allows for n^+ line patterns to be formed during the active area definitions. These serve as silicon-level interconnects.

The isolation characteristics of the field oxide are best understood by examining the structure of a completed MOS chip section. A poly gate enhancement-mode MOSFET can be fabricated using the sequence of Fig. 5.26. The first step is the growth of the thin MOSFET gate oxide with thickness x_{ox} as illustrated in Fig. 5.26(a). Next a p-type ion implant (e.g., BF_2) is performed to induce a positive threshold voltage shift for an E-mode

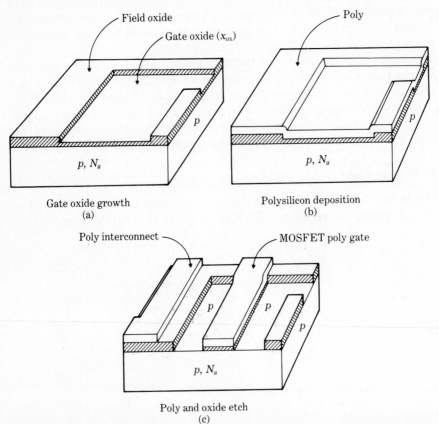

FIGURE 5.26 Enhancement-mode MOSFET fabrication sequence for etched field oxide isolation.

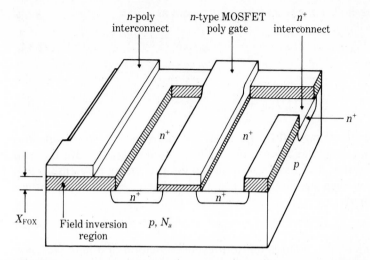

FIGURE 5.27 Final enhancement-mode MOSFET in an etched field oxide isolation process.

transistor. The gate polysilicon layer is then deposited over the entire surface, giving the layering shown by Fig. 5.26(b). The poly layer is also used as an interconnect level. The poly patterning is obtained from a masking step, and then the gate oxide is etched to leave the basic structure of Fig. 5.26(c). This shows a MOSFET pattern in the active area with a poly interconnect line running over the adjacent field region.

The final step needed to complete the basic MOSFET is to perform an n-type doping with either a diffusion or an ion implantation. This results in the completed MOSFET-interconnect arrangement in Fig. 5.27. The drain and source n^+ wells have been patterned using the gate (and FOX) as a mask. This is called a *self-aligned* MOSFET since the gate pattern allows for automatic alignment of the n^+ regions. Overlap capacitance is thus minimized. A direct consequence of this step is that the polysilicon layer is doped n-type unless a mask is provided prior to the doping step. (Undoped poly lines may be used as resistors.)

The purpose of the field oxide is to prevent the formation of inversion layers under the poly interconnects. The region of concern in the present example is shown in Fig. 5.27. It is seen that the polysilicon interconnect forms a MOS system with the field oxide such that

$$C_{\text{FOX}} = \frac{\varepsilon_{\text{ox}}}{X_{\text{FOX}}}$$

(5.3-1)

is the field capacitance per unit area. The process flow is set to give $X_{\text{FOX}} \gg x_{\text{ox}}$. Consequently, $C_{\text{FOX}} \ll C_{\text{ox}}$, indicating a reduced level of MOS field coupling into the substrate.

The *field threshold voltage* may be constructed as

$$(V_{T0})_{FOX} = V_{FB,FOX} + 2|\phi_F| + \frac{1}{C_{FOX}}\sqrt{2q\varepsilon_{Si}N_a(2|\phi_F|)}, \qquad (5.3\text{-}2)$$

where

$$V_{FB,FOX} = \Phi_{IS} - \frac{1}{C_{FOX}}(Q_{FOX} + Q_{ss,FOX}) \qquad (5.3\text{-}3)$$

is the field flatband voltage. Φ_{IS} represents the difference in work functions between the interconnect and substrate, while Q_{FOX} and $Q_{ss,FOX}$ are the field oxide charge densities. The basic constraint on the processing is to ensure that $(V_{T0})_{FOX}$ is greater than the largest voltage in the system. For example, with a single voltage supply,

$$(V_{T0})_{FOX} > V_{DD} \qquad (5.3\text{-}4)$$

represents the minimum requirement to prevent formation of a field inversion layer.

The value of the threshold voltage may be increased by two techniques. The first is to provide a field acceptor implant with a dose $D_{I,FOX}$ that increases the field threshold voltage by an amount

$$(\Delta V_{T0})_{FOX} = \frac{qD_{I,FOX}}{C_{FOX}}. \qquad (5.3\text{-}5)$$

The field implant must precede the initial field oxide growth. The resulting p^+ under the FOX is sometimes referred to as a *channel stop*.

An alternate approach is to apply a body bias voltage V_B to the p-type substrate. Assuming that V_B is negative with respect to ground, this induces a body bias threshold voltage shift of

$$(\Delta V_T)_{FOX} = \gamma_{FOX}(\sqrt{2|\phi_F| + |V_B|} - \sqrt{2|\phi_F|}), \qquad (5.3\text{-}6)$$

where

$$\gamma_{FOX} = \frac{1}{C_{FOX}}\sqrt{2q\varepsilon_{Si}N_a} \qquad (5.3\text{-}7)$$

is the field body bias factor. Using this technique requires that V_B be accounted for in the circuit design since all MOSFET threshold voltages will increase. It should be noted that the threshold voltage shifts described in the above formulas are proportional to X_{FOX}.

The main drawback of etched field oxide isolation is that $X_{FOX} \gg x_{ox}$ implies the existence of large oxide "steps" at the boundaries between active areas and field regions. Transitions of this type can lead to problems in depositing a conformal (uniform) layer of poly over the isolation terrain. For example, a crack may occur in the poly layer that could lead to chip failure. This gives motivation for examining isolation techniques that recess the field oxide into the silicon, yielding a more planar overall working surface.

5.3.2 LOCOS

The *local oxidation of silicon* (LOCOS) is a popular approach for achieving *recessed oxide isolation* (ROI). As implied by its name, LOCOS allows for the growth of an isolation oxide only in specified field regions. Active areas are shielded from the oxidizing ambient during this processing step. Surface silicon atoms are consumed during the FOX growth so that the isolation oxide is automatically recessed into the wafer. Denoting the field oxide thickness by X_{FOX}, eqn. (5.1-9) shows that the field recesses an amount

$$x_{Si} \simeq 0.46 X_{FOX} \tag{5.3-8}$$

relative to the original silicon surface. Consequently, an oxide step from the field to an active area is at most about $0.54 X_{FOX}$. This results in chip surfaces that are more planar than those obtained using a nonrecessed isolation oxide.

The technique of local oxidation relies on the use of a silicon nitride layer to inhibit oxide growth in active areas. The basic sequence is illustrated in Fig. 5.28. First, a stress-relief oxide layer and a nitride layer are patterned to define the desired active area geometry, shown in Fig. 5.28(a). Next an acceptor field implant is performed to increase the value of $(V_{T0})_{FOX}$. The resulting p^+ regions are shown in Fig. 5.28(b). The field implants are self-aligned to the active areas by the nitride masking.

Localized growth of the field oxide is achieved using the fact that Si_3N_4 does not react with O_2 and only slowly oxidizes in steam. A typical FOX growth might start with an initial growth in chlorinated O_2, followed by a long steam oxidation that produces most of the field oxide. The wafer would then be subjected to a dry O_2 anneal/growth step to improve the quality of the steam-grown oxide. The field regions (not covered by the nitride) are directly oxidized by this sequence of steps. The resulting structure appears as shown in Fig. 5.28(c).

The "lifting" of the nitride edges is easily understood. During the FOX growth, some of the oxidizing gas will diffuse through the stress-relief oxide to the underlying silicon. This induces an oxide growth around the edges of the nitride pattern, which in turn lifts the nitride layer. The transition from the stress-relief oxide to the FOX is called the *bird's beak* region because of its shape. The formation of the bird's beak in LOCOS results in a smaller active (planar) area than that originally defined by the nitride patterning. The loss of active area to the bird's beak is termed *encroachment*. The amount of encroachment increases with increasing thickness of the stress-relief oxide layer, since more oxygen can diffuse under the nitride pad.

MOSFETs are made in the active areas by first removing the nitride (Fig. 5.28d) and then etching the stress oxide down to the silicon (Fig. 5.28e). This allows for a controlled growth of the critical gate oxide with thickness x_{ox}, as illustrated by Fig. 5.28(f). The working value of the field oxide thickness X_{FOX} is also established at this point since the next layering will be the polysilicon gate.

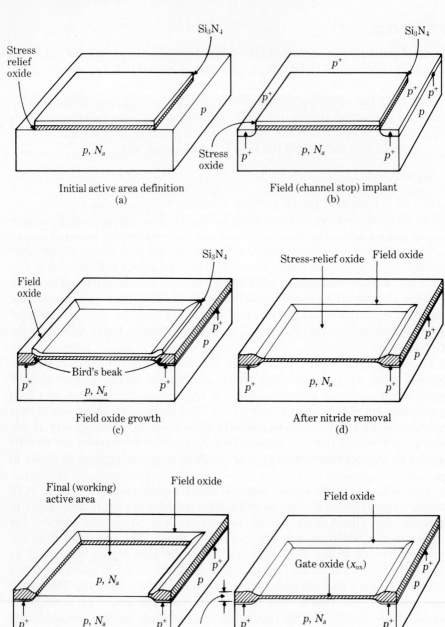

Initial active area definition
(a)

Field (channel stop) implant
(b)

Field oxide growth
(c)

After nitride removal
(d)

Stress oxide removal
(e)

MOSFET gate oxide growth
(f)

FIGURE 5.28   Basic LOCOS MOS isolation process flow.

LOCOS is a relatively complicated process to characterize and generally requires the use of numerical simulation studies. Although the details of the analysis are beyond the scope of the present treatment, three points deserve further comment.

First, it should be noted that the acceptor field implant used to increase $(V_{T0})_{FOX}$ must be relatively deep with a moderate dose level. This arises because the field oxide growth will consume the surface portions of the doped p^+ layer. The initial implant characteristics must be adjusted to account for this part of the processing. The dose level of the implant must be high enough to prevent the field from inverting. However, increasing the field implant density increases the MOSFET n^+/substrate depletion sidewall capacitance C_{j0sw}, as shown by eqn. (4.6-21). This consideration requires that only moderate implant doses (10^{12}–10^{13} [cm^{-2}]) be used.

The second point is concerned with the reaction of silicon nitride and steam by means of

$$Si_3N_4 + 6H_2O \rightarrow 3SiO_2 + 4NH_3.$$

The ammonia (NH_3) produced by this reaction tends to diffuse through the structure and can eventually make it to the underlying silicon. The ammonia will then react to form a layer of Si_3N_4 at the silicon surface. This is most prevalent around the edge of the nitride pattern and gives rise to what is sometimes called the "white ribbon" effect. The presence of this nitride can cause nonuniform gate oxide growths, so it must be removed with an additional processing step.

The last comment deals with the formation of the bird's beak and the subsequent active area encroachment. This is an important factor in chip layout since it requires that the initial active area definition be larger than the final working value. A typical order of magnitude might be 0.8 [μm] lateral encroachment (per side) using a 200 [Å] stress-relief oxide thickness; the actual value varies according to the process specifications. Encroachment is particularly crucial for VLSI circuits since it creates a fundamental limitation on high-density layouts.

5.3.3 Trench Isolation

LOCOS is a widely used technique since it provides the necessary isolation properties and is relatively straightforward to implement. However, the undesirable active area encroachment associated with the formation of the bird's beak has spurred development of new isolation schemes for use in high-density chip layouts. This section will discuss the use of "trenches" as an alternate approach to LOCOS.

Trench isolation uses a highly anisotropic silicon etch to create vertical-walled trenches around active areas. The trenches are filled with a dielectric to establish the boundaries needed to ensure electrical isolation. The process has the advantage that it exhibits only minor bird's beaking and negligible encroachment. In addition, trenches provide isolation structures that are almost completely recessed into the wafer.

The basic process flow involved in a trench isolated integrated circuit is illustrated in Fig. 5.29. Figure 5.29(a) shows a stress-relief oxide with a nitride layer used to define the active areas. The example chosen here also uses a CVD oxide layer as a mask in the trench patterning. Once the active areas are defined, the wafer is subjected to an anisotropic silicon etch (usually an RIE) that gives the initial formation of the trench structuring shown in Fig. 5.29(b). At this point, boron is implanted to create a p^+ channel-stop layer at the bottom of the trench. The oxide mask is then removed.

Ideally, isolation would be accomplished by filling the trench with an insulating dielectric such as silicon dioxide. Unfortunately, it is difficult to fill the trench using a CVD oxide because of void formation and other problems in trying to obtain uniform coverage. To overcome this, the wafer is first oxidized to grow oxide layers on the bottom and sidewalls; this is shown in Fig. 5.29(c). Then polysilicon is deposited over the structure, filling the trenches. The surface poly is etched, leaving the filled trenches illustrated in Fig. 5.29(d). The last step needed in the isolation process flow is to oxidize the wafer. This gives the final cross-section in Fig. 5.29(e).

This typical sequence illustrates some important points. First, note that there is a small amount of active area encroachment during the thermal oxidation of the trench walls in Fig. 5.29(c). Additional encroachment may occur during the poly oxide growth, which gives the surface oxide layer of Fig. 5.29(e) due to both silicon consumption and any bird's beak formation under the nitride. The magnitude of the encroachment is relatively small (particularly when compared with a standard LOCOS process), which makes this type of isolation a good candidate for high-density circuit layouts.

The use of polysilicon as a "filler" material for the trench provides good coverage but can introduce complications in the isolation characteristics. The basic problem is shown in Fig. 5.30(a), where an interconnect runs over the trench. Owing to normal coupling capacitances, the polysilicon inside the trench acts as an electrode with a floating potential.

There are three values of oxide thickness that enter into the analysis: the top oxide X_{TOX}, which is grown during the final oxidation step; the sidewall oxide X_{SOX}; and the bottom oxide X_{BOX}. Although the thickness of the trench sidewall and bottom oxides are usually different (due, for example, to nonuniform surface oxidation rates), $X_{BOX} = X_{SOX}$ will be assumed for simplicity. The important capacitances per unit area are thus given by

$$C_{TOX} = \frac{\varepsilon_{ox}}{X_{TOX}} \, [\text{F/cm}^2]$$

$$C_{SOX} = \frac{\varepsilon_{ox}}{X_{SOX}} = C_{BOX} \, [\text{F/cm}^2].$$

(5.3-9)

At the present time, typical values for the oxide layers are $X_{SOX} = 1500 \, [\text{Å}]$ and $X_{TOX} = 4000 \, [\text{Å}]$.

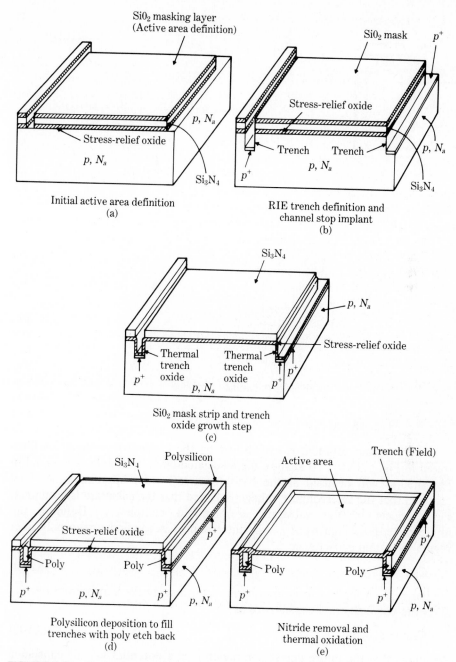

FIGURE 5.29 Basic sequence for trench isolated integrated circuits.

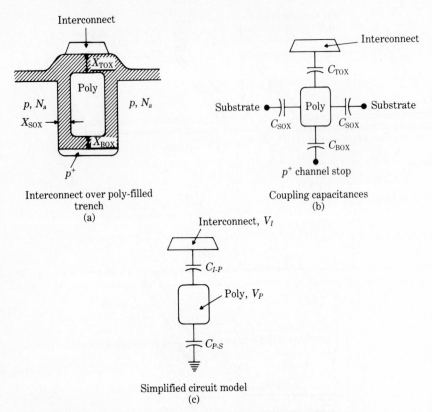

Interconnect over poly-filled
trench
(a)

Coupling capacitances
(b)

Simplified circuit model
(c)

FIGURE 5.30 Coupling problem with floating polysilicon filler.

The coupling problem is established by the capacitive connections illustrated in Fig. 5.30(b). To analyze the associated circuit problem, the model in Fig. 5.30(c) will be used as the basis for a low-order approximation. It is assumed that the interconnect voltage is V_I and that the substrate is grounded. Field effects in the substrate are ignored completely. Defining the interconnect-poly area by $A_{I\text{-}P}$, the interconnect is coupled to the poly filler region by a total capacitance

$$C_{I\text{-}P} = C_{TOX}A_{I\text{-}P}.$$ (5.3-10)

Similarly, denoting the total area of the trench sidewalls and bottom by A_{TR} gives the poly-substrate capacitance as

$$C_{P\text{-}S} = C_{SOX}A_{TR}.$$ (5.3-11)

Assuming that the entire polysilicon region is at a potential V_P, the capacitive voltage divider circuit gives

$$V_P = \frac{C_{I\text{-}P}}{C_{I\text{-}P} + C_{P\text{-}S}} V_I.$$ (5.3-12)

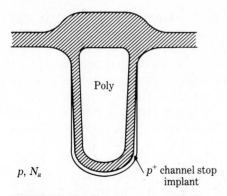

FIGURE 5.31 Example of nonideal trench cross section.

Minimum coupling is achieved by ensuring that $V_P \ll V_I$. This requires a large value for the capacitance ratio $C_{P\text{-}S}/C_{I\text{-}P}$.

The most important aspect of this analysis is that V_P must be kept below the value of the trench threshold voltage $(V_{T0})_{\text{BOX}}$ needed to invert the silicon at the bottom of the trench. This, of course, is required to preserve the isolation characteristics. Equation (5.3-12) shows that $V_P \ll V_I$ can be attained by satisfying the condition

$$\frac{C_{P\text{-}S}}{C_{I\text{-}P}} = \frac{A_{TR}}{A_{I\text{-}P}}\frac{X_{\text{TOX}}}{X_{\text{SOX}}} \gg 1. \tag{5.3-13}$$

The geometry of the trench structure automatically ensures that the capacitance ratio is large. This is seen by noting that $A_{TR} \gg A_{I\text{-}P}$ because the trench has both sidewall and bottom area contributions and surrounds an entire MOSFET active area. In addition, $X_{\text{TOX}} > X_{\text{SOX}}$ can be set by controlling the oxidation times for the two oxide growths.

The problems involved with trench isolation are much more complicated than the simplistic example discussed above. This additional complexity arises for many reasons. One is that the polysilicon is generally intrinsic so that the assumption of a single poly voltage V_P is not valid. An analysis must then account for the potential variation in both the polysilicon and the substrate. Another complicating factor is that the cross-sectional geometry of a realistic trench is not rectangular but assumes a profile similar to that illustrated in Fig. 5.31. These and other considerations need to be included to obtain a more accurate characterization of the trench isolation structures.

5.4 nMOS LOCOS Process Flow

A *process flow* describes the sequence of fabrication steps used to create an integrated circuit. Knowledge of the process flow allows for a complete electrical characterization of the layers used to construct the chip. It also provides the information needed for layout design.

This section deals with a process flow for fabricating a LOCOS-isolated nMOS integrated circuit with both enhancement-mode and depletion-mode transistors. Emphasis will be placed on the study of a depletion-load inverter on the chip. Extension of the concepts to other circuit configurations is straightforward. The sequences presented here are representative of realistic processes. However, only the basic steps are included in the discussion. Many variations are possible.

The process flow for fabricating a depletion-load nMOS inverter is shown by the drawings in Fig. 5.32. The starting point for the discussion is the initial active area patterning shown in Fig. 5.32(a). The drawing shows the active area geometry defined using the stress-relief oxide and nitride layers. In addition, a self-aligned p^+ channel stop field implant has been performed using the nitride patterning. For future reference it is noted that the E-mode driver MOSFET will be on the left side of the active area, while the D-mode load transistor will be placed on the right side.

The field oxide (FOX) used in the LOCOS isolation scheme is grown by subjecting the wafer to an oxidizing ambient. This results in the semirecessed field oxide structure shown in Fig. 5.32(b). Note the characteristic bird's beak formation around the nitride edges. The next steps are to remove the nitride and then etch the stress-relief oxide, which gives access to the silicon surface as in Fig. 5.32. Once this is accomplished, the critical MOSFET gate oxide with thickness x_{ox} is grown. Figure 5.32(d) shows the structure at this point in the process flow.

Two threshold adjustment ion implant steps are included in the fabrication sequence. The first is an acceptor implant used to set the threshold voltage $V_{T0} > 0$ of the enhancement-mode MOSFETs. Resist is used as a mask to selectively implant the E-mode active area; the patterning for this step is shown in Fig. 5.32(e). Implanted p-type threshold adjustment regions are not usually shown explicitly in cross-sectional drawings. Depletion-mode MOSFETs are made using the donor implant in Fig. 5.32(f). This step creates a patterned n-well, which eventually serves as the D-mode transistor channel. The use of donors ensures that the threshold voltage satisfies $V_{T0} < 0$.

Polysilicon is deposited over the gate oxide in the next step. The poly layer is patterned, and the gate oxide is etched. An n^+ implant is then performed to create self-aligned drain and source regions. The structure now appears as in Fig. 5.32(g), where the identities of the transistors become clear. Note that the n^+ implant uses the polysilicon layer as a mask, so the poly layer is doped heavily n-type unless a separate mask is provided to keep the layer undoped. This is important for computing both threshold voltages (in Φ_{GS}) and parasitic line resistances.

The last steps in the basic processing sequence are to coat the surface with CVD oxide and then pattern contact cuts where needed. Metal is deposited over the surface and patterned. The final inverter structure is shown in Fig.

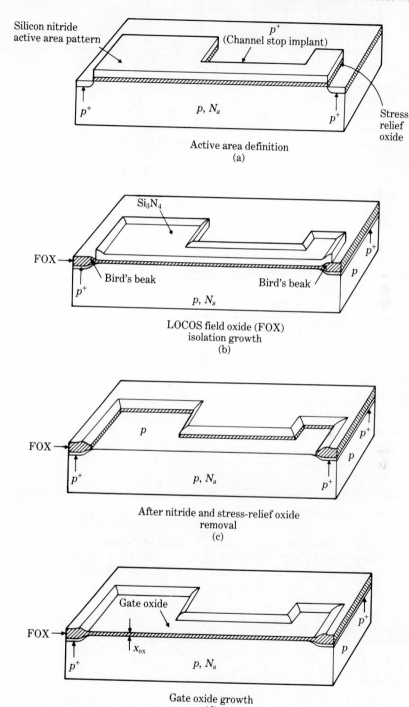

Silicon nitride
active area pattern

p^+
(Channel stop implant)

p^+

p, N_a

p^+

Stress
relief
oxide

Active area definition
(a)

Si_3N_4

FOX

Bird's beak

p^+

Bird's beak

p^+

p

p, N_a

LOCOS field oxide (FOX)
isolation growth
(b)

FOX

p

p^+

p^+

p

p^+

p, N_a

After nitride and stress-relief oxide
removal
(c)

Gate oxide

FOX

x_{ox}

p^+

p^+

p

p, N_a

Gate oxide growth
(d)

FIGURE 5.32 Basic nMOS LOCOS process flow for depletion-load inverter.

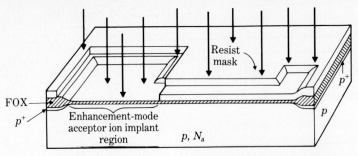

E-mode MOSFET threshold voltage adjustment ion implant
(e)

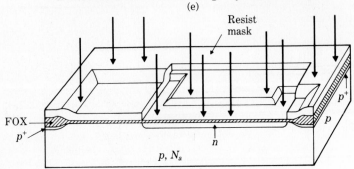

Depletion-mode MOSFET donor ion implant
(f)

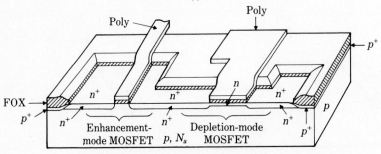

Poly deposition/patterning, oxide etch and n^+ implant
(g)

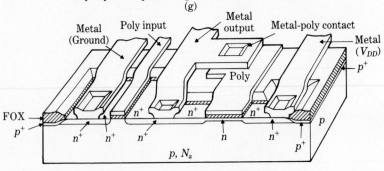

Completed inverter after metallization
(h)

Figure 5.32 (*contd.*)

5.32(h). Both metal/n^+ and metal/poly contacts have been made. The inverter input is at the poly level, while the output is chosen to be on the metal level. The power supply (V_{DD}) and ground connections to the circuit are obtained from the metal lines.

This sequence connects the source and gate of the depletion-load MOSFET by including oxide contact cuts for a metal interconnect. Since this is a common problem in depletion-load nMOS logic, alternate schemes have been developed to electrically connect n^+ regions to poly layers. Two of these techniques are included in the discussion here. In general, both will result in reduced real estate consumption.

The first n^+/poly technique is termed a *butting contact*. To effect this connection, the processing sequence is followed to obtain the poly patterning shown in Fig. 5.33(a). Note that the depletion MOSFET poly gate area over the field oxide has been reduced. The surface is coated with oxide, and then contact cuts are etched into the oxide layer. The structure of a butting contact

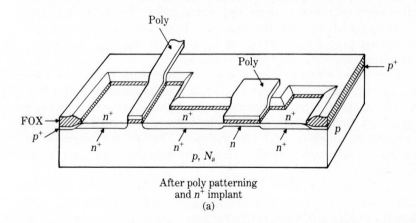

After poly patterning
and n^+ implant
(a)

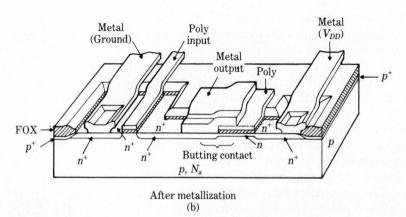

After metallization
(b)

FIGURE 5.33 Formation of a butting metal/poly/n^+ contact.

is illustrated in Fig. 5.33(b). Metal has been deposited such that it directly connects the poly gate to the n^+ region. The savings in real estate is evident from the metal/n^+ and n^+/poly contacts having merged into one. The main drawback of a butting contact is that the metal may crack because of the transition from the top of the poly to the n^+ layer.

Buried contacts achieve n^+/poly connections without using the metal layer. A typical sequence is illustrated in Fig. 5.34. The starting point for creating a buried contact is directly after the threshold adjustment implants have been performed. This is shown as Fig. 5.34(a). A buried contact relies on the fact that an n-type layer has been created for D-mode MOSFETs. The first step in creating the buried contact is to perform a selective etch into the gate oxide as in Fig. 5.34(b). Next the polysilicon is deposited and patterned so that the depletion MOSFET gate poly extends over the exposed n-region. Figure 5.34(c) gives the polysilicon patterning. The n^+ drain-source implant establishes the buried contact, since the polysilicon and n^+-wells are now electrically connected through the n-region. The final steps are to coat the wafer with oxide, etch contact cuts, and perform the metallization. Figure 5.34(d) gives the final structuring chosen for the example. Both the inverter input and output are on the poly level. The metal is used only for V_{DD} and ground connections in the gate.

The process flows described so far are characterized by having three interconnect layers; metal, poly, and n^+. These can be patterned to accommodate the wiring dictated by the circuit design. Although this may appear to allow a significant amount of freedom in the layout, a problem arises with using n^+-regions as interconnects. As discussed below, n^+ interconnects require excessive real estate, making them less attractive for chip implementations.

In the LOCOS process, an n^+-region is patterned as an active area with silicon nitride. Active area encroachment during the growth of the field oxide can significantly reduce the width of lines. Consequently, patterning an n^+ interconnect requires that the initial layout be larger than the final dimensions to account for bird's beak encroachment. Allowing for this encroachment increases the amount of chip area consumed. The most efficient interconnects are thus the metal and poly layers.

The layout of complicated logic systems is facilitated by having multiple interconnect layers available as needed. Owing to this consideration, process flows have been developed to provide additional conducting layers. Figure 5.35 shows the basic steps used for a double-poly scheme. This gives two polysilicon layers, denoted as Poly I and Poly II. Both may be used as MOSFET gates and interconnect lines. With the final metal layer added, this scheme provides three interconnect levels in addition to n^+ lines.

The sequence in Fig. 5.35 demonstrates how two polysilicon layers can be incorporated into the process flow such that either can be used for MOSFET gates. The starting point is the active area shown in Fig. 5.35(a). The (first) gate oxide with thickness x_{ox} is grown, and then the first poly layer (Poly I) is

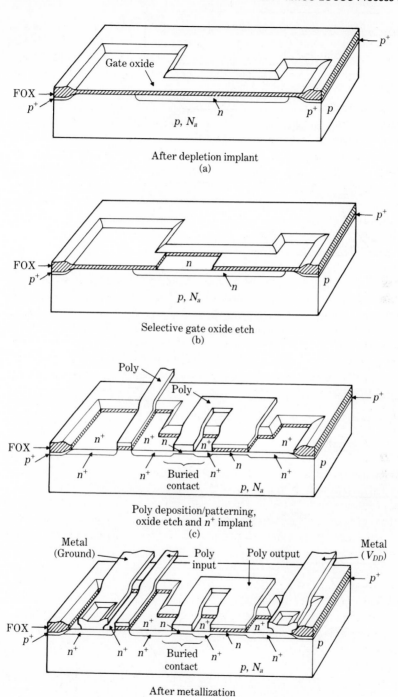

FIGURE 5.34 Formation of a buried poly/n^+ contact.

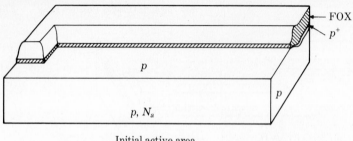

Initial active area
(a)

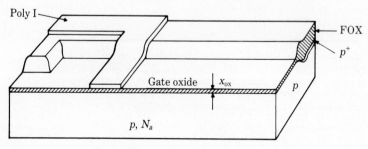

After gate oxide growth and first-level
poly (poly I) deposition/patterning
(b)

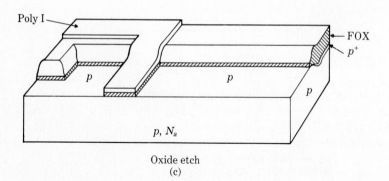

Oxide etch
(c)

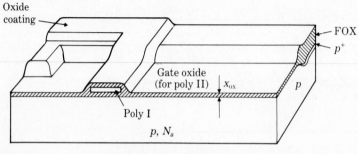

Oxidation for second poly
MOSFET gate oxide
(d)

FIGURE 5.35 Double-poly process flow.

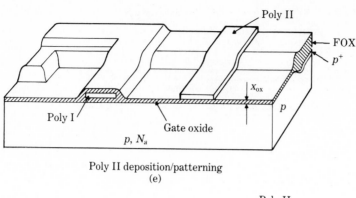

Poly II deposition/patterning
(e)

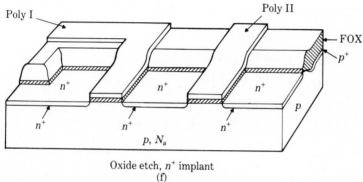

Oxide etch, n^+ implant
(f)

Figure 5.35   (*contd.*)

deposited and patterned. Figure 5.35(b) shows the structure at this point. The gate oxide is etched as drawn in Fig. 5.35(c) to allow access to the p-type silicon surface in preparation for the second poly layer. A second gate oxide is grown to a thickness x_{ox}. In an ideal process, the two gate oxides will be identical. As seen in Fig. 5.35(d), this results in a polyoxide growth over the Poly I layer. The second polysilicon layer, Poly II, is then deposited and patterned as illustrated in Fig. 5.35(e). Note that a capacitor can be formed between the Poly I and Poly II layers since there is an oxide layer between the two. After the gate oxide is etched, both poly layers are exposed. An n^+ drain-source implant gives the final structure, shown in Fig. 5.35(f). The metal deposition and patterning follows, which completes the process flow.

Other multilevel interconnect schemes are common. For example, single-poly, double-metal processes have been developed. Refractory metals are particularly useful in this regard.

EXAMPLE 5.4-1

The important parasitic interconnect capacitances are established during the fabrication cycle. Figure E5.1 shows the layering for a single-poly, single-metal

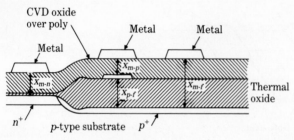

FIGURE E5.1

process. Assume oxide separations of

$$x_{p\text{-}f} = 0.60 \, [\mu\text{m}] \quad \text{(poly to field)};$$
$$x_{m\text{-}p} = 0.40 \, [\mu\text{m}] \quad \text{(metal to poly)};$$
$$x_{m\text{-}f} = 1.00 \, [\mu\text{m}] \quad \text{(metal to field)};$$
$$x_{m\text{-}n^+} = 0.42 \, [\mu\text{m}] \quad \text{(metal to } n^+)$$

as representative values. Compute the oxide capacitances per unit area for each case.

Solution

Since $\varepsilon_{ox} = (3.9)(8.85 \times 10^{-14})$, the interconnect capacitance is calculated using $C_{int} \simeq (3.45 \times 10^{-11})/x_{int} \, [\text{F/cm}^2]$. This gives

$$C_{p\text{-}f} \simeq 5.76 \times 10^{-9} \, [\text{F/cm}^2] = 0.0576 \, [\text{fF}/\mu\text{m}^2],$$
$$C_{m\text{-}p} \simeq 8.63 \times 10^{-9} \, [\text{F/cm}^2] = 0.0863 \, [\text{fF}/\mu\text{m}^2],$$
$$C_{m\text{-}f} \simeq 3.45 \times 10^{-9} \, [\text{F/cm}^2] = 0.0345 \, [\text{fF}/\mu\text{m}^2],$$
$$C_{m\text{-}n^+} \simeq 8.22 \times 10^{-9} \, [\text{F/cm}^2] = 0.0822 \, [\text{fF}/\mu\text{m}^2].$$

These values will be used as typical capacitance levels in later examples.   ■

5.5 Threshold Voltage Adjustment

The depletion-load nMOS fabrication sequence described in the last section has associated with it three distinct MOSFET threshold voltages. Multiple-threshold process flows introduce extra degrees of freedom, making them particularly useful in circuit design.

The basic threshold voltage is that obtained from the MOS system physics before ion implantation. From Section 1.1, this is given as

$$V_{T0} = V_{FB} + 2|\phi_F| + \frac{1}{C_{ox}}\sqrt{2q\varepsilon_{Si}N_a(2|\phi_F|)}, \tag{5.5-1}$$

where V_{FB} is the flatband voltage

$$V_{FB} = \Phi_{GS} - \frac{1}{C_{\text{ox}}}(Q_{\text{ss}} + Q_{\text{ox}}). \tag{5.5-2}$$

This is sometimes referred to as the "natural" threshold voltage of the wafer since it is set by the basic processing parameters including x_{ox} and N_a. It may be positive or negative, depending on the relative magnitude of each term.

Enhancement-mode MOSFETs are obtained by implanting acceptors with a dose $D_{I,a}$. This gives a positive shift to the natural threshold voltage, so

$$V_{T0,\text{E-mode}} = V_{T0} + \frac{qD_{I,a}}{C_{\text{ox}}} > 0. \tag{5.5-3}$$

Depletion-mode transistors are formed with a donor implant. Letting $D_{I,d}$ be the dose, this gives a negative shift such that

$$V_{T0,\text{D-mode}} = V_{T0} - \frac{qD_{I,d}}{C_{\text{ox}}} < 0. \tag{5.5-4}$$

Both $V_{T0,\text{E-mode}}$ and $V_{T0,\text{D-mode}}$ can be set by adjusting the dose of the implant. These serve as the fundamental circuit design values in the depletion-load process.

Zero-threshold MOSFETs are defined to have threshold voltages slightly greater than 0 [V]. These enhancement-mode devices require only small values of $V_{GS} > 0$ to induce conduction. Zero-threshold transistors are particularly useful for reducing threshold voltage losses and find wide application in certain types of switching and logic control networks.

Zero-threshold transistors can be fabricated by introducing another ion implantation step into the process flow. Alternatively, the natural threshold voltage V_{T0} may itself be slightly positive, which then gives a zero-threshold MOSFET without additional processing. To study this possibility, suppose that the gate is doped n-type with a donor density of $N_{d,\text{poly}}$. The flatband voltage is

$$V_{FB} \simeq -\left(\frac{kT}{q}\right)\ln\left(\frac{N_{d,\text{poly}}}{N_a}\right) - \frac{1}{C_{\text{ox}}}(Q_{\text{ss}} + Q_{\text{ox}}), \tag{5.5-5}$$

where Φ_{GS} from Problem 1.2 in Chapter 1 has been used. Since $V_{FB} < 0$, attainment of a zero-threshold device requires that

$$2|\phi_F| + \frac{1}{C_{\text{ox}}}\sqrt{2q\varepsilon_{\text{Si}}N_a(2|\phi_F|)} \geq |V_{FB}|. \tag{5.5-6}$$

Assuming that N_a is preset (to give a safe pn^+ reverse breakdown voltage), x_{ox} becomes the critical processing parameter. The value of x_{ox} needed to satisfy eqn. (5.5-6) depends on the oxide charges Q_{ss} and Q_{ox} and also on the doping levels in the gate and substrate. This may or may not be possible.

State-of-the-art processing lines have reached a significant level of complexity. Consequently, additional processing steps that enhance the circuit capabilities do not present any major technological problems. Economics is the important factor in the final decision.

5.6 nMOS Design Rules and Layout

Chip patterning represents the silicon implementation of a circuit design. Functionality of the chip depends on faithful reproduction of the pattern. *Design rules* are a set of specifications for the mask patterns used in the layout and provide geometric information such as the minimum widths and spacings for each layer. In addition, vertical alignment problems are addressed by specifications on the spacing between lines on different layers. Electrical characteristics are usually provided with the geometric patterning information. Items of interest include k' values, sheet resistances, and capacitance levels.

Design rules provide guidelines for chip layout at the mask level. The specific rules result from a detailed analysis of a given process flow and reflect the limitations of the processing line. In an industrial environment, design rule sets are highly proprietary because an enormous amount of information can be extracted from them. This section deals with the origins and application of design rules in chip mask design and layout.

5.6.1 Physical Basis of Design Rules

Formulation of a design rule set is based on many considerations. The most crucial items are (a) limitations in the lithography, (b) physics of the process flow, and (c) electrical characteristics of the final structures. A detailed discussion of design rule derivation is beyond the scope of this book. However, it is possible to introduce some of the factors that yield the final numbers.

The concept of a *minimum linewidth* is usually associated with limitations in the lithographic resolving power. A minimum linewidth (in units of microns [μm]) specifies the smallest width of a line that can be reliably patterned on a given layer. In terms of the interconnect example shown in Fig. 5.36, this design rule specifies the smallest value of w permitted. Other physical problems are also addressed by this design rule. For example, with a resistivity ρ [Ω-cm] and a layer thickness h, the sheet resistance of the interconnect is

$$R_s = \frac{\rho}{h} [\Omega],$$
(5.6-1)

so that

$$R = R_s n [\Omega]$$
(5.6-2)

gives the resistance of a path with length $\ell = nw$, where n is the number of squares of dimension ($w \times w$) in the direction of current flow. For long interconnects, n may become excessively large. Design rule sets may specify a larger value for min (w) in this case to reduce the overall resistance.

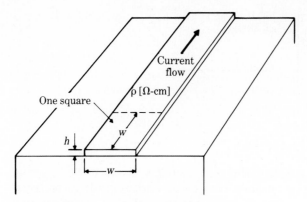

FIGURE 5.36 Interconnect geometry.

Minimum spacing rules for lines on a given layer also tend to originate from the lithography. In Fig. 5.37, two poly lines are separated by a distance d; the design rule for this layer will specify the minimum d permitted. This choice affects the electrical properties of the final structure since a parasitic coupling capacitance C_{coup} exists between the two lines. C_{coup} increases with decreasing separation distance d. The line-to-line spacing thus affects the performance of the circuit. Mutual inductance also enters into the problem since magnetic coupling is present.

Minimum spacing rules between lines on different layers are extremely important in layout. Restrictions arise because the layers must be stacked to form devices. If the vertical alignment of the layers is off, the fabricated structure may not behave properly. The complexity of the problem can be seen by noting that several layers must be applied and patterned to form an integrated circuit. Failure to correctly align even a single layer can result in a totally nonfunctional wafer.

Design rule specifications for spacing between lines of different layers are

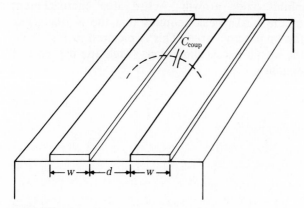

FIGURE 5.37 Interconnect spacing geometry.

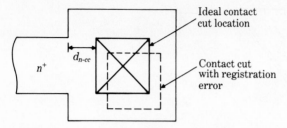

FIGURE 5.38   Registration error example.

chosen to compensate for *misalignment* or *registration* errors in the layering process. This is a complicated problem because it is not possible to perfectly align each new layer with existing wafer patterns. Rather, a registration tolerance must be allowed. This is usually determined by the lithographic equipment and the scheme used to place registration marks on the wafer for subsequent layers.

An important example of the alignment problem is shown in Fig. 5.38. A contact cut is needed in the oxide cover to access the n^+ patterned layer. The ideal location is shown by the solid line. The distance $d_{n\text{-}cc}$ between the edge of the n^+ boundary and the contact cut must be large enough to allow for registration errors when designing the contact cut mask. As long as the actual contact cut is within the n^+ borders, the connection made through the cut will work. The dashed line shows a case where a good cut will be made even though misalignment has occurred.

Spacings between lines on different layers also affect the electrical characteristics. Parasitic coupling capacitance is particularly important. Figure 5.39 shows a poly line and a metal line that are adjacent and thus experience field interactions. The spacing must be large enough to reduce this coupling to a negligible level.

MOSFET layout involves some special considerations. Consider first the active area definition in a LOCOS process. This mask sets the dimensions of the nitride pad used to inhibit oxide growth. Active area encroachment induced by the formation of the bird's beak requires that the nitride mask dimensions be larger than the final active area. This is pictured in Fig. 5.40, where the solid line shows the nitride mask dimensions, while the dashed line gives the final active area boundary.

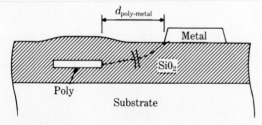

FIGURE 5.39   Poly-metal spacing example.

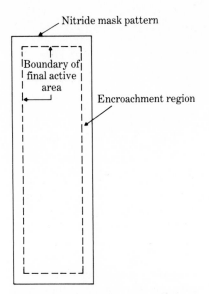

FIGURE 5.40 Nitride mask design.

The poly gate *overhang* distance d_{oh} is shown in Fig. 5.41. The distance d_{oh} is included to ensure that the n^+-regions will be separated in the event of a misalignment between the active area and the poly mask. d_{oh} is crucial in a self-aligned gate process since failure of the poly to cross the active area will result in shorted n^+-regions.

Another example is the mask defining the donor (n-type) implant in a depletion-mode MOSFET. The implanted region must be larger than the device active area to ensure that an n-layer is created if misalignment occurs. The outline of the depletion implant is shown in Fig. 5.42 by the dashed line. Similar problems arise for selectively implanted E-mode transistors.

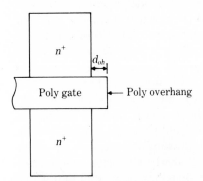

FIGURE 5.41 Poly gate overhang in a MOSFET.

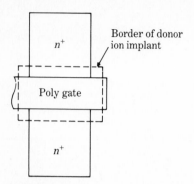

FIGURE 5.42 Threshold adjustment ion implant boundary spacings for a depletion-mode MOSFET layout.

5.6.2 Mask Drawings

Design rules are spacing and linewidth requirements for making masks. The active area encroachment problem just discussed shows that the mask geometries are different from the final device layout. The mask pattern set is sometimes referred to as the *drawn* layout, while the resulting chip patterns are termed the *actual* or *final* layouts. Encroachment, lateral diffusion, isotropic etching, and other physical processes require that a distinction be made.

An example of this was presented in the discussion of the overlap capacitance C_o in Section 4.6. Overlapping exists because of lateral doping effects in a self-aligned MOSFET structure. Denoting the drawn channel length by L', this gave

$$L' = L + 2L_o, \tag{5.6-3}$$

where L is the actual channel length. Figure 5.43 shows the superposed mask patterns for the gate and active area in solid lines. The final n^+ geometry (after processing) is indicated by the dashed lines. The boundary around the device edges is affected by both encroachment and lateral doping effects.

Device characteristics depend on the final geometric values. For example, L and W in I_D always refer to the length and width of the finished device. However, it is usually more convenient to present layouts in terms of mask drawings, particularly during the design phase. In this case, care should be exercised when obtaining geometric data for circuit simulation purposes.

Mask-level layout drawings usually show the ideal locations of the patterns. Multiple-layer drawings are common. In simple layouts, the regions are labeled (e.g., poly, n^+, etc.). More complicated drawings required coding to keep the layers separated. In single-color drawings, levels can be distinguished using different line types (solid, dashed, etc.) for boundaries. Alternatively, variations in shading or cross-hatching are common. Color coding is useful for CAD displays and plotters and is particularly nice for *stick*

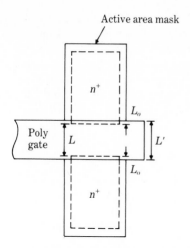

Active area mask

Poly gate

n^+

n^+

L_o

L

L'

L_o

FIGURE 5.43 MOSFET top view.

diagrams in VLSI system design. (A stick diagram displays only the routing of the patterns.)

5.6.3 Design Rule Sets

Design rules are process-specific, so minimum linewidths, spacings, dimensions, etc., are given in units of microns [μm]. The actual numbers provide a measure of the integration density since small values indicate higher packing densities. A design rule set can easily be 50 pages or longer. Simplified sets exist for applications that do not maximize packing density. An example of a basic nMOS design rule set is provided in Table 5.4. This is a generic list compiled for illustration only. Levels (masking layers) are denoted by number (01, 02, etc.) and name ("Active area," "Enhancement implant," etc.), with critical dimensions provided for each. A typical list of electrical parameters is also shown. Note in particular the variations in k' values for different-sized devices.

Increased interest in IC chip design for VLSI has prompted the development of *portable* design rules that can, in principle, be adapted to an arbitrary processing line. These are formulated in terms of a fundamental design length λ (giving them the name λ *design rules*). The length λ is chosen according to technology and is a measure of the minimum resolution of the lithography. A base set of λ design rules is provided in Table 5.5. Portability is incorporated since λ may be varied as needed. The main disadvantage of portable design rules is that most spacings are chosen as integer or half-integer multiples of λ for simplicity. This generally precludes the possibility of attaining a maximum packing density.

TABLE 5.4 Simplified nMOS Design Rule Set Example

Electrical Characteristics				
Parameter	Minimum	Typical	Maximum	Units
1. Enhancement mode				
k' ($L > 5\,[\mu\text{m}]$)	22	25	28	$[\mu\text{A/V}^2]$
k' ($L < 5\,[\mu\text{m}]$)	20	22	26	$[\mu\text{A/V}^2]$
γ	0.31	0.35	0.38	$[\text{V}^{1/2}]$
V_{T0}	0.7	0.9	1.1	$[\text{V}]$
2. Depletion mode				
k' ($L > 5\,[\mu\text{m}]$)	22	25	28	$[\mu\text{A/V}^2]$
k' ($L < 5\,[\mu\text{m}]$)	22	25	28	$[\mu\text{A/V}^2]$
γ	0.37	0.43	0.47	$[\text{V}^{1/2}]$
V_{T0}	-3.7	-3.4	-3.0	$[\text{V}]$

Layout Rules		
Layer Number	Name	Rules (Units of $[\mu\text{m}]$)
01	Active area	Minimum dimension — 3×4
		Minimum spacing — 3
02	Enhancement implant	Minimum overlap with active area (FOX) — 3
		Minimum overlap with active area (n^+) — 2
03	Depletion implant	Minimum overlap with active area (FOX) — 3
		Minimum overlap with active area (n^+) — 3
		Minimum spacing from E-mode MOSFET — 3
04	Buried contact	Minimum overlap — 4
		Minimum spacing to poly or device — 3
05	Poly	Mininum width — 3
		Minimum spacing — 3
		Minimum MOSFET gate overhang — 2
		Minimum spacing to n^+ — 3
		Minimum dimensions for buried contact — 3×4
06	Metal contact mask	Minimum dimensions — 3×3
		Minimum spacing to active area edge — 3
		Minimum spacing to poly edge — 3
07	Metal	Minimum width — 6
		Minimum spacing — 4
08	Passivation (overglass)	

TABLE 5.5 Simplied nMOS Portable Layout Rule Set Example

Layer Number	Name	Rules	
01	Active area	Minimum dimension	$2\lambda \times 2\lambda$
		Minimum spacing	2λ
02	Enhancement implant	Minimum overlap with active area (FOX)	1.5λ
		Minimum overlap with active area (n^+)	1.5λ
03	Depletion implant	Minimum overlap with active area (FOX)	1.5λ
		Minimum overlap with active area (n^+)	1.5λ
		Minimum spacing from E-mode MOSFET	2λ
04	Buried contact	Minimum overlap	2λ
		Minimum spacing to poly or device	2λ
05	Poly	Minimum width	2λ
		Minimum spacing	2λ
		Minimum MOSFET gate overhang	2λ
		Minimum spacing to n^+	1λ
		Minimum dimensions for buried contact	$2\lambda \times 4\lambda$
06	Metal contact mask	Minimum dimensions	$2\lambda \times 2\lambda$
		Minimum spacing to active area edge	2λ
		Minimum spacing to poly edge	2λ
07	Metal	Minimum width	3λ
		Minimum spacing	3λ
08	Passivation (overglass)		

5.6.4 Some Comments on Layout

Integrated circuit layout is usually classified as an art that is best learned by practice. If integration density is not important, layout becomes relatively staightforward. The complicating factor is that the geometry determines the circuit parameters such as parasitic capacitance and resistance; these in turn affect the performance and the overall design. Iterations are often required.

Computer-aided design (CAD) and computer-aided engineering (CAE) are mandatory for complete layout control. Varieties of workstations are available for graphics layout and schematic capture (where the layout can be used to generate a component list). Most layout software tools provide a design rule checker (DRC), which can be used to determine if any design rules have been violated. In addition, circuit compaction and interconnect routing capabilities are common.

The advent of VLSI has spurred the development of CAD/CAE techniques. Advances in these areas have rapidly changed the level of layout complexity possible. Circuit design is intimately related to layout, so maintaining pace with the state of the art is mandatory.

5.7 Processing Variations

Variations in the process flow result in nonuniformities on the finished wafers. Electrical and structural differences can be detected when comparing wafers in a given lot. For that matter, die characteristics vary across a single wafer. Precise control of the fabrication process is not possible.

Geometric variations are addressed by adhering to a design rule set in the layout. Electrical variations directly affect the circuit performance, so a circuit designer must work around the processing variables to ensure working circuits. This complicates matters since the design methodology must be extended and refined to a level that is compatible with the process limitations.

5.7.1 Generalized Modeling

Process variations are usually analyzed using statistical techniques. The problem can be understood by referring to the example in Fig. 5.44, which shows a histogram for threshold voltages in a given test group. The "target"

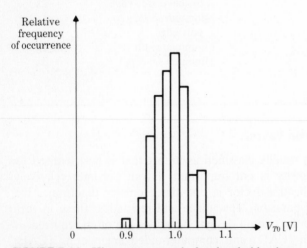

FIGURE 5.44 Histogram example for threshold voltage.

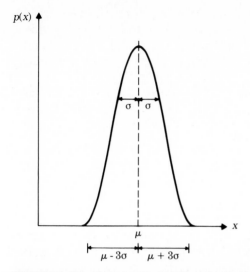

FIGURE 5.45 Gaussian probability distribution density.

value of the process is assumed to be $V_{T0} = +1$ [V]; this is termed the *nominal* or *working* value. Design rules generally provide nominal values for electrical characteristics.

Modeling the variations of a parameter is complicated and can be approached in several different ways. The simplest analysis is based on Gaussian distributions. For a given parameter x, the probability density $p(x)$ is given by

$$p(x) = \frac{1}{\sqrt{2\pi}\,\sigma} e^{-[(x-\mu)^2/2\sigma^2]}, \tag{5.7-1}$$

such that $p(x)\,dx$ gives the probability of finding the parameter value in the range x and $(x + dx)$. In this equation, μ is the mean or average value of the parameter and σ is the *standard deviation* of the distribution; σ measures the *spread* of the distribution. Figure 5.45 shows the general properties of a Gaussian.

Probability densities are used to compute probabilities. To calculate the probability P of finding the variable x between X_1 and X_2, differential contributions $p(x)\,dx$ in this range are summed by integrating:

$$P = \int_{X_1}^{X_2} p(x)\,dx. \tag{5.7-2}$$

Since the total probability of measuring x in $[-\infty, +\infty]$ is unity, the distribution is *normalized* to 1 by requiring

$$\int_{-\infty}^{+\infty} p(x)\,dx = 1. \tag{5.7-3}$$

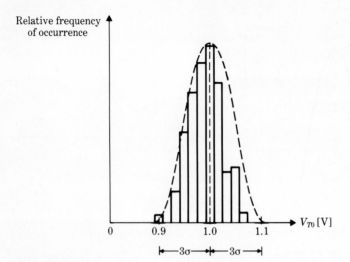

FIGURE 5.46   Adaptation of Gaussian distribution to threshold voltage histogram with large variations.

These equations indicate that the probability of measuring a value in a given range is proportional to the area under the curve.

It is useful to reference the probability to the values of μ and σ. An analysis of the problem shows that

$$\int_{\mu-3\sigma}^{\mu+3\sigma} p(x)\, dx \simeq 0.997, \tag{5.7-4}$$

which indicates that about 99.7% of the sample values lie in the interval $(\mu - 3\sigma, \mu + 3\sigma)$. The quantity 6σ is a convenient measure of the total spread.

To apply the Gaussian distribution to the threshold voltage problem in Fig. 5.44, let the nominal value $V_{T0} = 1$ [V] represent the mean. The standard deviation σ can be obtained by fitting the histogram to a Gaussian, as shown in Fig. 5.46. This allows for an estimation of $3\sigma \equiv \Delta V_{T0}$, so the threshold voltage can be assumed to lie in the range

$$V_{T0} = 1 \pm (\Delta V_{T0})\ [\text{V}]. \tag{5.7-5}$$

This becomes a circuit design constraint.

5.7.2   Analysis of Device Parameters

Process variations directly affect the device parameters used in circuit design. As an example, consider the threshold voltage spread just described. Variations in the gate oxide thickness x_{ox} are an important factor in this problem. To analyze this dependence, recall that

$$C_{ox} = \frac{\varepsilon_{ox}}{x_{ox}},$$

(5.7-6)

so taking differentials gives

$$dC_{ox} = -\frac{\varepsilon_{ox}}{x_{ox}^2} dx_{ox}.$$

(5.7-7)

This may be rewritten as

$$\frac{dC_{ox}}{C_{ox}} = -\frac{dx_{ox}}{x_{ox}},$$

(5.7-8)

where C_{ox} and x_{ox} are interpreted as nominal values.

To see the effect on V_{T0}, write

$$V_{T0} = V_{FB} + 2|\phi_F| + \frac{x_{ox}}{\varepsilon_{ox}}|Q_{B0}| + \frac{qD_I x_{ox}}{\varepsilon_{ox}}.$$

(5.7-9)

Assuming for the moment that only variations in x_{ox} are important, taking differentials yields

$$dV_{T0} = \frac{\partial V_{FB}}{\partial x_{ox}} dx_{ox} + \frac{1}{\varepsilon_{ox}}|Q_{B0}| \, dx_{ox} + \frac{qD_I}{\varepsilon_{ox}} dx_{ox}.$$

(5.7-10)

The flatband voltage term evaluates to (see eqn. 1.1-22)

$$\frac{\partial V_{FB}}{\partial x_{ox}} dx_{ox} = -\frac{1}{\varepsilon_{ox}}(Q_{ss} + Q_{ox}) \, dx_{ox},$$

(5.7-11)

so dV_{T0} can then be written in the form

$$dV_{T0} = [Q_{ss} + Q_{ox} - qD_I - |Q_{B0}|] \frac{dC_{ox}}{C_{ox}^2},$$

(5.7-12)

showing the dependence on variations in C_{ox}. This is useful because C_{ox} can be measured in the lab.

The general analysis technique can be extended to include more than one variable. For example, to compute the variation in V_{T0} due to changes in both x_{ox} and D_I, write

$$dV_{T0} = \frac{\partial V_{T0}}{\partial x_{ox}} dx_{ox} + \frac{\partial V_{T0}}{\partial D_I} dD_I.$$

(5.7-13)

This adds a term

$$+ \left(\frac{qD_I}{C_{ox}}\right) \frac{dD_I}{D_I}$$

(5.7-14)

to eqn. (5.7-13). For an arbitrary function $f = f(A, B, C, \dots)$, variations due to the parameters $A, B, C, \ldots$, can be computed by

$$df = \left(\frac{\partial f}{\partial A}\right) dA + \left(\frac{\partial f}{\partial B}\right) dB + \left(\frac{\partial f}{\partial C}\right) dC + \cdots. \tag{5.7-15}$$

To apply these results to a realistic situation, approximate

$$df \simeq \Delta f, \qquad dA \simeq \Delta A, \qquad dB \simeq \Delta B, \qquad \ldots, \tag{5.7-16}$$

where (Δf) represents the spread in f values due to changes (ΔA), (ΔB), etc., in the processing line.

5.7.3 Circuit Design Criteria

Process variations enter into the circuit design problem because they affect the voltages and currents. Consider a depletion-load nMOS inverter. The driver-load ratio for a given V_{OL} is calculated from

$$\beta_R = \frac{k_D'(W/L)_D}{k_L'(W/L)_L} = \frac{|V_{TL}(V_{OL})|^2}{[2(V_{OH} - V_{TD})V_{OL} - V_{OL}^2]}, \tag{5.7-17}$$

where

$$V_{TL}(V_{OL}) = V_{TOL} + \gamma_L(\sqrt{V_{OL} + 2|\phi_F|} - \sqrt{2|\phi_F|}). \tag{5.7-18}$$

Normal processing variations exist in the threshold voltages V_{TD} and V_{TL} and also in the k' values. To complicate matters more, power supply variations (as manifest in V_{OH}) should also be accounted for in the circuit design. A good design will account for these possibilities such that the range of V_{OL} output voltages is acceptable when interfaced with the other circuitry.

The simplest approach to designing around process variations is to use nominal values in the initial design cut and then vary the important parameters to track the circuit performance. Although a few hand calculations can be made, this generally requires a computer simulation of the circuit. For example, SPICE provides for a sensitivity analysis (.SENS) to determine tolerances. Experience is probably the best teacher for this type of problem.

Chip yield is also related to the problem. A processed wafer contains many die, only a fraction of which will operate correctly. The yield Y is defined to be the percentage of good die on the wafer. Defective chips either will be completely nonfunctional or may not meet electrical specifications for the product. In a new design, this may result from problems in processing, circuit design, or both. The importance of this consideration is immediately obvious: low yields give low (or negative) profits! Economics thus dictates that a circuit designer (who wants to stay employed) take variations into account.

5.8 CMOS Technologies

CMOS fabrication is complicated by the fact that both n-channel and p-channel MOSFETs are used. This requires that bulk regions of both polarities be available at the wafer surface. nMOS transistors are fabricated in grounded p-type bulk regions, while pMOS transistors are made in n-type bulk regions that are electrically connected to the power supply. The substrate wafer can be chosen either n-type or p-type. The CMOS process flow must include a step to create regions of the opposite polarity.

A possible approach to CMOS is to use a p-type wafer and selectively dope *n-tub* regions for placement of p-channel MOSFETs. The primary sequences for an inverter created in an n-tub LOCOS flow are shown in Fig. 5.47. The starting point is a p-type wafer. Ion implantation or diffusion is used to create n-tub regions; this is illustrated in Fig. 5.47(a). Allowance has been made for eventual connection of the n-tub region (the bulk electrode for pMOS transistors) to the power supply V_{DD} rail. The next step defines the active areas by nitride patterning, and then the LOCOS FOX is grown. Figure 5.47(b) shows the structure at this point. Note that the different widths for the two active areas have been patterned to give a symmetric inverter with $\beta_n = \beta_p$.

Gate oxide regrowth comes next, and then poly is deposited for the gate layer. Poly patterning followed by selective n^+ and p^+ implants gives the chip surface shown in Fig. 5.47(c). After a CVD oxide is applied and patterned, metal is deposited to give the output, V_{DD}, and ground connections shown in Fig. 5.47(d). This completes the main processing steps in the LOCOS flow.

Other approaches are possible. The direct complement of the n-tub process is the *p-tub* flow where an n-type wafer is used for pMOS transistors and p-tub regions are provided for n-channel MOSFETs. p-tub technology is ideal for retrofitting a pMOS (only) line for CMOS. *Twin-tub* processes have also been developed. In this approach, a thin CVD *epitaxial* layer of lightly doped silicon is grown on the substrate wafer. Then individual n-regions and p-regions are patterned in the epitaxial layer to provide MOSFET active areas.

A significant problem that can occur in CMOS circuits is *latch-up*. Figure 5.48(a) shows the origin of the problem in the n-tub inverter. Tracking the silicon layers from the power supply V_{DD} to ground shows the existence of a *4-layer p^+npn^+* structuring. This can be interpreted in different ways. Those familiar with power semiconductor devices will recognize this as the basis for a *silicon-controlled rectifier* (SCR). An SCR acts like a diode but is triggered into conduction using an external electrode. Alternately, the layering can be visualized as a pair of opposite polarity bipolar transistors. This view point is adopted here and gives the npn/pnp combination shown in Fig. 5.48(b).

Latch-up describes the condition in which current flow is diverted from the power supply to the ground electrode through the p^+npn^+ layering. Once current flow is initiated, the only way to stop it is to disconnect the power supply. Since no current is supplied to the circuit, the logic will not operate. In the worst-case scenario, the chip burns up from excessive Joule heating.

The bipolar transistor equivalent circuit shows the substrate current I from V_{DD}. If the base-emitter junction of the pnp transistor becomes forward biased, I begins to flow. This provides base current I_{Bn} to the npn transistor, which turns the device on. Collector current $I_{Cn} = I_{Bp}$ from the npn transistor

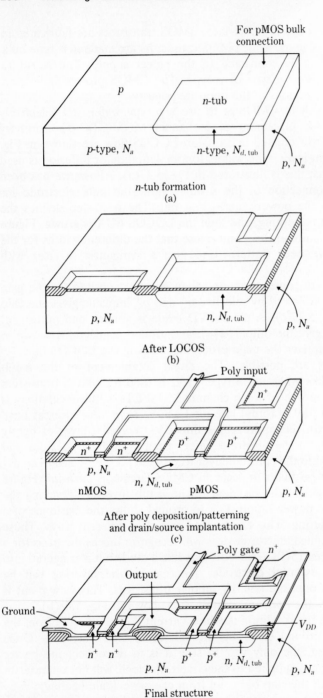

For pMOS bulk connection

p

n-tub

p-type, N_a n-type, $N_{d,\,tub}$

$p,\ N_a$

n-tub formation
(a)

$p,\ N_a$ $n,\ N_{d,\,tub}$ $p,\ N_a$

After LOCOS
(b)

Poly input

n^+

n^+ n^+ p^+ p^+

$p,\ N_a$

$n,\ N_{d,\,tub}$

nMOS   pMOS $p,\ N_a$

After poly deposition/patterning
and drain/source implantation
(c)

Poly gate n^+

Output

Ground

V_{DD}

n^+ n^+ p^+ p^+ $n,\ N_{d,\,tub}$

$p,\ N_a$ $p,\ N_a$

Final structure
(d)

FIGURE 5.47 CMOS process flow.

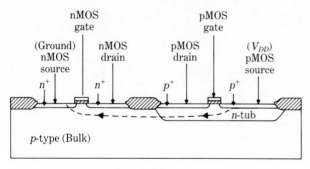

Inverter structure
(a)

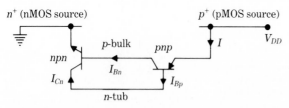

Equivalent bipolar transistor circuit
(b)

FIGURE 5.48 Latch-up in an *n*-tub CMOS inverter.

forces the *pnp* transistor to conduct more current. Feedback thus sustains the current flow and sinks current directly from V_{DD} to ground. This type of action leads to latch-up.

The bipolar analogy can be analyzed for the condition of latch-up. Denoting the common-base current gains by α_{npn} and α_{pnp}, the analysis gives [16]

$$I = \frac{I_{C0}}{1 - \alpha_{npn} - \alpha_{pnp}},$$ (5.8-1)

where I_{C0} is the reverse leakage current. The current gains are dependent on current such that $(d\alpha/dI) > 0$. Both α_{npn} and α_{pnp} increase with increasing current, so the condition

$$\alpha_{npn} + \alpha_{pnp} = 1$$ (5.8-2)

gives a singularity corresponding to latch-up current flow. After this is initiated, feedback sustains current flow.

Different approaches are available for dealing with latch-up. Using the bipolar transistor analogy shows that reducing the common-base current gains of the parasitic transistors can prevent the structure from ever reaching the singularity condition (eqn. 5.8-2). A direct way to implement this is to place heavily doped *guard rings* around the MOSFETs. Figure 5.49 shows both

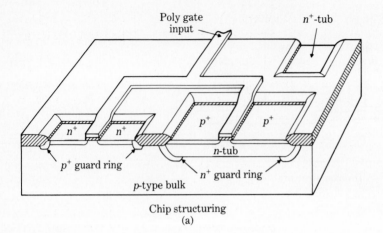

Chip structuring
(a)

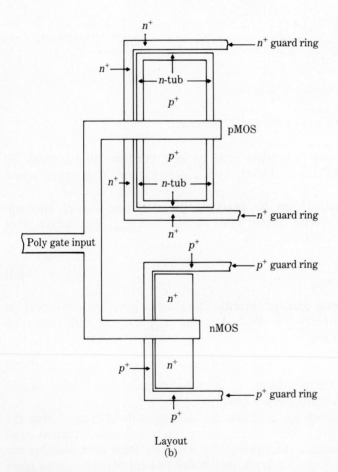

Layout
(b)

FIGURE 5.49 Guard rings for latch-up prevention.

perspective and top views of the guard ring placement. An n^+ ring is placed around the n-tub containing the pMOS device, while a p^+ ring surrounds the nMOS transistor.

Guard rings have the effect of changing the layering into a $p^+n^+p^+n^+$ structure in the lateral current flow direction. This reduces the effectiveness of the base and emitter regions in both transistors, which results in lower α values. (In terms of the device physics of bipolar transistors, the guard rings simultaneously reduce the emitter injection efficiency and the base transport factor.) Bias is controlled by tying the p^+ rings to 0 [V] and the n^+ rings to V_{DD}. Parasitic resistance and associated voltage drops are then eliminated, which keeps the bipolar transistors in cutoff.

Trench isolation in an n-tub or p-tub process automatically precludes the possibility of the latch-up by blocking the current flow path. No guard rings are necessary. Twin-tub also has this characteristic. However, the increased cost and complexity of twin-tub processing is significant.

Another CMOS technology is *silicon on sapphire* (SOS), which uses epitaxial silicon layers grown on a sapphire substrate. Sapphire is an insulator, so latch-up is not a problem. Moreover, the capacitance levels are greatly reduced, which allows for higher-speed circuits. Cost, however, is again a major factor against using SOS for everyday applications.

EXAMPLE 5.8-1

A CMOS process uses n-tub technology with the following specifications:

$$x_{ox} = 500 \, [\text{Å}], \quad \mu_n \simeq 580 \, [\text{cm}^2/\text{V-sec}], \quad \mu_p \simeq 218 \, [\text{cm}^2/\text{V-sec}],$$

$$N_a = 10^{15} \, [\text{cm}^{-3}], \quad N_{a,sw} = 5 \times 10^{15} \, [\text{cm}^{-3}],$$

$$N_{d,\text{tub}} = 10^{16} \, [\text{cm}^{-3}], \quad N_{d,sw} = 6 \times 10^{16} \, [\text{cm}^{-3}],$$

$$N_{d,n^+} = 10^{20} \, [\text{cm}^{-3}], \quad N_{a,p^+} = 10^{20} \, [\text{cm}^{-3}] \quad (n^+ \text{ and } p^+ \text{ regions}).$$

The base threshold voltages are set at $V_{T0n} = +0.70 \, [\text{V}]$, $V_{T0p} = -0.80 \, [\text{V}]$. Calculate the important circuit design parameters.

Solution
First,

$$C_{ox} = \frac{(3.9)(8.854 \times 10^{-14})}{0.05 \times 10^{-4}} \simeq 6.91 \times 10^{-8} \, [\text{F/cm}^2].$$

Next,

$$k_n' \simeq (580)(6.91 \times 10^{-8}), \qquad k_n' \simeq 40 \, [\mu\text{A/V}^2],$$

$$k_p' \simeq (218)(6.91 \times 10^{-8}), \qquad k_p' \simeq 15 \, [\mu\text{A/V}^2].$$

The body bias coefficients are

$$\gamma_n \simeq \frac{\sqrt{2(1.6 \times 10^{-19})(11.8)(8.854 \times 10^{-14})(10^{15})}}{6.91 \times 10^{-8}} \simeq 0.26 \, [\mathrm{V}^{1/2}] \quad (\mathrm{nMOS}),$$

$$\gamma_p \simeq \frac{\sqrt{2(1.6 \times 10^{-19})(11.8)(8.854 \times 10^{-14})(10^{16})}}{6.91 \times 10^{-8}} \simeq 0.84 \, [\mathrm{V}^{1/2}] \quad (\mathrm{pMOS}).$$

For an n-channel FET, the bulk Fermi potential is

$$2\,|\phi_{Fp}| \simeq 2(0.026) \ln \left(\frac{10^{15}}{1.45 \times 10^{10}} \right) \simeq 0.579 \, [\mathrm{V}] \quad (\mathrm{nMOSFETs}),$$

while a p-channel FET uses

$$2\,|\phi_{Fn}| \simeq 2(0.026) \ln \left(\frac{10^{16}}{1.45 \times 10^{10}} \right) \simeq 0.699 \, [\mathrm{V}] \quad (\mathrm{pMOSFETs}).$$

Next, examine the junction capacitances for the nMOS n^+/p-substrate and pMOS p^+/n-tub drain and source regions.

$$nMOS: \qquad \phi_o \simeq (0.026) \ln \left[\frac{10^{15} 10^{20}}{(1.45 \times 10^{10})^2} \right] \simeq 0.879 \, [\mathrm{V}],$$

and

$$C_{j0} \simeq \sqrt{ \frac{(1.6 \times 10^{-19})(11.8)(8.85 \times 10^{-14})}{2(10^{-15} + 10^{-20})(0.879)} } \simeq 9.75 \times 10^{-9} \, [\mathrm{F/cm}^2],$$

while the sidewall contribution quantities are

$$\phi_{osw} \simeq (0.026) \ln \left[\frac{(5 \times 10^{15}) 10^{20}}{(1.45 \times 10^{10})^2} \right] \simeq 0.921 \, [\mathrm{V}],$$

$$C_{j0sw} \simeq \sqrt{ \frac{(1.6 \times 10^{-19})(11.8)(8.854 \times 10^{-14})}{2(2 \times 10^{-16} + 10^{-20})(0.921)} } \simeq 2.13 \times 10^{-8} \, [\mathrm{F/cm}^2].$$

The sidewall capacitance must be multiplied by the junction depth x_{jn} for use in circuit calculations.

$$pMOS: \qquad \phi_o \simeq (0.026) \ln \left[\frac{10^{16} 10^{20}}{(1.45 \times 10^{10})^2} \right] \simeq 0.939 \, [\mathrm{V}],$$

and

$$C_{j0} \simeq \sqrt{ \frac{(1.6 \times 10^{-19})(11.8)(8.854 \times 10^{-14})}{2(10^{-16} + 10^{-20})(0.939)} } \simeq 2.98 \times 10^{-8} \, [\mathrm{F/cm}^2].$$

The sidewall contributions are

$$\phi_{osw} \simeq (0.026) \ln \left[\frac{(6 \times 10^{16}) 10^{20}}{(1.45 \times 10^{10})^2} \right] \simeq 0.985 \, [\mathrm{V}],$$

$$C_{j0sw} \simeq \sqrt{ \frac{(1.6 \times 10^{-19})(11.8)(8.854 \times 10^{-14})}{2(1.67 \times 10^{-17} + 10^{-20})(0.985)} } \simeq 7.23 \times 10^{-8} \, [\mathrm{F/cm}^2]. \quad \blacksquare$$

TABLE 5.6 Simplified CMOS Design Rule Set Example (*n*-tub)

Electrical Characteristics

Parameter	Minimum	Typical	Maximum	Units
1. *n*-channel MOSFETs				
k' ($L > 5$ [μm])	36	40	44	[μA/V^2]
k' ($L < 5$ [μm])	34	37	43	[μA/V^2]
γ	0.23	0.26	0.28	[V$^{1/2}$]
V_{T0}	0.7	0.9	1.1	[V]
2. *p*-channel MOSFET				
k' ($L > 5$ [μm])	15	18	21	[μA/V^2]
k' ($L < 5$ [μm])	14	16	20	[μA/V^2]
γ	0.76	0.84	0.93	[V$^{1/2}$]
V_{T0}	-1.0	-0.85	-0.7	[V]

Layout Rules

Layer Number	Name	Rules (Units of [μm])	
01	*n*-tub	Minimum dimensions	3×3
		Minimum spacing	3
02	Active area	Minimum dimension	3×4
		Minimum spacing	3
		Minimum distance to *n*-tub edge	9
03	nMOS threshold implant	Minimum overlap with active area (FOX)	3
		Minimum overlap with active area (n^+)	2
04	Poly	Minimum width	3
		Minimum spacing	3
		Minimum MOSFET gate overhang	2
		Minimum spacing to n^+	3
05	Metal contact mask	Minimum dimensions	3×3
		Minimum spacing to active area edge	3
		Minimum spacing to poly edge	3
06	Metal	Minimum width	6
		Minimum spacing	4
07	Passivation (overglass)		

5.9 CMOS Design Rules

Design rules for CMOS circuits are based on the same considerations as in nMOS designs. Lithography, processing limitations, and final electrical characteristics must be considered. The main differences in formulating a set of CMOS design rules are that (a) an opposite polarity tub mask must be included and (b) guard rings or some latch-up prevention becomes important. Both tend to increase the real estate requirement over that needed for an equivalent nMOS technology (i.e., for the same linewidth). However, as mentioned before, CMOS technologies have been developed to an advanced stage so that a direct comparison with NMOS is no longer meaningful.

CMOS design rules are process-specific. An illustrative set of rules is provided in Table 5.6. Portable design rules can also be formulated. A λ set example is listed in Table 5.7. An example of a CMOS inverter layout is shown in Fig. 5.50.

TABLE 5.7 Simplified CMOS Portable Layout Rule Set Example (n-tub)

Layer Number	Name	Rules	
01	n-tub	Minimum dimensions	$2\lambda \times 2\lambda$
		Minimum spacing	3λ
02	Active area	Minimum dimension	$3\lambda \times 2\lambda$
		Minimum spacing	2λ
		Minimum distance to n-tub edge	5λ
03	nMOS threshold implant	Minimum overlap with active area (FOX)	2λ
		Minimum overlap with active area (n^+)	1.5λ
04	Poly	Minimum width	2λ
		Minimum spacing	2λ
		Minimum MOSFET gate overhang	2λ
		Minimum spacing to n^+	2λ
05	Metal contact mask	Minimum dimensions	$3\lambda \times 2\lambda$
		Minimum spacing to active area edge	2λ
		Minimum spacing to poly edge	2λ
06	Metal	Minimum width	3λ
		Minimum spacing	3λ
07	Passivation (overglass)		

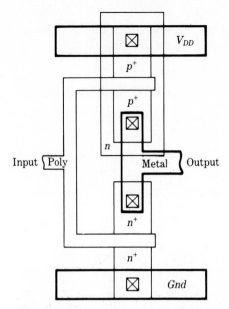

FIGURE 5.50   CMOS inverter layout in an *n*-tub technology (equal device size).

REFERENCES

(1) P. Antognetti, et al. (eds.), *Process and Device Simulation for MOS-VLSI Circuits* (NATO ASI Series), The Hague: Martinus Nijhoff, 1983.

(2) L. Chang, "A Derivative Method to Determine a MOSFET's Effective Channel Length and Width Electrically," *IEEE Electron Device Lett.*, vol. EDL-7, pp. 229–231, 1986.

(3) R. A. Colclaser, *Microelectronics Processing and Device Design,* New York: Wiley, 1980.

(4) L. Esaki and G. Soncini (eds.), *Large Scale Integrated Circuits Technology* (NATO ASI Series), The Hague: Martinus Nijhoff, 1982.

(5) S. K. Ghandhi, *VLSI Fabrication Principles,* New York: Wiley, 1983.

(6) A. B. Glaser and G. E. Subak-Sharpe, *Integrated Circuit Engineering,* Reading, MA: Addison-Wesley, 1977.

(7) L. A. Glasser and D. W. Dobberpuhl, *The Design and Analysis of VLSI Circuits,* Reading, MA: Addison-Wesley, 1985.

(8) D. J. Hamilton and W. G. Howard, *Basic Integrated Circuit Engineering,* New York: McGraw-Hill, 1975.

(9) G. De Mey, "Stochastic Geometry Effects in MOS Transistors," *IEEE J. Solid-State Circuits,* vol. SC-20, pp. 865–870, 1985.

(10) H. Mikoshiba, T. Homma, and K. Hamano, "A New Trench Isolation Technology as a Replacement for LOCOS," *IEDM Tech. Dig.*, pp. 578–581, 1984.

(11) A. D. Milne (ed.), *MOS Devices*, Edinburgh: Halstead Press, 1983.

(12) S. P. Murarka, *Silicides for VLSI Applications*, New York: Academic Press, 1983.

(13) R. D. Rung, "Trench Isolation Prospects for Application in CMOS VLSI," *IEDM Tech. Dig.*, pp. 574–577, 1984.

(14) B. Spinks, *Introduction to Integrated Circuit Layout*, Englewood Cliffs, NJ: Prentice-Hall, 1985.

(15) S. M. Sze, *VLSI Technology*, New York: McGraw-Hill, 1983.

(16) R. R. Troutman, *Latchup in CMOS Technology*, Hingham, MA: Kluwer Academic Publishers, 1986.

(17) N. Weste and K. Eshraghian, *Principles of CMOS VLSI Design*, Reading, MA: Addison-Wesley, 1985.

PROBLEMS

5.1 A thermal oxide layer is grown on a (100) silicon wafer using two steps. The first is a steam oxidation at 1000 [°C] for 60 minutes. This is followed by a dry O_2 growth at 1000 [°C] for 45 minutes.
(a) Calculate the oxide thickness x_{ox} after the steam oxidation step.
(b) Calculate the final oxide thickness x_{ox} after the dry O_2 growth is completed.

5.2 A *p*-tub CMOS process uses a 2-step boron diffusion into an *n*-type wafer that has a constant donor doping of 10^{15} [cm^{-3}]. The boron diffusion is described by the following diffusion schedule:

> Predeposit: 30 minutes at 900 [°C],
> Drive-in: 75 minutes at 1100 [°C].

(a) Calculate the depth x_j of the *p*-tub region after the diffusion.
(b) Plot the boron doping profile as a function of position x [μm] superposed on the *n*-type donor level of 10^{15} [cm^{-3}].
(c) Suppose that the drive-in is increased to 90 minutes. Find the *p*-tub depth (i.e., x_j) for this case.

5.3 The wafer in Problem 5.2 is subjected to additional heat treatment steps, which are summarized as follows:

> 30 minutes at 1000 [°C]
> 60 minutes at 950 [°C]
> 45 minutes at 1100 [°C].

Find the new location of the *pn* junction after these steps are

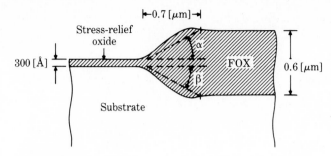

FIGURE P5.1

completed by using $\overline{Dt}$ in place of the drive-in (Dt) product. (Assume that the drive-in lasts 75 minutes as in the original problem statement.)

5.4 Phosphorus is implanted into silicon with an energy of 200 [keV], which results in a projected range of $R_p \simeq 0.25$ [μm] and a straggle of $\Delta R_p \simeq 0.08$ [μm]. The implant dose is 4×10^{14} [cm^{-3}], and the wafer doping is constant at a level of $N_a = 10^{15}$ [cm^{-3}] (p-type).
(a) Calculate the peak implant density N_p [cm^{-3}].
(b) What is the surface implanted doping level N_{ion} $(x = 0)$?
(c) Calculate the junction depth x_j where the pn junction is formed.

5.5 A LOCOS process yields the bird's beak cross section shown in Fig. P5.1. The field oxide is $X_{FOX} = 0.60$ [μm], and the stress-relief oxide has a thickness of 300 [Å]. The lateral encroachment region where the bird's beak is formed is 0.7 [μm]. Calculate the bird's beak angles α and β shown in the drawing.

5.6 A polysilicon layer is doped n-type during a self-aligned MOSFET process flow. The poly doping is approximated by $N_{d,poly} \simeq 10^{20}$ [cm^{-3}], and the electron mobility is $\mu_n \simeq 150$ [cm^2/V-sec]. The poly layer has a nominal thickness of 0.20 [μm].
(a) Determine the sheet resistance R_s [Ω] of the poly layer.
(b) Find the resistance R [Ω/μm] if the width of the line is 3 [μm].
(c) Find the percent increase in R [Ω/μm] if the thickness of the poly layer is reduced to 0.18 [μm].

5.7 A field oxide is grown to a thickness of $X_{FOX} = 0.60$ [μm] on a p-type substrate that has a doping density of $N_a = 10^{15}$ [cm^{-3}]. The FOX charge is $Q_{ss,FOX} = q(5 \times 10^{10})$ [C/cm^2] $\gg Q_{FOX}$.
(a) Calculate the value of $(V_{T0})_{FOX}$ assuming an n-type poly interconnect with a doping density of $N_{d,poly} = 10^{20}$ [cm^{-3}]. (Use the results of Problem 1.2 to calculate Φ_{IS}.).
(b) Find the dose D_I of the field acceptor ion implant needed to set $(V_{T0})_{FOX}$ to a value of +15 [V].

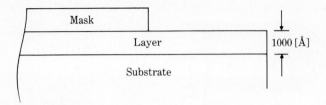

FIGURE P5.2

5.8 A wafer processing system using trench isolation results in the cross section of Fig. 5.30 with $X_{SOX} = 1500$ [Å], $X_{BOX} = 1700$ [Å], and $X_{TOX} = 4500$ [Å]. The trench itself is 1.4 [μm] wide and 5 [μm] deep.
(a) Calculate the poly-substrate capacitance per micron of trench.
(b) Calculate the interconnect-poly capacitance per micron of trench.

5.9 The MOSFET body bias coefficient is given by

$$\gamma = \frac{\sqrt{2q\varepsilon_{Si}N_a}}{C_{ox}}.$$

Suppose that the process gives nominal values of $x_{ox} = 500$ [Å] and $N_a = 10^{15}$ [cm^{-3}]. The oxide thickness is controlled to within 5%, while the acceptor doping N_a is known to have 8% variations.
(a) Calculate the percentage variation in γ.
(b) Find the minimum and maximum values of γ.

5.10 The layer shown in Fig. P5.2 is etched for a time $t_e = 1000$ [Å]/v_{ver} using different etching mechanisms. Assume that the masking layer is not affected.
(a) Provide scaled drawings of the etched layer cross section for degree of anisotropy values of $A = 0$, 0.25, 0.5, 0.75, and 1.0.
(b) Suppose that the etch time is increased to 1.5 t_e and that the substrate etch rates are identical to the etch rates for the layer. Sketch the profile if $A = 0.75$.

5.11 Cascaded CMOS inverters are shown in Fig. P5.3. A single stage of this layout was analyzed in Problem 4.10 of Chapter 4. Interconnect capacitances will now be added to the calculation. Assume interconnect levels of $C_{p-f} = 0.0576$ [fF/μm^2] and $C_{m-f} = 0.0345$ [fF/μm^2].
(a) Calculate the metal-field capacitance from the output of the first inverter to the metal-poly contact cut at the input to the second stage. Include only the metal-field lengths, i.e., ignore the regions where the metal overlaps with n^+, p^+, and poly.
(b) Find the input capacitance of the second stage as seen looking into the poly line. Then determine the sum ($C_{line} + G_G$).
(c) Compare the value of ($C_{line} + C_G$) to the value of C_{out} used in Problem 4.10. Does one contribution dominate?

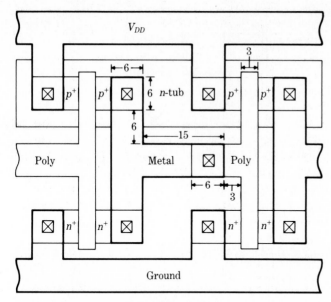

All dimensions in [μm]

FIGURE P5.3

*(d) Prepare a SPICE file that models both the inverter capacitances and the interconnect contributions. Study the transient response and compare the values of t_{HL} and t_{LH} [ns].

5.12 A certain process gives an active area encroachment of 0.4 [μm] per side and a lateral doping distance of $L_o = 0.3$ [μm] (gate overlap). Give the mask layouts for MOSFETs that have final (W/L) values of 0.5, 1, and 4. Assume that the minimum resolution of the system is 3 [μm].

5.13 *Stick diagrams* are simplified drawings that show the placement and routing of IC layers. Figure P5.4 shows a sample base set of stick diagram notation where each material layer is represented by a specific line type. (An alternate approach is to use a specific color for each layer. In some instances, the color chosen for each layer roughly corresponds to the color seen for the layer on a wafer illuminated by white light. Otherwise, the choice of colors has become somewhat arbitrary.)

Chip layout patterns can be described using stick diagrams. These are useful for initial planning. In a self-aligned MOS technology, the gate poly layer is used to mask the n^+ pattern in making nMOS transistors. Thus, assuming that a threshold adjustment implant is provided, drawing a poly stick that crosses an n^+ stick gives an E-mode MOSFET. A D-mode transistor is obtained by adding a depletion implant.

(a) Construct the stick diagrams for the three types of nMOS inverters.

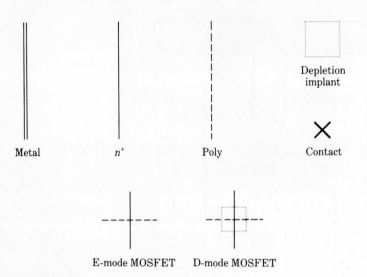

FIGURE P5.4

Then discuss the advantage of having a stick diagram vs. the information that is missing from the drawing.

(b) Create additional lines (or colors) to describe an n-tub CMOS process. Then construct the stick diagram for a CMOS inverter.

5.14 Consider a depletion load inverter in a process that exhibits threshold voltage variations of ΔV_{T0D} and ΔV_{T0L}.

(a) Starting with the design equation for β_R in eqn. (3.4-43), compute the change $\Delta \beta_R$ due to variations in V_{T0D} and V_{T0L}.

(b) Assume that $V_{OH} = 5$ [V], $\gamma = 0.37$ [V], $2|\phi_F| = 0.58$ [V], and

$$V_{T0D} = 0.9 \pm 0.1 \,[\text{V}], \qquad V_{T0L} = -3.3 \pm 0.1 \,[\text{V}]$$

give the threshold voltage ranges. A design value of $V_{OL} = 0.25$ [V] is set as a design specification. Use the expression derived in (a) to calculate the variation $\Delta \beta_R$. Then discuss the implications the analysis has on overall design acccuracy and chip yield.

CHAPTER  6

Combinational MOS Logic Circuits

This chapter initiates high-level MOS logic design by examining depletion load nMOS and CMOS circuit implementations of basic logic gates. Circuit properties are analyzed, and design considerations are presented. Various applications and extensions of the gate designs are also included. These are usually termed *random* or *combinational* logic designs.

The mathematics is based on the inverter equations developed in Chapters 3 and 4. This allows a relatively analytic presentation of the material and also provides a deeper understanding of the circuit properties. In general, the circuits are classified as *static* to indicate that no external clock is required.

6.1 nMOS NOR Gate

An nMOS NOR circuit is shown in Fig. 6.1. The 2-input NOR gate is constructed by simply paralleling two driver MOSFETs. The operation of the circuit is easily understood from qualitative arguments. If either $V_{in,A}$ OR $V_{in,B}$

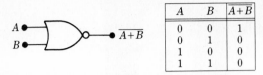

A	B	$\overline{A+B}$
0	0	1
0	1	0
1	0	0
1	1	0

Logic symbol and truth table
(a)

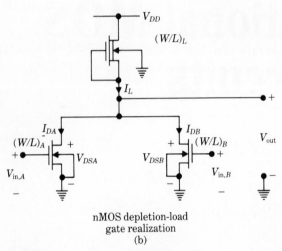

nMOS depletion-load
gate realization
(b)

FIGURE 6.1   Basic 2-input NOR gate in an nMOS technology.

is high, a conducting path to ground is created by the formation of a channel; the output voltage V_{out} will be at V_{OL}. The only time that $V_{out} = V_{OH}$ occurs is when both input voltages are low, since this will place both drivers in cutoff. The overall NOR function is actually an OR operation followed by a logic inversion (NOT). This arises because the NOR circuit is constructed using the inverter as a basis with the OR function provided by the two parallel drivers.

To analyze the circuit, first note that the use of a depletion load gives $V_{OH} = V_{DD}$, since this is almost identical to the case of a simple inverter. V_{OH} may be slightly less than the inverter output high voltage because the leakage currents of both drivers must be accounted for.

Calculating V_{OL} is more complicated because there are three combinations of input voltages that give a low output voltage. For arbitrary values of driver aspect ratios $(W/L)_A$ and $(W/L)_B$, this yields three distinct values for V_{OL}. If only one input voltage is high, the circuit reduces to the inverter analysis (since the other driver is in cutoff with $I_D = 0$). Using eqn. (3.4-22), the output low voltage for the depletion load NOR is found from

$$V_{OL} = (V_{OH} - V_{TD}) - \sqrt{(V_{OH} - V_{TD})^2 - \frac{1}{\beta_R}|V_{TL}(V_{OL})|^2}. \qquad \text{(6.1-1)}$$

It will be assumed that $V_{TD} = V_{TD,A} = V_{TD,B}$ since all enhancement-mode transistors have the same (nominal) threshold voltage in a given process. The value used for β_R depends on the input voltages. If $V_{in,A} = V_{OH}$ and $V_{in,B} = 0$, then

$$\beta_R = \frac{\beta_A}{\beta_L} = \frac{k'_D \left(\dfrac{W}{L}\right)_A}{k'_L \left(\dfrac{W}{L}\right)_L}, \tag{6.1-2}$$

since MOSFET A is providing the ratioed value for the voltage, i.e., $R_{on} = R_{on,A}$. Similarly, in the reverse situation where $V_{in,A} = 0$ and $V_{in,B} = V_{OH}$, MOSFET B is providing an on-resistance of $R_{on,B}$ so that

$$\beta_R = \frac{\beta_B}{\beta_L} = \frac{k'_D \left(\dfrac{W}{L}\right)_B}{k'_L \left(\dfrac{W}{L}\right)_L} \tag{6.1-3}$$

should be used. If the drivers are identical with $(W/L)_A = (W/L)_B$, then the V_{OL} levels are identical.

If both driver input voltages are at V_{OH}, then the calculations must be modified since

$$I_L = I_{DA} + I_{DB} \tag{6.1-4}$$

indicates that current is flowing through both of the driver MOSFETs. The drivers are nonsaturated while the load is saturated, so this may be written as

$$\beta_L |V_{TL}(V_{OL})|^2 = (\beta_A + \beta_B)[2(V_{OH} - V_{TD})V_{OL} - V_{OL}^2]. \tag{6.1-5}$$

Thus,

$$V_{OL} = (V_{OH} - V_{TD}) - \sqrt{(V_{OH} - V_{TD})^2 - \frac{\beta_L}{(\beta_A + \beta_B)}|V_{TL}(V_{OL})|^2} \tag{6.1-6}$$

gives the self-iterating equation that determines V_{OL} when $V_{in,A} = V_{in,B} = V_{OH}$. This is identical to eqn. (6.1-1), with β_R given by

$$\beta_R = \frac{\beta_A + \beta_B}{\beta_L} = \frac{k'_D \left[\left(\dfrac{W}{L}\right)_A + \left(\dfrac{W}{L}\right)_B\right]}{k'_L \left(\dfrac{W}{L}\right)_L}. \tag{6.1-7}$$

Since the effective driver β is now $(\beta_A + \beta_B)$, it is seen that the output voltage for this case will be smaller than that obtained when only one of the drivers is on. Consequently, this represents the best-case (i.e., lowest) V_{OL} value available from the circuit.

A simple DC design procedure for the 2-input NOR gate is obtained by noting the similarities between this gate and the basic inverter. In particular, when only a single input line is high, V_{OL} is determined directly from the inverter analysis. Since this represents a worst-case (highest) V_{OL} level, eqn. (3.4-43) may be used to write

$$\beta_R = \frac{|V_{TL}(V_{OL})|^2}{2(V_{OH} - V_{TD})V_{OL} - V_{OL}^2}, \tag{6.1-8}$$

where β_R is either β_A/β_L or β_B/β_L, depending on which driver is on. The simplest design choice is to set

$$\beta_A = \beta_B = \beta_R\beta_L \tag{6.1-9}$$

so that

$$\left(\frac{W}{L}\right)_A = \left(\frac{W}{L}\right)_B \tag{6.1-10}$$

indicates that the drivers are identical. This choice will work as long as the input voltage levels to both devices have approximately equal V_{OH} values. If the two input high voltages are different, then the (W/L) ratios must be adjusted accordingly, which gives rise to a nonsymmetric driver arrangement.

The design process is easily understood using the concept of equivalent MOSFET resistances. For a nonsaturated driver, the on-conductance was found as

$$G_{on} = \frac{1}{R_{on}} = \frac{k_D'}{2}\left(\frac{W}{L}\right)_D [2(V_{OH} - V_{TD}) - V_{OL}]. \tag{6.1-11}$$

Consequently, choosing $(W/L)_A = (W/L)_B$ is equivalent to setting the driver on-conductance to yield the desired output low voltage when only a single driver is providing a path to ground. If both drivers have high input voltages applied, then the equivalent driver resistance will be $(R_{on}/2)$ since the two are in parallel. This lowers the output voltage below the original (worst-case) design value, as expected from the analysis.

The switching times for the 2-input NOR gate may be computed using the analysis presented in Chapter 4. C_{out} is found from the basic capacitance circuit in Fig. 6.2 as

$$C_{out} \simeq K(V_{OH}, V_{OL})[C_{dbA} + C_{dbB} + C_{sbL}] + C_{GDA} + C_{GDB}$$
$$+ C_{GDL} + C_{line} + C_G, \tag{6.1-12}$$

where C_G is the input capacitance of the next stage. Note that C_{out} is larger than the inverter value by an amount

$$\Delta C_{out} \simeq K(V_{OH}, V_{OL})C_{dbB} + C_{GDB} + \Delta C_{line}, \tag{6.1-13}$$

with the first two terms corresponding respectively to the drain-bulk and gate-drain capacitances of driver B. ΔC_{line} represents the increased line

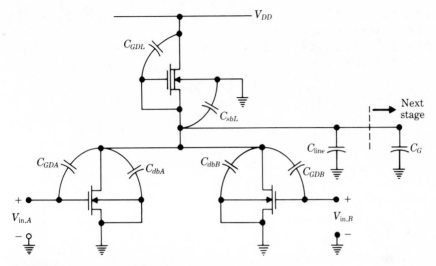

FIGURE 6.2 C_{out} calculation for a 2-input NOR gate.

capacitance due to the additional interconnect distance required for the second driver.

Recall now that all of the basic transient time intervals are proportional to the output node capacitance C_{out}. If a NOR gate and a NOT (inverter) gate are constructed with identical driver and load geometries, the above discussion shows that the NOR gate will exhibit a slower switching response when compared with the NOT function (assuming that the parasitic line capacitances are the same). When both types of gates are used in a digital system, this difference in timing must be accounted for. As a simple example, consider the two-gate configuration shown in Fig. 6.3. The input logic voltages are designated as $V_{\text{in},A}(t)$ and $V_{\text{in},B}(t)$. If both inputs A and B are switched simultaneously, the different load capacitance values will give unequal response times. In terms of the propagation delay times t_p, the logic signals will require respective time intervals of

$$(\Delta t)_C \simeq (t_p)_{\text{NOT}}, \qquad (\Delta t)_D \simeq (t_p)_{\text{NOT}} + (t_p)_{\text{NOR}} \qquad \text{(6.1-14)}$$

before the logic levels at the outputs C and D correctly reflect the input conditions. Thus, output C will become stable before the output at D. This type of timing problem may be important in a static logic design, particularly when the overall system response is considered. In an actual circuit design, it is possible to equalize the switching times of the two gates by varying the driver and load (W/L) values while maintaining β_R constant. For example, to speed up the NOR gate, $(W/L)_L$ can be increased to allow the load transistor greater current capabilities during the capacitor charge times. To preserve the value of V_{OL}, this requires that the driver aspect ratios be increased by the same amount. The resulting circuit exhibits faster switching times but will consume more chip area. The existance of unequal delay times gives motivation for the

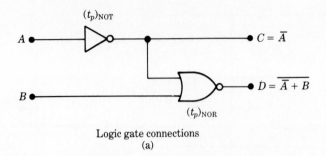

Logic gate connections
(a)

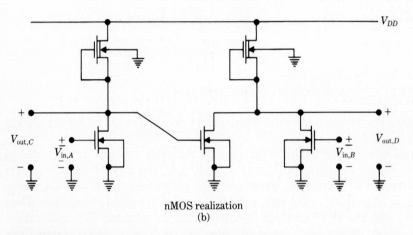

nMOS realization
(b)

FIGURE 6.3 Example of propagation delay problem.

introduction of the clocking pulses employed in dynamic logic circuits, which are studied in Chapter 8.

Although the discussion has been centered around the 2-input NOR gate, it is a simple matter to extend the circuit to an N-input NOR by paralleling N driver MOSFETs as drawn in Fig. 6.4. The analysis and design criteria established above remain generally valid for the N-input NOR gate. In particular, the worst-case (highest) V_{OL} voltage will arise from the smallest driver (W/L) value, while the best-case (lowest) V_{OL} value occurs when all of the inputs are simultaneously high. The simplest design approach for the N-input gate is to first compute the driver $(W/L)_D$ for an inverter and then set

$$(W/L)_D = (W/L)_A = (W/L)_B = \cdots = (W/L)_N. \tag{6.1-15}$$

A circuit designed in this manner will have the same basic properties discussed for the simpler 2-input gate. Care must be taken to ensure that the extra load capacitance does not severely degrade the transient performance of the circuit, since now the additional capacitance (over the inverter value) is given by $N(\Delta C_{out})$.

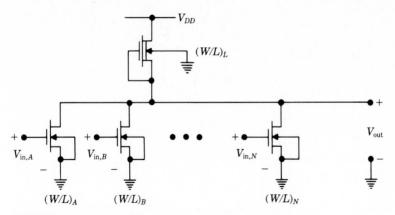

FIGURE 6.4 N-input NOR gate in an nMOS technology.

6.2   nMOS NAND Gate

The next gate arrangement to be examined is the 2-input NAND circuit shown in Fig. 6.5. This is obtained from the inverter by adding another driver in series. Conceptually, the operation of the circuit is straightforward. If either $V_{in,A}$ or $V_{in,B}$ is low, no direct path to ground exists since at least one of the drivers is in cutoff. The output will thus be at a level $V_{out} = V_{OH}$. The only time a conducting path exists through both drivers is when $V_{in,A}$ AND $V_{in,B}$ are both high; this results in an output voltage $V_{out} = V_{OL}$. Obviously, the NAND operation is obtained by combining the inverter NOT properties with the logical AND provided by the series-connected drivers.

To analyze the DC characteristics of the NAND circuit, first note that $V_{OH} = V_{DD}$ is still a good approximation. This is due to the use of a depletion-mode load transistor and also because the series driver arrangement restricts $I_{D,A}$ and $I_{D,B}$ to approximately the same leakage levels regardless of which transistor is in cutoff.

Calculating V_{OL} is more complicated. The basic starting point for the analysis is

$$I_L = I_{DA} = I_{DB}, \qquad (6.2\text{-}1)$$

with $V_{in,A} = V_{in,B} = V_{OH}$. Since both drivers will be nonsaturated,

$$\frac{\beta_L}{2}|V_{TL}(V_{OL})|^2 = \frac{\beta_A}{2}[2(V_{GS,A} - V_{TA})V_{DS,A} - V_{DS,A}^2]$$

$$= \frac{\beta_B}{2}[2(V_{OH} - V_{TB})V_{DS,B} - V_{DS,B}^2], \qquad (6.2\text{-}2)$$

with

$$V_{OL} = V_{DS,A} + V_{DS,B}. \qquad (6.2\text{-}3)$$

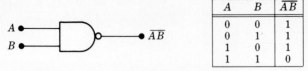

A	B	$\overline{AB}$
0	0	1
0	1	1
1	0	1
1	1	0

Logic symbol and truth table
(a)

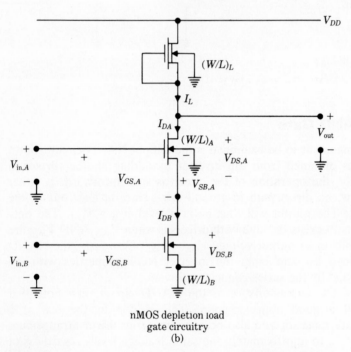

nMOS depletion load
gate circuitry
(b)

FIGURE 6.5   Basic 2-input NAND gate in an nMOS technology.

The complexity of this problem can be seen by noting

$$V_{GS,A} = V_{OH} - V_{DS,B} \tag{6.2-4}$$

and also that body bias effects arise in MOSFET A since

$$V_{TA}(V_{DS,B}) = V_{T0D} + \gamma(\sqrt{V_{DS,B} + 2|\phi_F|} - \sqrt{2|\phi_F|}). \tag{6.2-5}$$

Because of these relations (which originate from the presence of the extra node between the drivers), the solutions for $V_{DS,A}$ and $V_{DS,B}$ are quite difficult to find directly.

The easiest approach for calculating the output low voltage of the circuit is to note that since V_{OL} is assumed small, $V_{DS,B}$ is also quite small. This allows for two approximations to be made. First, $V_{GS,A} = V_{OH}$ is a reasonable first estimate for the gate-source voltage of MOSFET A. Second, the small value of

$V_{DS,B}$ implies that body bias effects in eqn. (6.2-5) will be negligible; thus, $V_{TA} = V_{TB} = V_{TD}$. Equation (6.2-2) may now be reduced to read

$$\frac{\beta_L}{2}|V_{TL}(V_{OL})|^2 \simeq \frac{\beta_A}{2}[2(V_{OH} - V_{TD})V_{DS,A} - V_{DS,A}^2]$$

$$= \frac{\beta_B}{2}[2(V_{OH} - V_{TD})V_{DS,B} - V_{DS,B}^2]. \tag{6.2-6}$$

Treating these as approximate quadratic equations gives the drain-source voltages as

$$V_{DS,A} \simeq (V_{OH} - V_{TD}) - \sqrt{(V_{OH} - V_{TD})^2 - \frac{1}{\beta_{R,A}}|V_{TL}(V_{OL})|^2},$$

$$V_{DS,B} = (V_{OH} - V_{TD}) - \sqrt{(V_{OH} - V_{TD})^2 - \frac{1}{\beta_{R,B}}|V_{TL}(V_{OL})|^2}, \tag{6.2-7}$$

where

$$\beta_{R,A} = \frac{\beta_A}{\beta_L}, \qquad \beta_{R,B} = \frac{\beta_B}{\beta_L} \tag{6.2-8}$$

are the respective driver-load ratios. Using eqn. (6.2-3), these two expressions may be added to give

$$V_{OL} \simeq 2(V_{OH} - V_{TD}) - \left\{ \sqrt{(V_{OH} - V_{TD})^2 - \frac{1}{\beta_{R,A}}|V_{TL}(V_{OL})|^2} \right.$$

$$\left. + \sqrt{(V_{OH} - V_{TD})^2 - \frac{1}{\beta_{R,B}}|V_{TL}(V_{OL})|^2} \right\}, \tag{6.2-9}$$

which is a self-iterating equation for V_{OL}. If the drivers are identical so that

$$\beta_{R,A} = \beta_{R,B} \equiv \beta_{R,AB}, \tag{6.2-10}$$

then the two square-root terms may be combined to yield the simpler equation

$$V_{OL} \simeq 2\left\{ (V_{OH} - V_{TD}) - \sqrt{(V_{OH} - V_{TD})^2 - \frac{1}{\beta_{R,AB}}|V_{TL}(V_{OL})|^2} \right\} \tag{6.2-11}$$

for approximating V_{OL}.

Equation (6.2-11) for the case of identical drivers is interesting since it contains the basic DC design parameters V_{OL} and $\beta_{R,AB}$. V_{OL} is plotted as a function of $\beta_{R,AB}$ in Fig. 6.6 using a set of typical processing parameters; the graph also shows $V_{OL}(\beta_R)$ for a simple inverter in dashed lines. To understand the difference between the two curves, note that the NAND gate equation in (6.2-11) has the same form as the inverter V_{OL} expression (which is given in eqn. 6.1.1 for the NOR gate), except for the factor of 2 on the right-hand side. For the NAND gate to produce low values of V_{OL}, the factor of 2 implies that the square-root term must be relatively close in magnitude to $(V_{OH} - V_{TD})$.

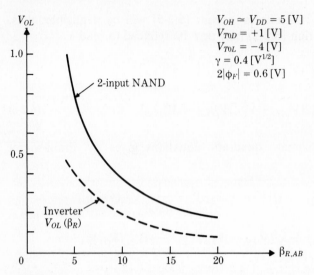

FIGURE 6.6 V_{OL} as a function of $\beta_{R,AB}$ as approximated by eqn. (7.2-11) for a 2-input NAND.

With regard to design parameters, this can be accomplished by choosing $\beta_{R,AB}$ to be a large number. Thus, the NAND gate $\beta_{R,AB}$ must be larger than the inverter driver-load ratio β_R used to achieve the same V_{OL} value.

The basic DC design equations for the ratioed circuit may be extracted from the relations in eqn. (6.2-6) by noting the constraint on the sum of the drain-source voltages given by eqn. (6.2-3). The simplest case is that where the drivers are chosen to be identical. Assuming

$$V_{DS,A} = V_{DS,B} \simeq \frac{1}{2} V_{OL}, \tag{6.2-12}$$

eqn. (6.2-6) may be rearranged to read

$$\beta_{R,AB} \simeq \frac{2 \,|V_{TL}(V_{OL})|^2}{\left[2(V_{OH} - V_{TD})V_{OL} - \dfrac{1}{2} V_{OL}^2\right]}, \tag{6.2-13}$$

or

$$\beta_{R,AB} \simeq 2\beta_R, \tag{6.2-14}$$

where β_R is the driver-load ratio for an inverter design with the same V_{OL} specification. This behavior can be seen from the graph in Fig. 6.6. In terms of layout considerations, eqn. (6.2-14) implies that the 2 NAND drivers will consume more than 4 times the chip area required for the driver MOSFET of an inverter.

This result may be clarified by the concept of equivalent resistances. The driver on-resistance R_{on} for an inverter is given by eqn. (6.1-11) of the previous section. Since the two NAND drivers are in series, the problem of establishing V_{OL} reduces to the requirement that

$$R_{on} = R_{on,A} + R_{on,B}. \qquad (6.2\text{-}15)$$

If the two drivers are identical, then $G_{on,A} = G_{on,B}$. Consequently, choosing

$$\left(\frac{W}{L}\right)_A = \left(\frac{W}{L}\right)_B \simeq 2\left(\frac{W}{L}\right)_D \qquad (6.2\text{-}16)$$

is equivalent to setting

$$R_{on,A} = R_{on,B} \simeq \frac{1}{2}R_{on} \qquad (6.2\text{-}17)$$

since the resistances are inversely proportional to the (W/L) value. It is noted in passing that the drivers can be designed to have different geometries so long as eqn. (6.2-15) is satisfied. The design equations for this case are found by choosing different values for $V_{DS,A}$ and $V_{DS,B}$ (as fractions of V_{OL}).

The evaluation of C_{out} for the transient analysis of the 2-input NAND gate is complicated by the presence of the additional node between the two drivers. The basic capacitance contributions are approximated as shown in Fig. 6.7. However, since the voltage V_X at the node between the two drivers varies with

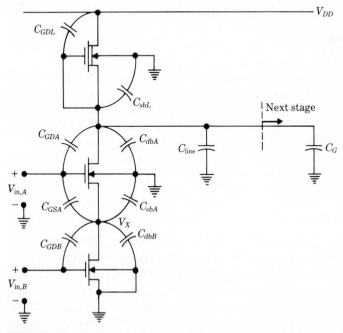

FIGURE 6.7 Basic capacitance contributions for the 2-input NAND gate.

the input switching conditions, only certain capacitors actually contribute to C_{out}. The resulting differences in switching times should be accounted for when the gate is used in a system design.

As an example, assume that initially $V_{\text{in},A} = V_{\text{in},B} = V_{OH}$. Then both devices are on and $V_X < V_{OL}$, i.e., only a small voltage exists across MOSFET B. Now suppose that $V_{\text{in},B}$ is switched to $0\,[\text{V}]$, while $V_{\text{in},A}$ remains at V_{OH}. Since MOSFET B is in cutoff, the output node charges to V_{OH}; V_X also increases toward this value. A quick examination of the circuit shows that every capacitor undergoes a voltage change. Thus,

$$C_{\text{out}} \simeq C_{GDA} + C_{GSA} + C_{GDB} + C_{GDL} + C_{\text{line}} + C_G$$
$$+ K(V_{OH}, V_{OL})(C_{dbA} + C_{sbA} + C_{dbB} + C_{sbL}) \tag{6.2-18}$$

is a reasonable approximation for the value of C_{out} that characterizes this possibility.

The situation is quite different if $V_{\text{in},B}$ is at V_{OH} while $V_{\text{in},A}$ is switched to $0\,[\text{V}]$. For this case, MOSFET B will be isolated from the output node since MOSFET A is in cutoff. Thus, V_X will remain at a low value. For the output to change to V_{OH}, the circuit then indicates that C_{GDA} and C_{dbA} must charge to this value. Consequently, these two capacitors represent the most important driver contributions to the output capacitance. This switching possibility is characterized by

$$C_{\text{out}} \simeq C_{GDA} + C_{GDL} + C_{\text{line}} + C_G + K(V_{OH}, V_{OL})(C_{dbA} + C_{sbL}), \tag{6.2-19}$$

which is obviously smaller than that in eqn. (6.2-18). This simple example illustrates the importance of the signal timing in determining the transient behavior of the gate. The worst-case conditions should always be examined, as these give an indication of the switching limitations.

This discussion may be easily extended to the case of an N-input NAND gate constructed using N drivers connected in series. The circuit for $N = 3$ is shown in Fig. 6.8. The design considerations for the 3-input structure are analogous to the simple 2-input NAND above. In particular, if all drivers are chosen to be identical, then the driver-load ratio $\beta_{R,ABC}$ becomes

$$\beta_{R,ABC} \simeq 3\beta_R, \tag{6.2-20}$$

which is equivalent to setting

$$R_{\text{on},A} + R_{\text{on},B} + R_{\text{on},C} \simeq R_{\text{on}}. \tag{6.2-21}$$

The drawbacks of using multiple-input NAND gates become immediately obvious. First, the driver aspect ratios become excessively large; for example, N identical drivers will each require a gate area of $(WL) = N(W/L)_D L^2$. This implies very expensive real estate costs (when compared to the amount of logic actually implemented). Also, the (worst-case) value of C_{out} increases significantly with each additional driver, which in turn degrades the transient response of the circuit. Owing to these observations, it is very rare to use

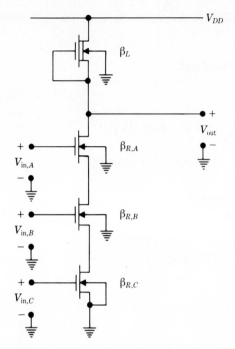

For identical drivers, $\beta_{R,ABC} = \beta_{R,A} = \beta_{R,B} = \beta_{R,C} \approx 3\beta_R$

FIGURE 6.8   3-input NAND gate in nMOS.

NAND gates with more than 3 inputs. Rather, static random logic design usually centers around using NOT and NOR gates in conjunction with the complex logic implementation schemes presented in the next section.

EXAMPLE 6.2-1

A 2-input nMOS NAND gate is described by the parameters

$$\beta_{R,AB} = 12, \quad V_{T0D} = +1\,[\text{V}], \quad V_{T0L} = -3.5\,[\text{V}], \quad V_{DD} = +5\,[\text{V}].$$

Assume that $\gamma_L = 0.40\,[\text{V}^{1/2}]$ and $2|\phi_{F,L}| = 0.6\,[\text{V}]$. Calculate V_{OL} for the circuit.

Solution
Equation (6.2-11) gives the self-iterating expression

$$V_{OL} = 2\left(4 - \sqrt{16 - \frac{1}{12}|V_{TL}(V_{OL})|^2}\right),$$

where

$$V_{TL}(V_{OL}) = -3.5 + 0.4(\sqrt{V_{OL} + 0.6} - \sqrt{0.6}).$$

This can be solved using hand calculations as summarized in the following table:

V_{OL} Guess	$|V_{TL}|$	Right-Hand Side
0.5	3.39	0.243
0.243	3.42	0.251
0.251	3.44	$0.251 \leftarrow V_{OL} \approx 0.251\,[V]$

The following is a BASIC program that performs this iteration:

```
10 READ VDD, VTOD, VTOL, GAMMA, FERMI, BR
20 DATA 5, 1, -3.5, 0.4, 0.6, 12
30 VOL=.1*VDD:REM INITIAL GUESS
40 VTL=VTOL+GAMMA*(SQR(VOL+FERMI)-SQR(FERMI))
50 VX=2*((VDD-VTOD)-SQR((VDD-VTOD)^2-
   (1/BR)*(ABS(VTL))^2))
60 IF ABS(VX-VOL) < .001 THEN 100
70 VOL=VX:GOTO 40
100 PRINT "VOL=";VOL

RUN
VOL=.2508307
```

The results of a SPICE simulation for the VTC are shown below and in Fig. E6.1. This corresponds to the case where both inputs are tied together and simultaneously switched.

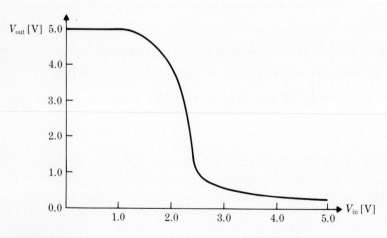

FIGURE E6.1

```
DMODE INVERTER
VDD 3 0 DC 5VOLTS
MD1 2 1 4 0 EMODE L=5U W=30U
MD2 4 1 0 0 EMODE L=5U W=30U
MLOAD 3 2 2 0 DMODE L=10U W=5U
VS 1 0 DC PULSE
.MODEL EMODE NMOS VTO=1 GAMMA=0.40 KP=2.5E-5
.MODEL DMODE NMOS VTO=-3.5 GAMMA=0.40 KP=2.5E-5
.DC VS 0 5 0.1
.PLOT DC V (2)
.END
```

6.3 Complex Static nMOS Logic

The treatment thus far has examined three basic nMOS circuits: the NOT, NOR, and NAND gates. With regard to logic design, this set of gates is more than sufficient for the implementation of arbitrary logic functions. The chip designer, however, must be more careful, since items such as the layout geometries and real estate minimization become critical factors. The true power of random static nMOS digital circuits arises when complex logic gates are introduced. These configurations allow for relatively complicated logic functions to be realized using a minimum number of transistors. This in turn reduces the overall chip area consumption, allowing for higher-density circuit integration.

The idea behind complex logic circuits is really quite simple. If the load is ignored, then drivers in series may be viewed as providing the AND operation, while parallel driver branches give OR capabilities. The load device is responsible for inverting the logic and is taken as the origin of the NOT function. To implement a complex logic function in nMOS, AND and OR gates are first used to construct the basic switching scheme. Each input requires a driver MOSFET. The chip area is minimized by noting that only one load device is used to provide the overall NOT function in the circuit. This single load device is connected between the power supply and the output node, making it common to all parallel driver branches. All that is really being described here is modifying the basic NOR circuit to allow for series-connected (AND) drivers in the branches. However, it is useful to view this process as constructing AOI (AND-OR-INVERT) gates, particularly when the logic function is complicated.

As an example of a complex logic gate, consider the logic diagram shown in Fig. 6.9(a). The AOI structuring of the function is immediately obvious: the output F is achieved using only AND, OR, and NOT gates (the NOR gate at the output is an OR cascaded into an inverter). The nMOS realization of this logic is obtained by applying the techniques described above and is drawn in Fig. 6.9(b). To understand the construction process, first note that the OR operation $(E + F)$ is obtained by paralleling MOSFETs E and F; the subsequent use of the AND operation with A is achieved by adding MOSFET

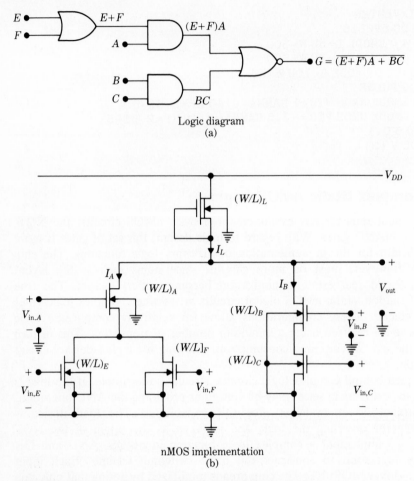

FIGURE 6.9 Example of an nMOS complex logic circuit.

A in series with the parallel-connected E and F drivers. Similarly, series connecting MOSFETs B and C give the logical AND operation of BC. The two driver branches are then paralleled, giving $[(E + F)A + BC]$. Inversion is provided by the load transistor.

The analysis of this circuit follows the same lines established in the previous two sections. Consider first the value of V_{OH}. This output voltage occurs when both driver branches are admitting only leakage current. For example, if both $V_{in,A}$ and $V_{in,B}$ are low, then $V_{out} = V_{OH}$. Since the circuit employs a depletion-mode MOSFET load, $V_{OH} = V_{DD}$ is still a reasonable approximation. However, there are two leakage current paths, so it is expected that V_{OH} will be slightly lower than that found for a simple inverter.

The output low voltage V_{OL} is measured when at least one of the branches is providing a conduction path to ground. With arbitrary driver aspect ratios, there are seven possible values of V_{OL} corresponding to the seven input voltage combinations that will pull the output node down toward ground potential. In particular, it is seen that the following input voltage specifications will result in $V_{\text{out}} = V_{OL}$:

$$V_{\text{in},A} = V_{\text{in},E} = V_{OH},$$

$$V_{\text{in},A} = V_{\text{in},F} = V_{OH},$$

$$V_{\text{in},A} = V_{\text{in},E} = V_{\text{in},F} = V_{OH},$$

$$V_{\text{in},B} = V_{\text{in},C} = V_{OH},$$

$$V_{\text{in},A} = V_{\text{in},E} = V_{\text{in},B} = V_{\text{in},C} = V_{OH},$$

$$V_{\text{in},A} = V_{\text{in},F} = V_{\text{in},B} = V_{\text{in},C} = V_{OH},$$

$$V_{\text{in},A} = V_{\text{in},E} = V_{\text{in},F} = V_{\text{in},B} = V_{\text{in},C} = V_{OH}. \tag{6.3-1}$$

For the first three combinations, the output low voltage is determined from

$$I_L = I_A, \tag{6.3-2}$$

while the fourth possibility requires using

$$I_L = I_B. \tag{6.3-3}$$

The remaining three input combinations are based on

$$I_L = I_A + I_B \tag{6.3-4}$$

since both branches will be conducting current. The worst-case value of V_{OL} will result when only a single branch is providing a conducting path to ground. In this case, either eqn. (6.3-2) or (6.3-3) will apply. The analysis depends on the specific (W/L) values for the circuit, with the driver branch employing the smallest (W/L) MOSFET aspect ratios giving the worst-case conditions.

The calculations can best be illustrated by reference to a specific circuit design. Assume that $(W/L)_A$ and $(W/L)_E$ represent the smallest driver geometries. The worst-case value of V_{OL} will occur when only $V_{\text{in},A}$ and $V_{\text{in},E}$ are high. The problem reduces to a 2-input NAND gate equivalent, so that by analogy with eqn. (6.2-9),

$$V_{OL} \simeq 2(V_{OH} - V_{TD}) - \left[\sqrt{(V_{OH} - V_{TD})^2 - \frac{1}{\beta_{R,A}} |V_{TL}(V_{OL})|^2} \right.$$

$$\left. + \sqrt{(V_{OH} - V_{TD})^2 - \frac{1}{\beta_{R,E}} |V_{TL}(V_{OL})|^2} \right] \tag{6.3-5}$$

constitutes the self-iterating equation for the largest V_{OL} value of the circuit. The respective driver-load ratios in this equation are defined by the usual relations

$$\beta_{R,A} = \frac{\beta_A}{\beta_L}, \qquad \beta_{R,E} = \frac{\beta_E}{\beta_L}. \tag{6.3-6}$$

The preceding analysis may be understood by again recalling that the driver on-conductances are proportional to (W/L). Consequently, the largest value of driver path resistance to ground will occur when the smallest geometry devices are the only ones that are turned on. The ratioed nature of static nMOS logic then implies that this will give the worst-case output low voltage.

The design process for this circuit follows the general guidelines established for the simpler gates. First, specifying V_{OL} (or NM_L) sets the driver-load ratio β_R for an inverter. An estimate of C_{out} then allows for the calculation of $(W/L)_L$ and, hence, the driver aspect ratio $(W/L)_D$. The objective of this approach is to design the worst-case V_{OL} of the complex logic circuit around the original inverter specifications. With regard to the series-connected drivers B and C, the NAND gate discussion of the previous section gives the simplest design choice as

$$\left(\frac{W}{L}\right)_B = \left(\frac{W}{L}\right)_C \simeq 2\left(\frac{W}{L}\right)_D. \tag{6.3-7}$$

The remaining branches may be designed in the same manner. Since transistors E and F are in parallel, they can be chosen to have identical geometries. The worst-case driver resistance for this branch occurs when only one of these MOSFETs is conducting. Since MOSFET A is included in a series connection, the choice of

$$\left(\frac{W}{L}\right)_A = \left(\frac{W}{L}\right)_E = \left(\frac{W}{L}\right)_F \simeq 2\left(\frac{W}{L}\right)_D \tag{6.3-8}$$

will provide the same maximum on-resistance as that used for an inverter. This design approach thus specifies that every driver in the circuit should have an aspect ratio of approximately $2(W/L)_D$. The circuit will operate as desired as long as every input voltage is able to reach a maximum level of V_{OH}. If some of the inputs do not meet this requirement, then their (W/L) ratios must be increased accordingly. This possibility was discussed in Chapter 3 and is easily extended to the present case.

An alternate design procedure is obtained by noting that specification of V_{OL} allows the individual drain-source voltages to be set. Once these choices are made, the current equations (6.3-2)–(6.3-4) give the required β_R values. As an example, consider the branch with drivers B and C and suppose that $V_{DS,B} = \frac{1}{3}V_{OL}$ and $V_{DS,C} = \frac{2}{3}V_{OL}$ are chosen. Using eqn. (6.3-3) gives

$$\frac{\beta_L}{2}|V_{TL}(V_{OL})|^2 = \frac{\beta_B}{2}\left[2(V_{GS,B} - V_{TB})\left(\frac{1}{3}V_{OL}\right) - \left(\frac{1}{3}V_{OL}\right)^2\right]$$

$$= \frac{\beta_C}{2}\left[2(V_{GS,C} - V_{TC})\left(\frac{2}{3}V_{OL}\right) - \left(\frac{2}{3}V_{OL}\right)^2\right], \tag{6.3-9}$$

which may be solved for $\beta_{R,B}$ and $\beta_{R,C}$. Note that MOSFET C has

$$V_{GS,C} = V_{OH}, \qquad V_{TC} = V_{T0D} \tag{6.3-10}$$

so that $\beta_{R,C}$ may be computed directly. MOSFET B, on the other hand, is seen to have

$$V_{GS,B} = V_{OH} - \frac{2}{3}V_{OL},$$

$$V_{TB} = V_{TOD} + \gamma\left[\sqrt{\frac{2}{3}V_{OL} + 2|\phi_F|} - \sqrt{2|\phi_F|}\right] \tag{6.3-11}$$

so that this approach automatically takes into account the driver body bias and the fact that the input voltage to this transistor is not the gate-source voltage required in the current expressions. Although this design technique requires additional intermediate calculations, it is more accurate than the simplified inverter-based procedure.

The basic capacitors that contribute to C_{out} of the circuit may be approximated as shown in Fig. 6.10. The value of C_{out} needed in the analysis will depend on the switching event under consideration. The worst-case transient conditions will occur when C_{out} is taken as the sum of all contributions. Note that the actual numerical values of the elements are determined by the DC circuit design. Consequently, iterations will generally be required in the overall design process when using the procedure discussed in Section 4.7. Obviously, this can get quite complicated, so computer design aids are highly desirable, if not essential.

The discussion in this section has centered around the AOI circuit in Fig. 6.9, which implements the logic function

$$G = \overline{A(E + F) + BC}. \tag{6.3-12}$$

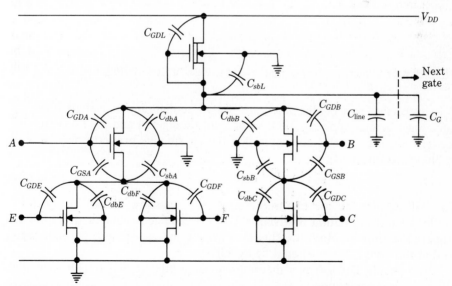

FIGURE 6.10 Basic capacitance contributions for the AOI gate in Fig. 6.9.

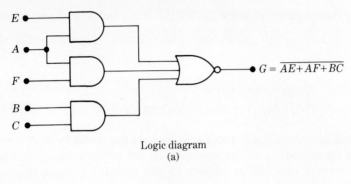

Logic diagram
(a)

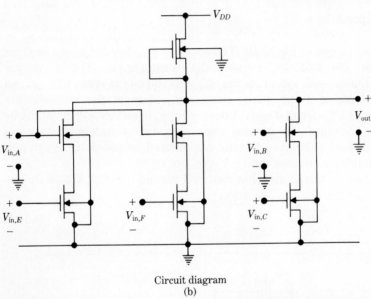

Circuit diagram
(b)

FIGURE 6.11 Alternate representation of the logic function in Fig. 6.9.

It is instructive to examine alternate nMOS configurations that perform the same logic by first applying the laws of Boolean algebra and then constructing the new circuits. To this end, note that the distributive property may be used in the first term to give

$$G = \overline{AE + AF + BC}. \qquad (6.3\text{-}13)$$

This will require the use of three 2-input AND gates as shown in Fig. 6.11. While the transformation is trivial, the new expression requires an additional (superfluous) driver. Both the chip area and output capacitance are increased, so that this configuration should be avoided.

Next, apply DeMorgan's theorems to G in eqn. (6.3-13). This gives the alternate representations

$$G = (\overline{AE})(\overline{AF})(\overline{BC}) \qquad (6.3\text{-}14)$$

and

$$G = (\bar{A} + \bar{E})(\bar{A} + \bar{F})(\bar{B} + \bar{C}), \tag{6.3-15}$$

which are equivalent to the original function. The logic diagrams and nMOS circuits for these expressions are illustrated in Figs. 6.12 and 6.13, respectively. It is seen by inspection that the original AOI configuration in Fig. 6.9 is a more efficient realization of this particular logic function. This comment applies to both the chip area consumption and the overall switching speed.

The simple example above demonstrates an important point: AOI structuring automatically implements complex static nMOS logic functions that are compact and that exhibit low capacitance values. The AOI structuring is easily designed into a circuit by ensuring that the logic is performed in proper order from the input to the output (first AND, then OR, and finally INVERT). This type of complex logic technique has the advantage that it is straightforward to implement and thus allows for relatively quick designs with acceptable characteristics. In addition, the structure permits a direct one-to-one relationship between the logic diagram and the circuitry. These points have helped to make the AOI technique quite popular in realistic working environments.

6.4 nMOS XOR, XNOR, and Associated Circuits

A useful example of complex logic arises in the *exclusive-OR* (XOR) function. The gate-level logic symbol and truth table are given in Fig. 6.14(a). The XOR output is defined by

$$F = A \oplus B$$
$$= A\bar{B} + \bar{A}B, \tag{6.4-1}$$

as is easily verified from the truth table. Taking the complement of the 2-input

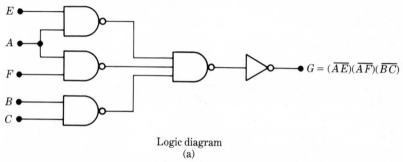

Logic diagram
(a)

FIGURE 6.12 Equivalent logic configuration of the circuit in Fig. 6.9.

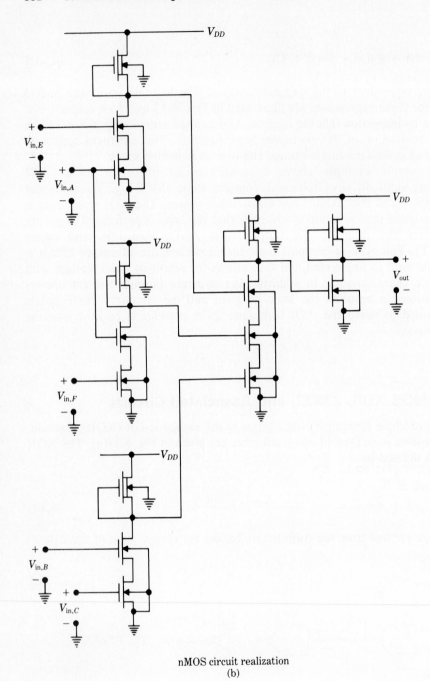

nMOS circuit realization
(b)

Figure 6.12 (*contd.*)

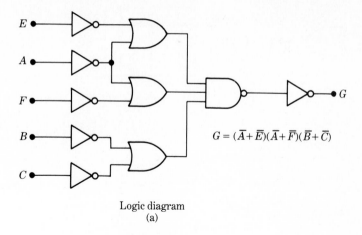

$$G = (\overline{A} + \overline{E})(\overline{A} + \overline{F})(\overline{B} + \overline{C})$$

Logic diagram
(a)

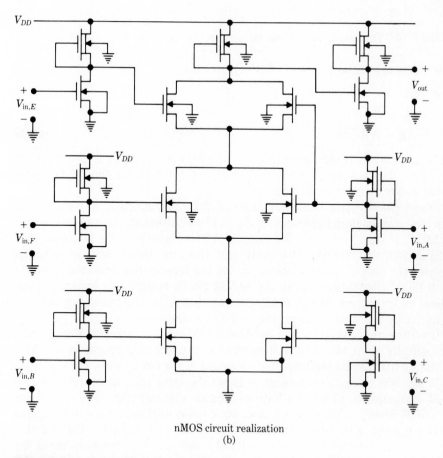

nMOS circuit realization
(b)

FIGURE 6.13   Equivalent product-of-sums logic circuit for Fig. 6.9.

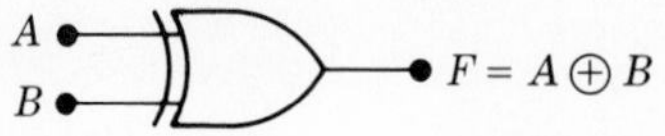

A	B	F
0	0	0
1	0	1
0	1	1
1	1	0

Exclusive OR (XOR)
(a)

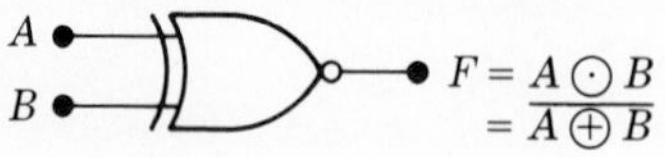

A	B	F
0	0	1
1	0	0
0	1	0
1	1	1

Exclusive NOR (XNOR)
(equivalence or equality)
(b)

FIGURE 6.14 The exclusive-OR and exclusive-NOR functions.

XOR yields the *exclusive-NOR* (XNOR) function

$$F = A \odot B$$
$$= \overline{A \oplus B}$$
$$= AB + \bar{A}\bar{B}. \qquad (6.4\text{-}2)$$

This is also known as the *equivalence* or *equality* function since a logic 1 results only when the two inputs are equal. Figure 6.14(b) shows the XNOR gate symbol and associated truth table.

Direct gate-by-gate implementation of the XOR function in eqn. (6.4-1) gives the logic diagram illustrated in Fig. 6.15(a). Although this can be realized in nMOS, it does not have the desired AOI structuring. An alternate gate configuration that follows the AOI patterning is shown in Fig. 6.15(b). Applying the nMOS AOI techniques of the last section then gives the circuit in Fig. 6.15(c). Comparing this to the nMOS circuit required to implement the original logic diagram of Fig. 6.15(a) shows that AOI structuring has reduced the transistor count and, hence, the real estate consumption.

Inverting the output of the original XOR logic diagram in Fig. 6.15(a) automatically gives the XNOR function. This procedure results in an AOI structuring that can be implemented by means of the circuit shown in Fig. 6.16. While this is a viable arrangement, it is not the most efficient in terms of real estate consumption. Figure 6.17(a) provides an alternate approach to realizing the XNOR function. This scheme does not exhibit AOI patterning but requires only three gates. The corresponding nMOS circuit is illustrated in Fig. 6.17(b). The overall real estate requirements for the two may be compared using the simple inverter-based design procedure discussed earlier. When this is done, it

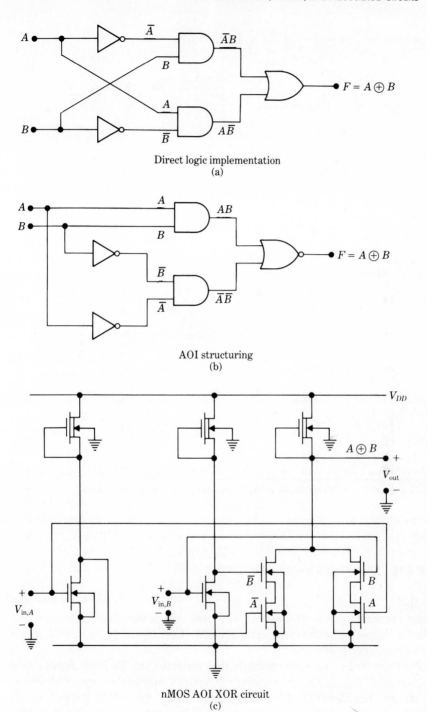

Direct logic implementation
(a)

AOI structuring
(b)

nMOS AOI XOR circuit
(c)

FIGURE 6.15   Evolution of nMOS XOR circuit.

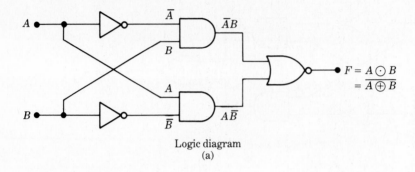

Logic diagram
(a)

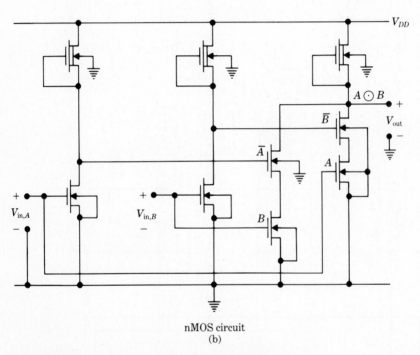

nMOS circuit
(b)

FIGURE 6.16 Direct implementation of XNOR.

is seen that both circuits require the same amount of driver area. However, the alternate implementation in Fig. 6.17(b) uses one fewer load transistor and also has a simpler interconnect arrangement. Consequently, it constitutes a more efficient silicon design.

Integration techniques are methods that increase the packing density in a given chip area. While this terminology is usually applied to layout schemes, many circuit functions can be integrated using the basic properties of MOSFETs. An interesting example of such a technique is illustrated in Fig. 6.18(a) for an nMOS XNOR circuit. The gate-source voltages for the

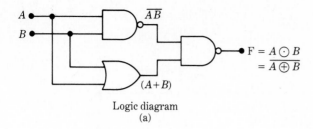

Logic diagram
(a)

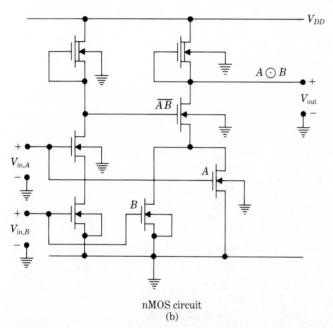

nMOS circuit
(b)

FIGURE 6.17   Alternate implementation of the nMOS XNOR function.

MOSFET $M1$ and $M2$ are given by

$$V_{GS,1} = V_B - V_A, \qquad V_{GS,2} = V_A - V_B. \tag{6.4-3}$$

Realization of the equality function occurs because $V_A = V_B$ places both $M1$ and $M2$ into cutoff, which gives $V_{out} = V_{OH}$. If the two inputs are not equal, then either $M1$ or $M2$ is conducting. This lowers the output voltage to V_{OL}. The actual value of V_{OL} depends on the path to ground provided by the circuitry that generates V_A and V_B.

Figure 6.18(b) shows the case where inverters are added to provide the XNOR input voltage V_A and V_B. Although it may appear that this changes the logic, it is easily verified that the output still has the properties needed to give the equivalence function. To demonstrate the DC analysis of this circuit,

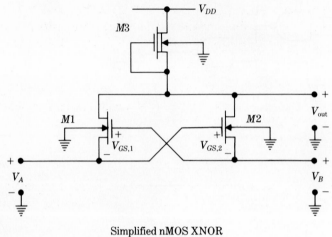

Simplified nMOS XNOR
(a)

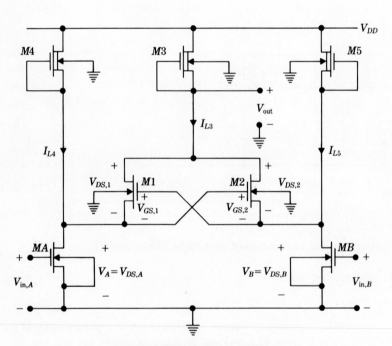

Simplified XNOR with input drivers
(b)

FIGURE 6.18   Compact nMOS XNOR circuit.

suppose that the inverter inputs are specified as $V_{in,A} = V_{DD}$ and $V_{in,B} = 0$. This places MOSFET MA in the nonsaturated mode, while MB is cutoff. Assuming that $V_B = V_{DD} \gg V_A = V_{DS,A}$, the relations in eqn. (6.4-3) give

$$V_{GS,1} = V_{DD} - V_{DS,A} > 0, \qquad V_{GS,2} = -V_{GS,1} < 0, \tag{6.4-4}$$

indicating that $M1$ is conducting in the nonsaturated mode, while $M2$ is in cutoff. The value of $V_{out} = V_{OL}$ for this input combination is thus established by the current I_{L3}, which flows through $M1$ and MA.

V_{OL} may be computed by first writing the KVL relation

$$V_{OL} = V_{DS,1} + V_{DS,A}. \tag{6.4-5}$$

The drain-source voltages needed in this equation are found from analyzing the current flow paths. Consider first the load current I_{L3}. Assuming that $M3$ is saturated,

$$I_{L3} = \frac{\beta_L}{2} |V_{TL}(V_{OL})|^2$$

$$= \frac{\beta_1}{2} [2(V_{GS,1} - V_{T1})V_{DS,1} - V_{DS,1}^2], \tag{6.4-6}$$

where the second step follows from applying KCL at the output node. In this equation, β_1 represents the device transconductance parameter for $M1$, and β_L is used to describe the depletion load MOSFET $M3$. Body bias effects are included by means of

$$V_{TL}(V_{OL}) = V_{T0L} + \gamma_L(\sqrt{V_{OL} + 2|\phi_F|} - \sqrt{2|\phi_F|}) \tag{6.4-7}$$

and

$$V_{T1}(V_{DS,A}) = V_{T0D} + \gamma(\sqrt{V_{DS,A} + 2|\phi_F|} - \sqrt{2|\phi_F|}), \tag{6.4-8}$$

where V_{T0D} and V_{T0L} denote the zero-body bias threshold voltages for the E-mode and D-mode transistors, respectively.

The current that flows through load transistor $M4$ is given by

$$I_{L4} = \frac{\beta_L}{2} |V_{TL}(V_{DS,A})|^2, \tag{6.4-9}$$

where it has been assumed that $M3$ and $M4$ have identical geometries (so that β_L describes both). Now the input driver MA must sink a total current of $(I_{L3} + I_{L4})$. Using the expressions above then yields

$$\frac{\beta_L}{2} [|V_{TL}(V_{DS,A})|^2 + |V_{TL}(V_{OL})|^2] = \frac{\beta_A}{2} [2(V_{DD} - V_{T0D})V_{DS,A} - V_{DS,A}^2], \tag{6.4-10}$$

where β_A is the conduction factor for MA.

The basic relationships in eqns. (6.4-5), (6.4-6), and (6.4-10) provide the information needed to find V_{OL}. Interest will be directed toward obtaining an iterative numerical technique as a method for extracting V_{OL}. This can be accomplished by first treating eqn. (6.4-10) as a quadratic in $V_{DS,A}$, so that

$$V_{DS,A} = (V_{DD} - V_{T0D})$$
$$- \sqrt{(V_{DD} - V_{T0D})^2 - \frac{1}{\beta_{R,A}}[|V_{TL}(V_{OL})|^2 + |V_{TL}(V_{DS,A})|^2]}, \quad \text{(6.4-11)}$$

where

$$\beta_{R,A} = \frac{\beta_A}{\beta_L} \quad \text{(6.4-12)}$$

is the driver-load ratio for $MA/M4$. Similarly, eqn. (6.4-6) may be solved for $V_{DS,1}$ in the form

$$V_{DS,1} = (V_{DD} - V_{DS,A} - V_{T1})$$
$$- \sqrt{(V_{DD} - V_{DS,A} - V_{T1})^2 - \frac{1}{\beta_{R,1}}|V_{TL}(V_{OL})|^2}, \quad \text{(6.4-13)}$$

with

$$\beta_{R,1} = \frac{\beta_1}{\beta_L} \quad \text{(6.4-14)}$$

being the driver-load ratio for the $M1/M3$ combination. These two equations are to be used with the constraint

$$V_{OL} = V_{DS,1} + V_{DS,A} \quad \text{(6.4-15)}$$

to iteratively compute V_{OL}. The procedure is relatively straightforward. A first guess for V_{OL} is used to compute $V_{DS,A}$ from eqn. (6.4-11). These two voltages are substituted into eqn. (6.4-13) to find a value for $V_{DS,1}$. The sum $(V_{DS,1} + V_{DS,A})$ is then compared with the original guess for V_{OL}. If the two agree, then a solution has been found. Otherwise, the sum can be used as a second guess for V_{OL}, and the procedure is repeated until agreement is reached. This generally converges quite rapidly regardless of the initial guess for V_{OL} and is easily implemented on a system with programming capabilities.

A design methodology for the simplified XNOR circuit may be developed from the preceding analysis. Since the circuit is ratioed, interest centers around obtaining a specified value for the output low voltage. The simplest procedure involves choosing $V_{DS,1}$ and $V_{DS,A}$ to be fractions of the desired V_{OL} value. Once these are established, the current equations (6.4-6) and (6.4-10) may be used to find the required driver-load ratios.

As an example, suppose that

$$V_{DS,1} = \frac{1}{2}V_{OL} = V_{DS,A} \quad \text{(6.4-16)}$$

is chosen. Rearranging eqn. (6.4-6) gives

$$\beta_{R,1} = \frac{|V_{TL}(V_{OL})|^2}{\left[V_{DD} - \frac{1}{2}V_{OL} - V_{T1}\left(\frac{1}{2}V_{OL}\right)\right]V_{OL} - \frac{1}{4}V_{OL}^2}, \tag{6.4-17}$$

while eqn. (6.4-10) is written

$$\beta_{R,A} = \frac{\left|V_{TL}\left(\frac{1}{2}V_{OL}\right)\right|^2 + |V_{TL}(V_{OL})|^2}{(V_{DD} - V_{T0D})V_{OL} - \frac{1}{4}V_{OL}^2}. \tag{6.4-18}$$

These give the important driver-load ratios. As in previous circuits, β_L is dictated by the transient response specifications for t_{LH} or t_p. This then gives $(W/L)_L$, which may be used for all of the load transistors ($M3$, $M4$, and $M5$). The results of eqn. (6.4-17) may be used to find $(W/L)_1$. To obtain a symmetric circuit, the choice $(W/L)_2 = (W/L)_1$ is made for $M2$. Finally, the input drivers MA and MB are taken to have $(W/L)_A = (W/L)_B$ as given by eqn. (6.4-18).

An important application of XOR logic is the construction of binary adders. The simplest of these is a *half-adder*, which produces the binary sum of two inputs A_0 and B_0 by means of the logical expression

$$S_0 = A_0 \oplus B_0. \tag{6.4-19}$$

To account for the case where both inputs are logical 1's, the carry C_0 is generated using

$$C_0 = A_0 B_0. \tag{6.4-20}$$

The logic diagram for a half-adder is shown in Fig. 6.19(a), with Fig. 6.19(b) giving the functional block equivalent. The nMOS implementation of a half-adder is straightforward. For example, the XNOR circuit in Fig. 6.17(b) can be modified to a half-adder by adding two inverter stages.

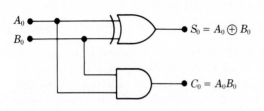

Basic XOR logic diagram
(a)

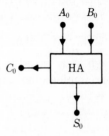

Functional block equivalent
(b)

FIGURE 6.19 A half-adder.

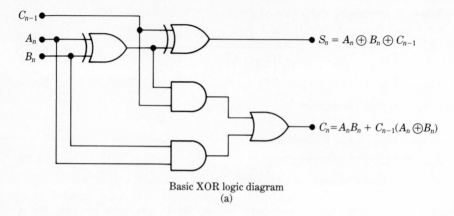

Basic XOR logic diagram
(a)

A_n	B_n	C_{n-1}	S_n	C_n
0	0	0	0	0
0	0	1	1	0
0	1	0	1	0
0	1	1	0	1
1	0	0	1	0
1	0	1	0	1
1	1	0	0	1
1	1	1	1	1

Truth table
(b)

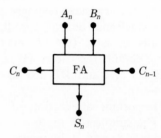

Functional block equivalent
(c)

FIGURE 6.20   A full-adder.

A *full-adder* allows for the inclusion of the carry generated by a preceding adder stage. Denoting the inputs to the nth-stage full-adder by A_n, and B_n, a full-adder performs binary addition using the logic

$$S_n = A_n \oplus B_n \oplus C_{n-1}, \tag{6.4-21}$$

where C_{n-1} is the carry bit from the previous $(n-1)$th stage. The XOR logic diagram, truth table, and functional block symbol for a full-adder is shown in Fig. 6.20. The output carry C_n of this stage is obtained from

$$C_n = A_n B_n + C_{n-1}(A_n \oplus B_n). \tag{6.4-22}$$

Although it is possible to construct an nMOS full-adder from a direct gate-by-gate implementation of the logic provided in Fig. 6.20(a), an AOI approach is more efficient. The logic diagram in Fig. 6.21(a) shows the structuring that is ideally suited for a complex nMOS gate design. The resulting circuit is illustrated in Fig. 6.21(b). Note that the bulk electrode connections have been omitted from the schematic. This is a common convention used to simplify the drawing. It implies that all of the bulk electrodes of the MOSFETs are grounded (or, alternately, set at some

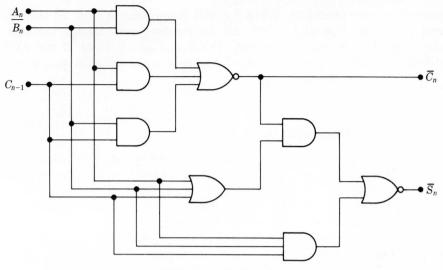

Full adder using AOI structuring
(a)

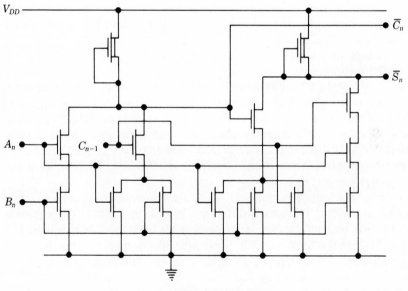

nMOS AOI full adder
(bulk electrodes grounded)
(b)

FIGURE 6.21   nMOS AOI full-adder.

common negative potential). When a circuit is analyzed, it must be remembered that body bias effects will still be present, even though the bulk connections are not explicitly shown. The design and analysis of the AOI full-adder follows the same procedures discussed in the previous section and will not be detailed here.

6.5 CMOS NAND and NOR Gates

CMOS logic circuits may be constructed by extending nMOS techniques. In standard CMOS structuring, each input is connected to both an n-channel and a p-channel MOSFET. The complementary nature of this scheme gives CMOS the unique property that the logic performed by the nMOS devices is duplicated by the pMOS transistors (or vice versa).

Figure 6.22(a) gives the circuit for a 3-input CMOS NAND gate. The logic formation can be seen by noting that the nMOS transistors are in series; this is identical to the structuring used for nMOS. Since the p-channel transistors must be arranged in a complementary manner, they are connected in parallel. A moment's reflection will verify that this would yield the NAND function in a positive-logic pMOS-only design (where the n-channel MOSFETs are replaced by a single load element). The 3-input NOR gate shown in Fig. 6.22(b) uses three n-channel MOSFETs in parallel to create the logic. This requires that the pMOS transistors be in series, as would be required for a NOR design in straight pMOS technology.

The complementary arrangements of the n-channel and p-channel MOSFETs in both gates gives output voltage levels of $V_{OH} = V_{DD}$ and $V_{OL} = 0$ [V]. This is true regardless of the device β values, as CMOS operation is ratioless. Stable input voltages prohibit the formation of a direct current path between V_{DD} and ground. Consequently, power supply current is required only when the inputs are switched. Generalized CMOS gates thus have the same low-power properties found for the simple CMOS inverter.

Implementing a logic function in CMOS always requires more transistors than needed in nMOS. The actual difference in chip real estate consumption between the two technologies depends on the MOSFET (W/L) values. The size of the MOSFETs used in CMOS circuits also establishes the transient switching properties. Note that C_{out} will have more contributions in a CMOS logic gate than those found for an equivalent nMOS implementation.

The NAND and NOR schematics illustrate that all nMOS transistors have grounded bulk electrodes, while the pMOS bulk regions are connected to V_{DD}. Including bulk connections in the circuit diagrams gets somewhat cumbersome, so it is common to introduce the alternate simplified MOSFET symbols shown in Fig. 6.23. These will be used interchangeably in the discussions. Although the bulk connections are not shown explicitly, body bias effects will still be present in some devices. A complete CMOS analysis should always take this into account.

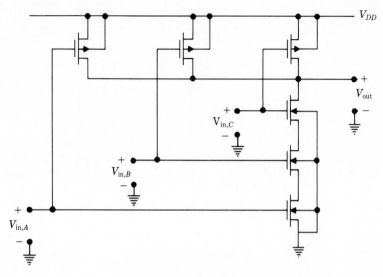

3-input CMOS NAND gate
(a)

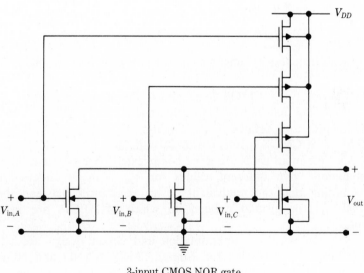

3-input CMOS NOR gate
(b)

FIGURE 6.22 Basic CMOS logic gates.

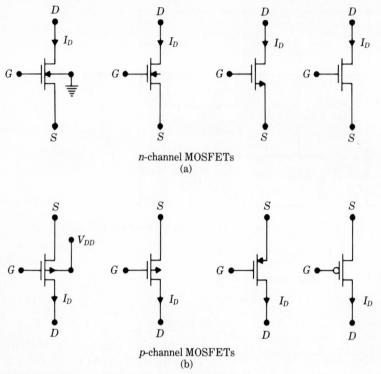

FIGURE 6.23 Simplified equivalent MOSFET symbols for CMOS logic circuits.

The ratioless property of CMOS logic allows the design discussion to center around establishing the desired VTC properties. Of particular interest is the logic threshold voltage V_{th} defined by the condition $V_{in} = V_{th} = V_{out}$. Recall from Section 3.5 that a CMOS inverter with $\beta_n = \beta_p$ and $V_{T0n} = |V_{T0p}|$ has $V_{th} = (V_{DD}/2)$. Multiple-input gates are more complicated since different input switching combinations give different values for V_{th}. This is understood qualitatively by noting that the equivalent device resistances between the output node and V_{DD} or ground depend on which MOSFETs are in a conducting mode.

Problems associated with V_{th} determination and circuit design are addressed in the following subsections for 2-input CMOS NAND and NOR gates. It is mentioned in passing that multiple-input nMOS gates also have V_{th} values that depend on the input switching combination. However, since this type of logic is ratioed, the design usually centers around V_{OL} instead of V_{th}.

6.5.1 2-Input NAND Analysis

Consider the 2-input CMOS NAND gate shown in Fig. 6.24, where V_{out} is initially at $V_{OH} = V_{DD}$. This output condition can arise from any of the

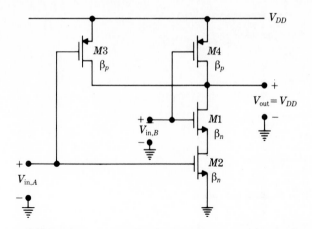

FIGURE 6.24 2-input CMOS NAND gate with a logic 1 output.

following input combinations:

$$V_{in,A} = 0, \qquad V_{in,B} = 0,$$
$$V_{in,A} = V_{DD}, \qquad V_{in,B} = 0, \qquad \qquad \textbf{(6.5-1)}$$
$$V_{in,A} = 0, \qquad V_{in,B} = V_{DD}.$$

An output transition of $V_{out} \rightarrow V_{th}$ may be induced by increasing input voltages that are initially zero up to V_{th}. Since there are three possibilities for this switching, three distinct values of V_{th} exist for the circuit. These may be computed by examining the current flow characteristics for each case. The analysis here will assume that both n-channel MOSFETs have a device transconductance of β_n. Similarly, the pMOS transistors are assumed to be identical with β_p.

SIMULTANEOUS SWITCHING

The first case to be studied is that where both $V_{in,A}$ and $V_{in,B}$ are initially zero. The two inputs are then simultaneously switched toward logic 1 states. The logic threshold voltage V_{th} is computed using the circuit arrangement shown in Fig. 6.25(a).

To analyze the circuit characteristics, note that the nMOS gate-source voltages are given by

$$V_{GS,1} = V_{th} - V_{DS,2}, \qquad V_{GS,2} = V_{th}. \qquad \qquad \textbf{(6.5-2)}$$

Using the output KVL equation

$$V_{th} = V_{DS,1} + V_{DS,2} \qquad \qquad \textbf{(6.5-3)}$$

in $V_{GS,1}$ then shows that

$$V_{GS,1} = V_{DS,1}. \qquad \qquad \textbf{(6.5-4)}$$

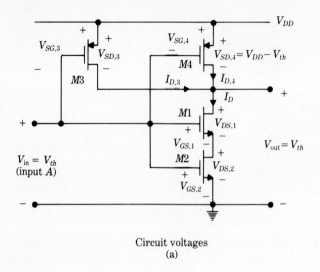

Circuit voltages
(a)

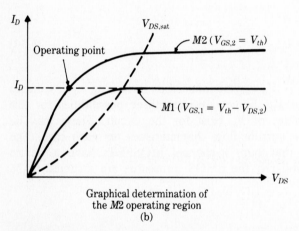

Graphical determination of
the $M2$ operating region
(b)

FIGURE 6.25   CMOS NAND under simultaneous input switching.

This demonstrates that MOSFET $M1$ is saturated with

$$I_D = \frac{\beta_n}{2}(V_{th} - V_{T0n} - V_{DS,2})^2 \qquad (6.5\text{-}5)$$

Although body bias effects will be present in $M1$, they will be ignored for simplicity.

MOSFET $M2$ has $V_{GS,2} > V_{GS,1}$. Combining this with the fact that β_n describes both transistors implies that $M2$ must be nonsaturated, with

$$I_D = \frac{\beta_n}{2}[2(V_{th} - V_{T0n})V_{DS,2} - V_{DS,2}^2]. \qquad (6.5\text{-}6)$$

This line of reasoning may be verified graphically by referring to the $I - V$

plot in Fig. 6.25(b). Since a common current I_D flows through both transistors, it is seen that $M1$ saturated requires that $M2$ must be in the nonsaturated mode because of the greater gate-source voltage. (The argument is valid only within the limits of the square-law MOSFET model and when channel-length modulation can be ignored.)

Equation (6.5-5) may be solved as

$$V_{DS,2} = (V_{th} - V_{T0n}) - \sqrt{\frac{2I_D}{\beta_n}}. \tag{6.5-7}$$

Substituting into eqn. (6.5-6) then gives

$$V_{th} - V_{T0n} = 2\sqrt{\frac{I_D}{\beta_n}} \tag{6.5-8}$$

as the first equation needed to find V_{th}.

The pMOS transistors have source-gate voltages of

$$V_{SG,3} = V_{SG,4} = (V_{DD} - V_{th}). \tag{6.5-9}$$

Since

$$V_{SD,3} = V_{SD,4} = (V_{DD} - V_{th}), \tag{6.5-10}$$

both $M3$ and $M4$ are saturated with equal drain current levels. Thus, the total current from V_{DD} to ground is given by

$$I_D = I_{D,3} + I_{D,4} = \beta_p(V_{DD} - V_{th} - |V_{T0p}|)^2. \tag{6.5-11}$$

This constitutes the second equation needed to find V_{th}. Substituting it into eqn. (6.5-8) and rearranging gives the threshold voltage for this switching combination as

$$V_{th} = \frac{V_{T0n} + 2\sqrt{\frac{\beta_p}{\beta_n}}(V_{DD} - |V_{T0p}|)}{1 + 2\sqrt{\frac{\beta_p}{\beta_n}}}. \tag{6.5-12}$$

Comparing this result with eqn. (3.5-34) for a simple CMOS inverter shows that there is an extra factor of 2 multiplying the square-root terms. For the case where $V_{T0n} = |V_{T0p}| = V_{T0}$ and $\beta_n = \beta_p$, this expression reduces to

$$V_{th} = \frac{2V_{DD} - V_{T0}}{3}. \tag{6.5-13}$$

Consequently, if the CMOS NAND is designed using the inverter criteria, a nonsymmetrical VTC will result when both inputs are simultaneously switched.

SINGLE-INPUT SWITCHING

An example of single-input switching is illustrated by the circuit in Fig. 6.26. Initially, $V_{in,A} = V_{DD}$ and $V_{in,B} = 0$. The B input is then increased to V_{th},

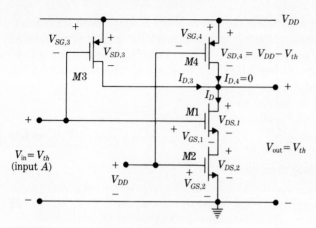

FIGURE 6.26 CMOS NAND circuit for single-input switching.

which establishes the voltage levels shown in the drawing. The gate-source voltages of the n-channel MOSFETs are obtained as

$$V_{GS,1} = V_{th} - V_{DS,2}, \qquad V_{GS,2} = V_{DD}. \tag{6.5-14}$$

Using KVL at the output nodes shows that $V_{DS,1} = V_{GS,1}$, so $M1$ is saturated. $M2$ is nonsaturated, giving

$$I_D = \frac{\beta_n}{2}(V_{th} - V_{T0n} - V_{DS,2})^2$$

$$= \frac{\beta_n}{2}[2(V_{DD} - V_{T0n})V_{DS,2} - V_{DS,2}^2]. \tag{6.5-15}$$

The first ($M1$) equation may be written as

$$V_{DS,2} = (V_{th} - V_{T0n}) - \sqrt{\frac{2I_D}{\beta_n}}. \tag{6.5-16}$$

Substituting this into the second ($M2$) equation then yields the V_{th} condition

$$\frac{4I_D}{\beta_n} = 2(V_{DD} - V_{T0n})(V_{th} - V_{T0n}) + 2\sqrt{\frac{2I_D}{\beta_n}}(V_{th} - V_{DD}) - (V_{th} - V_{T0n})^2. \tag{6.5-17}$$

Now consider the pMOS devices. Since $V_{SG,4} = 0$, $M4$ is in cutoff. MOSFET $M3$ is easily seen to be in saturation. Consequently, the second equation needed for V_{th} is

$$\sqrt{\frac{2I_D}{\beta_n}} = \sqrt{\frac{\beta_p}{\beta_n}}(V_{DD} - |V_{T0p}| - V_{th}), \tag{6.5-18}$$

which can be substituted into eqn. (6.5-17). Assuming that $V_{T0n} = |V_{T0p}| =$

V_{T0}, the algebra results in a quadratic equation for V_{th} of the form

$$\left[1 + 2\left(\sqrt{\frac{\beta_p}{\beta_n}} + \frac{\beta_p}{\beta_n}\right)\right]V_{th}^2$$

$$- \left\{\left[4\left(\frac{\beta_p}{\beta_n}\right) + 2 + 2\sqrt{\frac{\beta_p}{\beta_n}}\right](V_{DD} - V_{T0}) + 2V_{T0} + 2V_{DD}\sqrt{\frac{\beta_p}{\beta_n}}\right\}V_{th}$$

$$+ \left[2\left(\frac{\beta_p}{\beta_n}\right)(V_{DD} - V_{T0})^2 + 2\left(V_{T0} + \sqrt{\frac{\beta_p}{\beta_n}}\right)(V_{DD} - V_{T0}) + V_{T0}^2\right] = 0.$$

$$(6.5\text{-}19)$$

Although this is somewhat messy, V_{th} can be determined by substituting the circuit parameters and then using the quadratic solutions. If $\beta_n = \beta_p$, the equation may be solved to give

$$V_{th} = (V_{DD} - 0.6V_{T0}) - \frac{1}{5}\sqrt{5V_{DD}^2 - 10V_{DD}V_{T0} + 4V_{T0}^2},\qquad (6.5\text{-}20)$$

which allows for a quick determination of V_{th}.

The remaining case is where the initial input states are given by $V_{in,A} = 0$ and $V_{in,B} = V_{DD}$. This may be treated using the general techniques developed above and will not be covered in detail here.

The switching properties of the 2-input CMOS NAND can be extracted from the V_{th} expressions for the three input combinations. This results in the general property that V_{th} is the largest for the case of simultaneous input switching. A typical set of VTC curves is shown in Fig. 6.27. This graph also

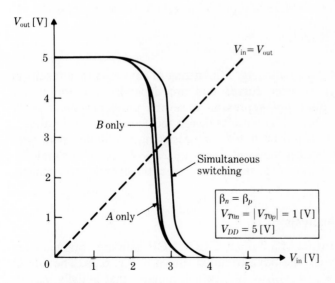

FIGURE 6.27 Switching characteristics for a 2-input CMOS NAND gate.

demonstrates that the value of V_{th} in a single switching case is dependent on which input voltage must be changed to induce the output transition.

DESIGN

The 2-input CMOS NAND gate can be designed around the value of V_{th} for the case of simultaneous switching. Rearranging eqn. (6.5-12) gives

$$\sqrt{\frac{\beta_p}{\beta_n}} = \frac{(V_{th} - V_{T0n})}{2(V_{DD} - V_{th} - |V_{T0p}|)}, \tag{6.5-21}$$

which may be used to determine the ratio (β_p/β_n) for a desired V_{th}. For example, if $V_{T0n} = |V_{T0p}|$, a value of $V_{th} = (V_{DD}/2)$ requires that $\beta_n = 4\beta_p$. This particular choice of V_{th} may be undesirable, since the single-input switching cases will have values less than $(V_{DD}/2)$. An alternate design procedure is to choose the average V_{th} for the three combinations to be approximately one-half of the power supply level.

Real estate consumption is often the most important consideration in high-density CMOS designs. In this situation, it is common to use minimum-size MOSFETs where $(W/L)_n = (W/L)_p$ represents the smallest layout permitted by the fabrication processes. Recalling that $\mu_n \simeq (2.5)\mu_p$, the pMOS-nMOS ratio is $(\beta_p/\beta_n) \simeq 0.4$. In a CMOS inverter where $V_{T0n} = |V_{T0p}| = V_{T0}$, this design choice gives

$$V_{th} \simeq \frac{0.368V_{T0} + 0.632V_{DD}}{1.632}, \tag{6.5-22}$$

as seen directly from eqn. (3.5-34). Similarly, a 2-input minimum-area NAND gate with equal-magnitude nMOS and pMOS threshold voltages yields

$$V_{th} \simeq \frac{1.264V_{DD} - 0.264V_{T0}}{2.264} \tag{6.5-23}$$

for the case of simultaneous switching. Minimizing of chip area thus results in nonsymmetrical voltage transfer curves that are dependent on the power supply voltage. This does not represent a severe problem unless noise sensitivity is critical. As such, it is a useful design technique to keep in mind.

As a final point, it should be noted that designing a CMOS circuit with $\beta_n \neq \beta_p$ gives nonsymmetrical transient times, i.e., $t_{LH} \neq t_{HL}$. This should be accounted for in the system timing, since it will affect the overall performance.

6.5.2 2-Input NOR Analysis

The switching characteristics of a 2-input CMOS NOR gate are obtained from the same type of analysis. The basic circuit is shown in Fig. 6.28. The output voltage is specified as $V_{out} = V_{OH} = V_{DD}$, which requires that initially $V_{in,A} = 0 = V_{in,B}$. There are three possible input combinations that can induce the

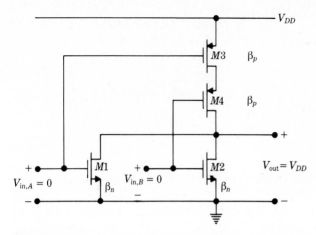

FIGURE 6.28   2-input CMOS NOR gate for V_{th} analysis.

output transition $V_{out} \to V_{th}$. Each is characterized by a distinct value for V_{th}. Two of these are discussed below.

SIMULTANEOUS SWITCHING

The simplest case is where both inputs are simultaneously increased to a level of $V_{in,A} = V_{in,B} = V_{th}$. The resulting circuit currents and voltages are provided in Fig. 6.29. To analyze this case, first note that the nMOS transistors have terminal voltages of

$$V_{GS,1} = V_{GS,2} = V_{th} = V_{DS,1} = V_{DS,2}. \tag{6.5-24}$$

Both devices are therefore saturated with equal current levels. The total drain

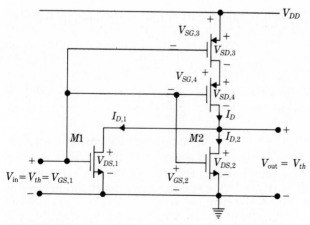

FIGURE 6.29 V_{th} circuit for simultaneous switching.

current is then given by

$$I_D = I_{D,1} + I_{D,2}$$
$$= \beta_n (V_{th} - V_{T0n})^2, \tag{6.5-25}$$

where it has been assumed that β_n describes both n-channel MOSFETs. This may be rearranged in the form

$$V_{th} - V_{T0n} = \sqrt{\frac{I_D}{\beta_n}}, \tag{6.5-26}$$

which constitutes the first equation needed to compute V_{th}.

The source-gate voltages of the pMOS devices are seen to be

$$V_{SG,3} = V_{DD} - V_{th},$$
$$V_{SG,4} = V_{DD} - V_{th} - V_{SD,3}. \tag{6.5-27}$$

Since KVL at the output gives

$$V_{DD} - V_{th} = V_{SD,3} + V_{SD,4}, \tag{6.5-28}$$

these show that $M3$ is nonsaturated while $M4$ is conducting in the saturated mode. Consequently,

$$I_D = \frac{\beta_p}{2} [2(V_{DD} - V_{th} - |V_{T0p}|)V_{SD,3} - V_{SD,3}^2]$$

$$= \frac{\beta_p}{2} (V_{DD} - V_{th} - |V_{T0p}| - V_{SD,3})^2, \tag{6.5-29}$$

where β_p is the device transconductance for the p-channel MOSFETs. Eliminating $V_{SD,3}$ yields

$$\sqrt{\frac{2I_D}{\beta_p}} = V_{DD} - V_{th} - |V_{T0p}|, \tag{6.5-30}$$

which is the second required equation. Substituting this relation into the nMOS expression (6.5-26) and rearranging finally gives

$$V_{th} = \frac{V_{T0n} + \frac{1}{2}\sqrt{\frac{\beta_p}{\beta_n}}(V_{DD} - |V_{T0p}|)}{1 + \frac{1}{2}\sqrt{\frac{\beta_p}{\beta_n}}} \tag{6.5-31}$$

as the logic threshold voltage when the inputs are simultaneously switched. There is now a factor of $\frac{1}{2}$ multiplying the square-root terms. This is to be compared with the factor of 2 found in eqn. (6.5-12) for the NAND gate. For the special case where $V_{T0n} = |V_{T0p}| = V_{T0}$ and $\beta_n = \beta_p$, the equation reduces to

$$V_{th} = \frac{V_{DD} + V_{T0}}{3}. \tag{6.5-32}$$

This demonstrates that designing the 2-input CMOS NOR gate using the inverter guidelines does not give $V_{th} = (V_{DD}/2)$.

SINGLE SWITCHING

Figure 6.30 shows the circuit when $V_{in,A} = 0$ is maintained while $V_{in,B}$ is increased to V_{th}. Since $V_{GS,1} = 0$, MOSFET $M1$ is in cutoff. $M2$ is described by $V_{GS,2} = V_{th} = V_{DS,1}$, so that it is operating in the saturated region. Consequently, the current is

$$I_D = I_{D,1} = \frac{\beta_n}{2}(V_{th} - V_{T0n})^2. \tag{6.5-33}$$

This provides one of the equations needed to find V_{th}.

The source-gate voltages of the pMOS transistors are

$$V_{SG,3} = V_{DD}, \qquad V_{SG,4} = V_{DD} - V_{th} - V_{SD,3}. \tag{6.5-34}$$

Since

$$V_{DD} - V_{th} = V_{SD,3} + V_{SD,4}, \tag{6.5-35}$$

it is easily verified that $V_{SG,4} = V_{SD,4}$; thus, $M4$ is saturated. $M3$ is nonsaturated, so the current relations are

$$I_D = \frac{\beta_p}{2}[2(V_{DD} - |V_{T0p}|)V_{SD,3} - V_{SD,3}^2]$$

$$= \frac{\beta_p}{2}(V_{DD} - V_{th} - |V_{T0p}| - V_{SD,3})^2. \tag{6.5-36}$$

The second ($M4$) equation may be solved to give

$$V_{SD,3} = (V_{DD} - |V_{T0p}| - V_{th}) - \sqrt{\frac{2I_D}{\beta_p}}. \tag{6.5-37}$$

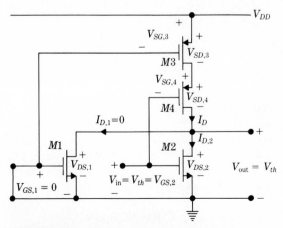

FIGURE 6.30 V_{th} for a single-input switching.

Substituting this into the first ($M3$) equation then yields

$$2\left(\frac{2I_D}{\beta_p}\right) = (V_{DD} - |V_{T0p}|)^2 - V_{th}^2 - 2\sqrt{\frac{2I_D}{\beta_p}} V_{th}, \qquad \text{(6.5-38)}$$

which is the remaining equation needed to compute V_{th}.

Combining eqns. (6.5-33) and (6.5-38) by eliminating I_D gives a quadratic equation for V_{th} in the form

$$\left[1 + 2\left(\frac{\beta_n}{\beta_p} + \sqrt{\frac{\beta_n}{\beta_p}}\right)\right] V_{th}^2 - 2V_{T0n}\left[2\left(\frac{\beta_n}{\beta_p}\right) + \sqrt{\frac{\beta_n}{\beta_p}}\right] V_{th}$$

$$- \left[(V_{DD} - |V_{T0p}|)^2 - 2\left(\frac{\beta_n}{\beta_p}\right) V_{T0n}^2\right] = 0. \qquad \text{(6.5-39)}$$

Once the circuit parameters are specified, the equation may be used to evaluate the value of V_{th} for this switching case. If $V_{T0n} = |V_{T0p}| = V_{T0}$ and $\beta_n = \beta_p$, the solutions reduce to

$$V_{th} = 0.6V_{T0} + 0.2\sqrt{5V_{DD}^2 - 10V_{DD}V_{T0} + 4V_{T0}^2}. \qquad \text{(6.5-40)}$$

The remaining switching case can be analyzed directly and will not be presented here.

The voltage transfer curves shown in Fig. 6.31 illustrate the general dependence of the CMOS NOR switching properties. For this gate, the lowest V_{th} value is the combination in which the inputs are simultaneously switched. This is opposite to that found for the NAND gate.

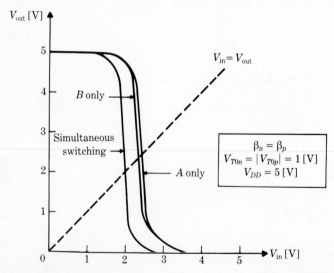

FIGURE 6.31 Switching characteristics for a 2-input CMOS NOR gate.

DESIGN

One approach to designing the 2-input CMOS NOR gate is to use eqn. (6.5-31), which describes V_{th} for the case of simultaneous switching. Rearranging gives

$$\sqrt{\frac{\beta_p}{\beta_n}} = \frac{2(V_{th} - V_{T0n})}{(V_{DD} - V_{th} - |V_{T0p}|)}, \tag{6.5-41}$$

which may be used to determine (β_p/β_n) for a desired V_{th}. As an example, suppose that $V_{T0n} = |V_{T0p}|$ and that the design goal is $V_{th} = (V_{DD}/2)$. The equation then gives $\beta_p = 4\beta_n$. This is exactly the inverse of that found for the NAND gate. An alternate design choice might be to choose the device ratios to give an average value of $(V_{DD}/2)$ for the three input combinations.

If minimum-size MOSFETs are desired, the logic threshold expression reduces to

$$V_{th} \simeq \frac{0.684V_{T0} + 0.316V_{DD}}{1.316}, \tag{6.5-42}$$

where it has been assumed that the MOSFET threshold voltages are equal to V_{T0}.

6.5.3   Comparison of CMOS NAND and NOR Gates

Recall from Sections 6.1 and 6.2 that the ratioed nature of nMOS logic makes NOR gates more real estate–efficient than NAND gates. This is not true in CMOS, since the logic is ratioless. Instead, the relative chip area requirements for CMOS NAND and NOR gates depend on the approach used to design the circuits.

The smallest layout geometries are obtained by using minimum-sized devices in which $(W/L)_n = (W/L)_p = (W/L)_{min}$ represents the smallest MOSFET dimensions allowed by the processing. For this case, it is obvious that the device real estate requirements depend only on the number of MOSFETs in the circuit, not on the particular logic function. Consequently, this design approach allows both NAND and NOR gates to be used as needed.

The situation is more critical when the circuits are designed for specific switching thresholds. To compare the CMOS NAND and NOR gates, suppose that a choice is made to set $V_{th} = (V_{DD}/2)$ when simultaneous input switching occurs. Assuming 2-input logic, this requires that a NAND gate be designed with $\beta_n = 4\beta_p$, while the NOR gate needs $\beta_p = 4\beta_n$. Using $\mu_n \simeq (2.5)\mu_p$ then gives the NAND requirement that

$$(W/L)_p \simeq \frac{2.5}{4}(W/L)_n. \tag{6.5-43}$$

Similarly, a NOR gate design will be based on

$$(W/L)_p \simeq (4)(2.5)(W/L)_n. \tag{6.5-44}$$

For the NAND gate, the logical choice is to set $(W/L)_p = (W/L)_{min}$. This gives

$$(W/L)_n \simeq (1.6)(W/L)_{min} \qquad \text{(NAND)} \qquad \text{(6.5-45)}$$

for the n-channel MOSFETs. The NOR gate would be chosen to have $(W/L)_n = (W/L)_{min}$, so the p-channel transistors are specified by

$$(W/L)_p \simeq (10)(W/L)_{min} \qquad \text{(NOR)}. \qquad \text{(6.5-46)}$$

Since each gate has two n-channel and two p-channel MOSFETs, this analysis shows that NAND gates will consume less area than NOR gates. Thus, a CMOS design based on this V_{th} specification should use NAND-based logic as much as possible to conserve chip area. Problems of this type should always be examined before implementing CMOS logic circuits.

6.6 Complex CMOS Logic

Complex logic gates can be realized in CMOS using AOI structuring techniques. An example is shown in Fig. 6.32. This circuit provides the logic function

$$F = \overline{A(E + F) + BC} \qquad \text{(6.6-1)}$$

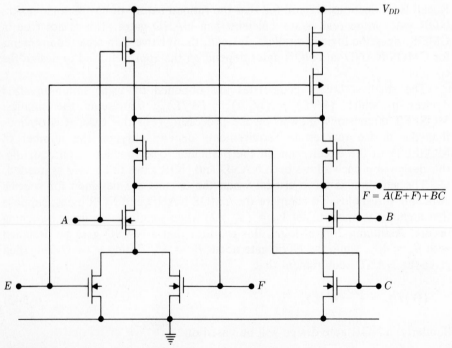

FIGURE 6.32   Example of CMOS AOI complex logic.

and constitutes the CMOS equivalent of the nMOS gate in Fig. 6.9 of Section 6.3. The circuit follows standard CMOS patterning in which each input is connected to the gate of one nMOS and one pMOS transistor. The analysis and design of the circuit is based on the generalized techniques used in the previous section to discuss CMOS NAND and NOR gates. Extending this type of AOI structuring to arbitrary circuit functions is straightforward, since the design is based on nMOS logic with the *p*-channel MOSFETs added in a complementary manner. OAI (OR-AND-INVERT) logic also constitutes a viable option since the circuits are ratioless.

6.7   Transmission Gate CMOS Logic

Complex CMOS combinational logic is often implemented using *bidirectional transmission gates* (or *pass gates*). These gates are constructed using one nMOS and one pMOS transistor in the parallel connected arrangement shown in Fig. 6.33(a). The transmission gate acts as a bidirectional switch that is controlled

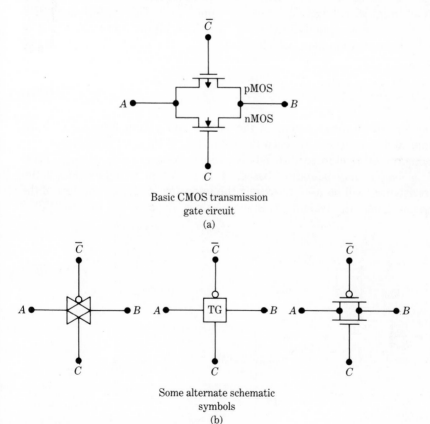

Basic CMOS transmission
gate circuit
(a)

Some alternate schematic
symbols
(b)

FIGURE 6.33   CMOS transmission gate.

by the gate signal C. When $C = 1$, both MOSFETs are on, allowing the signal to pass through the gate. This may be expressed by writing

$$C = 1 \Rightarrow A = B.$$

On the other hand, $C = 0$ places both MOSFETs in cutoff, creating a high impedance path between A and B and prohibiting all but leakage currents from flowing. Alternate schematic symbols for CMOS transmission gates are shown in Fig. 6.33(b).

This section will first analyze the basic circuit properties of the CMOS transmission gate. Once the characteristics are established, the discussion will turn to the use of these gating structures for implementing logic functions.

6.7.1 CMOS Transmission Gate Characteristics

The circuit properties of the CMOS transmission gate (TG) may be understood by referring to the example provided in Fig. 6.34. The input node (A) is set to a voltage $V_{in} = V_{DD}$. The output node (B) is connected to a capacitor C_{out}, which is at an initial voltage $V_{out} = 0$ when time $t = 0$. The control signal C is at a logic 1 state corresponding to $V_C = V_{DD}$. This allows a current I to flow through the transmission gate, which charges C_{out}. The process is described by the standard relation

$$V_{out}(t) = \frac{1}{C_{out}} \int_0^t I(\tau)\, d\tau. \tag{6.7-1}$$

The complementary arrangement of the nMOS and pMOS transistors allows the output node to eventually reach the value $V_{out} = V_{DD}$.

Solving the integral in eqn. (6.7-1) is tedious and will not be pursued here. Instead, a simple approximation based on the concept of equivalent drain-source resistances will be used to extract the important circuit properties of the transmission gate. Equivalent resistance values depend on the MOSFET

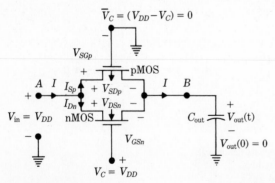

FIGURE 6.34 Transmission gate circuit for equivalent resistance calculations.

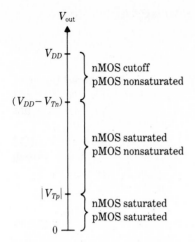

FIGURE 6.35 Operation modes of the MOSFETs used in the CMOS transmission gate.

operational modes, so that the charging process must be studied in terms of the MOSFET terminal voltages.

Consider first the nMOS transistor. It is seen that

$$V_{DSn} = V_{DD} - V_{\text{out}}, \qquad V_{GSn} = V_{DD} - V_{\text{out}} = V_{DSn}, \qquad \text{(6.7-2)}$$

so the MOSFET is conducting in the saturation mode until $V_{\text{out}} = (V_{DD} - V_{Tn})$ is reached. When the output node attains this value, the inversion layer loses its support from the gate voltage. The current flow thus goes to zero for values of V_{out} greater than this level.

The pMOS transistor has terminal voltages of

$$V_{SDp} = V_{DD} - V_{\text{out}}, \qquad V_{SGp} = V_{DD}. \qquad \text{(6.7-3)}$$

The source-gate voltage V_{SGp} is a constant once the input and control voltages V_{in} and V_C are set. The operational mode of the pMOS transistor depends on V_{out}. Initially, $V_{\text{out}} = 0$, so the device is saturated. Once V_{out} increases to $|V_{Tp}|$, the p-channel MOSFET goes into the nonsaturated mode and remains there until $V_{\text{out}} = V_{DD}$ is reached. The nMOS and pMOS operational characteristics as functions of V_{out} can be summarized using the voltage plot in Fig. 6.35.

To analyze the DC transmission properties of the TG, first note that the total current I through the gate is given by

$$I = I_{Dn} + I_{Sp}. \qquad \text{(6.7-4)}$$

Since the voltage across the TG is

$$V_{DSn} = V_{SDp} = (V_{DD} - V_{\text{out}}), \qquad \text{(6.7-5)}$$

the total equivalent resistance at any given set of voltages is

$$R_{eq} = \frac{(V_{DD} - V_{out})}{I}. \tag{6.7-6}$$

Alternately, the resistance of the individual transistors can be defined by

$$R_n = \frac{(V_{DD} - V_{out})}{I_{Dn}},$$

$$R_p = \frac{(V_{DD} - V_{out})}{I_{Sp}}. \tag{6.7-7}$$

Since the two are in parallel,

$$R_{eq} = \frac{R_n R_p}{R_n + R_p} \tag{6.7-8}$$

provides the relationship for the overall TG resistance.

Consider first the operation of the TG when $V_{out} < |V_{Tp}|$ and the MOSFETs are saturated. The equivalent resistances are

$$R_n = \frac{2(V_{DD} - V_{out})}{\beta_n(V_{DD} - V_{out} - V_{Tn})^2},$$

$$R_p = \frac{2(V_{DD} - V_{out})}{\beta_p(V_{DD} - |V_{Tp}|)^2}. \tag{6.7-9}$$

Note that the p-channel MOSFET is characterized by $V_{BSp} = 0$, so $V_{Tp} = V_{T0p}$, i.e., there are no body bias effects present. The nMOS transistor is more complicated since

$$V_{SBn} = V_{out} \tag{6.7-10}$$

implies that body bias must be included using

$$V_{Tn} = V_{T0n} + \gamma(\sqrt{2|\phi_F| + V_{out}} - \sqrt{2|\phi_F|}). \tag{6.7-11}$$

When V_{out} reaches $|V_{T0p}|$, the p-channel MOSFET goes into the nonsaturated region. This changes the equivalent p resistance to

$$R_p = \frac{2}{\beta_p[2(V_{DD} - |V_{T0p}|) - (V_{DD} - V_{out})]}. \tag{6.7-12}$$

R_n is still given by eqn. (6.7-9) since the n-channel MOSFET remains saturated.

The final transition occurs when C_{out} charges to a value $V_{out} = (V_{DD} - V_{Tn})$. The nMOSFET requires a minimum gate-source voltage of $V_{GSn} = V_{Tn}$ to create the conduction channel. Consequently, $V_{out} > (V_{DD} - V_{Tn})$ implies that the nMOS transistor will be in cutoff, which is effectively a high-impedance state in the charging analysis. The nonsaturated p-channel MOSFET controls the current flow during this portion of the charging event.

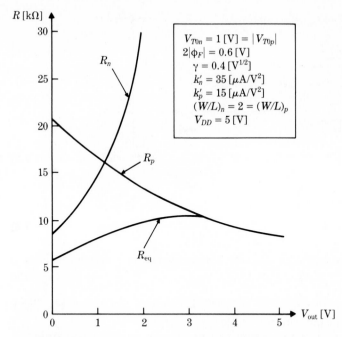

FIGURE 6.36 CMOS transmission gate equivalent resistances.

The preceding analysis allows for calculating R_n, R_p, and R_{eq} as functions of the output voltage V_{out} across the capacitor. An example of TG behavior is shown by the values plotted in the graph of Fig. 6.36. It is seen that R_n increases with increasing V_{out}. R_p simultaneously decreases in a manner that keeps R_{eq} relatively constant. Using the concept of equivalent resistances shows that the magnitude of I depends on the process and layout parameters. In particular, $(W/L)_n$ and $(W/L)_p$ constitute the basic design variables.

This approach may be extended to provide for a simple first-order model by viewing the event as charging C_{out} through R_{eq}. This gives the well-known behavior

$$V_{out}(t) \simeq V_{DD}(1 - e^{-t/R_{eq}C_{out}}), \tag{6.7-13}$$

which is characterized by the time constant

$$\tau_{TG} = R_{eq}C_{out}. \tag{6.7-14}$$

Although this represents only a simplistic viewpoint, it is seen that the transient performance (i.e., the charging time) can be improved by increasing both $(W/L)_n$ and $(W/L)_p$. Large (W/L) values decrease the equivalent drain-source resistances, which allows greater current flow through the gate. The tradeoff, of course, is that additional chip area is needed to implement the design.

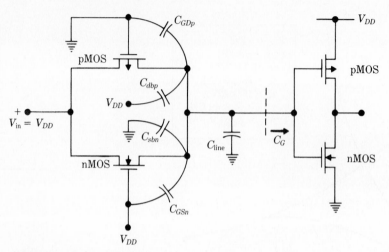

FIGURE 6.37 Approximation of C_{out} for a transmission gate driving an inverter circuit.

The bidirectional property of the TG is evident since the MOSFET arrangement allows for current flow in the opposite direction.

The capacitance C_{out} in the circuit can be approximated to first order by using the contributions shown in Fig. 6.37. The total value is

$$C_{out} \simeq C_{GSn} + C_{GDp} + K(V_{DD}, 0)[C_{sbn} + C_{dbn}] + C_{line} + C_G, \qquad \text{(6.7-15)}$$

where C_G depends on the gate following the TG. With the inverter,

$$C_G = C_{Gn} + C_{Gp}. \qquad \text{(6.7-16)}$$

This level of approximation for C_{out} is useful for understanding the dependence of the TG circuit performance on the device and layout parameters.

EXAMPLE 6.7.1-1

Find a simple approximation for the maximum value of R_{eq}.

Solution
A basic approximation for R_{eq} may be obtained by examining the point where the nMOS transistor goes into cutoff. This occurs when

$$V_{out} = V_{DD} - V_{Tn}(V_{out})$$
$$= (V_{DD} - V_{T0n}) - \gamma(\sqrt{2|\phi_F| + V_{out}} - \sqrt{2|\phi_F|}).$$

For V_{out} greater than this value, $R_n \to \infty$, so

$$R_{eq} \simeq R_p = \frac{2}{\beta_p[2(V_{DD} - |V_{T0p}|) - (V_{DD} - V_{out})]}.$$

Note that, if body bias is ignored,

$$R_{eq} \simeq \frac{1}{\beta_p \left[V_{DD} - |V_{T0p}| - \frac{1}{2} V_{Tn} \right]}.$$

■

EXAMPLE 6.7.1-2

A CMOS transmission gate is designed with $(W/L)_n = 2 = (W/L)_p$ in a process described by

$$k_n' = 25 \, [\mu A/V^2], \quad V_{T0n} = +1 \, [V], \quad \gamma_n = 0.37 \, [V^{1/2}], \quad 2|\phi_{Fp}| = 0.6 \, [V],$$
$$k_p' = 10 \, [\mu A/V^2], \quad V_{T0p} = -1 \, [V], \quad \gamma_p = 0.52 \, [V^{1/2}], \quad 2\phi_{Fn} = 0.7 \, [V].$$

Estimate the value of R_{eq} by evaluating $R_{eq} \simeq R_n$ at $V_{out} = 0$. Then perform a SPICE simulation to extract R_{eq}.

Solution
First calculate $R_{eq} \simeq R_n$ using eqn. (6.7-9) with $V_{out} = 0$:

$$R_{eq} \simeq \frac{2(5)}{2(25 \times 10^{-6})(5 - 1)^2} \simeq 12.5 \, [k\Omega].$$

To compare this approximation with a SPICE simulation, the load capacitance was set at $C_{out} = 100 \, [fF]$ and then a pulse was applied to the input of the TG. The SPICE file is shown below and the voltage plots are given in Fig. E6.2.

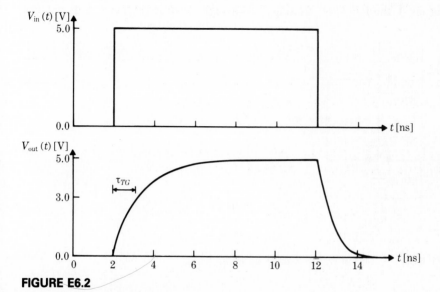

FIGURE E6.2

```
CMOS TG SIMULATION
VDD 3 0 DC 5VOLTS
MN 1 3 4 0 MN L=5U W=10U
MP 4 0 1 3 MP L=5U W=10U
VS 1 0 DC PULSE (0 5 2NS 1PS 1PS 10NS)
CLOAD 4 0 1.0E-13
.MODEL MN NMOS VTO=1.0 GAMMA=0.37 KP=2.5E-5 PHI=0.60
.MODEL MP PMOS VTO=-1.0 GAMMA=0.54 KP=1.0E-5 PHI=0.70
.TRAN 0.1NS 15NS
.PLOT TRAN V(4)
.END
```

To extrapolate the value of R_{eq}, use eqn. (6.7-13) to note that when $t = \tau_{TG}$,

$$V_{out} = V_{DD}[1 - e^{-1}] \simeq 3.16\,[\text{V}].$$

The time constant is then estimated by reading $\tau_{TG} \simeq 1.3\,[\text{ns}]$, so

$$R_{eq} \simeq \frac{1.3 \times 10^{-9}}{100 \times 10^{-15}} \simeq 13\,[\text{k}\Omega].$$

The approximation is quite good in this case. ∎

6.7.2 TG Logic Implementation

Transmission gates may be used to construct logic circuits by noting that signal paths can be set by the gate control lines. The simplest example of this type of control circuitry is the 2-input multiplexer shown in Fig. 6.38. The inputs A and the B are connected to TGs that are controlled by the switching signal S. When $S = 0$, TGA is on (conducting) while TGB is in a high-impedance state; the output is then $F = A$. Setting S to a logic 1 turns on TGB and switches

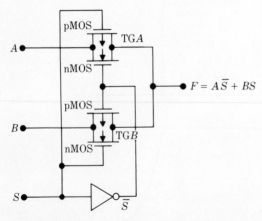

FIGURE 6.38 A 2-input transmission gate multiplexer.

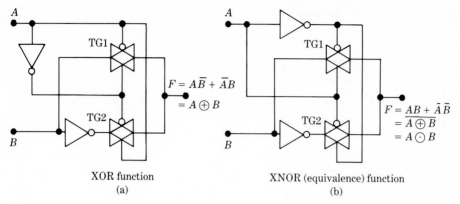

$$F = A\bar{B} + \bar{A}B$$
$$= A \oplus B$$

$$F = AB + \bar{A}\bar{B}$$
$$= \overline{A \oplus B}$$
$$= A \odot B$$

XOR function
(a)

XNOR (equivalence) function
(b)

FIGURE 6.39 CMOS XOR and XNOR functions using transmission gates.

TGA off. $S = 1$ thus gives $F = B$. The multiplexer operation can be summarized by writing

$$F = A\bar{S} + BS. \tag{6.7-17}$$

This example demonstrates how TGs can be used to control the data paths in a logic circuit. The circuit can be extended to the case of an n-input multiplexer.

Creating logic circuits using TGs generally requires that both a signal and its complement be available. These are used to directly control the conduction properties of transmission gates. This type of design approach provides for the XOR and XNOR gates in Fig. 6.39(a) and 6.39(b), respectively.

Consider first the XOR circuit. TG1 is controlled by A and $\bar{A}$ and transmits B. Pass gate TG2 is also controlled by A and $\bar{A}$ (in an inverse manner), but it transmits $\bar{B}$. The output can thus be viewed as B and its complement multiplexed by the switch control signal A. The output is then

$$F = A\bar{B} + \bar{A}B$$
$$= A \oplus B, \tag{6.7-18}$$

which is the XOR function. The XNOR (equivalence) circuit in Fig. 6.39(b) is obtained by similar construction. In this case, the connections of the control signal to TG1 are reversed from the scheme used in the XOR circuit. This complements each A factor in eqn. (6.7-18) so that the output is

$$F = AB + \bar{A}\bar{B}$$
$$= \overline{A \oplus B}$$
$$= A \odot B. \tag{6.7-19}$$

The XOR function can also be implemented using only 6 transistors. This type of gate is shown in Fig. 6.40(a). One transmission gate is used, and a CMOS inverter is included at the output. To understand the operation of this circuit, note that B acts as the TG control signal. When $B = 0$, the TG turns

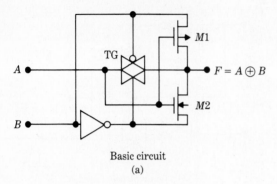

Basic circuit
(a)

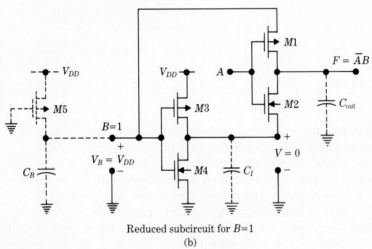

Reduced subcircuit for $B=1$
(b)

FIGURE 6.40 Six-transistor CMOS XOR.

on and A is transmitted through to the output. The inverter MOSFETs $M1$ and $M2$ are both off with this value of B since the polarity across the pair will prevent current flow.

The circuit operation for $B = 1$ may be understood by referring to the subcircuit in Fig. 6.40(b). In this case, the TG is off, so it is not shown in the diagram. An interesting point about the circuit is that the input voltage $V_B = V_{DD}$ acts as a power supply for the inverter MOSFETs $M1$ and $M2$. The voltage $V = 0$ produced by the first inverter provides the necessary ground connection (through $M4$).

The subcircuit also illustrates the important design considerations for the gate. The simplest case, where the XOR network is driven by an inverter, has been assumed. Transistor $M5$ constitutes the pMOS portion of the driving network. The output node capacitance C_{out} at F will be charged or discharged depending on the value of A. For a charging event with $A = 0$, MOSFETs $M1$ and $M5$ will be conducting to supply the necessary current. The actual rise time t_{LH} is determined by C_{out} and the values chosen for $(W/L)_1$ and $(W/L)_5$.

Discharging C_{out} is accomplished though $M2$ and $M4$; the value of the fall time t_{HL} is thus set by the design choices for $(W/L)_2$ and $(W/L)_4$. Note that the series-connected pMOS transistors $M1$ and $M5$ controlling t_{LH} are chosen in the same manner as for a CMOS NOR gate, while the nMOS series transistors $M2$ and $M4$ act like a CMOS NAND in determining t_{HL}.

The complete design of the circuit requires careful consideration of the overall system properties. This is seen by including the capacitances C_B and C_I shown in the reduced subcircuit. These contributions are important when the B input is switched from a logic 0 to a logic 1 state. C_B is approximated by

$$C_B \simeq C_{out,5} + [C_{G3} + C_{G4}] + C_{GS1} + K(V_{DD}, 0)C_{sb1} + C_{line,B}, \qquad (6.7\text{-}20)$$

where $C_{out,5}$ is the output capacitance contribution from the stage containing MOSFET $M5$. The charging transient for this capacitance is controlled primarily by $(W/L)_5$. The inverter capacitance, denoted by C_I, is approximately

$$C_I \simeq C_{GS2} + K(V_{DD}, 0)C_{sb2} + C_{GD3} + C_{GD4}$$
$$+ K(V_{DD}, 0)(C_{db3} + C_{db4}) + C_{line,I}. \qquad (6.7\text{-}21)$$

Switching B from 0 to 1 discharges this capacitor through $M4$, so $(W/L)_4$ is important in determining this transient time. It is noted in passing that $(W/L)_3$ of $M3$ establishes the charging time for this capacitor.

An XNOR circuit that uses only 6 MOSFETs is shown in Fig. 6.41. This has been constructed from the XOR circuit by simply reversing the B and $\bar{B}$ connections to the TG.

A half-adder circuit is easily obtained by recalling that this function uses inputs A_0 and B_0 to give the sum

$$S_0 = A_0 \oplus B_0, \qquad (6.7\text{-}22)$$

with the carry value

$$C_0 = A_0 B_0. \qquad (6.7\text{-}23)$$

One possible implementation of a half-adder based on a double TG XOR circuit is shown in Fig. 6.42. This uses $\bar{A}_0$ and $\bar{B}_0$ through a 2-input NOR gate

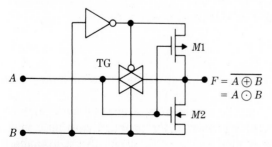

FIGURE 6.41   Six-transistor CMOS equivalence (XNOR) function.

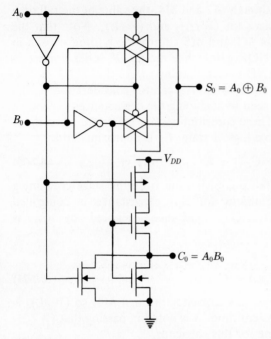

$S_0 = A_0 \oplus B_0$

$C_0 = A_0 B_0$

V_{DD}

FIGURE 6.42 A basic half-adder circuit.

to provide the carry as verified by the standard DeMorgan relation

$$C_0 = \overline{\overline{A}_0 + \overline{B}_0}$$
$$= A_0 B_0.$$

(6.7-24)

A full-adder is constructed in a similar manner. First, recall that the inputs A_n, B_n, and C_{n-1} to a full-adder generate the sum and carry outputs

$$S_n = A_n \oplus B_n \oplus C_{n-1},$$
$$C_n = A_n B_n + C_{n-1}(A_n \oplus B_n).$$

(6.7-25)

The sum S_n can be obtained using TG XOR logic in the form

$$S_n = (A_n \oplus B_n)\bar{C}_{n-1} + (\overline{A_n \oplus B_n})C_{n-1}.$$

(6.7-26)

Implementation of this function requires that $(A \oplus B)$ and its complement be generated, which is then XORed with C_{n-1}.

The output carry function C_n can be obtained by a direct combinational AOI CMOS circuit. A more elegant approach to the problem is to note that $(A \oplus B)$ and $(\overline{A \oplus B})$ must be generated for S_n, so these signals are available for deriving C_n without additional circuitry. Since the carry may be expressed

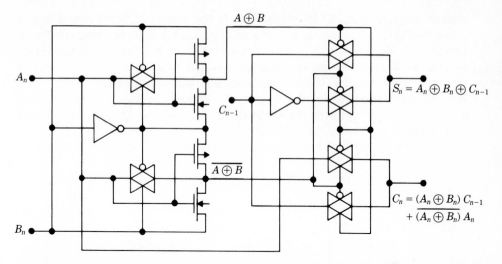

FIGURE 6.43 CMOS full-adder using transmission gate logic configuration.

in the alternate form

$$C_n = A_n B_n + A_n \bar{B}_n C_{n-1} + \bar{A}_n B_n C_{n-1}$$
$$= (A_n \oplus B_n) C_{n-1} + \overline{(A_n \oplus B_n)} A_n, \tag{6.7-27}$$

this term can be obtained using a TG multiplexer circuit.

Figure 6.43 shows a full-adder constructed using eqns. (6.7-26) and (6.7-27). The input circuits that produce $(A \oplus B)$ and $(\overline{A \oplus B})$ are symmetrical and minimize the number of MOSFETs. Although the sum S_n, can be generated using a single TG circuit with an inverter output, the carry C_n in eqn. (6.7-27) requires a more general dual-TG multiplexer. Multiplexing circuits have thus been used for both portions of the output circuitry. Proper (W/L) design for the MOSFETs then provides for more symmetrical rise and fall times at both outputs.

6.8 Nonstandard CMOS Gates

Standard CMOS AOI logic requires that each input be connected to the gate of both an nMOS and a pMOS transistor to provide the full complementary operation of the circuit. Following this general rule ensures the low power consumption characteristics, since only one type of MOSFET is conducting for a stable input voltage. The main drawback of this type of circuit design is that two MOSFETs are needed for every input. Implementing logic in standard CMOS can thus require a large number of transistors. This increases the real estate consumption and affects the integration level possible on the chip.

Nonstandard CMOS circuits are constructed using the fact that only one type of transistor (e.g., nMOS) is needed to form the logic. Since both nMOS and pMOS devices are available from the processing, one type of transistor is chosen to implement the logic function. A nonstandard CMOS circuit uses a single MOSFET of opposite polarity as a load. This is to be contrasted with regular CMOS, which uses opposite polarity MOSFETs in a complementary manner to achieve low power consumption and ratioless logic. Nonstandard CMOS circuits require fewer transistors but will dissipate DC power with certain combinations of input voltages. In addition, the logic becomes ratioed so that the output voltages are set by device geometries. These tradeoffs may be acceptable depending on the level of chip integration desired.

Figure 6.44 provides examples of nonstandard CMOS arrangements. The

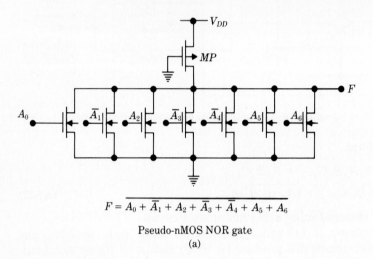

$$F = \overline{A_0 + \overline{A}_1 + A_2 + \overline{A}_3 + \overline{A}_4 + A_5 + A_6}$$

Pseudo-nMOS NOR gate

(a)

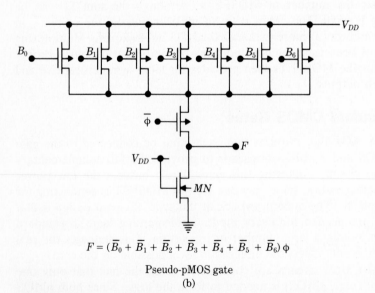

$$F = (\overline{B}_0 + \overline{B}_1 + \overline{B}_2 + \overline{B}_3 + \overline{B}_4 + \overline{B}_5 + \overline{B}_6)\, \phi$$

Pseudo-pMOS gate

(b)

FIGURE 6.44 Nonstandard CMOS logic.

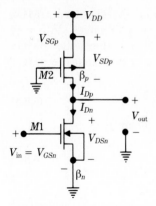

FIGURE 6.45 Pseudo-nMOS inverter circuit.

circuit in Fig. 6.44(a) illustrates a *pseudo-nMOS* approach in which *n*-channel MOSFETs are configured to implement the NOR function

$$F = \overline{(A_0 + \bar{A}_1 + A_2 + \bar{A}_3 + \bar{A}_4 + A_5 + A_6)}. \tag{6.8-1}$$

The *p*-channel MOSFET *MP* has its gate connected to ground and is used as a load device. *MP* conducts power supply current whenever the output is low ($F = 0$) because of the bias.

A *pseudo-pMOS* circuit is illustrated in Fig. 6.44(b). This gives an output

$$F = (\bar{B}_0 + \bar{B}_1 + \bar{B}_2 + \bar{B}_3 + \bar{B}_4 + \bar{B}_5 + \bar{B}_6)\phi, \tag{6.8-2}$$

as is easily verified using the rules of pMOS logic formation. The nMOS transistor *MN* is connected as a load, with its gate at the power supply level V_{DD}. The load transistor conducts *DC* current when the inputs are set to give $F = 1$.

The properties of nonstandard CMOS arrangements are most easily understood by analyzing a typical circuit. The pseudo-nMOS inverter of Fig. 6.45 is chosen for this purpose. The nMOS transistor *M1* is defined with terminal voltages of

$$V_{DSn} = V_{out}, \qquad V_{GSn} = V_{in}, \tag{6.8-3}$$

while the *p*-channel MOSFET *M2* has

$$V_{SDp} = V_{DD} - V_{out}, \qquad V_{SGp} = V_{DD}. \tag{6.8-4}$$

These allow for the calculation of critical VTC points.

Consider first the output high voltage V_{OH}. This may be computed by setting $V_{in} = 0 \,[\text{V}]$, placing *M1* in cutoff. Since the pMOS load transistor *M2* is biased on,

$$V_{OH} \simeq V_{DD} \tag{6.8-5}$$

is a reasonable approximation for the output voltage.

Now suppose that $V_{in} = V_{DD}$, so that $V_{out} = V_{OL}$. In standard CMOS logic, the ratioless property of the complementary pair gives $V_{OL} = 0$. However, the nonstandard circuit is more complicated since the pMOS transistor is conducting current. To analyze this case, first note that the nMOS transistor terminal voltages satisfy

$$V_{DSn} = V_{OL} < (V_{GSn} - V_{Tn}),$$ (6.8-6)

so that $M1$ is nonsaturated. The drain current is thus given by

$$I_{Dn} = \frac{\beta_n}{2}[2(V_{DD} - V_{Tn})V_{OL} - V_{OL}^2].$$ (6.8-7)

The p-channel MOSFET $M2$ voltages are

$$V_{SDp} = V_{DD} - V_{OL}, \qquad V_{SGp} = V_{DD}.$$ (6.8-8)

Assuming that $V_{OL} \ll V_{DD}$ as required for a good inverter and also that $V_{OL} < |V_{Tp}|$, these relations imply that $M2$ is saturated, with

$$I_{Dp} = \frac{\beta_p}{2}(V_{DD} - |V_{Tp}|)^2.$$ (6.8-9)

V_{OL} may now be computed by setting $I_{Dn} = I_{Dp}$. Rearranging gives

$$V_{OL}^2 - 2(V_{DD} - V_{Tn})V_{OL} + \left(\frac{\beta_p}{\beta_n}\right)(V_{DD} - |V_{Tp}|)^2 = 0,$$ (6.8-10)

which is a quadratic for V_{OL}. Solving gives

$$V_{OL} = (V_{DD} - V_{Tn}) - \sqrt{(V_{DD} - V_{Tn})^2 - \frac{\beta_p}{\beta_n}(V_{DD} - |V_{Tp}|)^2}$$ (6.8-11)

as the output low voltage. Assuming that $V_{T0n} = |V_{T0p}| = V_{T0}$ and noting that there are no body bias effects in the simple inverter, the solution may be reduced to

$$V_{OL} = (V_{DD} - V_{T0})\left[1 - \sqrt{1 - \frac{\beta_p}{\beta_n}}\right].$$ (6.8-12)

It is obvious that the driver-load ratio (β_n/β_p) must be greater than or equal to one for physical solutions.

To understand the implications of the ratioed behavior, note that setting $I_{Dn} = I_{Dp}$ provides the design equation

$$\frac{\beta_n}{\beta_p} = \frac{(V_{DD} - |V_{Tp}|)^2}{2(V_{DD} - V_{Tn})V_{OL} - V_{OL}^2}$$ (6.8-13)

for the driver-load ratio (β_n/β_p). The device aspect ratios are given by

$$\frac{(W/L)_n}{(W/L)_p} = \frac{k_p' \beta_n}{k_n' \beta_p}$$ (6.8-14)

so that the MOSFET geometries can be determined for a desired value of V_{OL}. The condition that $(\beta_n/\beta_p) > 1$ corresponds to the requirement that the gain of the stage be sufficient to invert the input signal.

As an example of the design constraints, suppose that $V_{DD} = 5\,[\text{V}]$ and $V_{T0n} = |V_{T0p}| = 1\,[\text{V}]$. For a design value of $V_{OL} = 0.2\,[\text{V}]$, $(\beta_n/\beta_p) = 10$ is needed. Raising the acceptable output low voltage to $V_{OL} = 0.4\,[\text{V}]$ lowers the minimum driver-load ratio to about 5.25. Since $(k_p/k_n) \simeq (1/2.5)$, a value of $V_{OL} = 0.4\,[\text{V}]$ requires that $(W/L)_n/(W/L)_p \simeq 2.1$. The ratioed nature of the logic clearly indicates that pseudo-nMOS does not conserve real estate unless there are a large number of inputs. In addition, nonzero V_{OL} levels must be acceptable in the circuit.

The inverter threshold voltage V_{th} may be computed by setting $V_{in} = V_{out} = V_{th}$. Equations (6.8-3) show that the n-channel MOSFET $M1$ is saturated. Assuming $V_{th} > |V_{Tp}|$, eqn. (6.8-4) places the pMOS transistor $M2$ in the nonsaturated mode. Equating currents then gives

$$\frac{\beta_n}{2}(V_{th} - V_{Tn})^2 = \frac{\beta_p}{2}[2(V_{DD} - |V_{Tp}|)(V_{DD} - V_{th}) - (V_{DD} - V_{th})^2]. \quad \textbf{(6.8-15)}$$

Rearranging provides the alternate design equation

$$\frac{\beta_n}{\beta_p} = \frac{[2(V_{DD} - |V_{Tp}|) - (V_{DD} - V_{th})](V_{DD} - V_{th})}{(V_{th} - V_{Tn})^2}, \quad \textbf{(6.8-16)}$$

which allows setting (β_n/β_p) for a desired inverter threshold voltage value. The resulting output low voltage V_{OL} would then be computed from eqn. (6.8-11). If the inverter is already designed, V_{th} may be found by writing eqn. (6.8-15) in the quadratic form

$$\left(1 + \frac{\beta_n}{\beta_p}\right)V_{th}^2 - 2\left(|V_{Tp}| + V_{Tn}\frac{\beta_n}{\beta_p}\right)V_{th} + \left[V_{Tn}^2\frac{\beta_n}{\beta_p} + 2\,|V_{Tp}|V_{DD} - V_{DD}^2\right] = 0.$$

$$\textbf{(6.8-17)}$$

Substituting appropriate values allows for calculation of V_{th}. The special case where $V_{Tn} = |V_{Tp}| = V_{T0}$ can be solved to give

$$V_{th} = V_{T0} + \sqrt{V_{T0}^2 - \frac{[V_{DD}^2 - 2V_{T0}V_{DD} - V_{T0}^2(\beta_n/\beta_p)]}{1 + (\beta_n/\beta_p)}}. \quad \textbf{(6.8-18)}$$

Pseudo-pMOS circuits can be analyzed using the same techniques. The results are analogous to those found for the pseudo-nMOS circuit in that the logic is ratioed and requires careful consideration of the driver-load ratio.

REFERENCES

(1) M. Annatatone, *Digital CMOS Circuit Design*, Hingham, MA: Kluwer Academic Publishers, 1986.

(2) M. I. Elmasry (ed.), *Digital MOS Integrated Circuits,* New York: IEEE Press (Wiley), 1981.

(3) O. G. Folberth and W. G. Grobman (eds.), *VLSI Technology and Design,* New York: IEEE Press (Wiley), 1984.

(4) L. A. Glasser and D. W. Dobberpuhl, *The Design and Analysis of VLSI Circuits,* Reading, MA: Addison-Wesley, 1985.

(5) M. J. Howes and D. V. Morgan (eds.), *Large Scale Integration,* Bath, UK: Wiley, 1981.

(6) C. A. Mead and L. Conway, *Introduction to VLSI Systems,* Reading, MA: Addison-Wesley, 1980.

(7) A. D. Milne, *MOS Devices,* Edinburgh: Halstead Press, 1983.

(8) A. Mukherjee, *Introduction to nMOS & CMOS VLSI System Design,* Englewood Cliffs, NJ: Prentice-Hall, 1986.

(9) D. A. Pucknell and K. Eshraghian, *Basic VLSI Design,* Sydney: Prentice-Hall, 1985.

(10) N. Weste and K. Eshraghian, *Principles of CMOS VLSI Design,* Reading, MA: Addison-Wesley, 1985.

PROBLEMS

Assume nMOS parameters of

$$V_{T0D} = +0.9\,[\text{V}], \quad V_{T0L} = -3.3\,[\text{V}], \quad V_{DD} = 5\,[\text{V}],$$
$$\gamma = 0.37\,[\text{V}^{1/2}], \quad 2\,|\phi_F| = 0.58\,[\text{V}], \quad k'_D = 25\,[\mu\text{A}/\text{V}^2] = k'_L$$

unless otherwise stated.

For CMOS problems, use

$$V_{T0n} = +0.8\,[\text{V}], \quad V_{T0p} = -0.8\,[\text{V}], \quad V_{DD} = 5\,[\text{V}],$$
$$k'_n = 40\,[\mu\text{A}/\text{V}^2], \quad k'_p = 16\,[\mu\text{A}/\text{V}^2], \quad \gamma_n = 0.26\,[\text{V}^{1/2}],$$
$$\gamma_p = 0.84\,[\text{V}^{1/2}], \quad 2\,|\phi_{Fp}| = 0.58\,[\text{V}], \quad 2\phi_{Fn} = 0.69\,[\text{V}]$$

for the representative process.

6.1 A 2-input nMOS NOR gate is designed with different driver geometries. Input A is described by $\beta_{R,A} = 5$, while input B has $\beta_{R,B} = 8$. The maximum value of the input voltage is specified to be $V_{\text{in}} = 5\,[\text{V}]$.
   (a) Calculate the best-case value of $V_{OL}\,[\text{V}]$.
   (b) Calculate the worst-case value of $V_{OL}\,[\text{V}]$.

6.2 Consider a 4-input nMOS NOR gate that is designed to have identical drivers. Denote the driver-load ratio by β_R for each input.

(a) Design the circuit by finding the nearest integer value of β_R that will guarantee a worst-case value of $V_{OL} = 0.4\,[\text{V}]$.

(b) Repeat the calculation for a worst-case value of $V_{OL} = 0.25\,[\text{V}]$.

*(c) Simulate the circuit in part (a) using SPICE. Examine input switching for the cases where only a single input is switched and when all inputs are changed at the same time.

6.3 A 2-input NOR gate is designed with $(W/L)_D = 4$ and $(W/L)_L = 0.5$. The output capacitance is estimated to be $C_{out} = 180\,[\text{fF}]$.

(a) Calculate the value of t_{LH}. Include body bias effects by averaging the output voltage.

(b) Calculate the value of t_{HL} when only a single driver is switched on.

(c) Now calculate the value of t_{HL} for the case when both drivers are simultaneously switched on.

6.4 A 2-input nMOS NOR gate is shown in Fig. P6.1. All dimensions are in

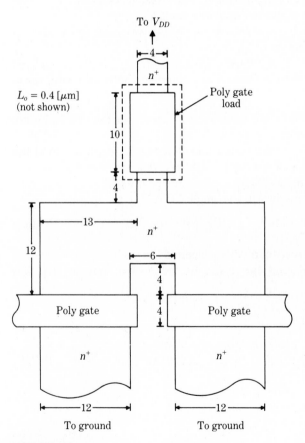

FIGURE P6.1

units of microns [μm]. The processing gives capacitance levels of

$$C_{ox} = 0.691\,[\text{fF}/\mu\text{m}^2], \quad C_o = 0.276\,[\text{fF}/\mu\text{m}],$$
$$C_{j0} = 0.0975\,[\text{fF}/\mu\text{m}^2], \quad \phi_o = 0.879\,[\text{V}],$$
$$C_{jsw} = 0.209\,[\text{fF}/\mu\text{m}], \quad \phi_{osw} = 0.957\,[\text{V}].$$

Gate overlap is not shown explicitly in the drawing but should be included in the calculations.

(a) Calculate the value of V_{OL} when both drivers are conducting. Use this value in the remaining portions of the problem.

(b) Calculate the total zero-bias capacitance $(C_{dbA} + C_{dbB} + C_{sbL})$ due to the n^+ region at the output node.

(c) Calculate $K(V_{OH}, V_{OL})$ and then find the average value of the n^+ output capacitance.

(d) Find the maximum value of $(C_{GDA} + C_{GDB})$ for the drivers.

(e) Find the maximum value of C_{GDL} of the load.

(f) Calculate the maximum, minimum, and average values of C_{out} that arise from the contributions shown (ignore C_{line} etc.)

(g) Calculate t_{HL} and t_{LH} when both inputs are switched from 0 [V] to V_{DD} [assume maximum C_{out}].

*(h) Prepare a full SPICE simulation file for (i) C_{out} = constant and (ii) SPICE-computed capacitances. Then compare the results with those in (g).

6.5   Two n-channel MOSFETs are connected as shown in Fig. P6.2. Ignore body bias in this problem.

(a) Assume that $M1$ is nonsaturated and that $M2$ is saturated. Find the aspect ratio (W/L) of a single MOSFET that will have I_D vs. V_{GS} characteristics identical to the series-connected pair.

(b) What operational mode will the single MOSFET be operating in?

6.6   Consider the 2-input nMOS NAND gate in Fig. 6.5 with identical drivers.

(a) Calculate the driver-load ratio needed to achieve a value of $V_{OL} = 0.25$ [V]. Round the result to the nearest integer that has

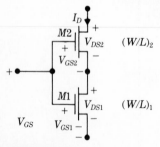

FIGURE P6.2

$V_{OL} \leq 0.25$ [V], and then calculate the actual value of V_{OL} after the rounding.

(b) Repeat for the case $V_{OL} = 0.10$ [V]. Compare the increase in driver real estate for the two cases if $L = 4$ [μm] is set for the driver MOSFETs and the load aspect ratios are identical for the two circuits.

6.7 A 2-input nMOS NAND gate is designed for an output low voltage of $V_{OL} = 0.20$ [V].
   (a) Calculate the value of $\beta_{R,AB}$ for the case of identical drivers.
   (b) Suppose that $V_{DS,A} = (V_{OL}/3)$ is chosen. Find $\beta_{R,A}$ and $\beta_{R,B}$ for this case.
   (c) Instead set $V_{DS,B} = (V_{OL}/3)$ and calculate the driver-load ratios.
   (d) Compare the results for all three cases. Comment on the primary effects each design set has on the overall gate switching characteristics.

6.8 The 2-input NAND gate shown in Fig. P6.3 has identical driver MOSFETs described by β_D. The input to the top driver is set at a voltage

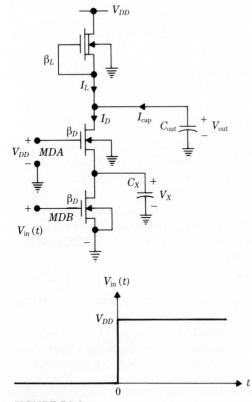

FIGURE P6.3

$V_{in,A} = V_{DD}$ for times of interest. The input to driver B is set at $V_{in,B} = V_{in}(t)$, which is pulsed at $t = 0$ from 0 [V] to V_{DD} as shown in the drawing. The problem deals with the discharge time t_{HL} for the circuit.

(a) Find a self-iterating equation for the intermediate node voltage V_X that is valid for times $t \le 0$.

(b) Now consider time $t \ge 0$. Use KCL to relate the currents, and then determine the conducting state of MOSFETs MDA and MDB during the discharge event. Derive an equation for t_{HL} by integrating the current through driver MDB. Assume integration limits of $V_{DSB}(0) = V_X$ and $V_{DSB}(t_{HL}) = \frac{1}{2}V_{OL}$, and assume that $I_L \ll I_D$.

Suppose that $\beta_D = 175$ [μA/V^2] and $\beta_L = 12.5$ [μA/V^2], while $V_{DD} = 5$ [V]. The capacitances are estimated by $C_{out} = 150$ [fF] and $C_X = 35$ [fF]. The other parameters are provided at the beginning of the problem listing.

(c) Calculate V_X ($t \le 0$).

(d) Find the value of V_{OL} for the NAND gate.

(e) Calculate t_{HL} using the expression derived in (b). Comment on the affect C_X has in determining t_{HL} for this calculation.

* (f) Simulate the circuit by performing a transient SPICE analysis.

6.9 The layout for a 2-input NAND gate is shown in Fig. P6.4. All dimensions are in units of microns [μm], and the depletion capacitance levels are the same as those given in Problem 6.4.

(a) Calculate V_{OL} for the circuit.

(b) Find the zero-bias n^+ capacitance ($C_{dbA} + C_{sbL}$).

(c) Calculate $K(V_{OH}, V_{OL})$, and then use this to find the average n^+ capacitance $K(V_{OH}, V_{OL})(C_{dbA} + C_{sbL})$.

(d) Refer to Fig. 6.7. Calculate the zero-bias value of ($C_{sbA} + C_{dbB}$). Then find the maximum values of ($C_{GSA} + C_{GDB}$). Calculate the maximum voltage at V_X, and then average the capacitances at this node by using $K(V_X, V_{OL})$ for the depletion contributions.

(e) Find the worst-case value of ($C_{GDA} + C_{GDL}$).

* (f) Prepare a SPICE file that simulates the switching of the NAND gate when (i) both inputs are switched and (ii) when the input to A is held at V_{DD} while the B input is switched. Use both SPICE-modeled capacitances and C_{out} levels calculated above.

6.10 Consider the 3-input CMOS NAND gate drawn in Fig. 6.22. The n-channel MOSFETs have β_n, while β_p gives the conduction properties for the p-channel devices.

Suppose that the inputs are connected, so $V_{in,A} = V_{in,B} = V_{in,C} = V_{in}$. Find an expression for the gate threshold voltage V_{th} under these conditions. Compare your answer with the 2-input NAND gate result.

6.11 A 2-input CMOS NAND has the structuring shown in Fig. 6.24 and is designed with $\beta_n = 2\beta_p$.

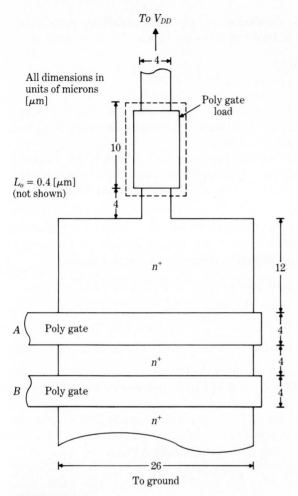

FIGURE P6.4

(a) Calculate the value of V_{th} when the inputs are simultaneously switched.

(b) Use eqn. (6.5-19) to calculate the value of V_{th} when only $M1$ is switched, with $V_{GS,2}$ held at V_{DD}.

*(c) Simulate all three switching possibilities using SPICE, and compare the relative values of V_{th}.

6.12 The 2-input CMOS NOR gate in Fig. 6.28 is designed with $\beta_n = 3\beta_p$.

(a) Calculate the value of V_{th} for the case of simultaneous input switching.

(b) $V_{GS,1}$ is grounded as shown in Fig. 6.30. Find V_{th} when $V_{GS,2}$ is increased from 0 [V] toward a logic 1 value.

*(c) Perform a SPICE simulation of all three switching possibilities. Discuss the results in terms of choosing β_n and β_p for V_{th} values.

6.13 Consider the logic function

$$F = AB + BC + AC.$$

(a) Design an nMOS AOI circuit that implements this function with a worst-case V_{OL} of 0.25 [V]. Assume a minimum linewidth of 4 [μm], and avoid superfluous transistors.

(b) Now implement the function using standard AOI CMOS. Compare the real estate requirements of the transistors if the 4 [μm] linewidth is applied to the CMOS process. (Use minimum size MOSFETs for simplicity.)

6.14 A basic CMOS NAND gate is shown in Fig. P6.5, where all dimensions are in [μm]. The overlap capacitance for all devices is specified as $C_o = 0.276$ [fF/μm]. The depletion capacitances are as follows:

$$nMOS: \quad C_{j0} = 0.0975 \,[\text{fF}/\mu m^2], \quad \phi_o = 0.879 \,[\text{V}],$$
$$C_{jsw} = 0.107 \,[\text{fF}/\mu m], \quad \phi_{osw} = 0.921 \,[\text{V}].$$
$$pMOS: \quad C_{j0} = 0.293 \,[\text{fF}/\mu m^2], \quad \phi_o = 0.939 \,[\text{V}],$$
$$C_{jsw} = 0.362 \,[\text{fF}/\mu m], \quad \phi_{osw} = 0.985 \,[\text{V}].$$

(These values correspond to the calculations in Example 5.8-1 with $x_j = 0.5$ [μm].) Note that $L_o = 0.4$ [μm] is not shown explicitly in the layout.

(a) Calculate the zero-bias $n^+ - p$ substrate capacitance $C_{d,n}$ at the output (i.e., at the drain of the A-input nMOS transistor). Then find the average value of this contribution using $K(V_{OH}, V_{OL})C_{d,n}$.

(b) Now calculate the zero-bias capacitance $C_{d,p}$ seen at the pMOS drain regions due to the p^+/n-tub junction. Average this value using $K(V_{OH}, V_{OL})$.

(c) Calculate the zero-bias and average values of the n^+ junction capacitance $C_{n,A-B}$ seen between the gates of nMOS transistors A and B.

(d) The poly-to-field capacitance is specified to be $C_{p-f} = 0.0576$ [fF/μm^2]. The poly lines are 4 [μm] wide. Calculate the total capacitance C_A seen looking into the A input line. Include the MOSFET gate capacitance in your calculation by assuming that the poly field and C_G are in parallel. Repeat for the input capacitance seen looking into the B line. Ignore the overhang contributions on the pMOS transistors and the upper nMOS device.

(e) Estimate the maximum value of C_{out} due to the contributions shown.

(f) Calculate the value of V_{th} for the case where both inputs are switched at the same time.

*(g) Prepare a SPICE file which includes the depletion capacitances by

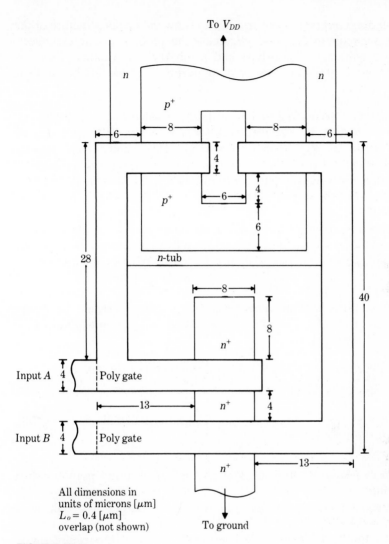

To V_{DD}

n

n

p^+

6

8

8

6

4

p^+

6

4

6

28

n-tub

40

8

8

n^+

Input A 4 Poly gate

13

n^+

n^+ 4

Input B 4 Poly gate

13

n^+

All dimensions in
units of microns [μm]
$L_o = 0.4$ [μm]
overlap (not shown)

To ground

FIGURE P6.5

specifying areas (AD, AS, etc.). Simulate the transient response of
the circuit using the .TRAN statement for the simultaneous switching
case.

6.15 Consider the NAND gate shown in Fig. P6-5 (introduced for Problem
6.14). Restructure the layout to reduce the input poly gate capacitance.
Then apply the layout patterning to a CMOS NOR gate. The use of stick
diagrams (introduced in Problem 5.13 of the previous chapter) may be
useful in the initial layout.

6.16 Use stick diagrams (see Problem 5.13) to show the layout structure of the half-adder circuit in Fig. 6-42. Examine the problems of interconnects and the requirement for p-type and n-tub regions. Comment on any general layout problems observed in performing the TG-based logic layout.

6.17 A CMOS TG is connected to a capacitance of value $C_{out} = 250$ [fF].
   (a) The transistors are designed with $(W/L)_n = 1 = (W/L)_p$. Calculate the value of R_{eq} when $V_{out} = 0$ [V], 2 [V], and 4 [V]. Assume $V_C = V_{DD}$.
   (b) Average the values in part (a) to obtain R_{eq}. Then estimate the time needed for the capacitor to charge from $V_{out} = 0$ [V] to a value $V_1 = 0.9 V_{DD}$.
   (c) Suppose that $(W/L)_n = (W/L)_p$ is increased to a value of 2. Calculate the charge time interval from 0 [V] to $0.9 V_{DD}$ for this case.
   (d) Finally, choose $(W/L)_n = 1$ and $\beta_p = \beta_n$. What is the percentage decrease in charge time for this case when compared with the value found in part (b)?
  *(e) Perform SPICE simulations on all three cases. Discuss the design factors of $(W/L)_n$ and $(W/L)_p$ as related to (i) circuit performance and (ii) layout.

6.18 Figure 6.38 shows the CMOS TG implementation of the function

$$F = AS + B\bar{S}.$$

   (a) Construct the CMOS AOI circuit which implements this function. Then compare transistor count between the two.
   (b) Discuss the design tradeoffs between the TG and AOI circuits.

6.19 Consider a 3-input NOR gate designed in CMOS using pseudo-nMOS structuring.
   (a) Design the circuit so that the worst-case output low voltage is $V_{OL} = 0.25$ [V].
   (b) Calculate V_{th} for the NOR gate when all three inputs are simultaneously switched from 0 [V] towards a logic 1 voltage.
   (c) Compare the real-estate consumption of this circuit with a standard CMOS complementary implementation. Assume $k'_n = 2.5 k'_p$.

6.20 Consider the simplified nMOS XNOR gate shown in Fig. 6.18(b).
   (a) Design the circuit for $V_{OL} = 0.35$ [V] by finding the β_R values. Assume equal drain-source voltages across the logic and driver MOSFETs.
   (b) Redesign the circuit for $V_{OL} = 0.35$ [V], but with $V_{DS,A} = 0.20$ [V] $= V_{DS,B}$.
   (c) Again set $V_{OL} = 0.35$ [V], only now choose $V_{DS,A} = 0.15$ [V] $= V_{DS,B}$. Compare the design of the two cases.

(d) Provide two different approaches for obtaining the XOR function from the basic gate circuit.

*(e) Now set $\beta_{R,1} = 6$ and $\beta_{R,A} = 4$. Calculate V_{OL} analytically, and then simulate the DC and transient characteristics using SPICE. Assume $C_{out} = 150\,[\text{fF}]$.

* (f) Choose a CMOS XNOR circuit, design it and compare the DC characteristics with the nMOS circuit.

CHAPTER 7

Bistable Logic Elements

Bistable logic circuits have two stable states and are commonly used for latches, memory elements, and timing circuits. This chapter discusses the basic bistable circuits in nMOS and CMOS designs. In addition, considerations for clocked circuits are introduced.

7.1 nMOS *SR* Flip-Flop

An *SR* flip-flop is shown in Fig. 7.1. The NOR gate implementation of Fig. 7.1(a) allows for construction of the function table in Fig. 7.1(b). Inputs are labeled S (set) and R (reset), and the outputs are Q and $\bar{Q}$. The operation of this bistable circuit can be understood by recalling that a logic 1 at any NOR input results in a logic 0 output. When $S = 0 = R$, the outputs are not affected and remain in the existing state. Setting the circuit with $S = 1$ and $R = 0$ drives Q to 1. Conversely, resetting the circuit with $R = 1$ and $S = 0$ results

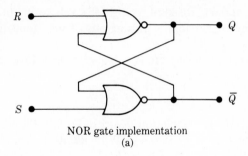

NOR gate implementation
(a)

S	R	Q	$\bar{Q}$	
0	0	Q	$\bar{Q}$	(Unchanged)
1	0	1	0	
0	1	0	1	
1	1	0	0	(Not permitted)

Function table
(b)

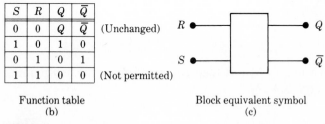

Block equivalent symbol
(c)

FIGURE 7.1 *SR* flip-flop.

in $Q = 0$. When $S = 1$ and $R = 1$, both outputs are 0. This contradicts the labeling of the outputs as Q and $\bar{Q}$, so this input combination is not permitted in normal operation. The functional block symbol for the *SR* flip-flop is given in Fig. 7.1(c).

The circuit diagram for the nMOS *SR* flip-flop is shown in Fig. 7.2. This constitutes a gate-by-gate implementation of the NOR-based logic. An alternate viewpoint is that the circuit consists of two cross-coupled inverters ($M1$, $M3$ and $M2$, $M4$) with MOSFETs MS and MR added to control the

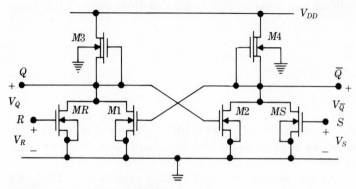

FIGURE 7.2 nMOS *SR* flip-flop.

switching. The input voltages V_S and V_R are kept at logic 0 levels to maintain an existing flip-flop state. The state is changed by pulsing one of the inputs to a logic 1 (V_{DD}) level.

The DC analysis of the circuit in a stable state is straightforward. First, the output high voltage is

$$V_{OH} \simeq V_{DD} \tag{7.1-1}$$

since the load transistors $M3$ and $M4$ are depletion devices. To find the value of V_{OL} at either Q or $\bar{Q}$, assume that $V_S = 0 = V_R$. With identical inverters (i.e., a symmetrical circuit), $(W/L)_1 = (W/L)_2 = (W/L)_D$ and $(W/L)_3 = (W/L)_4 = (W/L)_L$ so the driver-load ratio is

$$\beta_R = \frac{\beta_D}{\beta_L} = \frac{k'_D \left(\dfrac{W}{L}\right)_D}{k'_L \left(\dfrac{W}{L}\right)_L} \tag{7.1-2}$$

for either output. The value of V_{OL} is then computed from the self-iterating inverter equation (see eqn. 3.4-22):

$$V_{OL} = (V_{OH} - V_{TD}) - \sqrt{(V_{OH} - V_{TD})^2 - \frac{1}{\beta_R}|V_{TL}(V_{OL})|^2}. \tag{7.1-3}$$

It is important to keep in mind that Q and $\bar{Q}$ are complements, so when one output is V_{OH}, the other is at V_{OL}. Also, the value of V_{OL} computed from eqn. (7.1-3) is valid only in a stable state when the switching transistors MS and MR are off.

DC design is accomplished by using the inverter relation (see eqn. 3.4-43):

$$\beta_R = \frac{|V_{TL}(V_{OL})|^2}{2(V_{OH} - V_{TD})V_{OL} - V_{OL}^2}, \tag{7.1-4}$$

which sets β_R for a desired V_{OL}. Switching MOSFETs MS and MR can be chosen with (W/L) values identical to the inverter drivers $M1$ and $M2$. The best-case (i.e., smallest) V_{OL} level is then computed from eqn. (7.1-3) using the NOR relation

$$\beta_R = \frac{2k'_D \left(\dfrac{W}{L}\right)_D}{k'_L \left(\dfrac{W}{L}\right)_L}. \tag{7.1-5}$$

This is valid only when MOSFET pairs $(M1, MR)$ or $(M2, MS)$ are simultaneously conducting in the nonsaturated mode and occurs only when the circuit is being switched.

The complete design process requires setting values for MOSFET geometries. Since ratioed logic requires $\beta_R > 1$, $(W/L)_D > (W/L)_L$ must be

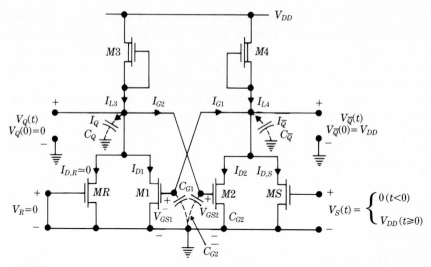

FIGURE 7.3   Setting $Q = 1$ for the *SR* flip-flop transient.

generally satisfied. The basic tradeoff for this type of circuit is between the real estate consumption and the charging time t_{LH}. The latter constraint arises because a small $(W/L)_L$ value limits the load current that is available to charge the output capacitance to V_{OH}.

The transient characteristics of the circuit may be examined by using the situation illustrated in Fig. 7.3. Initially, the output levels are $V_Q = 0$ and $V_{\bar{Q}} = V_{DD}$. Driver $M1$ is conducting in the nonsaturated mode, while $M2$ is in cutoff. At time $t = 0$, the set voltage V_S is pulsed from 0 to V_{DD} in an assumed steplike transition. V_R is kept at ground so that MR remains in cutoff. Turning on MS allows drain current $I_{D,S}$ to flow. This discharges the capacitance $(C_{\bar{Q}} + C_{G1})$ at the node $\bar{Q}$, so $V_{GS,1}$ is reduced. The current $I_{D,1}$ through $M1$ is correspondingly lowered, allowing the capacitance $(C_Q + C_{G2})$ at node Q to charge through the load MOSFET $M3$. Eventually, $V_{GS,1} < V_{TD}$ will be satisfied, placing $M1$ in cutoff. When this occurs, load current I_{L3} will charge $V_Q = V_{GS2}$ to V_{DD}. Since MOSFET $M2$ is strongly conducting, $V_{\bar{Q}} = V_{OL}$ will be attained.

Transient switching is described by setting up the differential equations of current flow. First, note that the capacitance $(C_{\bar{Q}} + C_{G1})$ discharges according to

$$I_{D2} + I_{D,S} - I_{L4} = -(C_{\bar{Q}} + C_{G1})\frac{dV_{\bar{Q}}}{dt}, \tag{7.1-6}$$

where $I_{D,S}$ is a function of V_S and $V_{DS,S} = V_{\bar{Q}}$, while I_{D2} depends on $V_{GS2} = V_Q$ and $V_{DS2} = V_{\bar{Q}}$. The load current I_{L3} is a function of V_Q. Charging $(C_Q + C_{G2})$ proceeds according to

$$I_{L3} - I_{D1} = (C_Q + C_{G2})\frac{dV_Q}{dt}, \tag{7.1-7}$$

where I_{L3} depends on V_Q and I_{D1} is a function of both $V_{DS1} = V_Q$ and $V_{GS1} = V_{\bar Q}$. These constitute a set of coupled differential equations for the voltages V_Q and $V_{\bar Q}$ and must be solved simultaneously subject to the proper initial conditions. The dependence of the MOSFET currents on the terminal voltages makes this an extremely complicated analytic problem.

A simpler approach can be used to obtain a low-order estimate for the switching times for the circuit. Assume that the switching event can be described in two steps. The first step is the discharging of $(C_{\bar Q} + C_{G1})$ from V_{DD} down to V_{TD} through the set MOSFET MS only. Using the analysis in Section 4.1, this requires a time

$$t_{\text{dis},\bar Q} \simeq \tau_D \left\{ \frac{2V_{TD}}{(V_{OH} - V_{TD})} + \ln \left[\frac{2(V_{OH} - V_{TD})}{V_{TD}} - 1 \right] \right\}, \qquad \text{(7.1-8)}$$

where the time constant is

$$\tau_D = \frac{(C_{\bar Q} + C_{G1})}{\beta_D(V_{OH} - V_{TD})}. \qquad \text{(7.1-9)}$$

The second step in the simplified analysis is to compute the time required for I_{L3} to charge $(C_Q + C_{G2})$ from V_{OL} to $V_1 = 0.9V_1$. M_1 is assumed to be in cutoff during this step since now $V_{GS,1} < V_{TD}$. From Section 4.2, this time interval is approximated by

$$t_{ch,Q} \simeq \tau_L \left\{ \frac{2(V_{DD} - |V_{TL}| - V_{OL})}{|V_{TL}|} + \ln \left[\frac{2|V_{TL}| - (V_{DD} - V_1)}{(V_{DD} - V_1)} \right] \right\}, \qquad \text{(7.1-10)}$$

where

$$\tau_L = \frac{(C_Q + C_{G2})}{\beta_L |V_{TL}|} \qquad \text{(7.1-11)}$$

is the load time constant. The total switching time is then approximated by summing these two intervals to obtain

$$t_{LH} \simeq t_{\text{dis},\bar Q} + t_{ch,Q}. \qquad \text{(7.1-12)}$$

Although many features of the switching process have been ignored, this expression illustrates the basic dependence of the transients on the MOSFET aspect ratios $(W/L)_D$ and $(W/L)_L$.

The design equation for $(W/L)_L$ may be obtained using eqn. (7.1-2) with a specified charge time requirement. A simpler approach is to use a reasonable $(W/L)_L$ value and then to check the switching time to ensure that it is acceptable. Capacitances must be estimated for the first calculations. The actual values resulting from the design geometries will be different, so that iterations will be necessary. Although the complete transient design requires the use of a computer simulation, it is possible to model the capacitance contributions for an initial estimate in hand calculations. The simplest

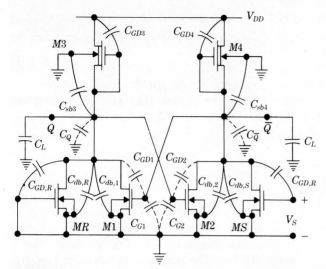

FIGURE 7.4 Capacitance contributions for *SR* flip-flop.

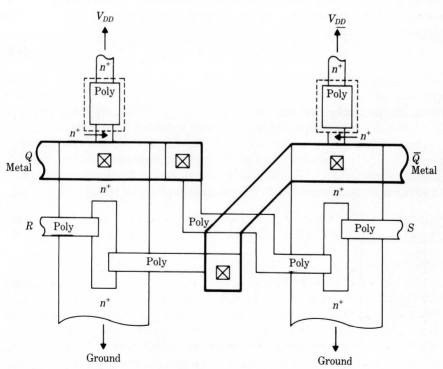

FIGURE 7.5 nMOS *SR* flip-flop layout.

equivalent circuit is shown in Fig. 7.4. This gives the approximate value

$$C_Q \simeq C_{GD,R} + C_{GD3} + K(V_{OH}, V_{OL})(C_{db,R} + C_{db1} + C_{sb3}) + C_L + C_{line}.$$

(7.1-13)

$C_{\bar{Q}}$ has the same contributions in a symmetrical design. It should be noted that this oversimplifies the problem since it ignores the gate-drain coupling capacitors C_{GD1} and C_{GD2} in MOSFETs $M1$ and $M2$, respectively.

A basic nMOS SR layout is shown in Fig. 7.5.

7.2 Clocked Flip-Flops

The inputs S and R to the basic SR flip-flop can be controlled by adding an externally applied clocking signal ϕ to the circuit. The clock itself is assumed to have the general characteristics illustrated in Fig. 7.6. ϕ ideally has a maximum of V_{DD} and a minimum of 0 [V]. The period of the clock is $T = 1/f$, where f is the frequency. The clock is assumed high for a time t_H. The *duty cycle* is defined as $(t_H/T) \times 100\%$ and gives the percentage time that $\phi = V_{DD}$.

The clocking of the SR flip-flop can be achieved by ANDing the inputs with ϕ. The logic diagram for this implementation is shown in Fig. 7.7(a). Inputs S and R are transmitted to the flip-flop circuitry only when $\phi = 1$. This is expressed by writing

$$S' = S\phi, \qquad R' = R\phi \tag{7.2-1}$$

for the inputs to the NOR circuitry. Outputs Q and $\bar{Q}$ become stable after the input voltages have had time to propagate through the circuitry. The block symbol for the clocked SR flip-flop is shown in Fig. 7.7(b). Note that the clock input line has a special notation to distinguish it from a logic input.

The nMOS circuitry for the clocked SR flip-flop is most easily implemented by using the AOI structuring implied by the logic diagram. This results in the schematic shown in Fig. 7.8. The layout of the clocked flip-flop simply requires the addition of clock input MOSFETs to the simpler case in Fig. 7.5.

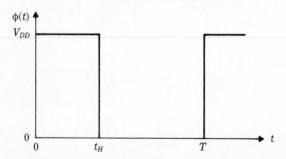

FIGURE 7.6 Ideal clock signal.

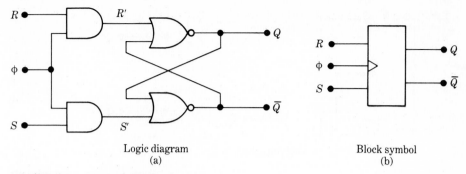

FIGURE 7.7 Clocked SR flip-flop.

A *D-type* flip-flop may be used for temporary bit storage or as a delay element. It is constructed from the basic clocked SR flip-flop by using a single data input D. D and $\bar{D}$ are connected to S and R, respectively, as in the logic diagram of Fig. 7.9(a). Since only a single input and its complement are used, the SR flip-flop problem of simultaneously having $S = 1$ and $R = 1$ does not occur. It is easily seen from the logic diagram that the output Q assumes the value of D at the data input when the clock condition $\phi = 1$ allows the signal to propagate through. Assuming that the D input is established, the output will be valid after the propagation delay time t_p of the circuit. The block logic symbol for the D-type flip-flop is shown in Fig. 7.9(b). Figure 7.9(c) provides the nMOS circuit as a straightforward modification of the simple clocked SR case.

A 2-input clocked *JK flip-flop* eliminates the problem of having an undefined input state of $S = 1 = R$ by utilizing feedback from the outputs.

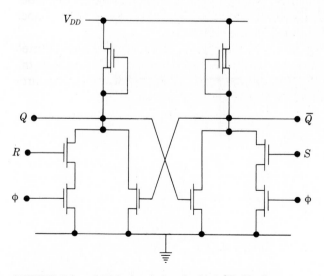

FIGURE 7.8 nMOS circuit for clocked SR flip-flop.

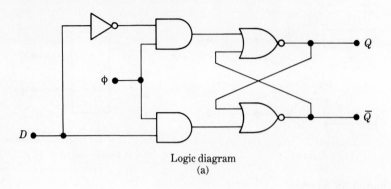

Logic diagram
(a)

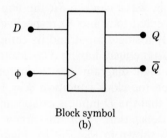

Block symbol
(b)

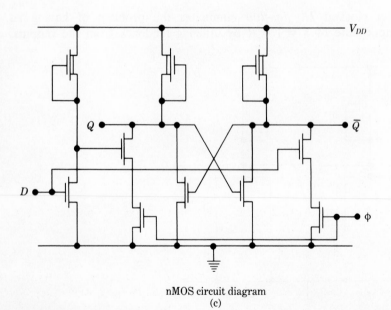

nMOS circuit diagram
(c)

FIGURE 7.9 *D*-type flip-flop.

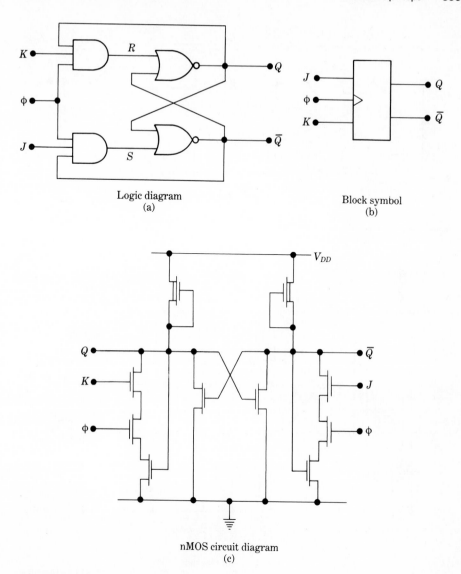

Logic diagram
(a)

Block symbol
(b)

nMOS circuit diagram
(c)

FIGURE 7.10 Basic *JK* flip-flop.

The basic AOI logic connections in Fig. 7.10(a) show that the inputs J and K are related to the internal S and R lines by

$$S = J\bar{Q}\phi, \qquad R = KQ\phi. \tag{7.2-2}$$

J now acts to set the output to $Q = 1$, while K is the reset line. The circuit behaves like an SR flip-flop as long as J and K are not simultaneously in logic 1 states. When $J = 1 = K$, the output Q changes (or *toggles*) states with the clock pulse. This means that if initially $Q = 1$, then $J = 1 = K$ will toggle the

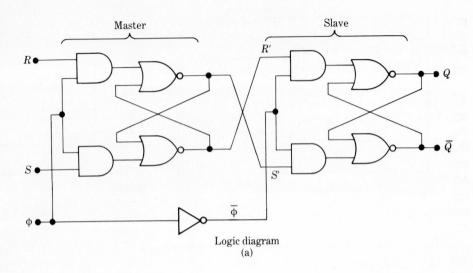

Logic diagram
(a)

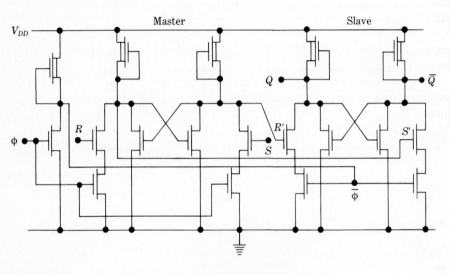

nMOS circuit diagram
(b)

FIGURE 7.11   An *SR* master-slave flip-flop.

output to $Q = 0$ and vice versa. The block symbol for a *JK* flip-flop is shown in Fig. 7.10(b), while the nMOS AOI implementation is given in Fig. 7.10(c).

Although the *JK* flip-flop appears to eliminate the problem associated with having simultaneous logic 1 input states, the feedback loop introduces a problem in the timing. This is seen by noting that the toggling of the output with $J = 1 = K$ continues as long as $\phi = 1$ is applied to the circuit. Oscillations result unless the clock time t_H satisfies $t_H < 2t_p$, where t_p is the propagation time of the circuit. This type of constraint on clock timing can be too restrictive for nMOS chip implementation, particularly since the exact $\phi(t)$ waveform depends on the total capacitive loading. A common approach to solving timing problems is to use a *master-slave* flip-flop.

Master-slave flip-flops are built by cascading two basic flip-flops together with opposite clock phases. The logic diagram for an *SR*-based master-slave arrangement is shown in Fig. 7.11(a), with the corresponding nMOS circuit in Fig. 7.11(b). The input *SR* flip-flop is operational when $\phi = V_{DD}$ and is termed the "master." During this portion of the clocking, inputs *S* and *R* permit data entry. When ϕ goes to 0, the master turns off and the "slave" circuit becomes active. Note that the master and slave portions are decoupled owing to the opposite clocking scheme. This allows the internal data lines *S'* and *R'* to be evaluated for final output as Q and $\bar{Q}$ without reference to the current input status at *S* and *R*.

The case where both inputs *S* and *R* are at stable logic 1 values is automatically accounted for since the master outputs give $S' = 0$ and $R' = 0$, indicating that the slave remains unchanged. The exception to this is when initially $S = 0 = R$ and both are switched toward logic 1 levels when $\phi = V_{DD}$. This creates a *race* condition between the two inputs. A race can lead to an indeterminate state entering the slave if only one input has reached the gate threshold voltage when ϕ goes to zero.

The *JK* master-slave flip-flop is a very useful circuit. It is constructed by cascading two basic *SR* flip-flop stages and then providing overall feedback from the slave outputs to the master inputs. The logic diagram is illustrated in Fig. 7.12(a), and the corresponding AOI nMOS circuit is drawn in Fig. 7.12(b). Decoupling the master and slave circuits by using ϕ and $\bar{\phi}$ allows for

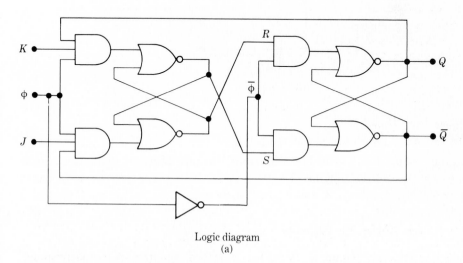

Logic diagram
(a)

FIGURE 7.12 *JK* master-slave flip-flop.

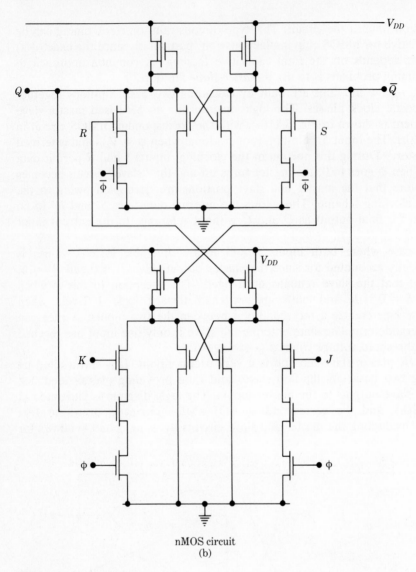

nMOS circuit
(b)

Figure 7.12   (*contd.*)

toggling when $J = 1 = K$ but eliminates the possibility of oscillations since only one stage is operational at a time.

7.3   The Schmitt Trigger

A Schmitt trigger circuit is useful for waveform shaping, particularly under startup or noisy conditions. The VTC of an ideal Schmitt trigger is illustrated

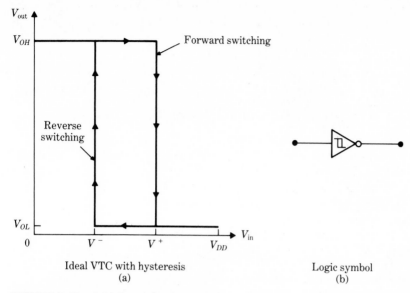

Ideal VTC with hysteresis
(a)

Logic symbol
(b)

FIGURE 7.13   Schmitt trigger inverter.

in Fig. 7.13(a), where it is seen that the curve exhibits a hysteresis loop. The circuit is characterized by two trigger voltages V^+ and V^-. The forward trigger voltage V^+ is important when the input is increased from $0\,[\text{V}]$ to V_{DD}. The reverse trigger voltage V^- describes the output switching as V_{in} is reduced from V_{DD} to $0\,[\text{V}]$. Since $V^- < V^+$, the Schmitt trigger acts as a threshold switch. The logic symbol for an inverter with a Schmitt trigger input is shown in Fig. 7.13(b).

The wave-shaping properties of a Schmitt trigger are illustrated by the example in Fig. 7.14. The first plot shows $V_{in}(t)$ with reference to the trigger voltages V^+ and V^-. During the time $t < t_1$, V_{in} is increasing, so the forward characteristics are important. Since $V_{in} < V^+$ in this time interval, the output is maintained at V_{OH}. At $t = t_1$, V_{in} crosses V^+, which drives the output to V_{OL}. V_{out} stays low until V_{in} decreases to V^- at time t_2. This transition is described by the reverse characteristics on the VTC, so that V_{out} increases to V_{OH}.

Figure 7.15 shows an nMOS implementation for a Schmitt trigger circuit. The forward switching properties may be understood by referring to the redrawn circuit in Fig. 7.16(a). Suppose that initially $V_{in} = 0$ and $V_{out} = V_{OH} = V_{DD}$. This places both driver MOSFETs $M1$ and $M2$ in cutoff. MOSFET $M3$ provides voltage feedback to the input circuit. It may be viewed as the load transistor in the inverter circuit formed by $M1$ and $M3$; V_I is the inverter output voltage. When $V_{in} = 0$, the gate of $M3$ is at $V_{OH} = V_{DD}$. Since

$$V_{GS3} \simeq V_{DD} - V_I$$

$$= V_{DS3}, \qquad\qquad\qquad\qquad\qquad\qquad\qquad \textbf{(7.3-1)}$$

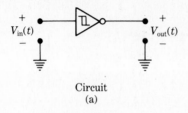

Circuit
(a)

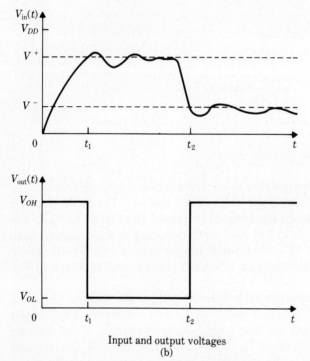

Input and output voltages
(b)

FIGURE 7.14   Waveform-shaping properties of a Schmitt trigger.

$M3$ acts like a saturated enhancement-mode load. The inverter output high voltage at this point is determined by (see eqn. 3.3-32)

$$V_{I,\max} = V_{DD} - V_{T3}, \tag{7.3-2}$$

where

$$V_{T3} = V_{T0} + \gamma(\sqrt{V_I + 2|\phi_F|} - \sqrt{2|\phi_F|}) \tag{7.3-3}$$

is the load threshold voltage. The value of $V_{I,\max}$ can be found by iterations.

The forward switching behavior is studied by increasing V_{in} as indicated by the horizontal portion of the Schmitt curve in Fig. 7.16(b). V_I decreases according to the saturated E-mode VTC shown in Fig. 7.16(c) until the point

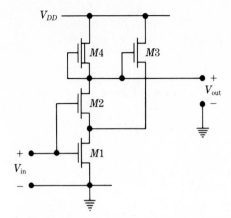

FIGURE 7.15   nMOS Schmitt trigger inverter.

where $M2$ turns on. This requires a gate-source voltage of

$$V_{GS2} = V_{T2}$$
$$= V_{T0} + \gamma(\sqrt{V_I + 2|\phi_F|} - \sqrt{2|\phi_F|})$$
$$= V_{T3}, \qquad (7.3\text{-}4)$$

corresponding to an input voltage of

$$V_{in} = V_{T2} + V_I. \qquad (7.3\text{-}5)$$

$M2$ turns on at this point, allowing I_{D2} to flow. V_{out} starts to fall to V_{OL} because

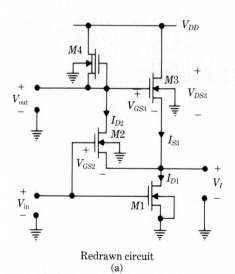

Redrawn circuit
(a)

FIGURE 7.16   Forward Schmitt trigger characteristics.

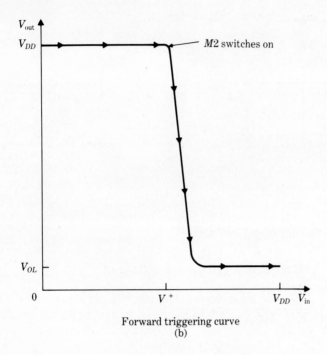

Forward triggering curve
(b)

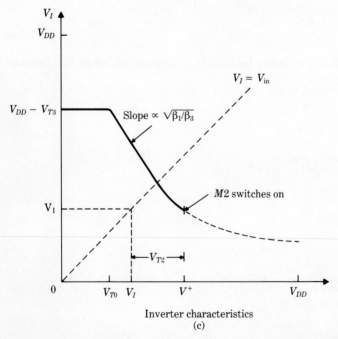

Inverter characteristics
(c)

Figure 7.16 (*contd.*)

of the current path through $M1$ and $M2$. Thus,

$$V^+ = V_{T2} + V_I \tag{7.3-6}$$

defines the forward trigger voltage V^+.

V^+ may be computed by evaluating the transistor currents. Consider first MOSFET $M1$. Noting $V_{in} = V_{GS1}$ and $V_I = V_{DS1}$, eqn. (7.3-5) gives

$$V_{DS1} = V_{GS1} - V_{T2}. \tag{7.3-7}$$

$M1$ does not experience body bias effects, so $V_{T1} = V_{T0}$. Since $V_{T2} > V_{T0}$, $M1$ is operating in the nonsaturated mode with

$$I_{D1} = \frac{\beta_1}{2}[2(V^+ - V_{T1})V_I - V_I^2]. \tag{7.3-8}$$

The load MOSFET $M3$ is saturated with

$$I_{S3} = \frac{\beta_3}{2}(V_{DD} - V_I - V_{T3})^2. \tag{7.3-9}$$

V^+ can be computed by setting $I_{D1} = I_{S3}$ and using eqn. (7.3-6) to relate V^+ to V_I. The complexity of the equations makes a numerical solution the easiest approach. A straightforward procedure is to iterate the current equation to find V_I and then use this value to compute V_{T2} and V^+.

A simpler analytic approximation can be obtained by ignoring body bias effects and assuming that $V_{T1} = V_{T2} = V_{T3} = V_{T0}$. For this case, eqn. (7.3-7) reduces to

$$V_{DS1} \simeq V_{GS1} - V_{T0}, \tag{7.3-10}$$

which places $M1$ on the edge of saturation with

$$I_{D1} \simeq \frac{\beta_1}{2}(V^+ - V_{T0})^2. \tag{7.3-11}$$

Using eqn. (7.3-6) gives I_{S3} in the form

$$I_{S3} = \frac{\beta_3}{2}(V_{DD} - V^+)^2, \tag{7.3-12}$$

so setting $I_{D1} = I_{S3}$ yields

$$V^+ \simeq \frac{V_{DD} + \sqrt{\dfrac{\beta_1}{\beta_3}}\,V_{T0}}{1 + \sqrt{\dfrac{\beta_1}{\beta_3}}} \tag{7.3-13}$$

as an approximation for the forward trigger voltage. Since body bias effects have been ignored, this will tend to underestimate V^+.

The important design parameter for V^+ is seen to be the driver-load ratio

$$\frac{\beta_1}{\beta_3} = \frac{(W/L)_1}{(W/L)_3}.$$

(7.3-14)

The design equation is obtained by rearranging eqn. (7.3-13) to read

$$\frac{(W/L)_1}{(W/L)_3} \simeq \left(\frac{V_{DD} - V^+}{V^+ - V_{T0}}\right)^2.$$

(7.3-15)

V^+ may be increased by decreasing (β_1/β_3). This result is understood by noting that V_I acts as a feedback voltage to keep $M2$ biased off. Decreasing (β_1/β_3) has the effect of making the $(M1, M3)$ inverter less sensitive to V_{in}, which in turn implies that V^+ will be larger.

Once $M2$ is biased on, the Schmitt output V_{out} rapidly drops to V_{OL} because of the current path established through $M1$ and $M2$. $M3$ is biased into cutoff, so the operational part of the circuit is a 2-input NAND. V_{OL} is found from the self-iterating equation

$$V_{OL} \simeq 2(V_{OH} - V_{T0}) - \left\{ \sqrt{(V_{OH} - V_{T0})^2 - \frac{1}{\beta_{R1}}|V_{T4}(V_{OL})|^2} \right.$$
$$\left. + \sqrt{(V_{OH} - V_{T0})^2 - \frac{1}{\beta_{R2}}|V_{T4}(V_{OL})|^2} \right\},$$

(7.3-16)

with the NAND driver-load ratios

$$\beta_{R1} = \frac{\beta_1}{\beta_4}, \qquad \beta_{R2} = \frac{\beta_2}{\beta_4}$$

(7.3-17)

as derived in Section 6.2. The NAND drivers in the Schmitt trigger should be designed with

$$(W/L)_1 > (W/L)_2$$

(7.3-18)

since MOSFET $M1$ must sink both I_{S3} and I_{D2} when $M2$ is switched on at $V_{\text{in}} = V^+$.

The reverse trigger voltage V^- is analyzed by decreasing V_{in} from V_{DD} toward 0 [V]. Since V_{out} is initially at V_{OL}, MOSFET $M3$ is off and the circuit behaves like a simple 2-input NAND gate. The reverse triggering is structured by the NAND VTC, as shown in Fig. 7.17. A reasonable approximation for V^- is

$$V^- \simeq V_{IH},$$

(7.3-19)

where V_{IH} is the input high voltage. The value of V_{IH} is set by β_{R1} and β_{R2}. Increasing these ratios lowers V_{IH}.

The overall design constraints revolve around setting the MOSFET aspect ratios. The subtle interplay that occurs is worth examining. First note that V^+ is established by (β_1/β_3) through eqn. (7.3-13). To make a dual-threshold

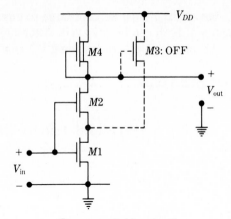

Reverse switching circuit
(a)

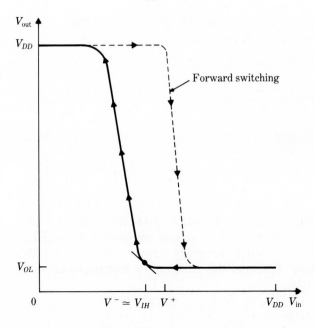

NAND VTC for reverse switching
(b)

FIGURE 7.17   Reverse Schmitt trigger characteristics.

Schmitt trigger,

$$V^+ > V^-$$ (7.3-20)

must be satisfied. V^- is designed by the NAND ratios (β_1/β_4) and (β_2/β_4) such that increasing these quantities lowers V^-. The interdependence arises from the fact that the design choice for β_1 affects both V^+ and V^-. (β_1/β_3) should be small to maintain a sufficient V^+, but (β_1/β_4) must be kept large to ensure that $V^- < V^+$. Simultaneous satisfaction of both can be achieved. Alternately, it is possible to make a single-threshold Schmitt trigger by designing $V^+ \simeq V^-$.

EXAMPLE 7.3-1

An nMOS process has $V_{T0} = +1$ [V], and $V_{DD} = 5$ [V] is used. Find the value of $(W/L)_1/(W/L)_3$ to set $V^+ = 4$ [V].

Solution
Using eqn. (7.3-15),

$$\frac{(W/L)_1}{(W/L)_3} = \left(\frac{5-4}{4-1}\right)^2 = \frac{1}{9},$$

so $(W/L)_3 = 9(W/L)_1$ is required. Note that this gives a driver-load (β_1/β_3) ratio less than unity in the saturated E-mode subcircuit of the forward trigger voltage analysis.   ■

EXAMPLE 7.3-2

The circuit designed in Example 7.3-1 was simulated on SPICE. The device aspect ratios were chosen to be

$$(W/L)_1 = (15/5), \quad (W/L)_2 = (40/5), \quad (W/L)_3 = (135/5),$$
$$(W/L)_4 = 0.5.$$

(The $(W/L)_2$ can be reduced, as suggested by eqn. 7.3-18. It lowers V^- here.)

The SPICE input file is as follows, and the results are plotted in Fig. E7.1. The internal algorithms used by SPICE generally do not converge in a DC analysis of a feedback circuit. To avoid this problem, a transient analysis is performed by pulsing the input.

```
SCHMITT TRIGGER ANALYSIS
VDD 3 0 DC 5VOLTS
M1 4 1 0 0 EMODE L=5U W=15U
M2 2 1 4 0 EMODE L=5U W=40U
M3 3 2 4 0 EMODE L=5U W=135U
M4 3 2 2 0 DMODE L=10U W=5U
VS 1 0 DC PULSE 0 5 2NS 2NS 2NS 3NS)
CLOAD 2 0 3.0E-14
.MODEL EMODE NMOS VTO=1 GAMMA=0.37 KP=2.5E-5
.MODEL DMODE NMOS VTO=-3 GAMMA=0.37 KP=2.5E-5
.TRAN 1NS 15NS
.PLOT TRAN V(2)
.END
```

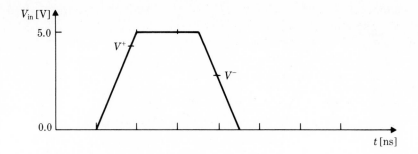

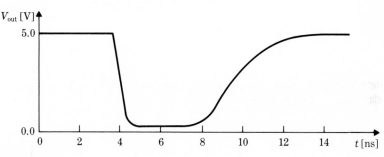

FIGURE E7.1

The SPICE simulation includes body bias. It is seen that the results give $V^+ \simeq 4.3\,[\text{V}]$. The higher value is expected. ∎

EXAMPLE 7.3-3

To find V^- for the Schmitt trigger, the $M1$-$M2$-$M4$ subcircuit is used (a 2-input NAND gate). Then V^- is approximately V_{IH} for this circuit. Using the parameters from the above example, the VTC was plotted (Fig. E7.2) using the DC SPICE analysis that follows. It is seen that $V^- \simeq 2.7\,[\text{V}]$.

```
SCHMITT TRIGGER V- INVERTER SUBCIRCUIT
VDD 3 0 DC 5VOLTS
M1 4 1 0 0 EMODE L=5U W=15U
M2 2 1 4 0 EMODE L=5U W=40U
M4 3 2 2 0 DMODE L=10U W=5U
VS 1 0 DC PULSE(0 5 2NS 2NS 2NS 30NS)
.MODEL EMODE NMOS VTO=1 GAMMA=0.37 KP=2.5E-5
.MODEL DMODE NMOS VTO=-3 GAMMA=0.37 KP=2.5E-5
.DC VS 0 5 0.1
.PLOT DC V(2)
.END
```

∎

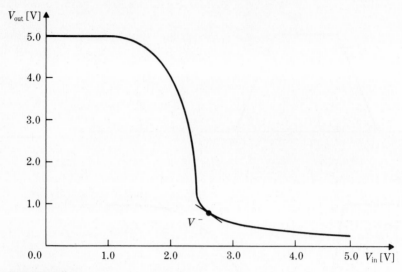

FIGURE E7.2

7.4 CMOS *SR* Flip-Flops

Static CMOS flip-flops can be constructed using the same techniques as for nMOS circuits. A CMOS *SR* flip-flop based on NOR gates is shown in Fig. 7.18. Standard CMOS structuring is used with complementary MOSFET pairs.

The operation of the circuit is similar to the nMOS equivalent, except that the pMOS transistors at the *S* and *R* inputs aid in switching. For example,

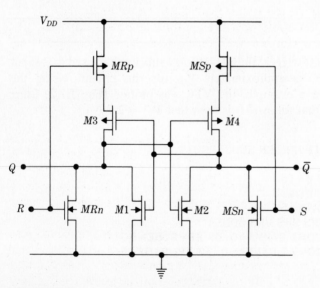

FIGURE 7.18 CMOS NOR *SR* flip-flop.

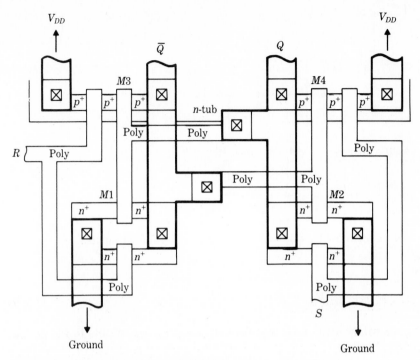

FIGURE 7.19   CMOS NOR *SR* flip-flop.

consider the case where the flip-flop is set by pulsing S from $0\,[\mathrm{V}]$ to V_{DD}. The input nMOS transistor MSn turns on, providing a path to ground for the output node $\bar{Q}$. MSp is simultaneously switched off, which decouples $\bar{Q}$ from the power supply. The capacitance at $\bar{Q}$ thus has a direct discharge path to ground for switching of the flip-flop state.

The most efficient layout of the CMOS NOR *SR* flip-flop uses minimum-sized MOSFETs, as discussed in Section 6.5.3. In this scheme, all transistors are designed with the same aspect ratio of $(W/L)_{\min}$. One possible layout for the circuit is shown in Fig. 7.19.

The real estate requirement can be reduced by using a nonstandard CMOS arrangement in which only nMOS transistors form the set and reset inputs. A NOR *SR* flip-flop circuit of this type is shown in Fig. 7.20. Comparing this with Fig. 7.18 shows that two pMOS transistors (MSp and MRp) have been eliminated. The price paid for this savings in real estate is increased power consumption and longer switching times.

A comparison of the two circuits can be made by using a simplified step pulse for $V_S(t)$ as shown in Fig. 7.21(a). It is assumed that $V_Q(0) = 0\,[\mathrm{V}]$ and $V_{\bar{Q}}(0) = V_{DD}$. Consider first the standard CMOS circuit of Fig. 7.18. Denoting the total capacitance at node $\bar{Q}$ by C_T gives the equivalent switching circuit shown in Fig. 7.21(b). Since transistor MSp is off, $I_p = 0$ and C_T discharges

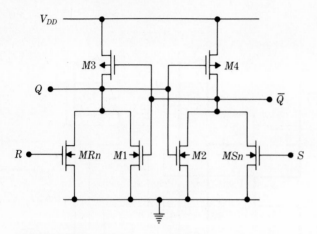

FIGURE 7.20   CMOS *SR* flip-flop with nonstandard set and reset inputs.

according to

$$I_{D,s}(V_Q, V_{\bar{Q}}) + I_{D2}(V_Q, V_{\bar{Q}}) = -C_T \frac{dV_{\bar{Q}}}{dt}, \tag{7.4-1}$$

where I_{D2} is initially zero. Note that the currents are proportional to the device aspect ratios (W/L).

The nonstandard flip-flop of Fig. 7.20 behaves in a slightly different manner. Since the pMOS device *MSp* has been removed, node $\bar{Q}$ maintains a conduction path to V_{DD} when V_S is switched to a logic 1 state. The equivalent discharge circuit is thus given by Fig. 7.21(c), with

$$I_{D,s}(V_Q, V_{\bar{Q}}) + I_{D2}(V_Q, V_{\bar{Q}}) - I_p(V_Q, V_{\bar{Q}}) = -C_T \frac{dV_{\bar{Q}}}{dt}. \tag{7.4-2}$$

If the MOSFET aspect ratios are identical for the two circuits, then the C_T values will be about equal. Equation (7.4-2) thus shows that the discharge current is reduced by an amount $I_p(V_Q, V_{\bar{Q}})$ when only nMOS switch transistors are used. As a result, both the switching time and the power dissipation are increased. These tradeoffs may be acceptable depending on the application.

The use of nonstandard switching can be extended further by adding input lines that perform logic using nMOS transistors. An example is given in Fig. 7.22, which sets $Q = 1$ according to the value of

$$S = A(B + C) \tag{7.4-3}$$

while a reset is achieved by

$$R = DF. \tag{7.4-4}$$

Proper care must be taken to ensure that S and R do not simultaneously evaluate to logic 1 states.

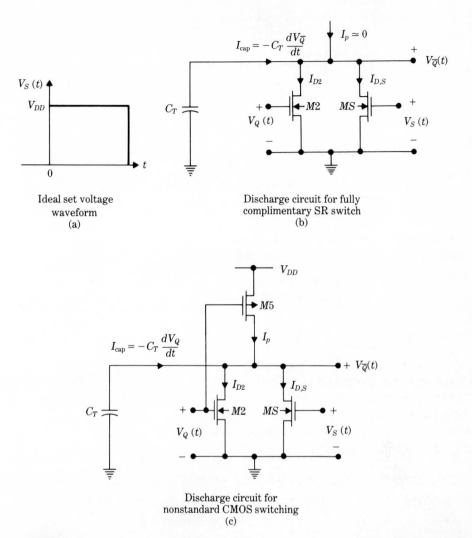

FIGURE 7.21 part (a): Ideal set voltage waveform (a)

FIGURE 7.21 part (b): Discharge circuit for fully complimentary SR switch (b)

FIGURE 7.21 part (c): Discharge circuit for nonstandard CMOS switching (c)

FIGURE 7.21 Comparison of discharging circuit for standard and nonstandard CMOS flip-flop switching.

CMOS allows for construction of an efficient NAND-based *SR* flip-flop using the logic diagram of Fig. 7.23(a). The designation of the inputs as the complements $\bar{S}$ and $\bar{R}$ is standard and is used because the resulting function table is the same as for the NOR design. Since CMOS circuits often generate signal complements, these inputs may be available without additional circuitry. Figure 7.23(b) shows the circuit diagram for the flip-flop. This may be constructed using minimum-sized MOSFETs or a symmetrical design can be achieved with equal β-values where $(W/L)_p \simeq 2.5(W/L)_n$. The main advantage of a symmetric design is equal rise and fall times at the outputs.

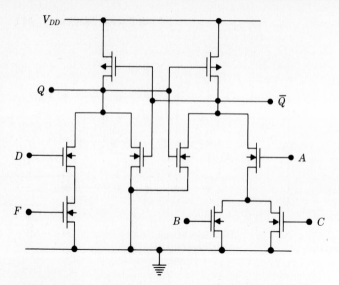

FIGURE 7.22 CMOS *SR* flip-flop with nMOS logic inputs.

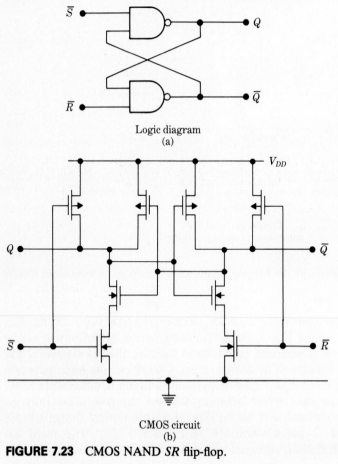

Logic diagram
(a)

CMOS circuit
(b)

FIGURE 7.23 CMOS NAND *SR* flip-flop.

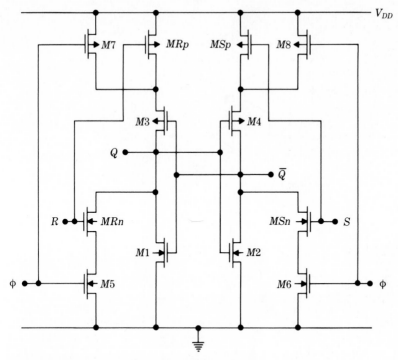

FIGURE 7.24   Clocked CMOS *SR* flip-flop.

7.5   Clocked CMOS Flip-Flops

CMOS flip-flops may be clocked using the same techniques introduced for nMOS circuits. An example is the *SR* flip-flop shown in Fig. 7.24. This is a direct CMOS implementation of the AOI logic diagram presented earlier in Fig. 7.7(a) in Section 7.2. Standard structuring has been used to provide for a complementary nMOS/pMOS pair of transistors at every input. The design and analysis of the circuit is straightforward and employs the general CMOS techniques developed in Chapter 6.

Transmission gates may be used to build CMOS flip-flops that are generally simpler and require fewer transistors than circuits designed with standard structuring. The TG-based equivalent of the *SR* flip-flop is shown in Fig. 7.25. The logic diagram in Fig. 7.25(a) shows cross-coupled inverters that are connected to the *S* and *R* inputs through TG*S* and TG*R*, respectively. The clocking signal controls the transmission gates and allows state changes when $\phi = V_{DD}$. During this time, *S* is directly connected to the *Q* output, while *R* is connected to $\bar{Q}$.

Figure 7.25(b) provides the CMOS circuit for the *TG*-based flip-flop. MOSFET pairs (*M*1, *M*3) and (*M*2, *M*4) are used for the flip-flop inverters. The (*W/L*) values for these devices determine the DC switching points and also establish the current flow levels for the transient characteristics.

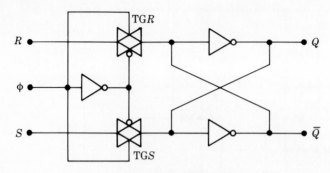

Logic diagram
(a)

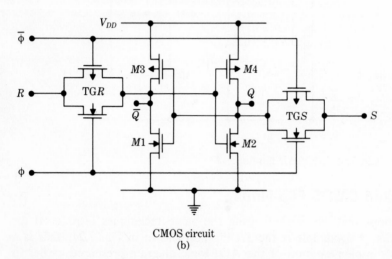

CMOS circuit
(b)

FIGURE 7.25 Clocked transmission gate input cross-coupled inverter flip-flop.

The overall design considerations for the circuit can be understood by examining a switching event. Figure 7.26(a) shows the charging circuit for setting the flip-flop with $V_S = V_{DD}$ when $\phi = 1$. Initial conditions at time $t = 0$ of $V_Q(0) = 0\,[\text{V}]$ and $V_{\bar{Q}}(0) = V_{DD}$ are assumed.

The S input voltage $V_S = V_{DD}$ introduces a transmission gate current of I_{TG}. When the current flow starts, $M4$ is in cutoff since $V_{SG4} = (V_{DD} - V_{\bar{Q}}) = 0\,[\text{V}]$. KCL thus gives the initial relation

$$I_{TG} = I_{D2} + I_Q + I_G, \tag{7.5-1}$$

which is valid as long as $(V_{DD} - V_{\bar{Q}}) < |V_{Tp}|$. I_{D2} is the non-saturated drain current

$$I_{D2} = \frac{\beta_2}{2}[2(V_{\bar{Q}} - V_{Tn})V_Q - V_Q^2] \tag{7.5-2}$$

through $M2$ that flows as soon as V_Q increases above $0\,[\text{V}]$. I_Q represents the

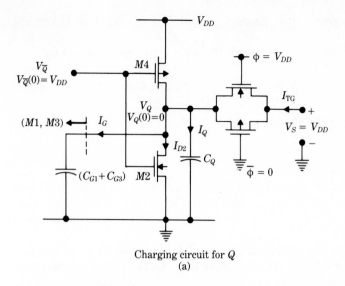

Charging circuit for Q
(a)

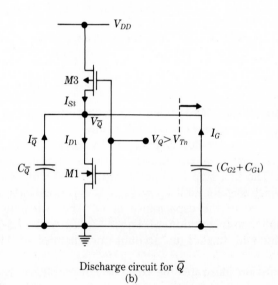

Discharge circuit for $\overline{Q}$
(b)

FIGURE 7.26   Switching for transmission gate SR flip-flop.

capacitor current

$$I_Q = C_Q \frac{dV_Q}{dt} \qquad (7.5\text{-}3)$$

into the equivalent capacitance C_Q at node Q, while I_G is the gate current

$$I_G = (C_{G1} + C_{G3}) \frac{dV_Q}{dt} \qquad (7.5\text{-}4)$$

used to charge the gate capacitances C_{G1} and C_{G3} of $M1$ and $M3$ in the opposite inverter pair.

The above contributions sum to give an initial charging described by

$$I_{TG} = \frac{\beta_2}{2}[2(V_{\bar{Q}} - V_{Tn})V_Q - V_Q^2] + (C_Q + C_{G1} + C_{G3})\frac{dV_Q}{dt}, \qquad (7.5\text{-}5)$$

where I_{TG} is approximated as

$$I_{TG} \simeq \frac{V_{DD} - V_Q}{R_{eq}}. \qquad (7.5\text{-}6)$$

R_{eq} is set by the (W/L) values for the transmission gate MOSFETs as described in Section 6.7. Note that pMOS transistor $M4$ will turn on when $(V_{DD} - V_{\bar{Q}}) = |V_{Tp}|$ is satisfied. This will provide additional current to charge the capacitance at Q.

The discharging of the complement node $\bar{Q}$ is treated in the same manner. Initially, $V_Q < V_{T1}$, so $M1$ is off. Consequently, no currents are flowing, so $V_{\bar{Q}} = V_{DD}$ is maintained. When V_Q reaches V_{T1}, $M1$ starts conducting in the saturated mode. Using KCL on the equivalent circuit in Fig. 7.26(b) gives

$$\frac{\beta_1}{2}(V_Q - V_{Tn})^2 = \frac{\beta_3}{2}[2(V_{DD} - V_Q - |V_{Tp}|)(V_{DD} - V_{\bar{Q}}) - (V_{DD} - V_{\bar{Q}})^2]$$

$$- (C_{\bar{Q}} + C_{G2} + C_{G4})\frac{dV_{\bar{Q}}}{dt} \quad (V_Q \geq V_{Tn}), \quad (7.5\text{-}7)$$

where it has been assumed that $M3$ is operating in the nonsaturated region. This is valid until $M1$ or $M3$ changes operational modes.

The initial switching of the flip-flop is described by using eqns. (7.5-5) and (7.5-7). Owing to the complexity of the equations, the solution technique will only be outlined here. The first step is to solve eqn. (7.5-5) with $V_Q = V_{DD}$. This will give $V_Q(t)$, which is valid until $V_Q = V_{Tn}$. At this point, $M1$ switches on, initiating the discharge of the capacitance at $\bar{Q}$. The switching is then described by the simultaneous solution of eqns. (7.5-5) and (7.5-7). The equations must be modified further to account for changes in MOSFET operational modes.

The preceding discussion illustrates the basic design problems involved in this circuit. For example, the aspect ratios of the transmission gates affect the switching times through R_{eq}. The values of (W/L) for these transistors must be large enough to provide the required current levels without introducing excessive capacitance.

The most involved design tradeoffs arise in choosing the aspect ratios for the inverter MOSFETs. A symmetric design is desirable to give equal output responses at Q and $\bar{Q}$. This requires that the nMOS transistors $M1$ and $M2$ be identical with $(W/L)_1 = (W/L)_2$ and also that the p-channel MOSFETs $M3$ and $M4$ have $(W/L)_3 = (W/L)_4$. However, this design choice does not minimize either t_{HL} or t_{LH}.

The tradeoffs can be seen by a qualitative interpretation of eqns. (7.5-5) and (7.5-7). First note that eqn. (7.5-5) indicates that a small value for $(W/L)_2$ of $M2$ would minimize t_{LH}, since more transmission gate current would be available for charging the capacitors. However, eqn. (7.5-7) shows that a large value of $(W/L)_1$ for $M1$ will allow the rapid discharge of output capacitance, as

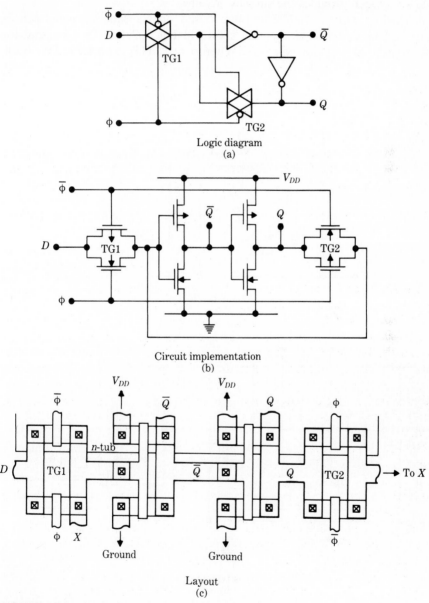

Logic diagram
(a)

Circuit implementation
(b)

Layout
(c)

FIGURE 7.27 CMOS transmission gate D-type flip-flop.

it provides increased I_{D1} levels. Clearly, setting $(W/L)_1 = (W/L)_2$ for a symmetric design will not result in minimum switching times. This is generally a reasonable tradeoff since symmetric outputs simplify the overall system timing. A simple design choice is to set $\beta_n = \beta_p$ for inverter thresholds of $V_{th} = (V_{DD}/2)$. $(W/L)_n$ can then be chosen according to the system timing requirements. Alternately, the same (W/L) value can be used for each MOSFET, which provides the simplest layout.

A clocked D-type CMOS flip-flop based on transmission gates can be constructed using the logic diagram shown in Fig. 7.27(a). Two transmission gates are used to gate the logic flow. When $\phi = 1$, TG1 is on while TG2 is off.

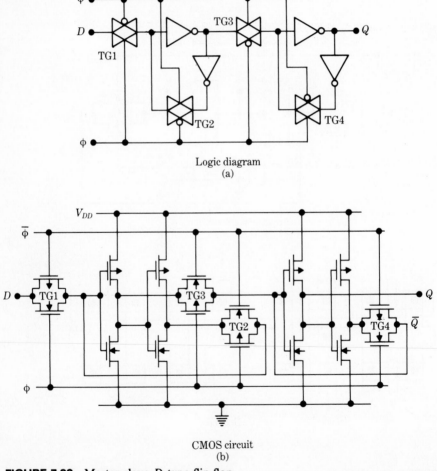

Logic diagram
(a)

CMOS circuit
(b)

FIGURE 7.28   Master-slave D-type flip-flop.

This allows data to enter the flip-flop from the D input such that $Q = D$. During the clock period when $\phi = 0$, TG1 is off and TG2 is on. TG2 is used to close the feedback path between the two inverters, which holds the flip-flop state established when $\phi = 1$. Figure 7.27(b) gives the circuit diagram, while Fig. 7.27(c) shows a representative layout. The simplest electrical design is obtained using $\beta_n = \beta_p$ for the inverter MOSFETs to set $V_{th} = (V_{DD}/2)$. The easiest layout employs identical (W/L) values for all transistors in the circuit.

A D-type master-slave flip-flop is built by extending the TG technique in a straightforward manner. The logic diagram in Fig. 7.28(a) shows two cascaded D-type flip-flops that are driven by opposite clock phases. During the time

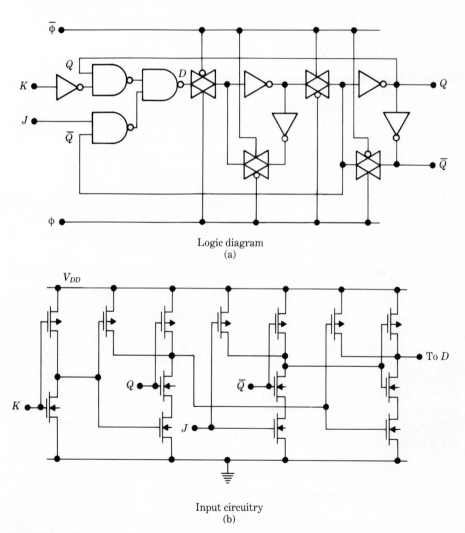

Logic diagram
(a)

Input circuitry
(b)

FIGURE 7.29 JK master-slave flip-flop.

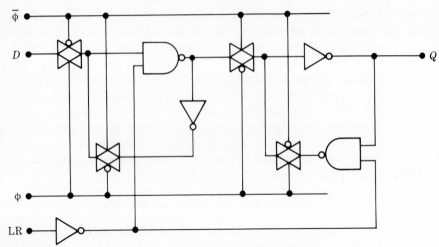

FIGURE 7.30 D-type master-slave flip-flop with clear (CLR).

when $\phi = 1$, input data is transmitted through TG1 of the master. When the clock falls to $\phi = 0$, TG2 establishes regenerative feedback in the master section to hold the state. This is transmitted to the slave flip-flop through TG3 during the same time interval. The next clocking period decouples the two stages and also turns on TG4 to provide feedback in the slave. The CMOS circuit realization of this flip-flop is shown in Fig. 7.28(b).

A JK master-slave flip-flop may be obtained from the D-type circuit by adding a small amount of input logic and then providing overall feedback from the slave to the master. Figure 7.29(a) shows an example of this type of construction. The inputs are now J and K such that

$$D = J\bar{Q} + \bar{K}Q. \tag{7.5-8}$$

The input logic is obtained using three NAND gates and an inverter. Consequently, the JK flip-flop circuit requires only the addition of the input circuitry drawn in Fig. 7.29(b).

As a final comment, it should be mentioned that extra control inputs that initially preset or clear (reset) the state of the flip-flop may be necessary in certain system applications. The TG-based D flip-flop can be easily modified for this capability by replacing the inverters with 2-input gates that accept the extra input. An example of a flip-flop with a clear input CLR is shown in Fig. 7.30. An input of CLR = 1 establishes the state $Q = 0$, while CLR = 0 allows the circuit to function normally.

7.6   CMOS Schmitt Trigger

A CMOS Schmitt trigger can be constructed by using the nMOS circuit in Fig. 7.16 of Section 7.3 as a guide. Since the forward triggering in the nMOS circuit is provided by MOSFETs $M1$, $M2$, and $M3$, the CMOS version of the Schmitt trigger is obtained by adding the pMOS complements of these transistors in place of the load device. The complete CMOS circuit is shown in Fig. 7.31(a). The complementary structuring makes it possible to implement a design where the forward trigger voltage V^+ and the reverse trigger voltage V^- satisfy

$$V^+ = \frac{1}{2} V_{DD} + \Delta V, \qquad V^- = \frac{1}{2} V_{DD} - \Delta V. \tag{7.6-1}$$

The voltage increment ΔV is determined by the device ratios. This type of VTC is shown in Fig. 7.31(b).

The forward switching characteristics are identical to those discussed for the nMOS circuit. Thus, V^+ can be approximated by (see eqn. 7.3-13)

$$V^+ \simeq \frac{V_{DD} + \sqrt{\frac{\beta_1}{\beta_3}} V_{T0n}}{1 + \sqrt{\frac{\beta_1}{\beta_3}}} \tag{7.6-2}$$

The important device design equation is then (see eqn. 7.3-15)

$$\frac{(W/L)_1}{(W/L)_3} \simeq \left(\frac{V_{DD} - V^+}{V^+ - V_{T0n}} \right)^2. \tag{7.6-3}$$

The reverse trigger voltage V^- is controlled by the pMOS transistors $M4$, $M5$, and $M6$. When V_{in} is decreased from V_{DD}, $M5$ is off while $M4$ and $M6$ act like a pMOS inverter. Denoting the output voltage of this inverter by V_{Ip} as shown in the schematic, reverse triggering takes place when the input voltage is reduced to a level

$$V^- = V_{Ip} - |V_{TS}| \tag{7.6-4}$$

since $M5$ will start conducting at this point.

To approximate V^-, assume that body bias can be ignored so that

$$V_{T4} = V_{T5} = V_{T6} = V_{T0p}. \tag{7.6-5}$$

Then $M4$ is conducting in the saturated mode with

$$I_{D4} = \frac{\beta_4}{2} (V_{DD} - V^- - |V_{T0p}|)^2 \tag{7.6-6}$$

while the feedback transistor $M6$ is also saturated such that

$$I_{S6} = \frac{\beta_6}{2} (V_{Ip} - |V_{T0p}|)^2$$

$$= \frac{\beta_6}{2} (V^-)^2. \tag{7.6-7}$$

Equation (7.6-4) was used in the last step. Equating currents $I_{D4} = I_{S6}$ then

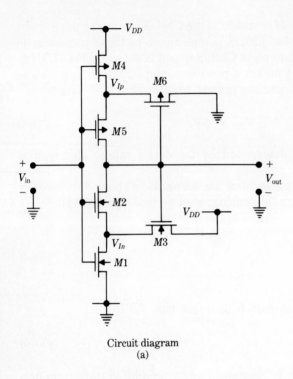

Circuit diagram
(a)

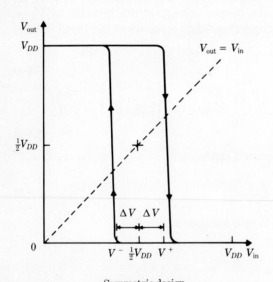

Symmetric design
transfer characteristics
(b)

FIGURE 7.31 A CMOS Schmitt trigger.

gives

$$V^- \simeq \frac{\sqrt{\frac{\beta_4}{\beta_6}}(V_{DD} - |V_{T0p}|)}{1 + \sqrt{\frac{\beta_4}{\beta_6}}}. \qquad (7.6\text{-}8)$$

Rearranging,

$$\frac{\beta_4}{\beta_6} = \frac{(W/L)_4}{(W/L)_6} \simeq \left(\frac{V^-}{V_{DD} - V^- - |V_{T0p}|}\right)^2, \qquad (7.6\text{-}9)$$

which is the reverse trigger voltage design equation. It is seen that V^- is set by the ratio (β_4/β_6).

To relate V^+ to V^-, assume that $V_{T0n} = |V_{T0p}| = V_{T0}$. Setting

$$\frac{\beta_1}{\beta_3} = \frac{\beta_4}{\beta_6} \equiv \beta_R \qquad (7.6\text{-}10)$$

for a symmetric nMOS/pMOS design and using eqn. (7.6-1) in conjunction with eqns. (7.6-2) and (7.6-8) yields

$$\Delta V \simeq \frac{V_{DD}(1 - \sqrt{\beta_R}) + 2\sqrt{\beta_R}\, V_{T0}}{2(1 + \sqrt{\beta_R})} \qquad (7.6\text{-}11)$$

as the voltage difference in the ideal symmetric VTC. Note that $\Delta V > 0$ must be satisfied for a realistic design. A design equation is obtained by writing this in the form

$$\sqrt{\beta_R} \simeq \frac{V_{DD} - 2(\Delta V)}{V_{DD} + 2(\Delta V) - 2V_{T0}}. \qquad (7.6\text{-}12)$$

This may be used to calculate the value of β_R for a specified trigger voltage separation. The CMOS Schmitt trigger discussed here has the nice property that the circuit can be designed to yield a symmetric hysteresis loop.

EXAMPLE 7.6-1

Design a symmetric CMOS Schmitt trigger with $\Delta V = 1.0\,[\text{V}]$. Assume that $V_{DD} = 5\,[\text{V}]$ and $V_{T0n} = 1\,[\text{V}] = |V_{T0p}|$.

Solution
The specification requires $V^\pm = \frac{1}{2}(5) \pm 1 = 3.5\,[\text{V}]$, $1.5\,[\text{V}]$. Using eqn. (7.6-3) gives the nMOS ratio

$$\frac{(W/L)_1}{(W/L)_3} = \left(\frac{5 - 3.5}{3.5 - 1}\right)^2 = 0.36.$$

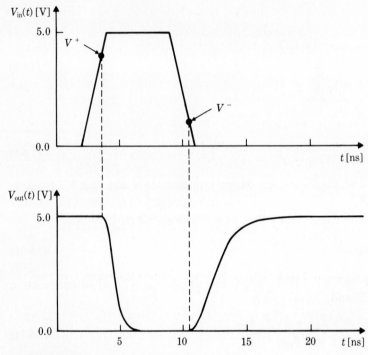

FIGURE E7.3

Similarly, eqn. (7.6-9) can be used to calculate the pMOS ratio

$$\frac{(W/L)_4}{(W/L)_6} = \left(\frac{1.5}{5 - 1.5 - 1}\right)^2 = 0.36,$$

as expected for a symmetric design. These can be verified using eqn. (7.6-12).

The SPICE simulation for the circuit is shown below and is plotted in Fig. E7.3. Integer rounding has been used in specifying the device ratios.

```
CMOS SCHMITT TRIGGER
VDD 3 0 DC 5VOLTS
M1 4 1 0 0 MN L=5U W=5U
M2 2 1 4 0 MN L=5U W=12.5U
M3 3 2 4 0 MN L=5U W=15U
M4 5 1 3 3 MP L=5U W=5U
M5 2 1 5 3 MP L=5U W=12.5U
M6 0 2 5 3 MP L=5U W=15U
VS 1 0 DC PULSE(0 5 2NS 2NS 2NS 5NS)
CLOAD 2 0 3.0E-14
.MODEL MN NMOS VTO=1 GAMMA=0.37 KP=2.5E-5
.MODEL MP PMOS VTO=-1 GAMMA=0.4 KP=1.0E-5
.TRAN 1NS 20NS
.PLOT TRAN V(2)
.END
```

REFERENCES

(1) M. Annaratone, *Digital CMOS Circuit Design,* Hingham, MA: Kluwer Academic Publishers, 1986.

(2) D. A. Hodges and H. G. Jackson, *Analysis and Design of Digital Integrated Circuits,* New York: McGraw-Hill, 1983.

(3) M. J. Howes and D. V. Morgan, *Large Scale Integration,* Bath, UK: Wiley, 1981.

(4) L. A. Glasser and D. W. Dobberpuhl, *The Design and Analysis of VLSI Circuits,* Reading, MA: Addison-Wesley, 1985.

(5) E. J. McCluskey, *Logic Design Principles,* Englewood Cliffs, NJ: Prentice-Hall, 1986.

PROBLEMS

Use the nMOS and CMOS process parameters listed at the beginning of the problem set in Chapter 6 unless otherwise specified.

7.1 An nMOS NOR-based *SR* flip-flop is designed using a driver-load ratio of $\beta_R = 6$. The outputs Q and $\bar{Q}$ are equally loaded by $C_Q = C_{\bar{Q}} = 240\,[\text{fF}]$.
   (a) Calculate the value of V_{OL} for the circuit that would be measured in a stable state.
   (b) The driver MOSFETs are designed with $(W/L)_D = 3$ with $L = 3\,[\mu\text{m}]$. The gate capacitance per unit area is given by $C_{ox} = 0.0691\,[\text{fF}/\mu\text{m}^2]$. Find the values of τ_L and τ_D for the circuit.
   (c) Use the approximations leading to eqn. (7.1-12) to estimate t_{LH} for the circuit.

7.2 An *SR* flip-flop is designed in nMOS using NAND gates.
   (a) Give the basic logic diagram that will implement the flip-flop. Provide the truth table for the circuit. (Use $\bar{S}$ and $\bar{R}$ as controls.)
   (b) Design the NAND gate to produce a value of $V_{OL} = 0.25\,[\text{V}]$. Then choose $(W/L)_L = 1$ and calculate the dimensions of the drivers if the minimum linewidth is specified to be $3\,[\mu\text{m}]$.
   (c) Draw stick diagrams for both the NOR-based and NAND-based flip-flops.

7.3 An nMOS NOR-based *SR* flip-flop design is specified to have an output low voltage of $V_{OL} = 0.2\,[\text{V}]$. The minimum linewidth is $3\,[\mu\text{m}]$, and the load transistors are set with $(W/L)_L = 3/6$.
   (a) Find the value of β_R needed to implement the design. Then round this to the nearest integer value that still has $V_{OL} \leq 0.2\,[\text{V}]$, and find $(W/L)_D$.
   (b) Assume that $C_Q = 150\,[\text{fF}]$, $C_{\bar{Q}} = 95\,[\text{fF}]$, and $C_{ox} = 0.0691\,[\text{fF}/\mu\text{m}^2]$. Estimate t_{LH} for the circuit.

*(c) Perform a SPICE simulation on the circuit using lumped-equivalent capacitances. Examine the effects of different load capacitances by simulating the transient response for both a set and a reset operation.

7.4 Design a clocked *SR* flip-flop using the circuit diagram shown in Fig. 7.8. Assume that V_{OL} is specified to have a maximum value of $V_{OL} = 0.4$ [V] and that the minimum linewidth is 3 [μm].

7.5 Extend the clocked *SR* flip-flop design in Problem 7.4 to create a *D*-type flip-flop.

7.6 This problem deals with the design on an nMOS Schmitt trigger based on the circuit shown in Fig. 7.16.
   (a) Find the value of (β_1/β_3) needed to set $V^+ = 2.75$ [V]. Then set $(W/L)_1 = 1$ and compute $(W/L)_3$.
   (b) Set $(W/L)_4 = 0.5$ for the depletion load transistor and assume that $(W/L)_1 = 2(W/L)_2$. Calculate the value of V_{OL} for the circuit.
   *(c) Run a SPICE transient simulation on the circuit and obtain values for V^+, V^-, and V_{OL}. Compare the results with the original design specification.

7.7 It is possible to obtain a set of Schmitt trigger design equations that include body bias effects. Consider the analysis of Section 7.3.
   (a) Show that the inverter voltage V_I satisfies the self-iterating equation

$$V_I = V^+ - V_{T2}(V_I)$$
$$= (V^+ - V_{T0}) - \gamma(\sqrt{V_I + 2|\phi_F|} - \sqrt{2|\phi_F|}).$$

   (b) Next show that

$$\frac{\beta_3}{\beta_1} = \frac{V_I[2(V^+ - V_{T0}) - V_I]}{(V_{DD} - V^+)^2}$$

   gives the ratio needed for designing the circuit to trigger at V^+.
   (c) A Schmitt trigger is built in nMOS with the specification of $V^+ = 3.0$ [V]. Iterate the equation in part (a) to find the value of V_I for this design. Then determine the ratio (β_3/β_1) for the circuit. Compare this value with that calculated from eqn. (7.3-13).
   (d) Plot (β_3/β_1) as a function of V^+ subject to the constraint in (a).
   *(e) Complete the circuit design choices, and simulate the circuit using SPICE.

7.8 Consider the basic *SR* flip-flop in a CMOS technology.
   (a) Design a NAND-based CMOS flip-flop using standard structuring.
   (b) Design a nonstandard flip-flop as shown in Fig. 7.20 that produces an output low voltage of $V_{OL} = 0.5$ [V] at Q and $\bar{Q}$. Compare the real estate requirements for the two. Then calculate the power dissipation for the nonstandard design when the inputs are stable.

7.9   The clocked TG-input flip-flop shown in Fig. 7.25 has three inputs: R, S, and ϕ. Discuss the design requirements on TGR, TGS, and the ϕ-input inverter circuit needed to equalize the propagation time through the system to the cross-coupled inverters.

7.10   Design a CMOS Schmitt trigger that has symmetric switching with $\Delta V = 1.0\,[V]$. *Simulate the triggering using SPICE.

7.11   Design a CMOS Schmitt trigger that has $V^- = 1.0\,[V]$ and $V^+ = 3.75\,[V]$. *Simulate the circuit using a SPICE transient analysis.

7.12   Derive the CMOS design equations that are equivalent to those found in Problem 7.7 for the nMOS circuit. Then use this set (which includes body bias) to redesign the circuit in Problem 7.11.

CHAPTER 8

Synchronous nMOS Logic

Designing a large digital network is complicated by the timing (propagation delays, etc.) and steering logic. Adding a clock waveform to the system controlling scheme allows for synchronization of events and data transfer.

Introducing clocks to control signal paths also allows for different circuit techniques to be developed. The *dynamic* circuits in this chapter are commonly used as the basis for complicated chip designs that require a structured timing for accurate control. New problems also arise, giving a different outlook on circuit design.

8.1 nMOS Pass Transistors

The most basic gating unit in nMOS logic is an *n*-channel MOSFET connected as a *pass transistor*. An example of this circuit is shown in Fig. 8.1(a), where the input to a depletion load inverter is controlled by the pass transistor *MP*. A

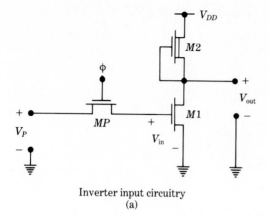

Inverter input circuitry
(a)

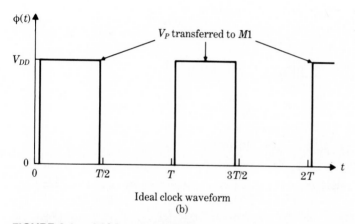

Ideal clock waveform
(b)

FIGURE 8.1 nMOS pass transistor.

clock signal ϕ is connected to the gate of *MP*. Data transfer can take place only when *MP* is in an active conduction mode with $\phi = 1$, as indicated in the clock waveform of Fig. 8.1(b). A control state of $\phi = 0$ puts *MP* in cutoff, so that the nodes V_P and V_{in} become isolated. Transistor *MP* is the nMOS equivalent of the CMOS transmission gate introduced in Section 7.7. The terms *pass gate* and *transmission gate* will be used interchangeably for both the nMOS and CMOS circuits.

The operational characteristics of the pass gate arise from the concept of charge transfer by flowing currents. Figure 8.2 illustrates how logic 1 and logic 0 voltage states are transferred to the inverter input through a pass transistor. In both cases, $\phi = 1$ establishes current flow through *MP*.

In Fig. 8.2(a), the pass gate input voltage is set to an ideal logic 1 level with $V_P = V_{DD}$. Assuming that the inverter input voltage V_{in} is initially $0\,[V]$, the current I charges the input capacitance C_{in}. The logic 1 state is thus transferred to the inverter input, which results in an output voltage of

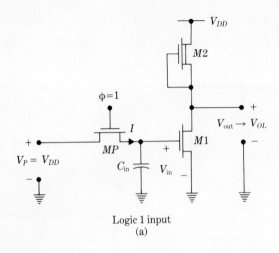

Logic 1 input
(a)

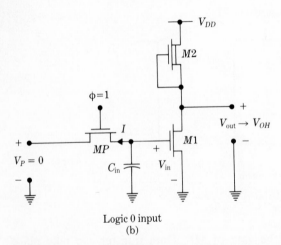

Logic 0 input
(b)

FIGURE 8.2   Charge transfer with nMOS pass transistor.

$V_{out} = V_{OL}$. Figure 8.2(b) shows the opposite case, where the pass gate input is grounded with $V_P = 0$ [V] while V_{in} is initially high. The input capacitance C_{in} discharges corresponding to a logic 0 input. This gives an inverter output voltage transition of $V_{out} \rightarrow V_{OH}$.

   This simple example shows the basic concepts involved in charge transfer of logic levels. When ϕ goes to 0 in the clock cycle, the input capacitance becomes isolated from V_P. The charge state of C_{in} corresponding to the value of V_{in} is stored until the next clock cycle. A node that holds a logic level by charge storage is sometimes referred to as a *soft node*. The existence of soft nodes gives rise to *dynamic logic* circuits, which are based on charge transfer and storage.

   nMOS pass transistors are widely used for controlling logic paths in synchronous logic circuits. The remaining portions of this section are devoted to a detailed study of their electrical characteristics.

8.1.1 Charging Analysis: Logic 1 Transfer

The first property to be examined is charge transfer by a MOSFET of a logic 1 voltage $V_P = V_{DD}$, as depicted in Fig. 8.2(a).

Consider the charging of C_{in} through pass transistor MP. The equivalent circuit for this situation is shown in Fig. 8.3(a), where an initial condition of $V_{in}(0) = 0\,[V]$ has been assumed. The pass transistor input is $V_P = V_{DD}$ corresponding to a logic 1 level. The gate of MP is at $\phi = V_{DD}$, which allows current I_{cap} to flow.

To analyze the current flow, first note that

$$V_{GS,P} = V_{DD} - V_{in}$$

$$= V_{DS,P}, \tag{8.1-1}$$

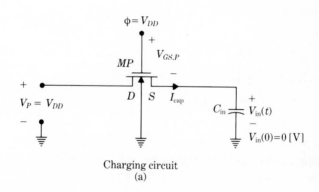

Charging circuit
(a)

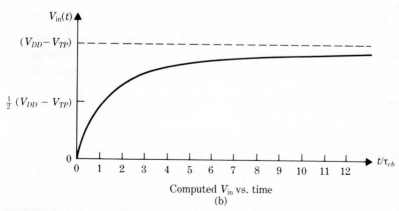

Computed V_{in} vs. time
(b)

FIGURE 8.3 Pass transistor charging characteristics.

so MP is always saturated. Since $I_{cap} = I_{D,P}$,

$$C_{in}\frac{dV_{in}}{dt} = \frac{\beta_P}{2}(V_{DD} - V_{in} - V_{TP})^2 \qquad (8.1\text{-}2)$$

is the differential equation for the capacitor voltage $V_{in}(t)$. V_{TP} is the threshold voltage of MP, while

$$\beta_P = (\mu_n C_{ox})\left(\frac{W}{L}\right)_P \qquad (8.1\text{-}3)$$

is the device transconductance parameter. Ignoring body bias effects for the moment, the solution to this equation with the initial condition $V_{in}(0) = 0\,[\text{V}]$ is

$$V_{in}(t) = (V_{DD} - V_{TP}) - \frac{(V_{DD} - V_{TP})}{\left[1 + \dfrac{\beta_P t}{2C_{in}}(V_{DD} - V_{TP})\right]}, \qquad (8.1\text{-}4)$$

as can be verified by direct substitution. It is convenient to introduce the charging time constant

$$\tau_{ch} \equiv \frac{2C_{in}}{\beta_P(V_{DD} - V_{TP})}. \qquad (8.1\text{-}5)$$

Then the solution may be written in the simplified form

$$V_{in}(t) = (V_{DD} - V_{TP})\left[\frac{(t/\tau_{ch})}{1 + (t/\tau_{ch})}\right], \qquad (8.1\text{-}6)$$

Figure 8.3(b) shows $V_{in}(t)$ as predicted by this expression.

The plot shows that the maximum voltage transferred through the pass transistor is

$$V_{in}(t \to \infty) = (V_{DD} - V_{TP})$$

$$= V_{max}. \qquad (8.1\text{-}7)$$

This illustrates the presence of a *threshold voltage loss* across the MOSFET as discussed in Section 3.3.1. Physically, the result may be understood by noting that V_{in} cannot exceed $(V_{DD} - V_{TP})$ because this is the largest source voltage that still allows for the channel to be formed.

Body bias is important to the overall operational characteristics since

$$V_{SB,P} = V_{in}. \qquad (8.1\text{-}9)$$

The complete expression for the threshold voltage is

$$V_{TP}(V_{in}) = V_{TOP} + \gamma(\sqrt{2|\phi_F| + V_{in}} - \sqrt{2|\phi_F|}), \qquad (8.1\text{-}10)$$

where V_{TOP} is the zero body bias threshold voltage of the pass transistor. (It is important that nMOS *pass transistor* quantities not be confused with pMOS values. Lowercase p subscripts will always be used to denote p-channel MOSFET parameters.) The value of $V_{\max}$ representing the highest voltage for V_{in} is computed by the self-iterating equation (see eqn. 3.3-32)

$$V_{\max} = V_{DD} - V_{TP}(V_{\max})$$
$$= (V_{DD} - V_{T0P}) - \gamma(\sqrt{2|\phi_F| + V_{\max}} - \sqrt{2|\phi_F|}). \qquad \textbf{(8.1-11)}$$

The result of this calculation may be used for $(V_{DD} - V_{TP})$ in eqn. (8.1-6) as an estimate that includes some body bias effects.

The presence of a threshold voltage drop across MP introduces some important circuit design considerations. First a cascaded chain of pass transistors will transmit a value of $V_{\max} = (V_{DD} - V_{TP})$ regardless of the number of MOSFETs in the chain. An example of this is shown in Fig. 8.4(a), where it is assumed that all transistors have the same V_{T0P} value. The first-pass MOSFET $MP1$ introduces a threshold voltage drop to give $V_{\max} = (V_{DD} - V_{TP})$. Maintaining a control voltage of $\phi = V_{DD}$ on devices $MP2$, $MP3$, and $MP4$ keeps these transistors in conduction. The analysis that led to eqn. (8.1-6) is valid for every transistor in the chain. As a result, the voltage $V_{\max}$ can propagate through the entire chain without further threshold voltage losses.

Next consider a pass transistor that is itself driven by a pass transistor. A 2-MOSFET network of this type is shown in Fig. 8.4(b). The first device $MP1$ has $V_P = V_{DD} = \phi$, which results in $V_{1,\max} = (V_{DD} - V_{TP1})$ being applied to the gate of $MP2$. The input to the second-pass transistor $MP2$ is set at V_{DD}. It is seen from the circuit that

$$V_{GS,P2} = V_{1,\max} - V_2, \qquad \textbf{(8.1-12)}$$

so that charging C_2 through the saturated MOSFET $MP2$ is described by

$$C_2\frac{dV_2}{dt} = \frac{\beta_{P2}}{2}(V_{1,\max} - V_2 - V_{TP2})^2. \qquad \textbf{(8.1-13)}$$

Comparing this with eqn. (8.1-2) shows that $V_2(t)$ has the form

$$V_2(t) = (V_{1,\max} - V_{TP2})\left[\frac{(t/\tau_{ch})}{1 + (t/\tau_{ch})}\right]. \qquad \textbf{(8.1-14)}$$

The maximum value for V_2 is thus

$$V_{2,\max} = V_{1,\max} - V_{TP2}, \qquad \textbf{(8.1-15)}$$

indicating another threshold voltage drop of V_{TP2} from $MP2$.

The results of this analysis must be used with care. A simple approximation for V_2 that ignores body bias is obtained by setting $V_{TP1} = V_{TP2} = V_{TP}$, so $V_{1,\max} = (V_{DD} - V_{TP})$. Equation (8.1-15) is then approximately

$$V_{2,\max} \simeq V_{DD} - 2V_{TP}. \qquad \textbf{(8.1-16)}$$

This gives the rule of thumb "one V_{TP} drop per MOSFET" when pass

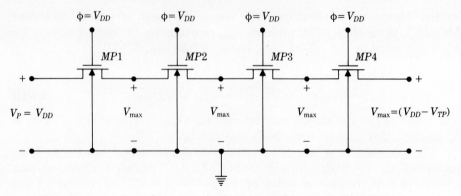

Cascaded pass transistor chain
(a)

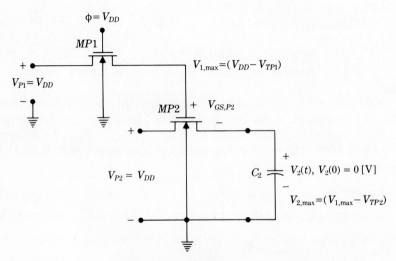

Pass transistor driving
another pass transistor
(b)

FIGURE 8.4 Pass transistor arrangements.

transistors drive other pass transistors. However, this oversimplifies the problem since the body bias levels on $MP1$ and $MP2$ are different. To see this, note that

$$V_{SB,P1} = V_{1,\max}$$

$$= V_{DD} - V_{TP1}(V_{1,\max}) \tag{8.1-17}$$

is the body bias on $MP1$ while

$$V_{SB,P2} = V_{2,\max}$$

$$= V_{1,\max} - V_{TP2}(V_{2,\max}) \tag{8.1-18}$$

must be used for $MP2$. To compute $V_{2,\max}$, $V_{1,\max}$ is first found using the self-iterating equation

$$V_{1,\max} = (V_{DD} - V_{T0P}) - \gamma(\sqrt{V_{1,\max} + 2\,|\phi_F|} - \sqrt{2\,|\phi_F|}).\qquad(8.1\text{-}19)$$

Once $V_{1,\max}$ is calculated, it can be substituted into

$$V_{2,\max} = (V_{1,\max} - V_{T0P}) - \gamma(\sqrt{V_{2,\max} + 2\,|\phi_F|} - \sqrt{2\,|\phi_F|})\qquad(8.1\text{-}20)$$

to solve for $V_{2,\max}$.

The existence of the threshold voltage drop when transferring a logic 1 state through a pass transistor represents a significant loss of voltage amplitude. Multiple nMOS threshold voltages may be designed into the fabrication process flow to compensate for this in circuit design. An example was discussed in Section 6.5. A "normal" E-mode V_{T0} is obtained by using an acceptor implant. A *natural* E-mode MOSFET has a value of V_{T0} slightly greater than 0 [V] and does not use an ion implant. This is also referred to as a *zero-threshold* transistor. (The ability to achieve this type of MOSFET depends on the values of x_{ox}, N_a, and the oxide charge densities.) It is seen that a natural E-mode transistor is well suited for use as a pass transistor since the magnitude of the threshold voltage loss will be reduced. Alternately, low threshold voltage MOSFETs may be fabricated by adding an extra implant step.

The switching time t_{LH} required for the pass transistor to charge up to a logic 1 output can be computed by rearranging eqn. (8.1-6) to read

$$t = \tau_{ch}\left[\frac{1}{1 - \dfrac{V_{in}(t)}{(V_{DD} - V_{TP})}} - 1\right].\qquad(8.1\text{-}21)$$

Defining $V_1 = 0.9V_{\max}$ as the final value of V_{in} for the rise time interval gives

$$t_{LH} = 9\tau_{ch}$$
$$= 9\left(\frac{2C_{in}}{\beta_P(V_{DD} - V_{TP})}\right).\qquad(8.1\text{-}22)$$

t_{LH} is proportional to C_{in} and inversely proportional to the pass transistor aspect ratio $(W/L)_P$. It should be noted that the rate of capacitor charging (dV_{in}/dt) decreases as time evolves. This occurs because the gate-source bias $V_{GS,P}$ in eqn. (8.1-1) decreases as V_{in} increases, reducing the current flow level.

8.1.2 Discharge Analysis: Logic 0 Transfer

Now consider the transfer of a logic 0 voltage level through a pass transistor as shown in Fig. 8.2(b). This involves discharging the capacitance C_{in} through MP as described with the equivalent circuit of Fig. 8.5(a). It is assumed that at time $t = 0$, $V_{in}(0) = (V_{DD} - V_{TP})$, corresponding to an initial logic 1 state.

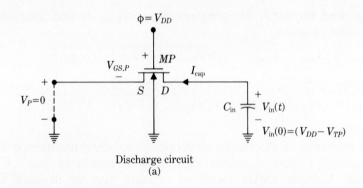

Discharge circuit
(a)

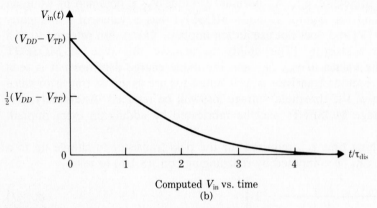

Computed V_{in} vs. time
(b)

FIGURE 8.5   Pass transistor discharge characteristics.

To analyze the circuit, note that

$$V_{GS,P} = V_{DD} - V_P$$

$$= V_{DD} \tag{8.1-23}$$

while

$$V_{DS,P} = V_{\text{in}}(t). \tag{8.1-24}$$

Since $V_{GS,P}$ is at the highest voltage in the system, $V_{GS,P} > V_{DS,P}$ is always true, indicating that MP remains nonsaturated. The discharge is then described by the differential equation

$$-C_{\text{in}} \frac{dV_{\text{in}}}{dt} = \frac{\beta_P}{2} [2(V_{DD} - V_{TP})V_{\text{in}} - V_{\text{in}}^2], \tag{8.1-25}$$

which is constructed by simply equating currents. Solving for $V_{\text{in}}(t)$ gives

$$V_{\text{in}}(t) = (V_{DD} - V_{TP}) \left(\frac{2e^{-t/\tau_{\text{dis}}}}{1 + e^{-t/\tau_{\text{dis}}}} \right), \tag{8.1-26}$$

where the discharge time constant is

$$\tau_{\text{dis}} = \frac{C_{\text{in}}}{\beta_P(V_{DD} - V_{TP})}.$$ (8.1-27)

$V_{\text{in}}(t)$ as predicted by this analysis is sketched in Fig. 8.5(b). Note that the charging and discharging time constants are related by

$$\tau_{\text{dis}} \simeq \frac{1}{2}\tau_{ch}.$$ (8.1-28)

This is only approximately true because there are no body bias effects present in the discharge circuit; in other words, $V_{SB,P} = 0$ gives $V_{TP} = V_{TOP}$.

The fall time t_{HL} may be computed by first solving eqn. (8.1-26) for t in the form

$$t = \tau_{\text{dis}} \ln \left(\frac{2(V_{DD} - V_{TP}) - V_{\text{in}}(t)}{V_{\text{in}}(t)} \right).$$ (8.1-29)

Defining the fall time t_{HL} as the interval required for V_{in} to reach a value $V_0 = 0.1(V_{DD} - V_{TP})$ gives

$$t_{HL} = \ln (19)\tau_{\text{dis}}$$

$$\simeq 2.94 \, \tau_{\text{dis}}.$$ (8.1-30)

Comparing this with the rise time t_{LH} in eqn. (8.1-22) using eqn. (8.1-28) yields

$$t_{LH} \simeq \frac{18}{\ln (19)} t_{HL}$$

$$\simeq 6.11 t_{HL},$$ (8.1-31)

which shows that C_{in} discharges much faster than it charges. This correlates with the circuit description, since the gate-source voltage $V_{GS,P} = V_{DD}$ remains constant during the discharge event. Note that the analysis gives the clock timing constraint

$$\left(\frac{T}{2} \right)_{\text{min}} > t_{LH} = 9\tau_{ch}$$ (8.1-32)

for the minimum clock period T needed to transfer a logic state through the pass transistor. Realistic clock frequencies are much smaller, since the propagation delays through the logic circuitry must be accounted for.

As a final comment, the transient switching times can be reduced by increasing the pass transistor aspect ratio $(W/L)_P$. In practice, however, minimum-size MOSFETs having $(W/L)_{\text{min}}$ are commonly used to reduce the real estate requirements. The amount of tradeoff between switching speed and chip area consumption depends on the value of C_{in}.

8.1.3   Charge Leakage from Soft Nodes

Dynamic logic circuits depend on the ability to store charge on circuit capacitances. Logic states are defined according to the voltage V associated with stored charge. The total charge Q stored on a capacitor C is

$$Q = CV \text{ [C]}. \tag{8.1-33}$$

A value of $Q = 0$ corresponds to a well-defined logic 0 charge state. Logic 1 charge states are determined by the voltage levels in the circuit. For example, the maximum charge stored at the output of a pass transistor is $C_{in}(V_{DD} - V_{TP})$, which then defines the largest logic 1 voltage on the node.

Storing an nMOS logic 0 charge state on a soft node is generally straightforward and is accomplished by using a pass transistor to isolate the node from voltage sources. This maintains a value of $Q = 0$ on the capacitor.

Logic 1 charge states are more difficult to control due to dynamic charge leakage off the soft node capacitance. Chip structuring results in unwanted conduction paths to ground. Although current flows are restricted to leakage levels, the loss of stored charge from a soft node can be significant. The degradation can reduce the charge to the point where it might be incorrectly interpreted as a logic 0 state. Pass transistors inherently possess leakage paths that must be considered in a circuit design.

Consider the basic circuit of Fig. 8.1, where a pass transistor drives a depletion load inverter, and assume that $\phi \to 0$ after the input capacitance C_{in} charges to a value of

$$V_{in} = V_{max} = (V_{DD} - V_{TP}). \tag{8.1-34}$$

The equivalent circuit for the charge leakage calculation is shown in Fig. 8.6, where the time origin has been chosen so that eqn. (8.1-34) is valid at $t = 0$.

The main paths for leakage current flow are through the pass transistor MP. Figure 8.7 provides a perspective view of a MOSFET structure to aid in

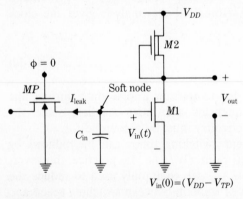

FIGURE 8.6   Circuit level model for soft node leakage.

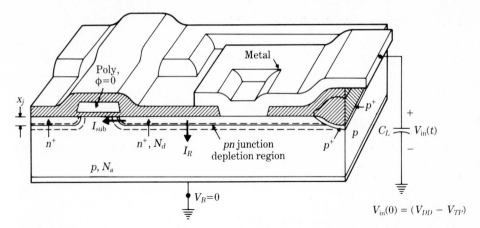

FIGURE 8.7 Chip perspective drawing of pass transistor *MP* for charge leakage calculation.

identifying the current components. The load capacitance C_L is given by

$$C_L = C_{GS1} + C_{line}, \qquad (8.1\text{-}35)$$

where C_{GS1} represents the input capacitance of the driver MOSFET $M1$, and C_{line} is the parasitic line contribution. Since both of these originate from the MOS layering, C_L can be approximated as a constant.

The total leakage current I_{leak} consists of two main components such that

$$I_{leak} = I_{sub} + I_R. \qquad (8.1\text{-}36)$$

I_{sub} is the subthreshold channel current that flows even though $V_{GS,P} < V_{TP}$. As discussed in Chapter 2, I_{sub} increases with decreasing channel length L, and is a strong function of the layout and fabrication technology. I_{sub} is also exponentially dependent on $(V_{GS,P} - V_{TP})$. Since it is assumed that MP is controlled by $\phi = 0\,[V]$, this exponential is small in the present problem. I_{sub} will thus be neglected in the following discussion. While this greatly simplifies the calculation, a realistic model should include I_{sub}, particularly in a small-geometry process.

I_R is the current across the reverse-biased pn junction formed by the n^+-well in the p-type substrate and is assumed to be the dominant contribution to I_{leak}. In general, current flow across a pn junction with an applied voltage V is described by

$$I = I_o(e^{qV/kT} - 1) + I_{dep}, \qquad (8.1\text{-}37)$$

where $V > 0$ for forward bias. (The properties of pn junctions are summarized in the Appendix.) The first term is the ideal Shockley current. For a step-profile junction with doping densities N_d and N_a, the reverse saturation current I_o assumes the form

$$I_o = qAn_i^2 \left[\frac{D_n}{L_n N_a} + \frac{D_p}{L_p N_d} \right], \tag{8.1-38}$$

where A is the cross-sectional diode area in the direction of current flow. D_n and D_p are the electron and hole diffusion coefficients (in the p and n sides, respectively), while

$$L_n = \sqrt{D_n \tau_n}, \qquad L_p = \sqrt{D_p \tau_p} \tag{8.1-39}$$

give the carrier diffusion lengths; τ_n and τ_p are the electron and hole lifetimes. The second term, I_{dep}, in eqn. (8.1-37) represents the currents due to trap states in the depletion region. The explicit form of I_{dep} depends on the bias.

Now consider the leakage problem shown in Fig. 8.7. Since $V_{\text{in}} > 0$, the capacitor voltage represents the level of reverse bias across the pn junction. I_R is thus obtained from eqn. (8.1-37) by setting $V = -V_{\text{in}}$. Assuming that $V_{\text{in}} \gg (kT/q)$,

$$I_R \simeq I_o + I_{\text{gen}}, \tag{8.1-40}$$

where I_{gen} is the generation current that originates in the depletion region. For a step-profile pn junction,

$$I_{\text{gen}} \simeq \frac{qAn_i}{2\tau_o} x_d, \tag{8.1-41}$$

where

$$\tau_o \simeq \frac{1}{2}(\tau_n + \tau_p) \tag{8.1-42}$$

is an effective minority carrier lifetime and x_d is the thickness of the depletion region. For the one-sided junction with $N_d \gg N_a$ and an applied reverse bias of $V_r(= V_{\text{in}})$,

$$x_d(V_R) \simeq \sqrt{\frac{2\varepsilon_{\text{Si}}}{qN_a}(\phi_o + V_r)}, \tag{8.1-43}$$

with

$$\phi_o = \left(\frac{kT}{q}\right) \ln \left(\frac{N_d N_a}{n_i^2}\right) \tag{8.1-44}$$

the built-in voltage. In room-temperature silicon, $I_{\text{gen}} \gg I_o$ is satisfied, so $I_R = I_{\text{gen}}$ will be assumed.

The above analysis shows that charge leakage may be approximated by the differential equation

$$-\frac{dQ}{dt} = I_{\text{leak}} \simeq I_{\text{gen}}, \tag{8.1-45}$$

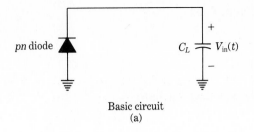

Basic circuit
(a)

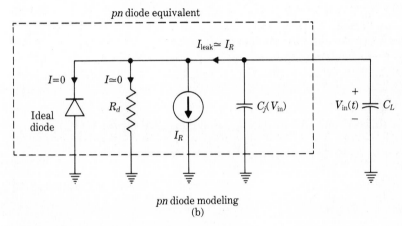

pn diode modeling
(b)

FIGURE 8.8 Device level leakage modeling.

where Q is the total charge stored on the soft node. The device level model for the problem is illustrated in Fig. 8.8. The basic problem of C_L discharging through a *pn* junction diode is shown in Fig. 8.8(a). In Fig. 8.8(b), the *pn* junction diode has been replaced by a device model consisting of an ideal diode in parallel with a large diode resistance R_d and a junction capacitance C_j. The model shows that Q has two contributions. It may be written in the form

$$Q = Q_j(V_{in}) + Q_L, \qquad (8.1\text{-}46)$$

where

$$Q_L = C_L V_{in} \qquad (8.1\text{-}47)$$

is the charge on the load capacitance C_L = constant, while

$$C_j(V_{in}) = \frac{dQ_j}{dV_{in}} \qquad (8.1\text{-}48)$$

defines the voltage-dependent depletion charge in the reverse-biased *pn* junction. The nonlinear depletion capacitance for a one-sided step-profile junction is written as (see eqn. 4.6-17)

$$C_j(V_{in}) = \frac{C_{j0}A}{\left(1 + \dfrac{V_{in}}{\phi_o}\right)^{1/2}},$$ (8.1-49)

with C_{j0} as the zero-bias capacitance per unit area

$$C_{j0} = \left.\frac{\varepsilon_{Si}}{x_d}\right|_{V_{in}=0}$$

$$\simeq \sqrt{\frac{qN_a\varepsilon_{Si}}{2\phi_o}}.$$ (8.1-50)

The complexity of the full problem is apparent. Noting that Q_j is a function of V_{in}, eqn. (8.1-45) gives

$$-\frac{dQ_j}{dV_{in}}\frac{dV_{in}}{dt} - C_L\frac{dV_{in}}{dt} \simeq I_{gen}.$$ (8.1-51)

Using eqns. (8.1-41) and (8.1-43) for I_{gen} and eqn. (8.1-49) for $C_j(V_{in})$,

$$-[C_L + C_j(V_{in})]\frac{dV_{in}}{dt} = I_{go}\left(1 + \frac{V_{in}}{\phi_o}\right)^{1/2},$$ (8.1-52)

where

$$I_{go} = \frac{qAn_i}{2\tau_o}\sqrt{\frac{2\varepsilon_{Si}\phi_o}{qN_a}}$$

$$= \frac{qAn_i\varepsilon_{Si}}{2\tau_o C_{j0}}.$$ (8.1-53)

A numerical analysis is required to find $V_{in}(t)$ as predicted by this equation.

It is possible to obtain analytic approximations for the charge leakage by using average or worst-case values for C_j and x_d. Consider first the technique of averaging. As shown in Section 4.6, the average depletion capacitance can be expressed by

$$C_{av} = K(V_1, V_2)C_{j0}A,$$ (8.1-54)

where

$$K(V_1, V_2) = \frac{2\phi_o}{(V_2 - V_1)}\left[\left(1 + \frac{V_2}{\phi_o}\right)^{1/2} - \left(1 + \frac{V_1}{\phi_o}\right)^{1/2}\right].$$ (8.1-55)

In the present problem, $V_1 = 0$ and $V_2 = (V_{DD} - V_{TP})$ for describing the entire discharging event. To compute the average depletion width $x_{d,av}$, the integral

$$x_{d,av} = \frac{1}{(V_2 - V_1)}\int_{V_1}^{V_2} x_d(V_{in})\,dV_{in}$$ (8.1-56)

is used. Substituting from eqn. (8.1-43) and integrating gives

$$x_{d,\text{av}} = H(V_1, V_2)x_{do},$$ (8.1-57)

where

$$x_{do} = \sqrt{\frac{2\varepsilon_{\text{Si}}\phi_o}{qN_a}}$$ (8.1-58)

is the zero-bias depletion width and

$$H(V_1, V_2) = \frac{2\phi_o}{3(V_2 - V_1)}\left[\left(1 + \frac{V_2}{\phi_o}\right)^{3/2} - \left(1 + \frac{V_1}{\phi_o}\right)^{3/2}\right].$$ (8.1-59)

With the average values, both C_{av} and $x_{d,\text{av}}$ are assumed constants in time. Equation (8.1-45) is now approximated by the simplified form

$$-\frac{dQ}{dt} \simeq -(C_{\text{av}} + C_L)\frac{dV_{\text{in}}}{dt} \simeq I_{g,\text{av}},$$ (8.1-60)

with

$$I_{g,\text{av}} = H(V_1, V_2)I_{go}.$$ (8.1-61)

Multiplying by dt and integrating gives

$$V_{\text{in}}(t) \simeq (V_{DD} - V_{TP}) - \frac{I_{g,\text{av}}}{(C_{\text{av}} + C_L)}t,$$ (8.1-62)

which approximates the leakage as being linear in time. The total time t for the charge leakage to yield a specific value of $V_{\text{in}} = V_{\text{low}}$ is

$$t \simeq (V_{DD} - V_{TP} - V_{\text{low}})\left(\frac{C_{\text{av}} + C_L}{I_{g,\text{av}}}\right).$$ (8.1-63)

This may be used to estimate the time interval allowed before the charge leakage yields an ambiguous logic level.

A worst-case analysis is also possible. This is obtained by maximizing the generation current using

$$I_{\text{gen}} = I_{go}\left(1 + \frac{V_{DD} - V_{TP}}{\phi_o}\right)^{1/2}$$ (8.1-64)

while minimizing the storage capacitance so that

$$C_{j,\text{min}} = \frac{C_{jo}A}{\left(1 + \dfrac{V_{DD} - V_{TP}}{\phi_o}\right)^{1/2}}.$$ (8.1-65)

Solving for $V_{\text{in}}(t)$ then yields

$$V_{\text{in}}(t) \simeq (V_{DD} - V_{TP}) - \frac{I_{go}t}{(C_{j,\text{min}} + C_L)}\left(1 + \frac{V_{DD} - V_{TP}}{\phi_o}\right)^{1/2},$$ (8.1-66)

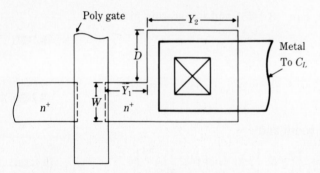

FIGURE 8.9 Top-view geometry for depletion capacitance calculation.

which may be used to estimate the worst-case (shortest) charge holding time.

The alert reader will have noticed that the above junction capacitance calculations ignored the difference in sidewall dopings. This is easily included in the analysis by using the pass transistor layout provided in Fig. 8.9. Following the discussion in Section 4.6, the junction capacitance is split into

$$C_j(V_{in}) = C_{bottom}(V_{in}) + C_{sw}(V_{in}).$$ (8.1-67)

The area of the bottom portion of the n^+-well is

$$A_{bottom} = W(Y_1 + Y_2) + DY_2,$$ (8.1-68)

so

$$C_{bottom}(V_{in}) = \frac{C_{j0}[W(Y_1 + Y_2) + DY_2]}{\left(1 + \dfrac{V_{in}}{\phi_o}\right)^{1/2}}.$$ (8.1-69)

The sidewall capacitance per unit length is

$$C_{jsw} = C_{j0sw}x_j,$$ (8.1-70)

where (see eqn. 4.6-21)

$$C_{j0sw} = \sqrt{\frac{q\varepsilon_{Si}}{2\left(\dfrac{1}{N_{a,sw}} + \dfrac{1}{N_d}\right)\phi_{osw}}}$$ (8.1-71)

and ϕ_{osw} is the sidewall built-in voltage. The sidewall capacitance is then seen to be

$$C_{sw} = \frac{2C_{jsw}(Y_1 + Y_2 + D + W)}{\left(1 + \dfrac{V_{in}}{\phi_{osw}}\right)^{1/2}}.$$ (8.1-72)

Assuming that

$$\phi_{osw} \approx \phi_o, \qquad\qquad\qquad (8.1\text{-}73)$$

the total zero-bias capacitance is given as

$$C_T = C_{j0}[W(Y_1 + Y_2) + DY_2] + 2C_{jsw}(Y_1 + Y_2 + D + W). \qquad (8.1\text{-}74)$$

The sidewall contributions can be accounted for by using this value of C_T in place of $(C_{j0}A)$ in all of the charge leakage equations.

The junction capacitance calculation is more complicated for a pass transistor that uses a buried contact. An example of this is shown in Fig. 8.10. Figure 8-10(a) provides the layout of a MOSFET where a buried poly-n^+ contact exists to connect the transistor to the load capacitance C_L. Recall from the process flow description in Section 5.4 that the buried contact is made before the n^+ drain-source implant. The contact relies on the n-type depletion implant to provide the necessary conductive properties. Taking a cross-sectional view of the device along the line $A\text{-}A'$ yields the doping shown in Fig.

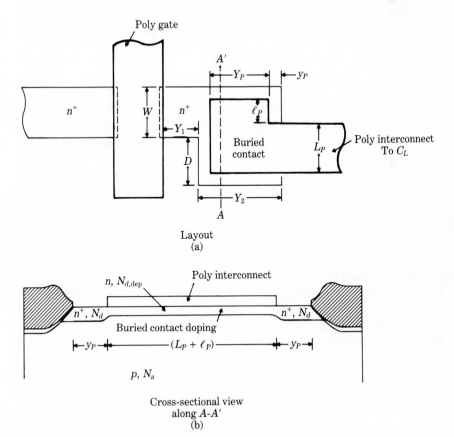

Layout
(a)

Cross-sectional view
along $A\text{-}A'$
(b)

FIGURE 8.10 Junction capacitance calculation for a buried contact.

8.10(b). It is different from the rest of the well. This changes the junction capacitance for this region.

A simple estimate of C_j for this case can be obtained by partitioning the well into proper areas. Denoting the buried contact capacitance per unit area by $C_{j0,\text{bur}}$, the total capacitance of the well is

$$C_T = C_{j0}[W(Y_1 + Y_2) + DY_2 - Y_P(\ell_P + L_P) - y_P L_P]$$
$$+ C_{j0,\text{bur}}[Y_P(\ell_P + L_P) + y_P L_P]$$
$$+ 2C_{jsw}[Y_1 + Y_2 + D + W]. \tag{8.1-75}$$

It is easily seen that the area used in the first term is the total bottom area of the well minus the area of the buried contact. The second term is then added to account for the difference in doping levels. Note that the sidewall contribution in the third term is not affected.

Finally, it should be noted that I_{gen} also depends on the specifics of the n-well, since x_d is a function of doping. The modifications for depletion depth and areas are straightforward and follow the same line of reasoning established above for the capacitances. The details are left as an exercise.

This series of examples illustrates the important points in pass transistor operation.

EXAMPLE 8.1-1

A pass transistor is connected to the gate of a driver MOSFET as shown in Fig. E8.1. The layout dimensions are in units of microns [μm], and the gate overlap is $L_o = 0.4\,[\mu\text{m}]$ (which is not explicitly shown).

Calculate the limiting values of C_{in} seen by the pass transistor. Assume that the circuit operates with $V_{DD} = 5\,[\text{V}]$. Use the depletion capacitance values from Example 4.6-1 and the parasitic field capacitances found in Example 5.4-1. Assume $V_{TOP} = 0.5\,[\text{V}]$, $\gamma = 0.4\,[\text{V}^{1/2}]$, and $2\,|\phi_F| = 0.6\,[\text{V}]$.

Solution
Each contribution will be calculated, then summed.

C_{metal}: The main contribution is the central metal-to-field capacitance

$$C_{\text{metal}} = C_{m\text{-}f}(\text{Area}) = (0.0345)(8^2) = 2.208\,[\text{fF}].$$

Parasitic contributions also exist in the contact regions from metal to n^+ and metal to poly. However, these are nonuniform and are neglected for simplicity.

C_{poly}: The poly interconnect capacitance (not including the gate capacitance of the driver) is

$$C_{\text{poly}} = C_{p\text{-}f}(\text{Area}) = (0.0576)(12^2 + 4^2) = 9.216\,[\text{fF}].$$

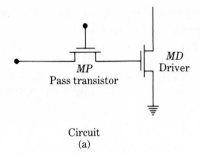

Circuit
(a)

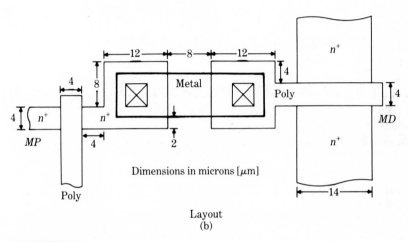

Dimensions in microns [μm]

Layout
(b)

FIGURE E8.1

C_G: The total gate capacitance of the driver is

$C_G = C_{ox}WL' = (0.0691)(14)(4) \simeq 3.870 \,[\text{fF}]$.

C_{n^+}: The depletion capacitance of the n^+-substrate pass MOSFET is found by first computing the zero-bias values. The bottom contribution is

$C_{d,\text{bottom}} = C_{j0}(\text{Area of bottom})$

$= (0.0975)[12^2 + (4)(4.4)] = 15.756 \,[\text{fF}],$

while the contribution from the sidewall is

$C_{d,sw} = C_{jsw}(\text{Perimeter})$

$= (0.209)[4(12) + 2(4.4)] \simeq 11.871 \,[\text{fF}].$

These represent the maximum values of depletion capacitance, so

$C_{d,\text{max}} = 15.756 + 11.871 = 27.627 \,[\text{fF}].$

The minimum values occur when the n^+-substrate is reverse-biased by the largest V_{in} possible. Assuming that V_{DD} is placed at the input of the pass transistor, this value is given by eqn. (8.1-11):

$$V_{max} = 4.5 - 0.4(\sqrt{0.6 + V_{max}} - \sqrt{0.6}),$$

which yields $V_{max} \simeq 3.96$ [V] by iterations. Noting the built-in potential values $\phi_o \simeq 0.879$ [V], $\phi_{osw} \simeq 0.957$ [V], the minimum capacitance values are

$$C_{d,min} = \frac{15.756}{\left(1 + \dfrac{3.96}{0.879}\right)^{1/2}} + \frac{11.871}{\left(1 + \dfrac{3.96}{0.957}\right)^{1/2}} \simeq 11.952 \text{ [fF]}.$$

Thus,

$$C_{in,max} \simeq 2.208 + 9.216 + 3.870 + 27.627 = 42.921 \text{ [fF]},$$

while

$$C_{in,min} \simeq 2.208 + 9.216 + 3.870 + 11.952 = 27.246 \text{ [fF]}. \qquad \blacksquare$$

EXAMPLE 8.1-2

Calculate the generation leakage current I_{go} through the $n^+\text{-}p$ substrate pass transistor junction in Example 8.1-1. Assume carrier lifetimes of $\tau_n = 0.1$ [μs] and $\tau_p = 1.0$ [μs].

Solution
The bottom and sidewall calculations will be separated. The leakage through the bottom is given by eqn. (8.1-53) as

$$I_{go,bottom} = \frac{(1.6 \times 10^{-19})(1.616 \times 10^{-6})(1.45 \times 10^{10})(11.8)(8.854 \times 10^{-14})}{2(0.55 \times 10^{-6})(9.75 \times 10^{-9})}$$

$$\simeq 0.365 \text{ [pA]}.$$

The sidewall leakage current is found using the same formula, modified for $C_{j0sw} = 4.18 \times 10^{-8}$ [F/cm^2] and $x_j = 0.5 \times 10^{-4}$ [cm]:

$$I_{go,sw} = \frac{(1.6 \times 10^{-19})(2.84 \times 10^{-7})(1.45 \times 10^{10})(11.8)(8.854 \times 10^{-14})}{2(0.55 \times 10^{-6})(4.18 \times 10^{-8})}$$

$$\simeq 14.971 \text{ [fA]}.$$

The sidewall current is much smaller than the leakage through the bottom because of the shallow junction depth.

The total zero-bias leakage current due to generation in the depletion region is

$$I_{go} \simeq 0.380 \text{ [pA]}. \qquad \blacksquare$$

EXAMPLE 8.1-3

Calculate the diode reverse saturation current I_o due to leakage across the n^+-p substrate junction, and compare this value with I_{go}. Assume mobilities of $\mu_n = 1275\,[\text{cm}^2/\text{V-sec}]$ (p-type substrate) and $\mu_p = 50\,[\text{cm}^2/\text{V-sec}]$ (n^+-region). Assume that the sidewall contributions are negligible.

Solution
The diffusion lengths are calculated from eqn. (8.1-39) as

$$L_n = \sqrt{(0.026)(1275)(0.1 \times 10^{-6})} \approx 1.82 \times 10^{-3}\,[\text{cm}],$$

$$L_p = \sqrt{(0.026)(50)(10^{-6})} \approx 1.14 \times 10^{-3}\,[\text{cm}],$$

where the Einstein relations $D = (kT/q)\mu$ have been used. The equilibrium minority carrier densities are

$$n_{po} \approx \frac{n_i^2}{N_a} = \frac{(1.45 \times 10^{10})^2}{10^{15}} \approx 2.10 \times 10^5\,[\text{cm}^{-3}],$$

$$p_{no} \approx \frac{n_i^2}{N_d} = \frac{(1.45 \times 10^{10})^2}{10^{20}} \approx 2.1\,[\text{cm}^{-3}].$$

Equation (8.1-38) then gives

$$I_o = (1.6 \times 10^{-19})(1.616 \times 10^{-6})\left[\frac{(33.15)(2.1 \times 10^5)}{(1.82 \times 10^{-3})} + \frac{(1.3)(2.1)}{(1.14 \times 10^{-3})}\right],$$

so

$$I_o \approx 0.99\,[\text{fA}].$$

By comparison, $I_o \ll I_{\text{gen}}$, so a total reverse current of

$$I_R \approx I_{\text{gen}}$$

is a good approximation. ∎

EXAMPLE 8.1-4

The voltage at the input to the driver MOSFET MD in Fig. E8.1 must be at least 2.50 [V] to be interpreted as a logic 1 input. Assuming that $V_{\text{max}} = 3.96$ [V] is originally at the input, calculate the holding time for the logic 1 state. Use average values of depletion capacitance and leakage current.

Solution
First look at depletion capacitance. The zero-bias bottom contribution was found as 15.756 [fF]. Then

$$(C_{d,\text{av}})_{\text{bottom}} = K(3.96, 2.50)(15.756).$$

Since

$$K(3.96, 2.50) = \frac{2(0.879)}{(3.96\text{-}2.50)}\left[\left(1 + \frac{3.96}{0.879}\right)^{1/2} - \left(1 + \frac{2.50}{0.879}\right)^{1/2}\right] \simeq 0.464,$$

this gives

$$(C_{d,\text{av}})_{\text{bottom}} \simeq 7.311\,[\text{fF}].$$

For the sidewall contributions, calculate

$$K(3.96, 2.50) = \frac{2(0.957)}{(3.96 - 2.50)}\left[\left(1 + \frac{3.96}{0.957}\right)^{1/2} - \left(1 + \frac{2.50}{0.957}\right)^{1/2}\right] \simeq 0.480,$$

so

$$(C_{d,\text{av}})_{sw} \simeq (0.480)(11.871) \simeq 5.698\,[\text{fF}].$$

Summing gives

$$C_{d,\text{av}} \simeq 13.009\,[\text{fF}].$$

The MOS capacitances sum to give the constant value

$$C_L \simeq 15.294\,[\text{fF}].$$

To average the currents, H from eqn. (8.1-59) must be calculated. For the bottom,

$$H(3.96, 2.5) = \frac{2(0.879)}{3(3.96 - 2.5)}\left[\left(1 + \frac{3.96}{0.879}\right)^{3/2} - \left(1 + \frac{2.50}{0.879}\right)^{3/2}\right] \simeq 2.159,$$

so

$$(I_{g,\text{av}})_{\text{bottom}} \simeq (2.159)(0.365) \simeq 0.788\,[\text{pA}].$$

For the sidewalls,

$$H(3.96, 2.5) = \frac{2(0.957)}{3(3.96 - 2.5)}\left[\left(1 + \frac{3.96}{0.957}\right)^{3/2} - \left(1 + \frac{2.50}{0.957}\right)^{3/2}\right] \simeq 2.089,$$

which gives

$$(I_{g,\text{av}})_{sw} \simeq (2.089)(14.971)\,[\text{fA}] \simeq 0.031\,[\text{pA}].$$

Note that $H > 1$ corresponding to the fact that I_g increases with reverse voltage. The total average generation current is

$$I_{g,\text{av}} \simeq 0.819\,[\text{pA}].$$

The holding time estimate for a logic 1 state is then obtained using eqn. (8.1-63) as

$$t \simeq (3.96 - 2.5)\left(\frac{28.303 \times 10^{-15}}{0.819 \times 10^{-12}}\right),$$

or $t \simeq 0.050\,[\text{sec}]$.   ■

EXAMPLE 8.1-5

Estimate the worst-case holding time for the circuit.

Solution
First, maximize the leakage (generation) current using eqn. (8.1-64) to

$$I_{\text{gen}} \simeq (0.365)\left(1 + \frac{3.96}{0.879}\right)^{1/2} + (0.014971)\left(1 + \frac{3.96}{0.957}\right)^{1/2} [\text{pA}]$$

$$\simeq 0.890 \,[\text{pA}].$$

The load capacitance is minimized by writing the depletion contributions as

$$C_{j,\min} = \frac{15.756}{\left(1 + \dfrac{3.96}{0.879}\right)^{1/2}} + \frac{11.871}{\left(1 + \dfrac{3.96}{0.957}\right)^{1/2}} \simeq 11.952 \,[\text{fF}],$$

which combines with the constant MOS capacitances to give

$$(C_{j,\min} + C_L) \simeq 27.246 \,[\text{fF}].$$

With these approximations, the holding time is

$$t \simeq (3.96 - 2.50)\left(\frac{27.246 \times 10^{-15}}{0.890 \times 10^{-12}}\right) \simeq 44.7 \,[\text{ms}].$$

The holding time decreases if V_{low} is increased. This makes it necessary to maintain dynamic operation of the circuit to avoid false logic inputs. ∎

8.2   2-Phase Clocking

Single-phase timing with one clock waveform $\phi(t)$ allows for the construction of basic clocked logic circuits. An example of this is a classical *sequential network* made up of flip-flops and combinational logic driven by one clock signal. While single-phase clocking is useful for certain types of nMOS static and quasi-static circuits, it can lead to problems in the system timing. This is particularly true at the chip level, as normal process variations permit specifying only nominal values for switching times.

2-phase clocking allows for a more controlled type of data transfer by using two clock waveforms $\phi_1(t)$ and $\phi_2(t)$. The most useful scheme is one in which the clocks are nonoverlapping, as illustrated in Fig. 8.11. This implies that

$$\phi_1(t)\phi_2(t) = 0 \qquad\qquad (8.2\text{-}1)$$

holds for arbitrary times t.

The extra degree of freedom introduced by using two non-overlapping clocks permits the design of systems in which data transfer in and out of gates

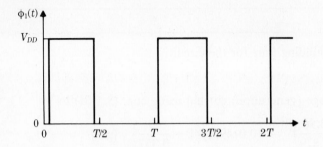

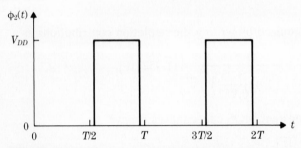

FIGURE 8.11 Ideal nonoverlapping 2-phase clocks.

or subsystems occurs only during specified time intervals. An example of this type of circuit is shown in Fig. 8.12. Pass gates are alternately controlled by ϕ_1 and ϕ_2, which sets the relative timing for bit transfer. When $\phi_1 = 1$, *MPA* and *MPB* conduct, allowing *A* and *B* to enter Gate 1. During the second part of the clock cycle, $\phi_1 = 0$ and $\phi_2 = 1$. *MPA* and *MPB* are both in cutoff, while

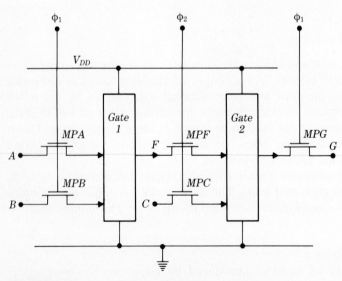

FIGURE 8.12 Basic 2-phase clocking.

MPF and *MPC* are on. The output *F* of gate 1 and input variable *C* are fed into gate 2 during this time. The result of these inputs will be available at node *G* during the next clock cycle when ϕ_1 returns to a value of 1.

A clock period *T* is defined as *one bit time* for the system timing. The data transfer is structured by the 2-phase clocking such that the delay per gate is set at one-half of a bit time ($T/2$). Use of this metric allows a relative system timing scale to be established.

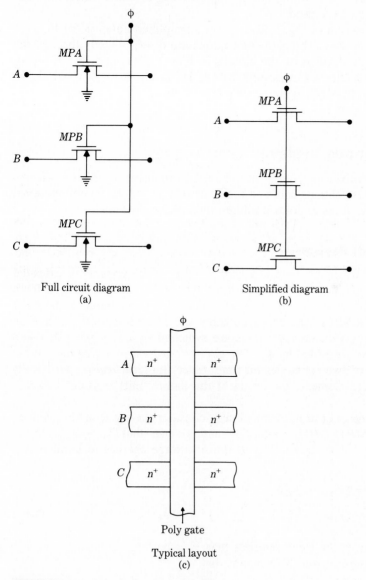

Full circuit diagram
(a)

Simplified diagram
(b)

Typical layout
(c)

FIGURE 8.13 Convention for multiple-pass transistor connections.

The 2-phase circuit operation should be compared with static combinational logic circuits that produce valid outputs after a propagation time t_p. Since the propagation time is set by the structuring and layout of each gate, system timing can become extremely complicated. 2-phase clocking schemes can be used to establish a basis for system control. Large-scale system design then becomes more tractable. The main drawback of this type of circuit is that the clock period must be long enough to allow for switching of the slowest gate in the chain. Thus, the overall system response will be slower than if purely combinational logic were used.

The circuit diagram of Fig. 8.12 employs a simplified notation for the pass transistor wiring in which the gate connections are drawn directly through the MOSFETs. This is clarified by the circuits in Figs. 8.13(a) and (b). A layout example for connected pass transistors is provided in Fig. 8.13(c), which shows how the notation correlates with the chip level view.

8.3   Synchronous Depletion Load Logic

Ratioed depletion load logic gates may be used to construct 2-phase synchronous systems. These circuits combine the basic features of static combinational logic with the data transfer control offered by 2-phase clocking.

8.3.1   Basic Shift Register

The simplest circuit of this type is a *shift register*, which consists of cascaded inverter stages separated by pass transitors. The general logic diagram for a 3-stage register is shown in Fig. 8.14(a), while the circuit implementation is provided in Fig. 8.14(b). The pass transistors are labeled as *MPX*, with X an integer. Pass gates with odd X values are switched by ϕ_1, while gates with even X values are controlled by ϕ_2. The circuit transfers (or shifts) data to the right at the rate of two states per bit time. Since the term *register* generically describes a storage element, the origin of the name "shift register" becomes clear.

To analyze the operation of the circuit, consider the input section defined by MOSFETS *MP*1, *MD*1, and *ML*1 and assume that $V_P = V_{DD}$. When $\phi_1 = 1$, the input capacitance $C_{\text{in},1}$ to the first stage charges to a value (see eqn. 8.1-11)

$$V_{\text{in},1} = V_{\text{max}} = (V_{DD} - V_{TP})$$
$$= (V_{DD} - V_{T0P}) - \gamma(\sqrt{V_{\text{max}} + 2|\phi_F|} - \sqrt{2|\phi_F|}). \tag{8.3-1}$$

Proper operation of the stage requires that V_{max} be large enough to act as a logic 1 input to the inverter. This means that

$$V_{\text{max}} > V_{IH} \tag{8.3-2}$$

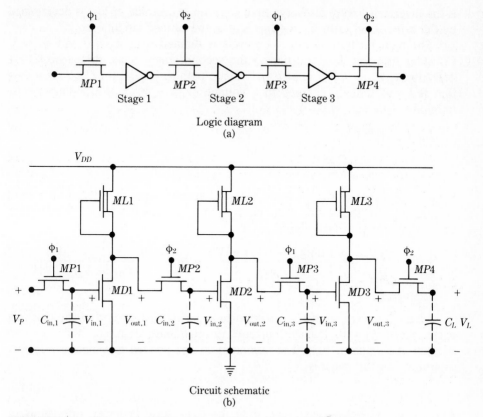

FIGURE 8.14 Three-stage shift register using depletion load inverters.

must be satisfied. If the capacitor is initially uncharged, the time required to set the input voltage is (see eqn. 8.1-22)

$$t_{LH,P} = 9\tau_{ch}$$

$$= \frac{18C_{in,1}}{\beta_P(V_{DD} - V_{TP})}. \qquad (8.3\text{-}3)$$

This induces the transition $V_{out,1} \rightarrow V_{OL}$ at the output of the first inverter stage. The time required for the output to fall to V_{OL} is estimated by (see eqn. 4.1-13)

$$t_{HL,1} = \tau_D\left\{\frac{2V_{TD}}{(V_{DD} - V_{TD})} + \ln\left[\frac{2(V_{OH} - V_{TD})}{V_{OL}} - 1\right]\right\}, \qquad (8.3\text{-}4)$$

where

$$\tau_D = \frac{C_{out,1}}{\beta_D(V_{OH} - V_{TD})} \qquad (8.3\text{-}5)$$

is the inverter (driver) discharge time constant. The value of V_{OL} is determined by the driver-load ratio $\beta_R = (\beta_{D1}/\beta_{L1})$ of the ratioed circuit.

The next portion of the clock cycle is defined with $\phi_1 = 0$ and $\phi_2 = 1$. Consider first the data transfer to the second stage. Since driver MOSFET $MD1$ is on, $V_{out,1} = V_{OL}$ allows a discharge path to ground for $C_{in,2}$. Assuming that this was initially charged to a voltage $(V_{DD} - V_{TP})$, the discharge time interval is (see eqn. 8.1-30)

$$t_{HL,P} \simeq 2.94\tau_{dis}$$

$$\simeq \frac{2.94 C_{in,2}}{\beta_P(V_{OH} - V_{TD})}, \tag{8.3-6}$$

which establishes a logic 0 input into the second-stage inverter. The output voltage $V_{out,2}$ of the second stage charges to V_{OH} through $ML2$. This switching time is estimated as (see eqn. 4.2-23)

$$t_{LH,2} = \frac{C_{out,2}}{\beta_{L2}|V_{TL}|}\left\{\frac{2(V_{DD} - |V_{TL}| - V_0)}{V_{TL}} + \ln\left[\frac{2|V_{TL}| - (V_{DD} - V_1)}{(V_{DD} - V_1)}\right]\right\}. \tag{8.3-7}$$

Charge leakage from $C_{in,1}$ is important during this portion of the clocking cycle. Since $\phi_1 = 0$, leakage currents through pass transistor $MP1$ tend to degrade the value of $V_{in,1}$. The minimum charge needed to maintain a logic 1 input level to this stage is

$$Q_{in,min} = C_{in,1}V_{IH}. \tag{8.3-8}$$

If $Q_{in,1}$ decays below this level, then the logic state becomes ambiguous or incorrect. Using eqn. (8.1-63) gives an estimate for the maximum clock period as

$$\left(\frac{T}{2}\right)_{max} \simeq \Delta V\frac{C_{in,1}}{I_{leak}}, \tag{8.3-9}$$

where

$$\Delta V \simeq (V_{DD} - V_{TP}) - V_{IH} \tag{8.3-10}$$

is the maximum allowed voltage decay. The problem of charge leakage thus establishes a minimum clock frequency of

$$f_{min} \simeq \frac{1}{T_{max}}$$

$$\simeq \frac{I_{leak}}{2(\Delta V)C_{in,1}}. \tag{8.3-11}$$

The maximum clock frequency is set by noting that $(T/2)_{min}$ must be long enough for the input capacitance to charge and for the logic to propagate through an inverter stage. A simple estimate for this time is

$$\left(\frac{T}{2}\right)_{\min} \approx t_{LH,P} + t_p, \tag{8.3-12}$$

where t_p is the inverter propagation time discussed in Section 4.3. The maximum frequency is then computed from

$$f_{\max} = \frac{1}{T_{\min}}. \tag{8.3-13}$$

These limits on the clock frequency must be observed to ensure proper circuit operation.

The most critical part of the circuit design is based on the fact that the threshold voltage loss through the pass transistor gives a maximum inverter input of

$$V_{\text{in}} = (V_{DD} - V_{TP}). \tag{8.3-14}$$

Since this must be greater than V_{IH}, relatively large values of β_R are needed.

The basic design equation is easily obtained by noting that when $V_{\text{in}} = (V_{DD} - V_{TP})$, the driver is nonsaturated with

$$I_{D1} = \frac{\beta_{D1}}{2}\{2[(V_{DD} - V_{TP}) - V_{TD}]V_{OL} - V_{OL}^2\}, \tag{8.3-15}$$

while the load MOSFET is saturated such that

$$I_{L1} = \frac{\beta_{L1}}{2}|V_{TL}(V_{OL})|^2. \tag{8.3-16}$$

Equating currents and rearranging gives the driver-load requirement

$$\beta_R = \frac{|V_{TL}(V_{OL})|^2}{2[(V_{DD} - V_{TP}) - V_{TD}]V_{OL} - V_{OL}^2}, \tag{8.3-17}$$

as expected from the generalized design equation (3.4-45). In this expression, $V_{TD} = V_{TOD}$ since there are no driver body bias effects. The equation illustrates the usefulness of having a technology that allows for small pass transistor threshold voltages V_{TP}, since this eases the requirement on β_R values.

The overall design is complicated by charge leakage from the soft input nodes. This is important at low clock frequencies. Charge leakage is accounted for by adding the design constraint that some decay of the V_{in} level be tolerated. Analytically this implies that the voltage difference ΔV defined in eqn. (8.3-10) must be large enough to ensure that an unambiguous logic 1 input state can be held for at least a time period $(T/2)$.

The problem can be illustrated in terms of the inverter voltage transfer characteristic (VTC), as shown in Fig. 8.15. Increasing β_R decreases V_{IH}.

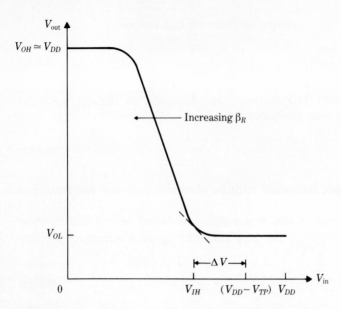

FIGURE 8.15   Inverter voltage transfer characteristic for ΔV design.

Recall from Section 3.4 that the value of V_{IH} can be estimated by first solving

$$V_{\text{out}} \simeq \frac{|V_{TL}(V_{\text{out}})|}{\sqrt{3\beta_R}} \tag{8.3-18}$$

for V_{out} and then using this in

$$V_{IH} \simeq V_{TD} + 2V_{\text{out}} + \frac{\eta(V_{\text{out}})\,|V_{TL}(V_{\text{out}})|}{\beta_R}, \tag{8.3-19}$$

where

$$\eta(V_{\text{out}}) = \frac{\gamma}{2\sqrt{V_{\text{out}} + 2\,|\phi_F|}}. \tag{8.3-20}$$

The design procedure is to first compute β_R using eqn. (8.3-17) and then check the resulting value of V_{IH}. The value chosen for ΔV depends on the leakage current level through the pass transistor and also on the input capacitance. The large β_R requirement in eqn (8.3-17) will aid in solving this problem since C_{GS} is increased accordingly. The design tradeoff arises in the real estate consumption.

A somewhat obvious final comment is that the simplest design for the shift register is obtained using identical pass transistors and inverter stages. This implies identical layout patterns to equalize the capacitance levels. An example is shown in Fig. 8.16. The layout uses the general patterning established for static logic circuits, with the pass transistors added in a straightforward manner.

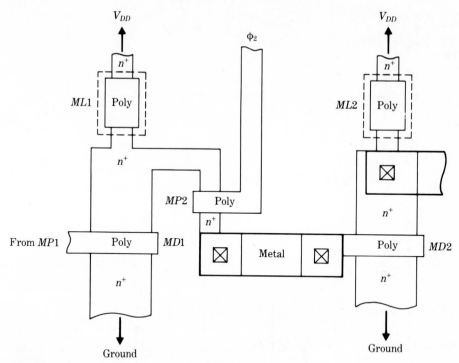

FIGURE 8.16 Shift register layout example.

8.3.2 Synchronous Complex Logic

Combinational and AOI logic gates can be included in the clocked circuits as implied by Fig. 8.12 in the previous section. A simple example of this type of circuit is shown in Fig. 8.17. The logic diagram is drawn in Fig. 8.17(a), with the corresponding nMOS circuit structuring in Fig. 8.17(b).

The operation of the circuit is straightforward. When $\phi_1 = 1$, $\phi_2 = 0$, the pass transistors allow inputs into the first gate. This has a standard AOI form to give the function

$$F1 = \overline{AB + C}. \tag{8.3-21}$$

During the second half of the clock cycle ($\phi_1 = 0$, $\phi_2 = 1$), $F1$ is transferred to the second gate where the exclusive-NOR operation is performed with $F2$. The circuit implements the XNOR function using the AOI configuration presented in Fig. 6.17 in Chapter 6. The value

$$G = F1 \odot F2 \tag{8.3-22}$$

is available at the output node G during the next half clock cycle when $\phi_1 = 1$ allows the transfer out of the chain.

The circuit chosen for this example illustrates the timing design permitted

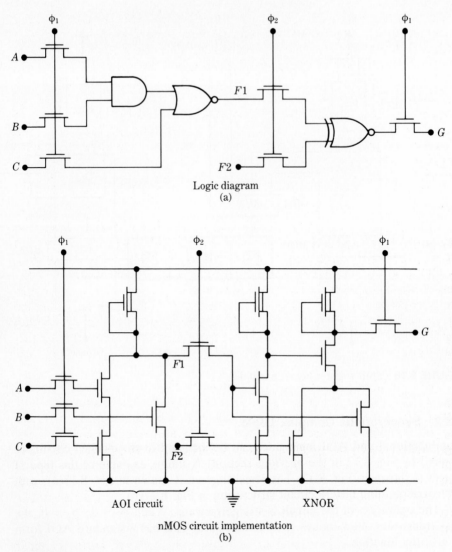

Logic diagram
(a)

nMOS circuit implementation
(b)

FIGURE 8.17   Clocked AOI and combinational logic gate arrangement.

by the 2-phase clocking. First note that the propagation delay for the AOI circuit will be different than that for the XNOR gate. Since the data transfer is systematically controlled by the clock signals ϕ_1 and ϕ_2, the difference in t_p values is unimportant so long as the half bit time $(T/2)$ is long enough to allow the gate outputs to stabilize before transfer. The tradeoff, of course, is that it takes longer to propagate the logic down the gate chain. Such a compromise is generally worth the price because it automatically solves many timing problems. This is particularly true for large, complex logic systems.

An extremely useful technique for controlling data transfer is to gate pass transistors with composite signals made up of both clock pulses and logic variables. An example of this type of circuit is the simple 2-line path controller of Fig. 8.18. The logic diagram in Fig. 8.18(a) shows how the clock ϕ_1 is combined with the select signal SEL to produce two auxiliary control signals $\phi_1 \cdot SEL$ and $\phi_1 \cdot \overline{SEL}$. These are used to gate pass transistors $MP0$ and $MP1$, respectively. With $SEL = 1$, the input data is routed to path P_0 when $\phi_1 = 1$. Alternately, a select input of $SEL = 0$ enables $MP1$ when $\phi_1 = 1$. This connects the input to path P_1. Data paths P_0 and P_1 can be directly routed into combinational logic circuits with outputs gated by ϕ_2. Figure 8.18(b) gives the circuit diagram for the path controller. The control signals $\phi_1 \cdot SEL$ and $\phi_1 \cdot \overline{SEL}$ are generated using NOR gate logic since this represents a minimum area layout in depletion load nMOS.

Another example of control signal generation is the clocked 2-to-4 line decoder of Fig. 8.19. The general block symbol is shown in Fig. 8.19(a). The inputs are denoted A_0 and A_1, with the internal switching controlled by clock ϕ_1 and a select signal SEL. The operation of the circuit is specified by the function table in Fig. 8.19(b). In essence, all output lines L_0, L_1, L_2, and L_3 are 0 whenever $SEL = 0$. With control signals such that $\phi_1 \cdot SEL = 1$, the circuit functions as a normal decoder. In this case, one output line is set at a logic 1 level depending on the input combination (A_0, A_1). The decoder outputs can be used to drive transmission gates or may serve as direct logic variables.

The NOR-based logic diagram for implementing the decoder is shown in Fig. 8.19(c). While the function table verification is straightforward, it is instructive to examine the decoder function using the NOR circuitry shown in Fig. 8.19(d). The logic implementation is based on the observation that any high NOR input causes a logic 0 output. The 3-input NOR gates each have one input driven by SEL. When $SEL = 0$, all gates are disabled with logic 0 outputs since ground paths are provided by the drivers. A select condition of $SEL = 1$ allows control by the inputs A_0, A_1, and their complements. The decoding process simply uses the proper combination of these signals to ensure that all drivers remain in cutoff in the desired gate line.

A final example of using controlled gating is illustrated by the dynamic selectively loadable register shown in Fig. 8.20(a). The control circuitry combines ϕ_1 and the load signal LD using the same logic as in the path controller of Fig. 8.18. The operation of the circuit is straightforward. Setting $LD = 1$ allows the register to be loaded with the input through $MP1$. This action takes place when $\phi_1 = 1$. After this portion of the clock cycle, the register state can be held using $LD = 0$, since this keeps $MP1$ in cutoff.

Consider now the mechanism used to hold the register state. Pass transistor $MP2$, which is controlled by $\phi_1 \cdot \overline{LD}$, is used to maintain charge levels in the dynamic circuit. The subcircuit for this part of the circuit operation is shown in Fig. 8.20(b). Suppose that the input capacitance C_{in} has been initially charged to a value of $V_{in} = V_{max} = (V_{DD} - V_{TP})$ through $MP1$.

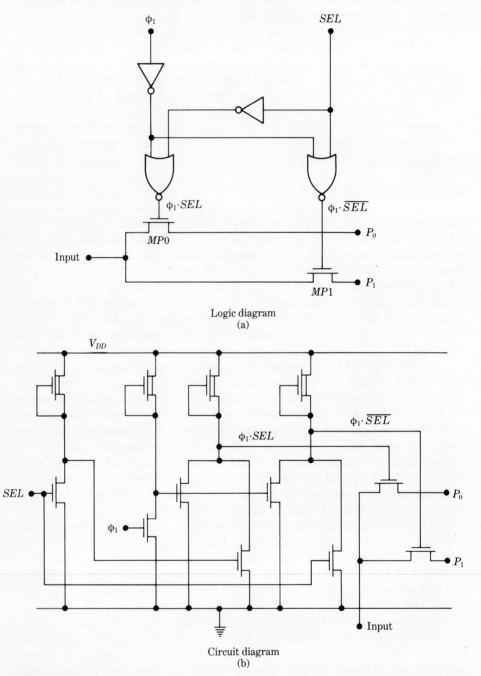

FIGURE 8.18   2-line path controller.

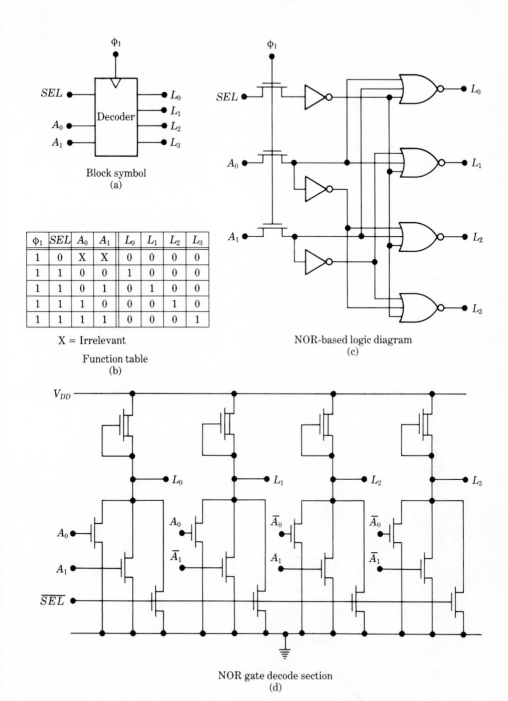

Block symbol
(a)

ϕ_1	SEL	A_0	A_1	L_0	L_1	L_2	L_3
1	0	X	X	0	0	0	0
1	1	0	0	1	0	0	0
1	1	0	1	0	1	0	0
1	1	1	0	0	0	1	0
1	1	1	1	0	0	0	1

X = Irrelevant

Function table
(b)

NOR-based logic diagram
(c)

NOR gate decode section
(d)

FIGURE 8.19 Control signal generation using a 2-to-4 decoder circuit.

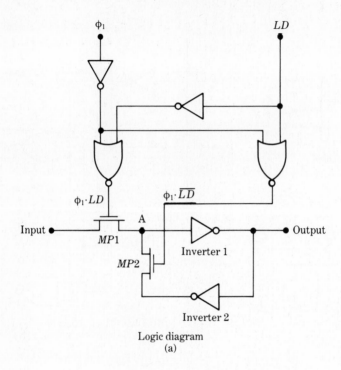

Logic diagram
(a)

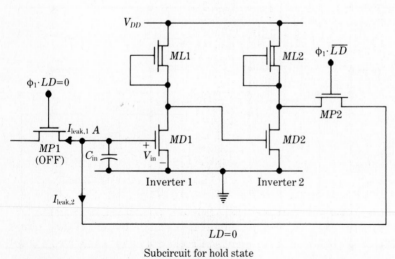

Subcircuit for hold state
(b)

FIGURE 8.20 Dynamic register with load control.

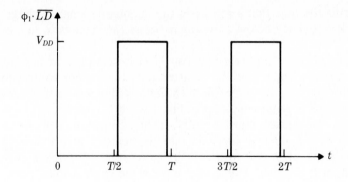

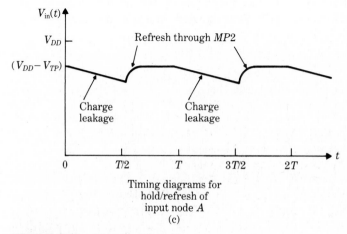

Timing diagrams for
hold/refresh of
input node A
(c)

FIGURE 8.20 (Continued)

When $\phi_1 = 0$, $MP2$ is in cutoff. Charge leakage then occurs through both $MP1$ and $MP2$ such that

$$I_{\text{leak}} = I_{\text{leak},1} + I_{\text{leak},2} \tag{8.3.23}$$

constitutes the total leakage current from the soft node A. This degrades the value of V_{in} with time. However, the voltage at A is periodically *refreshed* by the current supplied through $MP2$ when $\phi_2 = 1$ in the normal clock cycle. This current originates from the V_{DD} path established through load MOSFET $ML2$. The timing diagrams shown in Fig. 8.20(c) demonstrate how the refresh mode acts to restore the input voltage level. The technique of refreshing a soft node in dynamic circuits is widely used to reduce charge leakage problems.

8.3.3 Clocked Static Register Circuits

All of the nMOS logic circuits presented above have soft node inputs consisting of MOSFET driver gates connected to clocked pass transistors.

Charge leakage limits the time that a soft node can maintain a logic 1 voltage. Reducing the clock frequency below a minimum value thus results in loss of data.

It is often useful to have registers that can hold logic values at low (or zero) clock frequencies. These can be constructed using clock-controlled feedback in a double-inverter cascade. The technique is identical to that used for the clocked CMOS TG D-type flip-flop in Fig. 7.27 of Chapter 7.

An example of a single-phase register built from static inverters is shown in Fig. 8.21(a). When $\phi = 1$, data is admitted into the register circuit through $MP1$. A clock condition of $\phi = 0$ turns $MP2$ on, creating a feedback path that sustains the input logic level. This is true even if the clock is idled in the $\phi = 0$ state.

The 2-phase equivalent of this circuit is shown in Fig. 8.21(b). The operation is straightforward, with data allowed to enter the register when

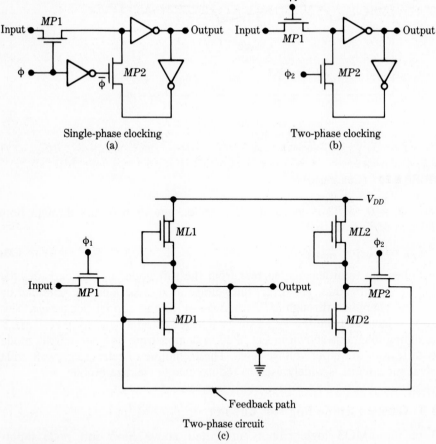

Single-phase clocking
(a)

Two-phase clocking
(b)

Two-phase circuit
(c)

FIGURE 8.21 Clocked static-design registers.

$\phi_1 = 1$, $\phi_2 = 0$. The hold clock condition is defined with $\phi_1 = 0$, $\phi_2 = 1$. The circuit diagram for the 2-phase case is provided in Fig. 8.21(c). It is seen that the input to the first inverter is no longer a soft node because of the feedback voltage provided through $MP2$. In particular, a logic 1 input high voltage at driver $MD1$ is maintained by the V_{DD} path through load MOSFET $ML2$ when $\phi_2 = 1$.

8.4 Dynamic Charge Sharing

Charge sharing occurs when the voltage held on a dynamic soft node is used to drive another soft node in the circuit. The effect of charge sharing depends or the ratio of the capacitances on the two nodes. Consider the circuit in Fig. 8.22(a). The input of the first inverter is set to $V_{in} = 0\,[V]$, which gives a first-stage output voltage of $V_{OH} = V_{DD}$. This is connected to the input of the second inverter through two pass transistors $MP1$ and $MP2$. $MP1$ is controlled by a clock waveform ϕ, while $MP2$ is connected to some logic variable A. This arrangement may or may not exhibit charge sharing problems, depending on the relative phasing of $\phi(t)$ and $A(t)$.

Charge sharing will occur if ϕ and A are phased as shown in Fig. 8.22(b). During the first half clock cycle $[0, T/2]$, $\phi = 1$ while $A = 0$. $MP1$ thus conducts current, which charges C_1 to a value

$$V_1 = V_{DD} - V_{TP}$$

$$\equiv V_o. \tag{8.4-1}$$

This is normal charging through a pass transistor since the power supply V_{DD} provides the necessary current. The total charge stored on C_1 during this time is

$$Q_T = C_1(V_{DD} - V_{TP})$$

$$= C_1 V_o. \tag{8.4-2}$$

$MP2$ is in cutoff, so no current flows to C_2 during this time. It is assumed that C_2 is initially uncharged at $0\,[V]$.

Charge sharing occurs during the second half clock cycle when the control signals are $\phi = 0$ and $A = 1$. Since $\phi = 0$, $MP1$ is in cutoff. This holds the charge Q_T on C_1 until $A = 1$ turns $MP2$ on and allows transfer to C_2. Although the situation is similar to normal transmission through a pass transistor, the amount of charge available for transfer to C_2 is limited since Q_T constitutes the total charge. Q_T must therefore be "shared" between the two capacitors.

The basic charge sharing problem may be understood using the reduced circuit in Fig. 8.23. The capacitor voltages are denoted by $V_1(t)$ and $V_2(t)$. The time origin has been shifted such that at $t = 0$,

$$V_1 = (0) = V_{DD} - V_{TP} = V_o, \qquad V_2(0) = 0 \tag{8.4-3}$$

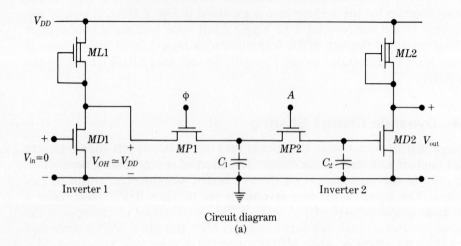

Circuit diagram
(a)

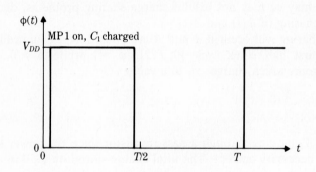

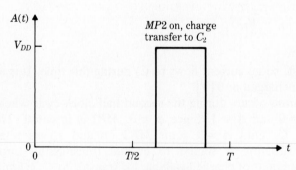

Pass transistor control
signals for charge
sharing example
(b)

FIGURE 8.22   nMOS charge sharing.

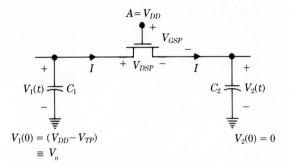

FIGURE 8.23 Charge sharing through a pass transistor.

specifies the initial conditions. MOSFET $MP2$ has terminal voltages of

$$V_{DSP}(t) = V_1(t) - V_2(t)$$

$$V_{GSP}(t) = V_{DD} - V_2(t). \tag{8.4-4}$$

MOSFET current I flows from C_1 to C_2 as shown, giving the charge transfer mechanism.

Before progressing into a detailed analysis, it is instructive to examine the problem using basic circuit theory. In the limit of $t \to \infty$, equilibrium is reached with $I \to 0$ and the final voltages satisfy $V_1 = V_2 = V_f$. C_1 and C_2 are effectively in parallel because of the inversion channel in $MP2$. Since the total charge available for transfer was Q_T,

$$Q_T = (C_1 + C_2)V_f \tag{8.4-5}$$

must be satisfied. However, eqn. (8.4-2) must also be satisfied so that equating the two gives

$$V_f = \frac{C_1}{C_1 + C_2} V_o$$

$$= \frac{1}{1 + (C_2/C_1)} V_o. \tag{8.4-6}$$

This illustrates the basic effect of charge sharing. The final voltage V_f on the two capacitors is determined by the ratio of capacitances (C_2/C_1). If $C_1 \gg C_2$, then $V_f \approx V_o$, indicating efficient charging of C_2. However, in the general case, V_f will reach only a fraction of V_o. For example, if $C_1 = 4C_2$, $V_f = 0.8V_o$.

Returning now to the original problem in Fig. 8.23, it is seen that the current I satisfies

$$I = -C_1 \frac{dV_1}{dt} = C_2 \frac{dV_2}{dt}. \tag{8.4-7}$$

Using eqns. (8.4-3) and (8.4-4) gives the MOSFET voltages as

$$V_{DSP}(0) = V_{DD} - V_{TP}, \qquad V_{GSP}(0) = V_{DD}, \tag{8.4-8}$$

so the transistor is initially at the edge of saturation. However, the time evolution of the terminal voltages must be studied to determine the state of the MOSFET during the charge transfer event.

The drain-source voltage has a rate of change

$$\frac{dV_{DSP}}{dt} = \frac{dV_1}{dt} - \frac{dV_2}{dt}$$

$$= -\left(1 + \frac{C_2}{C_1}\right)\frac{dV_2}{dt}, \tag{8.4-9}$$

as verified using eqn. (8.4-7). The rate of change of the gate-source voltage is easily seen to be

$$\frac{dV_{GSP}}{dt} = \frac{d}{dt}(V_{DD} - V_2)$$

$$= -\frac{dV_2}{dt}. \tag{8.4-10}$$

Comparing these two expressions shows that V_{DSP} decreases at a faster rate than V_{GSP}. Consequently, $V_{DSP} < (V_{GSP} - V_{TP})$ will be true, indicating that MP2 is nonsaturated during the charge transfer. The current I in eqn. (8.4-7) thus assumes the form

$$I = \frac{\beta_P}{2}[2(V_{DD} - V_2 - V_{TP})(V_1 - V_2) - (V_1 - V_2)^2]. \tag{8.4-11}$$

A full solution of the problem requires simultaneously finding $V_1(t)$ and $V_2(t)$. This is quite complicated and will not be pursued further.

The *RC* network shown in Fig. 8.24 constitutes a simpler model of the problem, which illustrates the important charge transfer properties. Capacitor C_1 is initially charged to a value V_o. At time $t = 0$, the switch is closed,

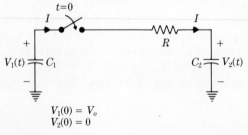

FIGURE 8.24   Simplified RC circuit for charge transfer analysis.

allowing current I to charge C_2. The linear resistor R is used as a low-order approximation for including the MOSFET conduction properties.

To find $V_1(t)$ and $V_2(t)$, the current flow is expressed as

$$\frac{dV_1}{dt} = -\frac{I}{C_1}, \qquad \frac{dV_2}{dt} = \frac{I}{C_2}, \tag{8.4-12}$$

so

$$\frac{d}{dt}(V_1 - V_2) = -\frac{I}{C_{eq}}, \tag{8.4-13}$$

where

$$C_{eq} = \frac{C_1 C_2}{C_1 + C_2} \tag{8.4-14}$$

is the equivalent series capacitance. Since $(V_1 - V_2)$ is the voltage across the resistor, the equation becomes

$$\frac{d}{dt}(V_1 - V_2) = -\frac{(V_1 - V_2)}{\tau}, \tag{8.4-15}$$

with

$$\tau = RC_{eq} \tag{8.4-16}$$

the time constant. Integrating gives the solution as

$$V_1(t) - V_2(t) = V_o e^{-t/\tau}. \tag{8.4-17}$$

Note that charge conservation enters the problem through the condition

$$Q_T = C_1 V_o, \tag{8.4-18}$$

as discussed at the beginning of this section.

$V_2(t)$ is computed by first differentiating eqn. (8.4-17) to give

$$\frac{dV_1}{dt} - \frac{dV_2}{dt} = V_o \frac{d}{dt}(e^{-t/\tau})$$

$$= -\left(1 + \frac{C_2}{C_1}\right)\frac{dV_2}{dt}. \tag{8.4-19}$$

The second step follows from eqn. (8.4-9). Rearranging and integrating then yields directly

$$V_2(t) = \left(\frac{C_1}{C_1 + C_2}\right)V_o(1 - e^{-t/\tau}). \tag{8.4-20}$$

$V_1(t)$ is found by substituting this into eqn. (8.4-17). The algebra gives

$$V_1(t) = V_o - \left(\frac{C_2}{C_1 + C_2}\right)V_o(1 - e^{-t/\tau}), \tag{8.4-21}$$

completing the analysis.

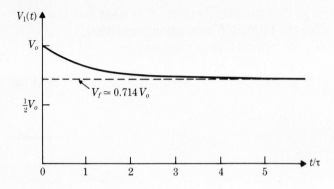

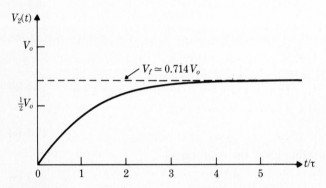

FIGURE 8.25   Plots of $V_1(t)$ and $V_2(t)$ for a capacitance ratio of $(C_2/C_1) = 0.4$.

Plots for $V_1(t)$ and $V_2(t)$ are shown in Fig. 8.25 for a capacitance ratio of $(C_2/C_1) = 0.4$. It is seen that both voltages have the expected behavior in that $V_1(t)$ decays from V_o while $V_2(t)$ is charged by the current flow. Asymptotically, the two attain the value

$$V_1(t \to \infty) = V_2(t \to \infty) = V_f$$

$$= \left(\frac{C_1}{C_1 + C_2}\right)V_o, \tag{8.4-22}$$

which agrees with the calculation in eqn. (8.4-6). With $(C_2/C_1) = 0.4$, $V_f \approx 0.714 V_o$.

The analysis here shows that charge sharing between two dynamic soft nodes separated by a pass transistor can degrade the logic 1 voltage level. The basic design criteria that will minimize the voltage reduction is to require that

$$C_1 \gg C_2 \tag{8.4-23}$$

be satisfied. This generally requires careful planning in the layout, since the capacitance contributions are sensitive functions of the geometry.

The problem can be illustrated by examining the first-try layout shown in

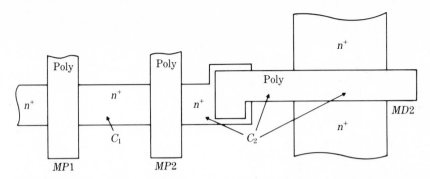

FIGURE 8.26   Layout problem for satisfying charge sharing capacitance requirement.

Fig. 8.26 for $MP1$, $MP2$, and $MD2$ as connected in the original circuit of Fig. 8.22. For this case, C_1 is primarily depletion capacitance from the n^+p junction. C_2, on the other hand, has contributions from depletion capacitance, parasitic poly line capacitance, and the MOS gate capacitance of $MD2$. It is very possible that $C_1 < C_2$ with the layout as shown. Reversing this inequality to yield an acceptable ratio might be accomplished by increasing the n^+ area between $MP1$ and $MP2$.

Finally, it should be noted that the reduction of a logic 1 voltage from charge sharing must be accounted for in designing the second-stage inverter. With regard to the original circuit in Fig. 8.22, this says that the driver-load ratio $\beta_{R2} = (\beta_{D2}/\beta_{L2})$ must be large enough to ensure that V_f is correctly interpreted as a logic 1 input level.

8.5   2-Phase Enhancement Load Logic

The two-phase (2ϕ) circuits examined up to this point have been based on static depletion load logic concepts. Dynamic enhancement load circuits provide an alternate approach to MOS logic. State-of-the-art designs tend to use techniques from both types of circuits. It is therefore worthwhile to study the details of 2ϕ circuits that employ E-mode MOSFETs as loads.

The general structure of a 2-phase clocked enhancement load inverter circuit is shown in Fig. 8.27. The inverter consists of driver and load MOSFETs MD and ML, with pass transistors $MP1$ and $MP2$ included as input and output gating devices. The input is controlled by a clock signal ϕ_{in}, while the output is timed according to ϕ_{out}.

An important difference between this configuration and depletion load circuits is that a clocking signal ϕ_L can be applied to the gate of load MOSFET ML as shown. Additional control is then provided for gating the power supply path from V_{DD}. ϕ_L may be phased with either ϕ_{in} or ϕ_{out}. Each results in a distinct type of circuit behavior.

Dynamic enhancement load logic is itself interesting for three reasons.

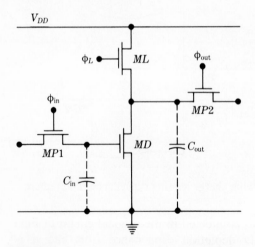

FIGURE 8.27   General 2-phase clocked enhancement load inverter arrangement.

First, for given values of W and L, an enhancement-mode MOSFET requires less real estate than a depletion-mode transistor. This is because no depletion implant design rule spacings are needed for the E-mode MOSFET. Second, it is possible to design ratioless enhancement load logic circuits that use minimum-size transistors throughout. Depletion load arrangements require specific driver-load values to achieve proper V_{OL} levels. Finally, the ability to switch the load MOSFET with a clock signal allows for control over the current consumption. Reduced power dissipation is thus possible.

One observation should be made concerning the clock amplitudes used in enhancement load circuits. It is seen from Fig. 8.27 that the load MOSFET ML can be operated in either the saturated or nonsaturated mode, depending on the amplitude of ϕ_L. If $\max(\phi_L) = V_{DD}$, then ML will be saturated. The highest output voltage is then restricted to (see eqn. 3.3-6)

$$V_{OH} = V_{DD} - V_{TL}(V_{OH}) \qquad (8.5\text{-}1)$$

because of the threshold voltage loss. Increasing $\max(\phi_L)$ to a value

$$\max(\phi_L) \geq V_{DD} + V_{TL}(V_{DD}) \qquad (8.5\text{-}2)$$

allows the load MOSFET to operate in the nonsaturated mode, which then gives

$$V_{OH} \simeq V_{DD}. \qquad (8.5\text{-}3)$$

The transmission gates also exhibit threshold voltage losses if $\max(\phi_{in}) = V_{DD} = \max(\phi_{out})$.

The amplitude of the clock waveforms can be adjusted to give the desired performance. The two critical levels are V_{DD} and $[V_{DD} + V_{TL}(V_{DD})]$. These are shown in the timing diagrams of Fig. 8.28, with V_ϕ denoting the actual

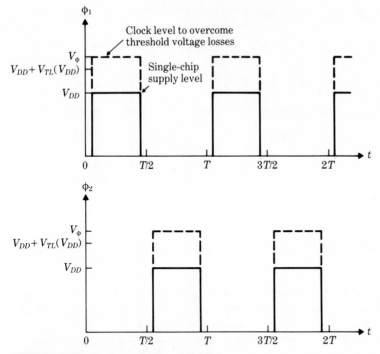

FIGURE 8.28 2-phase clock amplitudes.

value. Threshold voltage losses can be avoided using the large V_ϕ level drawn. However, this requires an extra on-chip voltage level.

The discussion here will assume the simplest case, where a single voltage level V_{DD} is used for both the power supply and clocks. The analysis is easily modified for a different choice of clock amplitude. The terminology "during ϕ_1" will sometimes be used to reference the time interval when $\phi_1 = 1$, $\phi_2 = 0$. Similarly, "during ϕ_2" will imply the half clock cycle when $\phi_1 = 0$, $\phi_2 = 1$ is valid.

8.5.1 2-Phase Ratioed Dynamic Logic

The first logic configuration to be examined is that where the clocks are chosen with $\phi_{in} = \phi_1$ and $\phi_{out} = \phi_2 = \phi_L$. The input is thus out of phase with the load and output clocking. Three stages of a shift register constructed using this scheme are shown in Fig. 8.29. This is termed a *ratioed* dynamic circuit since the driver-load ratio β_R establishes the value of V_{OL} for the inverters.

To understand the operation of the circuit, assume that an input voltage of $V_{in} = V_{DD}$ is applied to the first-pass transistor $MP1$. The resulting charge transfer sequence is shown by the subcircuits in Fig. 8.30. The circuits do not show MOSFETs that are placed into cutoff by a clock level of $\phi = 0$ to aid in following the charge-discharge paths.

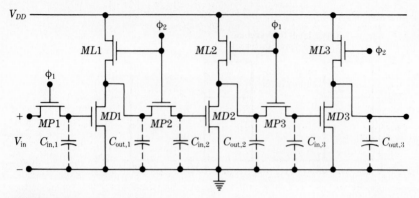

FIGURE 8.29 Dynamic 2ϕ ratioed shift register.

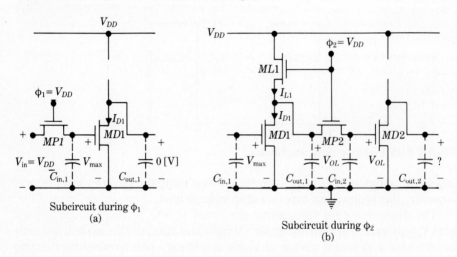

Subcircuit during ϕ_1
(a)

Subcircuit during ϕ_2
(b)

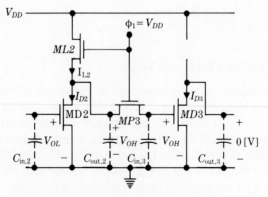

Subcircuit during next ϕ_1 cycle
(c)

FIGURE 8.30 Charge transfer through a ratioed shift register.

The first transfer is during ϕ_1 as shown in Fig. 8.30(a). The input voltage $V_{in} = V_{DD}$ charges $C_{in,1}$ through $MP1$. Including the threshold voltage loss gives the voltage

$$V_{max} = V_{DD} - V_{TP}(V_{max}) \tag{8.5-4}$$

on the input capacitor. Since this constitutes a logic 1 state, $V_{max} > V_{TD}$ is assumed to be satisfied; $MD1$ thus turns on. Any charge on the output capacitance $C_{out,1}$ is the lost to ground by the current I_{D1}, so the final voltage across the capacitor is 0 [V].

The next half clock cycle is defined by $\phi_1 = 0$, $\phi_2 = 1$. The charge transfer mechanisms during this time interval are illustrated by Fig. 8.30(b). Since $C_{in,1}$ acts as a soft node, $MD1$ remains in an active mode. ϕ_2 turns on the load MOSFET $ML1$, permitting current to flow from V_{DD}. The voltage across $C_{out,1}$ is V_{OL}, which is determined by the driver-load ratio $\beta_R = \beta_D/\beta_L$. V_{OL} may be computed using the techniques discussed in conjunction with eqn. (3.3-10) of Section 3.3.

Continuing with the analysis, note that pass transistor $MP2$ is biased on during ϕ_2. Any charge on $C_{in,2}$ is thus provided a discharge path to ground through $MP2$ and $MD1$, which sets the voltage across this capacitor at V_{OL}. Assuming $V_{OL} < V_{TD}$ as required for functional circuit, driver $MD2$ is biased into cutoff. The voltage across $C_{out,2}$ floats at an unknown value between 0 [V] and V_{DD} during this part of the clock cycle.

Figure 8.30(c) shows the circuit for the start of the next bit time when $\phi_1 = 1$. The voltage across $C_{in,2}$ is maintained at V_{OL}, so $MD2$ stays in cutoff. The clock voltage $\phi_1 = V_{DD}$ turns on load transistor $ML2$. Current thus flows to charge $C_{out,2}$ to V_{OH}, which is transferred to $C_{in,3}$ through pass gate $MP3$. $MD3$ conducts, allowing the voltage across $C_{out,3}$ to reach 0 [V].

The sequence of charge transfer described above brings out the important characteristics of the ratioed circuit. First, each inverter input acts as a soft node in the circuit operation. Thus, charge leakage gives a low-frequency limit for the clocks. Next, it is seen that the inverter outputs are not soft nodes since they gain voltage support from V_{DD} when the load MOSFETs are conducting. Consequently, there are no charge sharing problems in the pass transistors. Finally, note that power supply current flows only when a load MOSFET is biased on by a clock pulse. The overall power dissipation is thus lower than in a depletion load circuit.

Inverter design centers around setting β_R to give the desired V_{OL}, as expected for a ratioed behavior. Since it has been assumed that $\max(\phi) = V_{DD}$, the saturated E-mode load analysis of Section 3.3 is valid. The design equation is thus

$$\beta_R = \frac{[V_{DD} - V_{OL} - V_{TL}(V_{OL})]^2}{[2(V_{OH} - V_{TD})V_{OL} - V_{OL}^2]}, \tag{8.5-5}$$

which was given earlier as eqn. (3.3-33). Recall that $\beta_R > 1$ must be satisfied

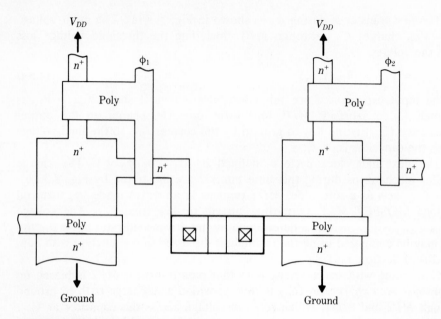

FIGURE 8.31 Ratioed shift register.

to yield an operational inverter. A layout example of the shift register is shown in Fig. 8.31.

Combinational logic can be implemented into the timing scheme using standard AOI structuring. Figure 8.32 provides an example of the technique. Pass transistors *MPA*, *MPB*, and *MPC* are controlled by ϕ_1. These gate the inputs to driver MOSFETs *MDA*, *MDB*, and *MDC* when $\phi_1 = 1$, with the input charge states held on the soft node capacitances. During ϕ_2, the load transistor *ML* and the output pass gate *MPF* are on. This allows transfer of the output state *F* to the next stage.

The combinational logic example above evaluates one logic function $F = \overline{AB + C}$ during ϕ_1. Since the output is not valid until the next half cycle when $\phi_2 = 1$, this may waste time if the clock period *T* is large. In this case it may be possible to perform more than one level of logic during the ϕ_1 time interval.

An example of a 2-level logic function is shown in Fig. 8.33. The first-level inputs are admitted during ϕ_1. The gate is configured to yield the combinational function *F*1 denoted in the diagram. This serves as an input into the second-level gate, along with X_0 and X_1. X_0 and X_1 may be gated by ϕ_1, although this is not shown explicitly in the circuit. The second-level logic gate produces an output *F*2, which constitutes the function transferred out of the 2-level network during ϕ_2.

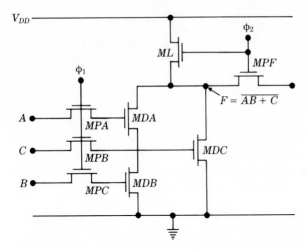

FIGURE 8.32 Basic AOI logic in dynamic 2ϕ ratioed circuit.

It is important to note that $F1$ and, hence, $F2$ are not valid logic levels until $\phi_2 = 1$. This arises because the load MOSFETs must be conducting to establish the final voltage levels in the ratioed scheme. For example, suppose that the input during ϕ_1 has $A = 0$. The voltage on the first-level output capacitance at $F1$ is not defined until $\phi_2 = 1$ allows the node to charge to a logic 1 level through $ML1$. Consequently, the overall output function $F2$ does not achieve a valid status until the ratioed evaluations are permitted during ϕ_2.

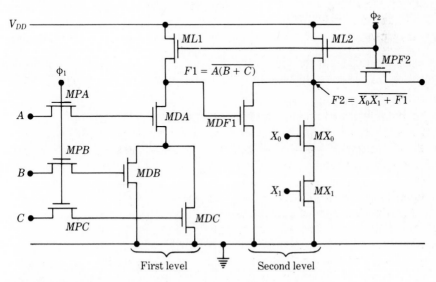

FIGURE 8.33 2-level logic formation in ratioed 2ϕ network.

8.5.2   Ratioless 2φ Logic

Ratioless logic can be achieved by choosing $\phi_{\text{in}} = \phi_1 = \phi_L$ and $\phi_{\text{out}} = \phi_2$ for the clocking in the generalized inverter of Fig. 8.27. This type of nMOS logic circuit is particularly interesting as it gives an output low voltage of $V_{OL} = 0\,[\text{V}]$. The term "ratioless" is used because this value is obtained from dynamic switching and is not affected by the driver-load ratio β_R. Minimum-area MOSFETs with $(W/L)_{\text{min}}$ can thus be used for every transistor in the circuit. Substantial real estate savings are possible using this type of circuitry.

Figure 8.34 shows a 3-stage ratioless shift register circuit. The only difference between this circuit and the ratioed configuration of Fig. 8.29 is the phasing of the load. In the present case, the load is in phase with the input pass transistors.

The charge transfer sequence through the shift register is illustrated by the subcircuits provided in Fig. 8.35. In Fig. 8.35(a), an input voltage $V_{\text{in}} = V_{DD}$ is applied. During ϕ_1, this is transmitted through $MP1$ to the input capacitance $C_{\text{in},1}$, resulting in V_{max} at the driver MOSFET $MD1$. V_{max} is still computed using eqn. (8.5-4).

The phasing of the load transistors shows that $ML1$ is conducting during this time interval. With $\phi_1 = V_{DD}$, $ML1$ is saturated while $MD1$ is nonsaturated. The output voltage across $C_{\text{out},1}$ is denoted by V_r. V_r is determined by the inverter β_R, so the voltage is indeed ratioed at this point. However, the output is not considered valid until the next half clock cycle, when transfer to the second stage occurs. V_r is thus classified as an intermediate dynamic logic state.

Charge transfer during ϕ_2 is illustrated by the subcircuit in Fig. 8.35(b). Consider the first-stage driver $MD1$. Since $V_{GS1} = V_{\text{max}}$, $MD1$ is biased into a conduction mode. The charge

$$Q_r = C_{\text{out},1} V_r, \tag{8.5-6}$$

which existed on $C_{\text{out},1}$ at the end of the ϕ_1 cycle, is lost to ground by the drain current flow I_{D1}. This establishes an output low voltage of $V_{OL} = 0\,[\text{V}]$ across

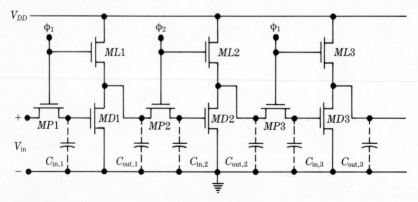

FIGURE 8.34   Ratioless shift register circuitry.

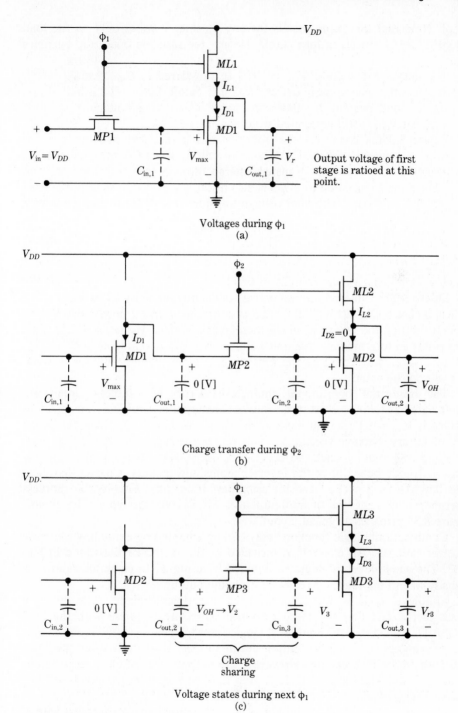

Output voltage of first stage is ratioed at this point.

Voltages during ϕ_1
(a)

Charge transfer during ϕ_2
(b)

Voltage states during next ϕ_1
(c)

FIGURE 8.35   Ratioless shift register charge transfer.

$C_{out,1}$. Note that this value results for any β_R since it relies only on the load MOSFET $ML1$ switching into cutoff. Hence, the ratioless operation becomes clear.

The output low voltage $V_{OL} = 0\,[\text{V}]$ is transferred to $C_{in,2}$ through $MP2$, which places the second-stage driver $MD2$ into cutoff. Since $ML2$ is biased into saturation, load current I_{L2} flows to charge $C_{out,2}$ to a voltage V_{OH}. It is obvious that V_{OH} is still computed from eqn. (8.5-1).

Figure 8.35(c) illustrates the voltages during the next half clock cycle, when $\phi_1 = 1$ is reestablished. $MD2$ is in cutoff, so the output node $C_{out,2}$ constitutes a soft node that transfers charge through pass transistor $MP3$ to $C_{in,3}$. Since the input node to $MD3$ is also a soft node, charge sharing between $C_{out,2}$ and $C_{in,3}$ occurs. The final voltage transferred to $MD3$ is thus given by

$$V_2 = V_3 = \left[\frac{1}{1 + (C_{in,3}/C_{out,2})}\right]V_{OH}, \tag{8.5-7}$$

which assumes that V_3 was initially at $0\,[\text{V}]$.

Driver $MD3$ is biased into an active conducting mode by V_3, while $\phi_1 = 1$ indicates that the load MOSFET $ML3$ is saturated. Current flows from V_{DD} to ground through the inverter, which then gives a voltage V_{r3} across $C_{out,3}$. This constitutes an intermediate dynamic voltage level and is analogous to V_r in Fig. 8.35(a). Note, however, that the difference in input voltages implies that the two will not be equal.

The important operational characteristics of the ratioless circuit are summarized as follows. First, the value of the output low voltage transferred to a stage is $V_{OL} = 0\,[\text{V}]$, regardless of the driver-load ratio. β_R does determine the temporary dynamic voltage V_r. This represents capacitor charge that must be gated to ground to achieve the value $V_{OL} = 0\,[\text{V}]$. Second, the clocking gives soft node behavior at the inverter inputs and outputs. Charge sharing is thus important for proper transfer using pass transistors. Finally, the ratioless operation allows the use of minimum-area MOSFETs throughout the circuit. Figure 8.36 provides a typical layout.

Combinational logic functions may be implemented by extending the basic inverter circuitry. An example is provided by the AOI gate illustrated in Fig. 8.37. The single-level 2ϕ logic is obtained by using ϕ_1 to clock all input pass transistors and the load, while ϕ_2 controls the output.

Multilevel logic is not generally used in ratioless circuits. The difficulty arises in the fact that the input pass transistors are in phase with the load MOSFET clocking. A multilevel logic arrangement would require that cascaded stages yield a stable output during the half clock cycle when the input and load MOSFETs are on. However, the output of a ratioless stage is not valid until after the input-load clock goes to 0, so a multilevel cascade will not function properly.

One way around this problem is to use ratioed gates cascaded into a

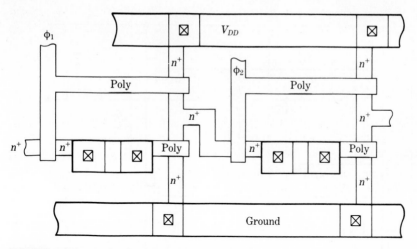

FIGURE 8.36 Ratioless shift register layout.

ratioless stage, as shown in Fig. 8.38. In this circuit, the input AOI gate performs the function

$$F = \overline{AC + B}. \tag{8.5-8}$$

Since the output F feeds the next stage, the voltage at this point must be stable while $\phi_1 = 1$. This can be accomplished by standard ratioed design. The

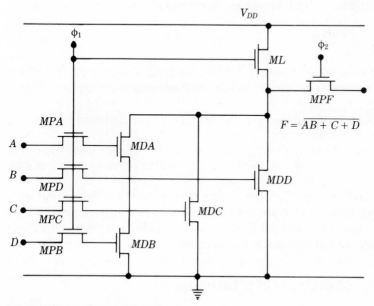

FIGURE 8.37 Ratioless AOI gate.

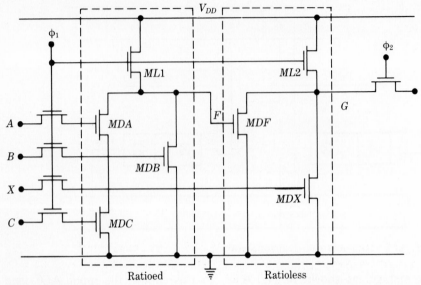

FIGURE 8.38   2-level logic using a ratioed gate cascaded into a ratioless circuit.

simplest design choice is to take

$$\left(\frac{W}{L}\right)_A = \left(\frac{W}{L}\right)_C = 2\left(\frac{W}{L}\right)_B \tag{8.5-9}$$

with the load MOSFET $ML1$ having an aspect ratio

$$\left(\frac{W}{L}\right)_{L1} = \frac{(W/L)_B}{\beta_R}, \tag{8.5-10}$$

where β_R is the equivalent inverter driver-load ratio computed from eqn. (8.5-5). Using a ratioless design in this stage will not produce acceptable logic 0 levels during ϕ_1.

The second stage is ratioless, so the MOSFETs are designed with aspect ratios of

$$\left(\frac{W}{L}\right)_F = \left(\frac{W}{L}\right)_X = \left(\frac{W}{L}\right)_{L2} = \left(\frac{W}{L}\right)_{min} \tag{8.5-11}$$

in a minimum-area layout. Ratioless design may be used for the second stage since the output G is a soft node connected to a pass transistor gated by ϕ_2. The use of this type of multilevel logic allows for additional logic calculations but increases the overall real estate requirement.

8.6   Voltage Bootstrapping Techniques

Bootstrapping generically refers to methods to "pull up" something. The term originates from the concept of "pulling up one's boots by the straps" and is

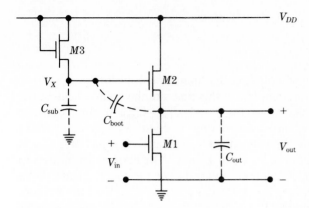

FIGURE 8.39 Enhancement load inverter with bootstrapped output.

loosely applied in electrical engineering to various situations. In context here, bootstrapping implies dynamic MOS circuit techniques to pull up voltages in a manner that overcomes threshold voltages losses.

The basic features of bootstrapping can be illustrated by examining the modified enhancement-load inverter shown in Fig. 8.39. MOSFET $M3$ is the bootstrapping transistor. As will be seen below, $M3$ allows the inverter to attain an output high voltage of $V_{OH} = V_{DD}$. In the simpler case of a saturated E-mode load, the gate of $M2$ would be connected to V_{DD}, giving

$$V_{OH} = V_{DD} - V_{T2}(V_{OH}). \tag{8.6-1}$$

The bootstrapped circuit, on the other hand, derives the output high voltage using dynamic charge storage to create a bootstrap voltage of

$$V_X \geq V_{DD} + V_{T2}(V_{DD}) \tag{8.6-2}$$

on the gate of $M2$. This allows full transmission of the power supply level V_{DD} to the output node through $M2$.

To analyze the circuit operation, assume initially for times $t < 0$ that $V_{in} = V_{DD}$ and $V_{out} = V_{OL}$. The basic inverter pair $M1$, $M2$ is ratioed. V_{OL} is thus found using current equations. To this end, note that the driver is nonsaturated with a drain current of

$$I_{D1} = \frac{\beta_1}{2}[2(V_{DD} - V_{T1})V_{OL} - V_{OL}^2]. \tag{8.6-3}$$

The gate of $M2$ is at voltage V_X such that

$$V_X = V_{DD} - V_{T3}(V_X). \tag{8.6-4}$$

Consequently, $M2$ is operating in the saturated mode with a drain current given by

$$I_{D2} = \frac{\beta_2}{2}(V_X - V_{OL} - V_{T2})^2. \tag{8.6-5}$$

Equating currents then gives an equation that may be solved for V_{OL}.

The important bootstrap parameters at this time are the voltages across the two parasitic capacitances shown as C_{boot} and C_{sub}. C_{boot} is the bootstrap capacitor that dynamically couples the output node to V_X. In the present circuit, C_{boot} is approximately the gate-source capacitance $C_{GS,2}$ of $M2$. C_{sub} is the substrate capacitance that couples V_X to ground. The main contributions to C_{sub} are from C_{line} and the n^+p junction capacitance at the source of $M3$. With the stable logic 1 input voltage, the voltage V_X across C_{sub} is given by eqn. (8.6-4), while C_{boot} has a voltage $(V_{OL} - V_X) < 0$.

The bootstrapping of the output occurs when V_{in} is switched from V_{DD} to $0\,[\text{V}]$. This causes V_{out} to increase from V_{OL}, corresponding to the charging of C_{out} through $M2$. The change in voltage $(dV_{out}/dt) > 0$ is coupled to node X by C_{boot}, which will be shown to increase V_X. This mechanism will then allow the output node to charge all the way up to V_{DD}.

Figure 8.40 shows the capacitive subcircuit used to analyze the transients, where i_b is the current into C_{boot} and i_s represents the current that charges C_{sub}. In the lowest order of approximation,

$$i_b \simeq i_s = i, \tag{8.6-6}$$

so

$$i \simeq C_{boot}\frac{d(V_{out} - V_X)}{dt} \simeq C_{sub}\frac{dV_X}{dt}. \tag{8.6-7}$$

This may be rearranged to give

$$C_{boot}\frac{dV_{out}}{dt} \simeq (C_{boot} + C_{sub})\frac{dV_X}{dt}, \tag{8.6-8}$$

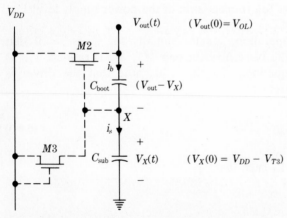

FIGURE 8.40 Bootstrapping capacitor subcircuit.

or

$$\frac{dV_X}{dt} \simeq \left(\frac{C_{boot}}{C_{boot} + C_{sub}}\right)\frac{dV_{out}}{dt}. \tag{8.6-9}$$

Since $(dV_{out}/dt) > 0$, this shows that $(dV_X/dt) > 0$. Consequently, $V_X(t)$ increases with time.

This equation may be used to predict the general behavior of the bootstrapping arrangement. Multiplying by dt and integrating gives

$$\int_{V_{DD}-V_{T3}}^{V_{X,max}} dV_X \simeq \left(\frac{C_{boot}}{C_{boot} + C_{sub}}\right)\int_{V_{OL}}^{V_{DD}} dV_{out}, \tag{8.6-10}$$

where it has been assumed that the output voltage changes from V_{OL} to V_{DD}. Solving for the maximum value of V_X gives

$$V_{X,max} \simeq V_{DD} - V_{T3} + r(V_{DD} - V_{OL}), \tag{8.6-11}$$

with

$$r = \frac{C_{boot}}{C_{boot} + C_{sub}} \tag{8.6-12}$$

the capacitance ratio. This demonstrates that the final value of V_X depends on the relative capacitance levels. The theoretical upper limit occurs when $C_{boot} \gg C_{sub}$, so $r = 1$. In this case,

$$V_{X,max} \simeq 2V_{DD} - V_{T3} - V_{OL}. \tag{8.6-13}$$

It is obvious that bootstrapping the gate of the load MOSFET $M2$ allows V_{out} to reach the full power supply value of V_{DD}.

The design problem centers around setting the values of the capacitances. To find the constraint on the relative values needed for C_{boot} and C_{sub}, note that the minimum value of V_X needed to bring V_{out} to V_{DD} is

$$V_{X,min} = V_{DD} + V_{T3}. \tag{8.6-14}$$

Using this as an upper limit on the left side of eqn. (8.6-10) gives the critical capacitance relation as

$$r \simeq \frac{2V_{T3}}{V_{DD} - V_{OL}}. \tag{8.6-15}$$

This may be rearranged to the form

$$\left(\frac{C_{boot}}{C_{sub}}\right)_{min} \simeq \frac{2V_{T3}}{V_{DD} - V_{OL} - 2V_{T3}}, \tag{8.6-16}$$

which gives the minimum value of C_{boot} relative to C_{sub} to bootstrap V_{out} to V_{DD}.

The dynamic properties of the bootstrapping technique are important to the overall operation. First, it is seen that the charge stored on C_{sub} is subject to leakage through $M3$. Consequently, some refreshing will be necessary. This can be accomplished by periodically switching $V_{in}(t)$.

Ensuring that C_{boot} is much larger than C_{sub} is critical in the bootstrapping technique. This can often be accomplished by careful layout of $M2$ and $M3$. Alternately, an extra transistor can be wired as a MOS capacitor that functions as C_{boot}. This is shown in Fig. 8.41. Figure 8.41(a) shows the simplified symbol for a MOS capacitor obtained by connecting the drain and source together. In Fig. 8.41(b), the MOS capacitor is inserted to act directly as C_{boot} in the circuit. In this case, the value of the capacitance is simply C_G.

Now note that eqn. (8.6-9) shows that V_X changes at a rate proportional to (dV_{out}/dt). The upper limit on the charge rate is

$$\left(\frac{dV_X}{dt}\right)_{max} = \frac{dV_{out}}{dt}, \qquad\qquad (8.6\text{-}17)$$

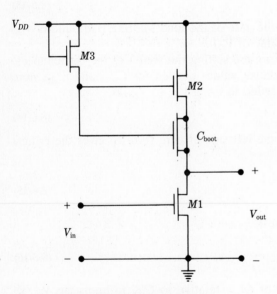

MOS capacitor
(a)

Use as a bootstrap capacitor
(b)

FIGURE 8.41   Bootstrapping with an MOS capacitor.

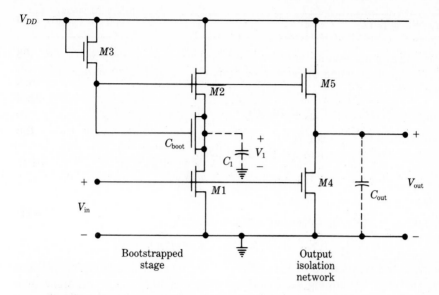

FIGURE 8.42   Isolation network for large-output capacitance.

which occurs when $r = 1$. If C_{out} is large, (dV_{out}/dt) is small, so the change in $V_{out}(t)$ will be relatively slow. In this case, the charge rate (dV_X/dt) can be increased by isolating the bootstrapping circuit from C_{out} as shown in Fig. 8.42. MOS capacitor C_{boot} now couples V_X to the drain of MOSFET $M1$, which is characterized by a small nodal capacitance C_1. In this case, a first-order approximation is

$$\frac{dV_X}{dt} \simeq r \frac{dV_1}{dt},$$

(8.6-18)

where

$$\frac{dV_1}{dt} \propto \frac{1}{C_1}.$$

(8.6-19)

Since C_1 will be much smaller than C_{out}, the bootstrap voltage V_X will rise much more rapidly than if it were connected directly to C_{out}. Of course, implementation of this circuit adds two transistors to the device count. Also note that this simple analysis ignores the overlap capacitance of $M5$, which gives weak coupling to C_{out}.

Although there are many possible variations for bootstrapped circuits, all tend to be based on the concepts presented here. Bootstrapping is a particularly useful technique to remember when designing high-density dynamic logic circuits.

EXAMPLE 8.6-1

The bootstrapping circuit can be understood by constructing an inverter as shown in Fig. 8.39 using discrete capacitors C_{boot} and C_{sub}. For this example, assume

$$C_{boot} = 70\,[fF], \qquad C_{sub} = 10\,[fF],$$

so the capacitance ratio is $r = 7/8$.

Assume that the devices are specified by

$$\left(\frac{W}{L}\right)_1 = \frac{35}{5} = 7, \qquad \left(\frac{W}{L}\right)_2 = \frac{5}{10} = 0.5 = \left(\frac{W}{L}\right)_3$$

and for $V_{DD} = 5\,[V]$ the important parameters are

$$k' = 25\,[\mu A/V^2], \qquad \gamma = 0.37\,[V^{1/2}], \qquad 2\,|\phi_F| = 0.6\,[V], \qquad V_{T0} = 1\,[V].$$

To analyze the bootstrapping arrangement, the inverter consisting of $M1$ and $M2$ is analyzed. Assuming saturated operation of $M2$, the calculations give $V_{OL} \simeq 0.05\,[V]$.

To find $V_{X,max}$ use eqn. (8.6-11). Including body bias on $M3$ gives

$$V_{X,max} \simeq (5 - 1 + 4.33) - 0.37(\sqrt{V_{X,max} + 0.6} - \sqrt{0.6}).$$

This self-iterates to give $V_{X,max} \simeq 7.56\,[V]$.

The SPICE file for analyzing the circuit is given below, with the plot

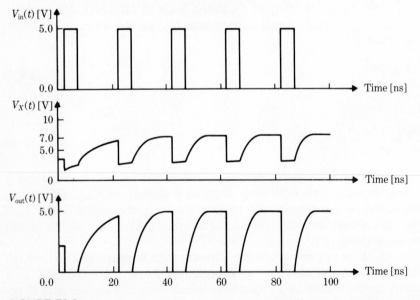

FIGURE E8.2

showing $V_{in}(t)$, $V_X(t)$, and $V_{out}(t)$ shown in Fig. E8.2. (The output capacitance affects only the switching times, so it has been chosen arbitrarily small.)

```
BOOTSTRAPPED NMOS CIRCUIT EXAMPLE
VDD 3 0 DC 5VOLTS
M1 2 1 0 0 EMODE L=5U W=35U
M2 3 4 2 0 EMODE L=10U W=5U
M3 3 3 4 0 EMODE L=10U W=5U
CSUB 4 0 1.0E-14
CBOOT 4 2 7.0E-14
VS 1 0 DC PULSE(0 5 2NS 1PS 1PS 5NS 20NS)
CLOAD 2 0 3.0E-15
.MODEL EMODE NMOS VTO=1 GAMMA=0.37 KP=2.5E-5
.TRAN 1NS 100NS
.PLOT TRAN V(2)
.END
```
■

8.7 Precharging

Precharging is a dynamic circuit technique that increases switching speed while simultaneously reducing the power dissipation. It is most useful when applied to nodes with large values of capacitance.

The concept is quite simple and can be understood by referring to the example in Fig. 8.43. When $\phi = 1$, MOSFET $M1$ acts as a pass gate between V_{DD} and the output node. This allows the output capacitance C_{out} to charge and is thus termed the *precharge* portion of the clocking cycle.

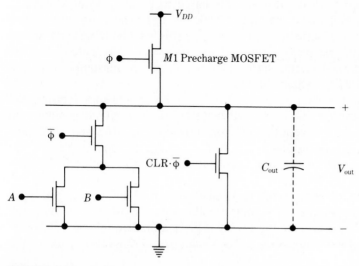

FIGURE 8.43 Precharge example.

Logic is performed when $\phi = 0$ and the inputs are admitted to the driver transistors. This constitutes the *evaluation* portion of the logic cycle. The output node remains at the precharge voltage level unless the input logic provides a discharge path to ground. For example, a clear input of CLR = 1 permits current flow out of C_{out}. This is called a *conditional discharge* since it occurs only when the input conditions specify a logic 0 output.

The advantages of precharging are seen from this example. Recall that the charging time is proportional to C_{out}. Large values of C_{out} thus result in long charging times. Since the output is precharged before the inputs are admitted, the logic propagation time is not affected. In particular, a logic 1 output state is immediately available, while a logic 0 output state requires a time (see eqn. 8.1-30)

$$t_{HL} \simeq \frac{2.94 C_{out}}{\beta_D (V_{DD} - V_{TD})},$$ (8.7-1)

assuming that $\phi = V_{DD}$. β_D and V_{TD} characterize the driver (or set of drivers) that is providing the path to ground.

The savings become even more apparent when it is noted that the precharge MOSFET $M1$ has a gate-source voltage of $V_{GS1} = (V_{DD} - V_{out})$. V_{GS1} thus decreases as C_{out} charges, reducing the current flow level with time. The inherent problem of large t_{LH} values associated with enhancement-load MOSFETs becomes irrelevant, as the logic propagation time is the only important factor.

Power dissipation is also reduced when precharging is used. The timing of the precharge event precludes the possibility of creating a direct path between V_{DD} and ground. Current is first used to precharge C_{out}. If the inputs are set so that $V_{out} \to 0$, then the capacitor will discharge to ground, giving dissipated power. On the other hand, if the output evaluates to a logic 1, no current flow path exists and the charge is held on C_{out}. This simply represents stored electric energy in the capacitor. Consequently, no power is dissipated when the capacitor is maintained in this state. The overall power consumption of the circuit is thus limited to dynamic usage.

Figure 8.44 provides an example of a dynamic ratioless shift register that performs logic using precharge and evaluate cycles. The circuit is similar to that shown in Fig. 8.34, except that evaluation MOSFETs $ME1$ and $ME2$ have been added. This introduces new soft nodes, which are denoted by $X1$ and $X2$. In addition, the capacitances are now split so that C_{X1} and C_{X2} become important.

The operation of the circuit is straightforward. During ϕ_1, the input voltage sets the charge on $C_{in,1}$. A logic 1 voltage level allows C_{X1} to discharge to 0 [V] through $MD1$, while a logic 0 input gives an indeterminate voltage on node $X1$. $ML1$ is conducting, so $C_{out,1}$ is precharged during this portion of the clock cycle.

When $\phi_2 = 1$, evaluation MOSFET $ME1$ is switched on. If the voltage on

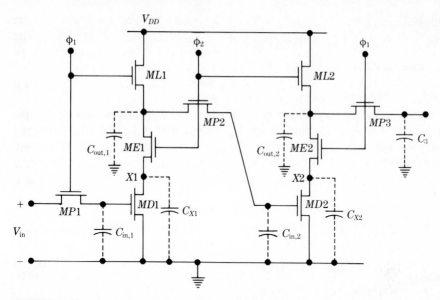

FIGURE 8.44 Dynamic ratioless shift inverter with precharge/evaluation modes.

$C_{in,1}$ is high $(>V_{TD})$, then $MD1$ conducts and allows $C_{out,1}$ and $C_{in,2}$ to discharge to a value of $V_{OL} = 0\,[\text{V}]$. On the other hand, a low voltage $(<V_{TD})$ on $C_{in,1}$ places $MD1$ in cutoff. This results in charge sharing from the precharged capacitor $C_{out,1}$ to C_{X1} and $C_{in,2}$. The subcircuit for this case of charge sharing is shown in Fig. 8.45.

The analysis of the charge sharing again revolves around setting capacitor ratios. Assuming clock levels with $\max(\phi) = V_{DD}$, $C_{out,1}$ is initially charged to

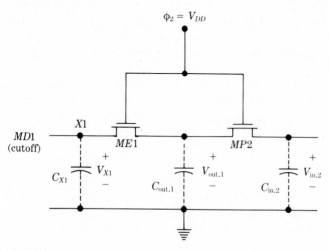

FIGURE 8.45 Charge sharing circuit during ϕ_2 with $V_{in} < V_{TD}$.

a voltage

$$V_{out,1} = V_{max} = V_{DD} - V_{TL1}(V_{max}), \tag{8.7-2}$$

which represents a total charge of

$$Q_T = C_{out,1}V_{max}. \tag{8.7-3}$$

The worst-case situation occurs when both C_{X1} and $C_{in,2}$ are initially uncharged with $V_{X1} = 0 = V_{in,2}$. Q_T must then be shared with C_{X1} and $C_{in,2}$ with the stipulation that the final voltage V_f across the three parallel-connected capacitors satisfy

$$V_f > V_{TD2} \tag{8.7-4}$$

to ensure that driver $MD2$ has a logic 1 level at its input.

The final voltage on the capacitor network is computed by using conservation of charge to write

$$Q_T = (C_{X1} + C_{out,1} + C_{in,2})V_f. \tag{8.7-5}$$

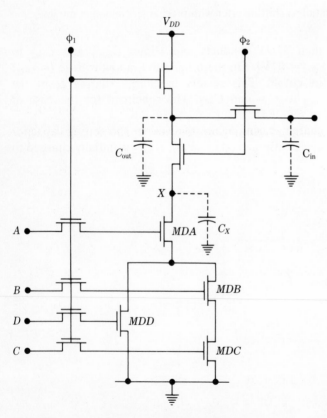

FIGURE 8.46 Combination single-level logic gate using precharge/evaluate modes.

Solving for V_f using eqn. (8.7-3) gives

$$V_f = \left(\frac{C_{out,1}}{C_{X1} + C_{out,1} + C_{in,2}} \right) V_{max}. \tag{8.7-6}$$

The value of $V_{in,2} = V_f$ at the gate of $MD2$ is thus dependent on the relative capacitance values. Maximizing V_f is accomplished by designing the circuit such that

$$C_{out,1} \gg C_{X1}, \qquad C_{out,1} \gg C_{in,2}. \tag{8.7-7}$$

Obviously, care must be taken in the layout to ensure that a reasonable amount of charge transfer is possible.

Single-level combinational logic may be implemented using this circuit as a basis. An example is shown in Fig. 8.46. The precharge and evaluate modes operate on the same principles discussed above. The only difference is that the logic evaluation during ϕ_2 involves multiple input states.

REFERENCES

(1) M. I. Elmasry (ed.), *Digital MOS Integrated Circuits*, New York: IEEE Press (Wiley), 1981.

(2) L. A. Glasser and D. W. Dobberpuhl, *The Design and Analysis of VLSI Circuits*, Reading, MA: Addison-Wesley, 1985.

(3) M. J. Howes and D. V. Morgan, *Large Scale Integration*, Bath, UK: Wiley, 1981.

(4) C. A. Mead and L. Conway, *Introduction to VLSI Systems*, Reading, MA: Addison-Wesley, 1980.

(5) J. Mavor, M. A. Jock, and P. B. Denyer, *Introduction to MOS LSI Design*, London: Addison-Wesley, 1983.

(6) A. Mukherjee, *Introduction to nMOS and CMOS VLSI System Design*, Englewood Cliffs, NJ: Prentice-Hall, 1986.

(7) S. Muroga, *VLSI System Design*, New York: Wiley, 1982.

(8) D. A. Pucknell and K. Eshraghian, *Basic VLSI Design*, Sydney: Prentice-Hall, 1985.

(9) B. Randell and P. C. Treleaven, *VLSI Architecture*, London: Prentice-Hall, 1983.

PROBLEMS

Assume nMOS parameters of $k' = 30 \, [\mu A/V^2]$ and

$$V_{T0} = +0.8 \, [V] \text{(E-mode)}, \quad V_{T0} = -3.3 \, [V] \text{(D-mode)}, \quad V_{DD} = 5 \, [V],$$
$$V_{T0} = +0.4 \, [V] \text{(pass)}, \quad 2|\phi_F| = 0.58 \, [V], \quad \gamma = 0.37 \, [V^{1/2}]$$

and capacitance levels of

$$C_{ox} = 0.691 \, [\text{fF}/\mu\text{m}^2], \quad C_o = 0.207 \, [\text{fF}/\mu\text{m}],$$

$$C_{j0} = 0.0975 \, [\text{fF}/\mu\text{m}^2], \quad \phi_o = 0.879 \, [\text{V}],$$

$$C_{jsw} = 0.209 \, [\text{fF}/\mu\text{m}], \quad \phi_{osw} = 0.957 \, [\text{V}],$$

and

$$C_{p-f} = 0.0576 \, [\text{fF}/\mu\text{m}^2], \quad C_{m-p} = 0.0863 \, [\text{fF}/\mu\text{m}^2],$$

$$C_{m-f} = 0.0345 \, [\text{fF}/\mu\text{m}^2], \quad C_{m-n^+} = 0.0822 \, [\text{fF}/\mu\text{m}^2],$$

for all problems unless otherwise stated.

8.1   Consider the chain of pass transistors shown in Fig. 8.4(a). Calculate the value of $V_{max} \, [\text{V}]$ transmitted through the chain for the process above.

8.2   Find the values of $V_{1,max}$ and $V_{2,max}$ shown in Fig. 8.4(b) for this process. Calculate the percent error in $V_{2,max}$ if body bias effects are ignored.

8.3   A pass transistor is connected to the input of an inverter as shown in Fig. 8.2. The input capacitance is approximated as a constant value of $C_{in} = 80 \, [\text{fF}]$. The pass transistor has an aspect ratio of $(W/L)_P = 2$. Assume that max $(\phi) = V_{DD}$.

   (a) Calculate the charging time constant τ_{ch} ignoring body bias effects. Does body bias increase or decrease this value? Use the worst-case possibility for the rest of the problem.

   (b) Plot $V_{in}(t)$ for the charging event. Then find t_{LH}.

   (c) Now calculate the discharging time constant τ_{dis} and then plot $V_{in}(t)$ for this case. Calculate t_{HL} for the circuit.

   (d) The leakage current through the pass transistor in a cutoff state is estimated to be $20 \, [\text{pA}]$. The inverter is designed to interpret the input voltage as a logic 1 level to a minimum value $V_{low} = 3.20 \, [\text{V}]$. Calculate the time t that a logic 1 can be held on the input node.

   (e) Now consider the inverter circuitry. Calculate the value of β_R needed to produce $V_{OL} = 0.30 \, [\text{V}]$ when $V_{in} = V_{low}$. Repeat for the case where V_{low} is reduced to $3.0 \, [\text{V}]$.

   * (f) Simulate the charging and discharging of C_{in} through the pass transistor using SPICE. Then specify the leakage current in the .MODEL parameters and examine the charge leakage from C_{in} as a function of time.

8.4   The pass transistor in Fig. P8.1 feeds the input of an inverter. The n^+/p-substrate zero-bias leakage current is $J_{go} = 1.13 \, [\text{fA}/\mu\text{m}^2]$. The circuit uses a buried contact to connect the n^+ and poly levels. The doping under the poly contact region increases the leakage current to $J_{go,bur} = 1.10 J_{go}$ and decreases the junction capacitance to $C_{j0,bur} = 0.91 C_{j0}$.

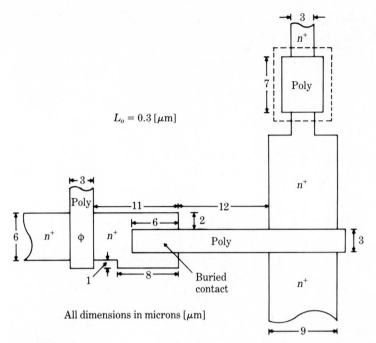

FIGURE P8.1

(a) Calculate the minimum value of the depletion capacitance due to the n^+/p-substrate region. Include the gate overlap region.

(b) Find the MOS capacitance due to the poly interconnect and the inverter gate. Then sum to obtain the minimum storage capacitance.

(c) Calculate the maximum value of the reverse leakage through the n^+-well to the substrate. Ignore sidewall contributions, but distinguish between the "normal" n^+p junction and that under the buried contact.

(d) Use the dimensions shown on the drawing (accounting for gate overlap) and find the value of β_R for the inverter. Then estimate V_{IH}.

(e) Calculate the holding time for a logic 1 charge state using the worst-case values of capacitance and current.

(f) Calculate the minimum and maximum values of the charge stored on the node in a logic 1 state. How many electrons does this correspond to?

8.5 Use eqns. (8.3-18) and (8.3-19) to obtain a plot of V_{IH} as a function of β_R for the process and circuit parameters provided at the beginning of the problem set.

8.6 Consider the charge sharing problem illustrated in Fig. 8.23.

(a) Plot V_f as a function of the capacitance ratio (C_2/C_1) in the range 1 to 20, assuming the initial conditions shown on the drawing.

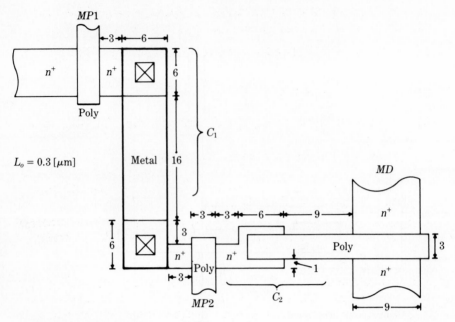

All dimensions in microns [μm]

FIGURE P8.2

- **(b)** Suppose that the initial voltages on the capacitors are set at $V_1(0) = 0\,[\text{V}]$ and $V_2(0) = (V_{DD} - V_{TP})$. Find V_f for capacitor values $C_1 = 10C_2$, $C_1 = C_2$, and $C_1 = 0.1C_2$. Discuss the characteristics of each case.
- **(c)** Now let $C_1 = 100\,[\text{fF}] = 5C_2$, and assume that $V_1(0) = (V_{DD} - V_{TP})$ while $V_2(0) = 1.2\,[\text{V}]$. Find the time constant for this case and then plot $V_1(t)$ and $V_2(t)$.
- ***(d)** Simulate the circuit in part (c) using SPICE.

8.7 Figure P8.2 shows the layout for two pass transistors $MP1$ and $MP2$. The pass MOSFETs are connected by a metal layer, while $MP2$ drives MOSFET MD by means of a buried contact. Assume that the buried contact region is described by the characteristics introduced in Problem 8.4.

- **(a)** Calculate the minimum value of C_1 (between the two pass transistors).
- **(b)** Calculate the maximum value of C_2.
- **(c)** Assume that C_2 is initially uncharged. What is the largest voltage that can be transferred to it when charge sharing occurs? Discuss the implications this has on the design of MD (which is the driver for a logic gate).

8.8 Design a ratioed half-adder circuit using 2-phase clocking. Then extend the design to a full-adder in multilevel logic. (Use E-mode loads.)

8.9 Implement an AOI full-adder in ratioless 2-phase logic. (Refer to Fig. 6.21a.)

8.10 Consider the bootstrapping circuit shown in Fig. 8.39. Assume that $C_{\text{sub}} = 50\,[\text{fF}]$.

(a) Calculate the value of V_{OH} if bootstrapping were not used. Include body bias.

(b) What is the smallest value of V_X that will give $V_{OH} = V_{DD}$?

(c) Assume that $V_{OL} = 0.25\,[\text{V}]$ is set by the circuit design. Find the minimum value of C_{boot} needed to bootstrap the output to V_{DD}. Ignore body bias for this calculation.

(d) Suppose that $C_{\text{boot}} = 125\,[\text{fF}]$ is set. Find the capacitance ratio r. Then use this in eqn. (8.6-11) to find a self-iterating equation for $V_{X,\text{max}}$. Solve the equation.

8.11 The bootstrapping circuit of Fig. 8.41 is used to give $V_{OH} = V_{DD}$. The layout of MOSFETs $M2$ and $M3$ is shown in Fig. P8.3 for this case. $M3$ is connected to the gate of $M2$ by a buried contact. Assume that the buried contact has the properties listed in Problem 8.4.

(a) Find the maximum value of C_{sub} as defined by the $M2 - M3$ layout. Assume that the poly gate contribution from $M2$ ends at the edge of the active area.

(b) Calculate the value of C_{boot} needed to establish a capacitor ratio of $r = 0.7$. Then find the dimensions of the MOSFET gate needed to implement the design. Assume a minimum linewidth of 3 [μm], and ignore the interconnect capacitance (poly-to-field) from $M2$ to C_{boot}.

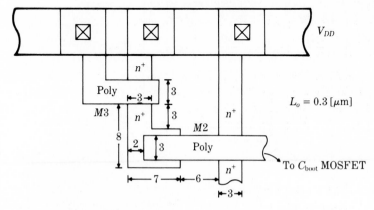

All dimensions in microns [μm]

FIGURE P8.3

(c) Assume that $r = 0.7$ has been established. Calculate the maximum voltage $V_{X,\text{max}}$ including body bias effects.

(d) Calculate the worst-case (i.e., highest) leakage current through the n^+-substrate junction.

(e) Combine the results from parts (b)–(d) and calculate the time that V_X can hold a high enough value to produce $V_{OH} = V_{DD}$ at the output.

* (f) Perform a SPICE simulation on the circuit. Investigate the voltage buildup of both V_X and V_{out} using a transient input.

*8.12 Design a bootstrapped network with output isolation as illustrated in Fig. 8.42. Simulate the response using SPICE, and track V_1, V_X, and V_{out} for different values of C_{out}.

8.13 Consider the precharging circuit shown in Fig. 8.43. The output capacitance is approximated as $C_{\text{out}} = 15 \,[\text{pF}]$. The clock has a frequency of $f = 5 \,[\text{MHz}]$ (with a 50% duty cycle) and $\max(\phi) = V_{DD}$.

(a) Calculate V_{OL}, V_0, and V_1 for the circuit.

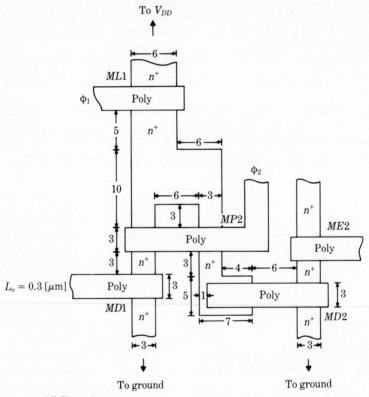

All dimensions
in microns [μm]

FIGURE P8.4

(b) Find the aspect ratio (W/L) of the precharge transistor needed to ensure that C_{out} reaches V_1 within a time $(T/2)$ (T = clock period).

(c) Find the minimum aspect ratio for an evaluate MOSFET that can discharge C_{out} to V_0 within one half of a clock cycle.

8.14 A ratioless shift register was shown in Fig. 8.44. Figure P8.4 provides a possible layout for the first stage of the circuit. Assume that the clock signals have a maximum value of $V_{DD} = 5\,[\text{V}]$.

(a) Calculate the zero-bias value of $C_{out,1}$ from the layout geometry. Then find the minimum value of this capacitor.

(b) Find the maximum value of C_{X1}.

(c) Find the maximum value of $C_{in,2}$ using a buried contact capacitance $C_{j0,bur} = 0.91\,C_{j0}$ as described in Problem 8.4.

(d) Calculate the total charge Q_T stored on $C_{out,1}$ during a precharge.

(e) Find the final voltage V_f in the worst case.

*8.15 Write a computer program that solves the charge leakage differential equation (8.1-52) for $V_{in}(t)$. Use C_L and $C_{j0}A$ as the inputs. (A finite difference approach is straightforward; alternately, just multiply by dt and increment over time.)

8.16 A better analytic approximation for charge leakage is obtained by writing eqn. (8.1-52) as

$$\frac{dV_{in}}{dt} \simeq -\frac{I_{go}}{C}\left(1 + \frac{V_{in}}{\phi_o}\right)^{1/2},$$

where C is a constant capacitance. Noting the simple integral

$$\int \frac{dx}{\sqrt{a + bx}} = \frac{2\sqrt{a + bx}}{b},$$

this can be integrated directly for $V_{in}(t)$.

(a) Find $V_{in}(t)$ by this approach, being sure to apply the initial condition $V_{in}(0) = (V_{DD} - V_{TP})$. Comment about the range of validity for the solution.

(b) Plot $V_{in}(t)$ from this approximation, and thus find an expression for the time t that a logic 1 state can be held on the soft node.

8.17 A dynamic random-access-memory ($DRAM$) cell is shown in Fig. P8.5. This cell uses four MOSFETs. Pass transistors $MP0$ and $MP1$ are used to allow data (charge) in and out and are controlled by the *word line* voltage, which chooses a cell in the chip array. The actual data storage takes place in the dynamic flip-flop formed by cross-coupled MOSFETs $M1$ and $M2$. The important capacitances are the storage capacitance C_{stor} and the *bit line* (data paths) capacitance C_{line}.

(a) Describe how data can be written to the cell using the word line select voltage and apply voltages to the bit lines.

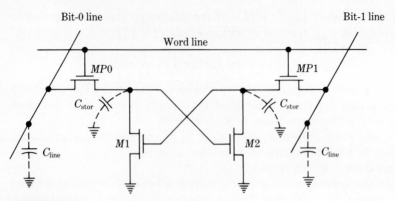

FIGURE P8.5

(b) Examine the storage capacitance C_{stor}. Find the general expression for estimating the capacitances that are used to store the charge. Where are the basic charge leakage paths in the circuit?

(c) Discuss the problem of charge sharing when a cell is read (i.e., the word line is brought high to access the storage capacitors).

8.18 A single-MOSFET DRAM cell is shown in Fig. P8.6. This has the problems of charge sharing and charge leakage associated with many dynamic circuits. For this problem, assume that the storage capacitor has a value $C_{cell} = 50$ [fF] and the bit line capacitance is $C_{line} = 0.9$ [pF].

(a) Find the maximum charge that is stored on the cell capacitance.

(b) Assume that the bit line is initially uncharged. Calculate the final voltage on the bit line after a cell with maximum charge stored is accessed. Why is this operation termed a *destructive read* event?

(c) The low voltage calculated in part (b) gives motivation for a precharge network for the bit line. Use E-mode transistors to design a network that would precharge C_{line} to a value close to $(V_{DD}/2)$. Keep the design simple.

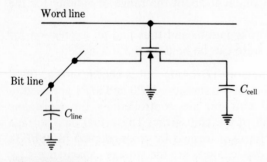

FIGURE P8.6

CHAPTER 9

Synchronous CMOS Logic

The concepts developed in the previous chapter are now applied to CMOS. This results in circuit structures that resemble the nMOS versions. However, the availability of nMOS and pMOS transistors allows for higher-level system design approaches to be developed. These are introduced and analyzed in detail.

9.1 Switching Properties of CMOS Transmission Gates

CMOS transmission gates (TGs) were introduced in Section 6.7 as a means of implementing static switch logic. Transmission gates are also useful for certain types of synchronous CMOS circuits, in which their transient characteristics become important. In addition, the analysis in this section is applicable to more general types of synchronous CMOS logic configurations.

Recall that a CMOS TG is constructed by paralleling an nMOS transistor

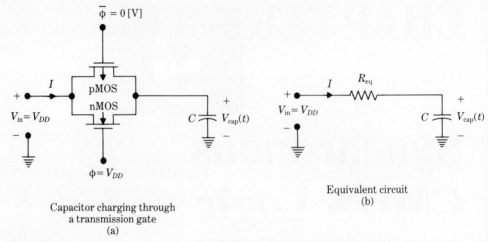

FIGURE 9.1   Basic transmission gate modeling scheme.

with a pMOS transistor. Figure 9.1(a) shows a circuit where the input voltage $V_{in} = V_{DD}$ is used to charge a capacitor C through a transmission gate. The TG is biased into the conducting mode by applying control signals of $\phi = V_{DD}$ to the nMOS gate and $\bar{\phi} = 0\,[V]$ to the pMOS gate. The current I flows as shown to charge the capacitor.

The simplest model for the TG is obtained by using the equivalent resistance

$$R_{eq} = \frac{R_n R_p}{R_n + R_p}$$

$$\simeq \text{Constant.} \qquad\qquad (9.1\text{-}1)$$

The lowest-order equivalent circuit for the capacitor charging is then the simple series RC shown in Fig. 9.1(b). Assuming that $V_{cap}(t = 0) = 0$, the capacitor voltage is described by

$$V_{cap}(t) = V_{DD}(1 - e^{-t/\tau_{TG}}). \qquad\qquad (9.1\text{-}2)$$

where

$$\tau_{TG} = R_{eq}C \qquad\qquad (9.1\text{-}3)$$

is the transmission gate time constant. A discharge event with $V_{in} = 0\,[V]$ and $V_{cap}(t = 0) = V_{DD}$ gives a capacitor voltage decay of

$$V_{cap}(t) = V_{DD}e^{-t/\tau_{TG}}. \qquad\qquad (9.1\text{-}4)$$

This level of modeling shows that R_{eq} and C are the important transient quantities. R_{eq} is a sensitive function of the MOSFET aspect ratios $(W/L)_n$ and $(W/L)_p$, while C depends on the TG geometries, the layout, and the load characteristics.

The switching properties important to synchronous CMOS logic can be extracted by studying the basic circuit shown in Fig. 9.2(a), where two inverters are connected by a transmission gate. Gate "bubbles" are used to distinguish pMOS transistors. This notation allows for quick recognition of the MOSFET polarities and is quite convenient to use. The TG is controlled by $\phi = 1$ and $\bar{\phi} = 0$, so it is in a conducting state.

Analysis of the transient characteristics can be based on the simplified equivalent circuit shown in Fig. 9.2(b). $V_{\text{in}}(t)$ is the input voltage to the system while $V_1(t)$ is the first-stage output. The TG has been replaced by an equivalent resistance R_{eq}. The second-stage input is modeled by the capacitance C_2. $V_2(t)$ is the voltage across C_2 and represents the second-stage input voltage.

Both charging and discharging transients are needed to characterize the circuits. These are discussed below, where the emphasis is on obtaining

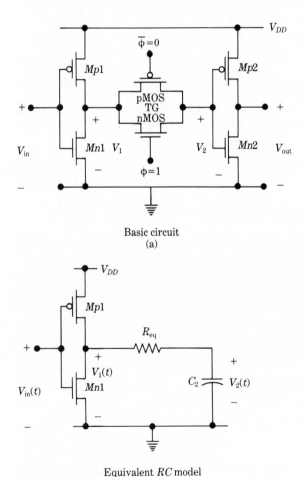

Basic circuit
(a)

Equivalent RC model
(b)

FIGURE 9.2 General switching circuit for synchronous logic.

expressions for $V_2(t)$ in each case. The analysis also provides the design parameters and gives insight into the switching performance expected in various synchronous logic schemes.

9.1.1 Discharging Analysis: Logic 0 Transfer

The first case to be studied is the transfer of a logic 0 state through the transmission gate. This is accomplished by discharging the capacitor C_2 from an initial voltage V_{DD} to 0 [V] through the TG and nMOS transistor $Mn1$. The system input voltage is taken to be $V_{in} = V_{DD}$. Since $Mp1$ is in cutoff, the subcircuit in Fig. 9.3 is sufficient to perform the analysis.

The first-order approximation for the discharging event is obtained by assuming that R_{eq} is negligible, so that $V_{DSn} = V_2$. The effects of R_{eq} will be included later using techniques developed in Chapter 4. Once this approximation is made, it is seen that the MOSFET starts out with

$$V_{DSn} = V_{DD}, \qquad V_{GSn} = V_{DD}, \tag{9.1-5}$$

so the transistor is initially saturated. Equating currents gives the differential equation

$$I_D = \frac{\beta_{n1}}{2}(V_{DD} - V_{Tn})^2$$

$$= -C_2\frac{dV_2}{dt} \tag{9.1-6}$$

for $V_2(t)$. Since $V_2(0) = V_{DD}$, the solution is easily seen to be

$$V_2(t) = V_{DD} - \frac{\beta_{n1}}{2C_2}(V_{DD} - V_{Tn})^2 t, \tag{9.1-7}$$

which indicates an initial linear decrease in the capacitor voltage. This expression is valid until a time t_o when the MOSFET enters the nonsaturated region with

$$V_{DSn} \approx V_2(t_o) = V_{DD} - V_{Tn}. \tag{9.1-8}$$

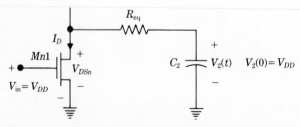

FIGURE 9.3   Subcircuit used for discharge calculation.

Solving gives this time as

$$t_o = \frac{2C_2 V_{Tn}}{\beta_{n1}(V_{DD} - V_{Tn})^2}.$$ (9.1-9)

For times $t > t_o$, the MOSFET is nonsaturated. The current equation is now given as

$$I_D = \frac{\beta_{n1}}{2}[2(V_{DD} - V_{Tn})V_2 - V_2^2]$$

$$= -C_2\frac{dV_2}{dt}.$$ (9.1-10)

This equation has the same form as eqn. (8.1-25) in Chapter 8. Using the initial condition $V_2(t_o) = (V_{DD} - V_{Tn})$ gives

$$V_2(t) = (V_{DD} - V_{Tn})\left[\frac{2e^{-(t-t_o)/\tau_n}}{1 + e^{-(t-t_o)/\tau_n}}\right],$$ (9.1-11)

where τ_n is the nMOS discharge time constant defined by

$$\tau_n = \frac{C_2}{\beta_{n1}(V_{DD} - V_{Tn})}.$$ (9.1-12)

Note that t_o can be written in the alternate form

$$t_o = \frac{2V_{Tn}}{(V_{DD} - V_{Tn})}\tau_n.$$ (9.1-13)

Equations (9.1-7) and (9.1-11) provide the analytic forms for $V_2(t)$. The total waveform $V_2(t)$ as a function of (t/τ_n) is plotted in Fig. 9.4.

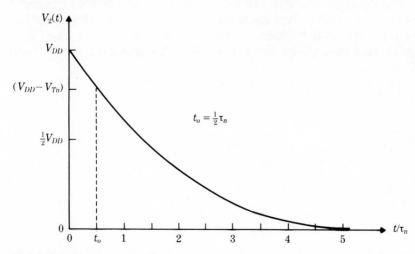

FIGURE 9.4 V_2 vs. t for discharge.

The effects of the transmission gate are included in the present analysis by using the equivalent resistance R_{eq}. For a first-order approximation, this simply requires that the time constant be replaced by

$$\tau_n \simeq C_2 \left[\frac{1}{\beta_{n1}(V_{DD} - V_{Tn})} + R_{eq} \right], \qquad (9.1\text{-}14)$$

which models R_{eq} as being in series with the MOSFET resistance. Increasing τ_n in this manner shows that longer discharge times are necessary. Note that C_2 includes contributions from the TG transistors, so τ_n is scaled by both resistance and capacitance factors.

The total discharge time t_{HL} from $V_2 = V_{DD}$ to a level $V_2(t_{HL}) = 0.1 V_{DD}$ may be computed directly from eqn. (9.1-11). The algebra gives

$$t_{HL} = \tau_n \left\{ \frac{2V_{Tn}}{(V_{DD} - V_{Tn})} + \ln \left[\frac{2(V_{DD} - V_{Tn})}{0.1V_{DD}} - 1 \right] \right\}, \qquad (9.1\text{-}15)$$

where eqn. (9.1-13) has been used. It is important to remember that the transmission gate parasitics increase the value of τ_n over that for a basic inverter. Consequently, adding the TG as a controlled gating element slows down the response of the circuit.

9.1.2 Charging Analysis: Logic 1 Transfer

Consider now the transfer of a logic 1 state through the transmission gate. The system input voltage is taken to be $V_{in} = 0 \,[V]$, so that $Mn1$ is off while $Mp1$ conducts to charge C_2. The subcircuit in Fig. 9.5 gives the charging path with an assumed initial condition of $V_2(0) = 0 \,[V]$. R_{eq} is assumed to be negligible in the first-order calculations but is included later by modifying the time constant. The equations that describe the charging may be obtained by simply complementing the results of the discharge analysis. This approach was used in Section 4.4 for the CMOS inverter discussion. For the present case, it is instructive to work through the details to reinforce the concepts.

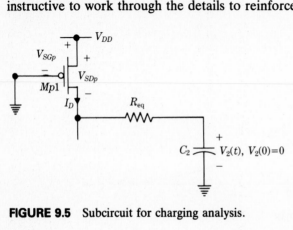

FIGURE 9.5 Subcircuit for charging analysis.

At the beginning of the charging process, the p-channel MOSFET $Mp1$ is saturated with

$$V_{SGp} = V_{DD} \simeq V_{SDp},\qquad(9.1\text{-}16)$$

ignoring the voltage across R_{eq}. The first portion of the charging cycle is thus described by

$$I_D = \frac{\beta_{p1}}{2}(V_{DD} - |V_{Tp}|)^2$$

$$= C_2\frac{dV_2}{dt},\qquad(9.1\text{-}17)$$

which has the solution

$$V_2(t) = \frac{\beta_{p1}}{2C_2}(V_{DD} - |V_{Tp}|)^2 t.\qquad(9.1\text{-}18)$$

This gives the charging behavior until a time t_o when the MOSFET enters the nonsaturated mode of operation. To compute the value of t_o, first note that

$$V_{SDp} = V_{DD} - V_2(t_o)$$

$$= V_{DD} - |V_{Tp}|,\qquad(9.1\text{-}19)$$

so eqn. (9.1-18) gives

$$t_o = \frac{2|V_{Tp}|}{(V_{DD} - |V_{Tp}|)}\tau_p,\qquad(9.1\text{-}20)$$

where τ_p is the pMOS charging time constant defined by

$$\tau_p = \frac{C_2}{\beta_{p1}(V_{DD} - |V_{Tp}|)}.\qquad(9.1\text{-}21)$$

$V_2(t)$ in the time interval $[0, t_o]$ may be written in the alternate form

$$V_2(t) = \frac{(V_{DD} - |V_{Tp}|)}{2\tau_p}t.\qquad(9.1\text{-}22)$$

As expected, the initial charging is approximately linear in time.

MOSFET $Mp1$ is nonsaturated for times $t > t_o$, so the charging is described by

$$I_D = \frac{\beta_{p1}}{2}\{2(V_{DD} - |V_{Tp}|)[V_{DD} - V_2(t)] - [V_{DD} - V_2(t)]^2\}$$

$$= C_2\frac{dV_2}{dt}.\qquad(9.1\text{-}23)$$

To solve this equation, first introduce a voltage $v(t)$ defined by

$$v(t) = V_{DD} - V_2(t).\qquad(9.1\text{-}24)$$

Since

$$\frac{dv(t)}{dt} = -\frac{dV_2}{dt}, \tag{9.1-25}$$

eqn. (9.1-23) may be written as

$$\frac{\beta_{p1}}{2}[2(V_{DD} - |V_{Tp}|)v(t) - v^2(t)] = -C_2\frac{dv}{dt}. \tag{9.1-26}$$

It is recognized that this equation has the same form as eqn. (9.1-10) in the discharge analysis. The solution for $v(t)$ is thus given by

$$v(t) = (V_{DD} - |V_{Tp}|)\left[\frac{2e^{-(t-t_o)/\tau_p}}{1 + e^{-(t-t_o)/\tau_p}}\right]. \tag{9.1-27}$$

Using eqn. (9.1-24), the capacitor voltage $V_2(t)$ in this time interval is given by

$$V_2(t) = V_{DD} - (V_{DD} - |V_{Tp}|)\left[\frac{2e^{-(t-t_o)/\tau_p}}{1 + e^{-(t-t_o)/\tau_p}}\right]. \tag{9.1-28}$$

Figure 9.6 shows the general behavior of $V_2(t)$ as predicted by eqns. (9.1-22) and (9.1-28).

The effects of the transmission gate may be approximated by modifying the time constant to

$$\tau_p \simeq C_2\left[\frac{1}{\beta_{p1}(V_{DD} - |V_{Tp}|)} + R_{eq}\right]. \tag{9.1-29}$$

C_2 again has contributions from the TG MOSFETs, in addition to extra C_{line} terms. The charging time t_{LH} defined by $V_2(t_{LH}) = 0.9V_{DD}$ is easily computed

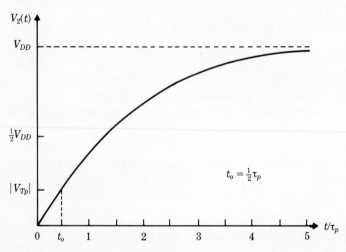

FIGURE 9.6 V_2 vs. t for charging.

from eqn. (9.1-28) as

$$t_{LH} = \tau_p \left\{ \frac{2\,|V_{Tp}|}{(V_{DD} - |V_{Tp}|)} + \ln \left[\frac{2(V_{DD} - |V_{Tp}|)}{0.1 V_{DD}} - 1 \right] \right\}.$$ (9.1-30)

It is seen by inspection that the pMOS charging is the exact complement of the nMOS discharge, as expected from the CMOS structuring.

The analysis clearly shows the important design parameters for the TG coupled inverters. First, the MOSFETs used in the transmission gate establish the value of R_{eq} and also contribute to the capacitance C_2. Next, the first-stage driver MOSFETs $Mp1$ and $Mn1$ act as the charge-discharge devices, so the values of β_{p1} and β_{n1} are the basic design variables that set the transient time intervals. Note that if $\beta_{p1} = \beta_{n1}$ and $V_{Tn} = |V_{Tp}|$, then $t_{HL} = t_{LH}$. Finally, the parasitic layout capacitances determine the working value of C_2.

EXAMPLE 9.1.2-1

Suppose that the circuit in Fig. 9.2 is designed with $(W/L) = 2$ for all transistors, and the power supply is $V_{DD} = 5\,[V]$. The process parameters are given as

$$k'_n = 40\,[\mu A/V^2], \quad V_{T0n} = 0.8\,[V], \quad \gamma = 0.26\,[V^{1/2}], \quad 2\,|\phi_{Fp}| = 0.58\,[V],$$
$$k'_p = 16\,[\mu A/V^2], \quad V_{T0p} = -0.8\,[V], \quad \gamma = 0.84\,[V^{1/2}], \quad 2\phi_{Fn} = 0.70\,[V].$$

C_2 is approximated as 150 [fF].

Calculate the values of t_{LH} and t_{HL} for the circuit. Use the technique developed in Example 6.7.1-1 to estimate R_{eq}.

Solution
First, estimate R_{eq}. This approach requires that V_{out} be found from

$$V_{out} = (5 - 0.8) - 0.26(\sqrt{V_{out} + 0.58} - \sqrt{0.58}),$$

which iterates to $V_{out} \simeq 3.85\,[V]$. Then

$$R_{eq} \simeq \frac{2}{2(16 \times 10^{-6})[2(5 - 0.8) - (5 - 3.85)]} \simeq 8620\,[\Omega]$$

is the estimate of the equivalent TG gate resistance.

t_{HL}: Using eqn. (9.1-14),

$$\tau_n \simeq (150 \times 10^{-15}) \left[\frac{1}{(40 \times 10^{-6})(2)(5 - 0.8)} + 8620 \right] \simeq 1.74\,[ns],$$

so eqn. (9.1-15) gives

$$t_{HL} \simeq (1.74) \left\{ \frac{2(0.8)}{(5 - 0.8)} + \ln \left[\frac{2(5 - 0.8)}{0.5} - 1 \right] \right\} \simeq 5.47\,[ns].$$

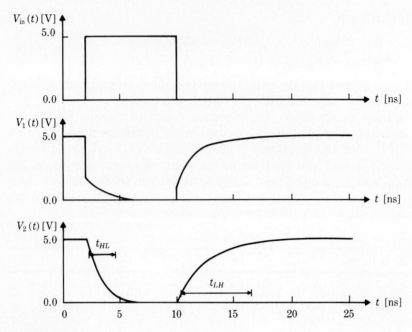

FIGURE E9.1

t_{LH}: The time constant is

$$\tau_p = (150 \times 10^{-15})\left[\frac{1}{(2)(16 \times 10^{-6})(5 - 0.8)} + 8620\right] \simeq 2.41 \,[\text{ns}],$$

so eqn. (9.1-30) gives

$$t_{LH} \simeq (2.41)(3.14) \simeq 7.57 \,[\text{ns}].$$

This problem was simulated using the SPICE file shown below. The resulting plots (Fig. E9.1) show that the simulated t_{HL} value is less than the 5.47 [ns] calculated above, while the value of t_{LH} is closer to the analytic result.

```
CMOS TG TRANSMISSION
VDD 3 0 DC 5VOLTS
MN1 2 1 0 0 MN L=5U W=10U
MP1 2 1 3 3 MP L=5U W=10U
MN 2 3 4 0 MN L=5U W=10U
MP 4 0 2 3 MP L=5U W=10U
VS 1 0 DC PULSE(0 5 2NS 1PS 1PS 8NS)
CLOAD 4 0 1.5E-13
.MODEL MN NMOS VTO=0.8 GAMMA=0.26 KP=4.0E-5 PHI=0.58
.MODEL MP PMOS VTO=-0.8 GAMMA=0.84 KP=1.6E-5 PHI=0.70
.TRAN 1NS 25NS
.PLOT TRAN V(2)
.PLOT TRAN V(4)
.END
```

9.1.3 Charge Leakage

Soft nodes that are driven through CMOS transmission gates exhibit charge leakage problems. Since charge leakage degrades logic 1 voltage levels, this effect can give a lower limit for the clock frequency used to control the gates. The charge leakage through nMOS pass transistors was analyzed in Section 8.1.3. Although the CMOS problem is similar, the analysis is complicated by the fact that the gate uses both an nMOS and a pMOS transistor.

The basic problem is shown in Fig. 9.7(a). The soft node stores the initial voltage $V_2(t = 0) = V_{DD}$ on capacitor C_2. The TG is biased off by grounding the gate of Mn while keeping V_{DD} on the gate of Mp. I_{leak} represents the leakage current that discharges the capacitor by means of

$$I_{\text{leak}} = -C_2 \frac{dV_2}{dt}. \tag{9.1-31}$$

As discussed below, current paths exist through both MOSFETs in the transmission gate.

Figure 9.7(b) shows the cross section for the transistors in a simplified n-tub CMOS technology. The load capacitance C_L represents the contributions that originate away from the TG sections of the chip. For example, the inverter arrangement in Fig. 9.2 gives

$$C_L = C_{Gn2} + C_{Gp2} + C_{\text{line}} \tag{9.1-32}$$

as a reasonable approximation. C_2 must be decomposed in this manner to account for pn junction depletion capacitances.

Two important pn junctions are shown in the drawing. The first is denoted pn junction 1 and is the p^+n junction formed in the n-tub portion of the p-channel MOSFET Mp. Note that the n-tub bulk is connected to V_{DD}. The second is labeled as pn junction 2 and is the n^+p junction formed in the p-type substrate for Mn. The substrate is assumed grounded with a bulk voltage $V_B = 0$.

The structure of the CMOS transmission gate shows the complexity of the leakage problem. Each MOSFET has reverse-biased pn junctions that admit leakage currents. In addition, subthreshold currents $I_{\text{sub},n}$ and $I_{\text{sub},p}$ should be included in the circuit analysis. A complete analysis of the charge leakage paths generally requires a computer simulation. However, it is possible to analytically study some aspects of the problem by simplified modeling. (*Surface leakage* is also ignored here.)

The model used here assumes that subthreshold currents are negligible in both devices. This approximation implies that the leakage currents through the reverse-biased pn junctions dominate. Figure 9.8 shows the equivalent circuit where junctions 1 and 2 are represented by diodes. Applying KCL gives the current I_L out of C_L as

$$I_L \simeq I_{R2} - I_{R1}, \tag{9.1-33}$$

where I_{R1} and I_{R2} are the reverse currents through diodes 1 and 2, respectively.

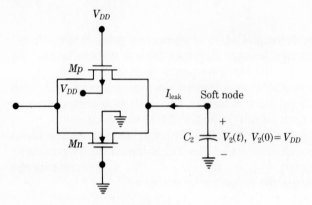

Soft node leakage circuit
(a)

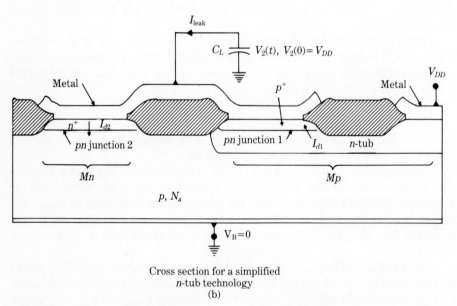

Cross section for a simplified
n-tub technology
(b)

FIGURE 9.7 Charge leakage problem for a CMOS transmission gate.

It is important to note that I_{R1} represents current flow from the power supply V_{DD}.

The lowest-order analysis of the leakage current is based on the simple RC circuit shown in Fig. 9.9. Each diode has been replaced by a parallel combination of a conductance G and a capacitance C_d. These elements are assumed to be linear, time-invariant quantities. Although this model greatly oversimplifies the problem, it does illustrate the overall effects of having two leakage current paths.

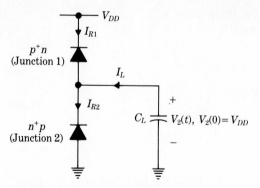

FIGURE 9.8   Diode model for charge leakage.

Interest centers on finding $V_2(t)$ with the initial condition $V_2(0) = V_{DD}$. Summing currents,

$$i = i_2 - i_{c1} - i_{g1}$$

$$= -(C_{d2} + C_L)\frac{dV_2}{dt}, \tag{9.1-34}$$

where

$$C_2 = C_L + C_{d2} \tag{9.1-35}$$

is the total capacitance. The current components are

$$i_2 = G_2 V_2,$$
$$i_{g1} = G_1(V_{DD} - V_2),$$
$$i_{c1} = C_{d1}\frac{d(V_{DD} - V_2)}{dt}, \tag{9.1-36}$$

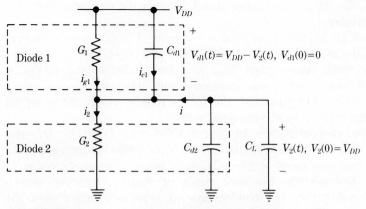

FIGURE 9.9   First-order RC approximation for charge leakage.

so eqn. (9.1-34) may be written in the form

$$C\frac{dV_2}{dt} + GV_2(t) = G_1 V_{DD}. \tag{9.1-37}$$

In this equation,

$$G = G_1 + G_2 \tag{9.1-38}$$

while

$$C = C_L + C_{d1} + C_{d2} \tag{9.1-39}$$

is the total capacitance. Solving this linear differential equation with the initial condition $V_2(0) = V_{DD}$ results in

$$V_2(t) = \frac{V_{DD}}{G}(G_2 e^{-t/\tau_d} + G_1), \tag{9.1-40}$$

where

$$\tau_d \equiv \frac{C}{G} \tag{9.1-41}$$

is the time constant. The voltage $V_{d1}(t)$ across diode 1 is

$$V_{d1}(t) = V_{DD} - V_2(t)$$

$$= V_{DD}\frac{G_2}{G}(1 - e^{-t/\tau_d}). \tag{9.1-42}$$

Note that as $t \to \infty$,

$$V_{d1} \to \frac{G_2}{G}V_{DD}, \qquad V_2 \to \frac{G_1}{G}V_{DD}. \tag{9.1-43}$$

The final values of V_{d1} and V_2 are thus dependent on the conductance ratios (G_1/G) and (G_2/G). In this limit, the circuit assumes the characteristics of a simple voltage divider.

Figure 9.10 shows V_{d1} and V_2 plotted as functions of (t/τ_d) for two different conductance ratios. In Fig. 9.10(a), $(G_1/G) = 0.7$ and $(G_2/G) = 0.3$. The soft node voltage V_2 thus decays from V_{DD} to a final value of $(G_1/G)V_{DD} = 0.7V_{DD}$. V_{d1}, on the other hand, increases from $0\,[\text{V}]$ up to a level of $(G_2/G)V_{DD} = 0.3V_{DD}$. For this case, the higher relative conductance for G_1 serves to keep the voltage V_2 at a reasonable level.

The opposite case is shown in Fig. 9.10(b) with $(G_1/G) = 0.3$ and $(G_2/G) = 0.7$. $V_2(t)$ decays exponentially from V_{DD}, while V_{d1} increases such that the two waveforms cross at $(V_{DD}/2)$. $V_2(t)$ has an asymptotic value of $0.3V_{DD}$, which is less than the final value of $V_{d1}(t)$. Physically this is understood by noting that the conductance to ground is larger than the conductance to V_{DD}. The level of charge leakage in this case is significant,

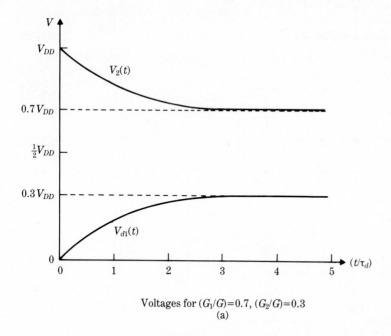

Voltages for $(G_1/G)=0.7$, $(G_2/G)=0.3$
(a)

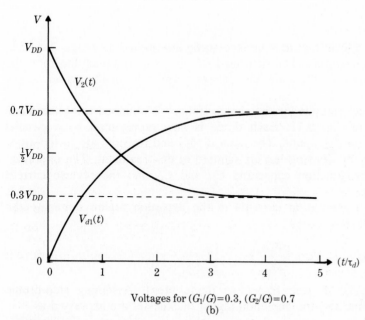

Voltages for $(G_1/G)=0.3$, $(G_2/G)=0.7$
(b)

FIGURE 9.10 $V_2(t)$ and $V_{d1}(t)$ for simplified RC model.

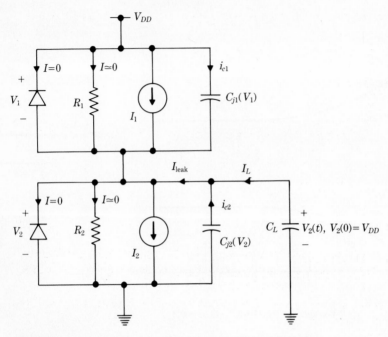

FIGURE 9.11   Diode models for charge leakage calculation.

since V_2 falls to a level that may be incorrectly interpreted as a logic 0 state at the input of the next gate. The simplified RC circuit model thus illustrates the possibilities when the soft node has leakage paths to both the power supply and ground.

A higher-level analysis for the charge leakage is obtained using the diode models shown in Fig. 9.11. Each diode is now represented by a parallel combination of four elements. The ideal diodes and resistors are included for completeness in the drawing but are ignored in the treatment. The important elements are the junction capacitors C_{j1} and C_{j2} and the reverse current sources I_1 and I_2.

Assume that generation currents in the depletion region dominate the reverse diode currents. Then

$$I_1 \simeq \frac{qA_1 n_i}{2\tau_o} x_{d1}, \qquad I_2 \simeq \frac{qA_2 n_i}{2\tau_o} x_{d2}, \tag{9.1-44}$$

where A_1 and A_2 are the respective diode areas. Assuming step-profile junctions for simplicity, the depletion region thicknesses are expressed as

$$x_{d1} = x_{do1}\left(1 + \frac{V_1}{\phi_{o1}}\right)^{1/2}, \qquad x_{d2} = x_{do2}\left(1 + \frac{V_2}{\phi_{o2}}\right)^{1/2}, \tag{9.1-45}$$

where ϕ_{o1} and ϕ_{o2} are the built-in voltages. With one-sided doping profiles,

$$x_{do1} \simeq \sqrt{\frac{2\varepsilon_{Si}\phi_{o1}}{qN_{d,\text{tub}}}}, \qquad x_{do2} \simeq \sqrt{\frac{2\varepsilon_{Si}\phi_{o2}}{qN_a}} \qquad \text{(9.1-46)}$$

give the zero-bias values.

The junction capacitances are given by

$$C_{j1} = \frac{C_{j01}A_1}{\left(1 + \dfrac{V_1}{\phi_{o1}}\right)^{1/2}}, \qquad C_{j2} = \frac{C_{j02}A_2}{\left(1 + \dfrac{V_2}{\phi_{o2}}\right)^{1/2}}, \qquad \text{(9.1-47)}$$

where

$$C_{j01} = \frac{\varepsilon_{Si}}{x_{do1}}, \qquad C_{j02} = \frac{\varepsilon_{Si}}{x_{do2}} \qquad \text{(9.1-48)}$$

are the zero-bias values per unit area. Sidewall contributions for both the generation currents and capacitances can be included in the usual manner. They are ignored in the present analysis to simplify the equations.

The total leakage current I_{leak} is seen from Fig. 9.11 to consist of two terms such that

$$I_{\text{leak}} = I_L + i_{c2}. \qquad \text{(9.1-49)}$$

Since C_L is approximately a constant,

$$I_L = -C_L \frac{dV_2}{dt} \qquad \text{(9.1-50)}$$

gives the relation for the first term. The current i_{c2} flowing out of $C_{j2}(V_2)$ is complicated by the fact that depletion capacitance is nonlinear. The capacitance is defined by

$$C_{j2}(V_2) = \frac{dQ_{j2}(V_2)}{dV_2}, \qquad \text{(9.1-51)}$$

so the chain rule gives i_{c2} as

$$i_{c2} = -\frac{dQ_{j2}}{dt} = -C_{j2}(V_2)\frac{dV_2}{dt}. \qquad \text{(9.1-52)}$$

Using C_{j2} from eqn. (9.1-47) yields

$$i_{c2} = -\frac{C_{j02}A_2}{\left(1 + \dfrac{V_2}{\phi_{o2}}\right)^{1/2}}\left(\frac{dV_2}{dt}\right), \qquad \text{(9.1-53)}$$

which illustrates the complete functional form.

Applying KCL to the circuit gives the current flow as

$$I_{leak} = (I_2 - I_1) - i_{c1}, \tag{9.1-54}$$

where

$$i_{c1} = C_{j1}(V_1)\frac{dV_1}{dt} \tag{9.1-55}$$

is the current through C_{j1}. Combining the expressions above and noting that

$$V_1 = V_{DD} - V_2 \tag{9.1-56}$$

results in the differential equation

$$\frac{dV_2}{dt} = -\frac{qn_i}{2\tau_o C_T(V_2)}[A_2 x_{d2}(V_2) - A_1 x_{d1}(V_{DD} - V_2)] \tag{9.1-57}$$

for the voltage $V_2(t)$. In this expression,

$$C_T(V_2) = C_L + C_{j2}(V_2) + C_{j1}(V_{DD} - V_2) \tag{9.1-58}$$

represents the total voltage-dependent capacitance. Although this equation is sufficiently complex to require numerical solutions, it is possible to extract some general information concerning the behavior of $V_2(t)$.

Recall that the initial conditions on the voltages are $V_2(0) = V_{DD}$ and $V_1(0) = [V_{DD} - V_2(0)] = 0\,[V]$. As long as

$$I_2 > I_1 \tag{9.1-59}$$

is valid, $(dV_2/dt) < 0$, indicating that $V_2(t)$ decreases with time. Eventually the generation currents on the right-hand side of eqn. (9.1-57) become equal in magnitude and $(dV_2/dt) = 0$. This condition is expressed by

$$A_2 x_{d2}(V_2) = A_1 x_{d1}(V_{DD} - V_2). \tag{9.1-60}$$

The value of V_2 at this point is found using eqns. (9.1-45) and (9.1-46), which result in

$$V_2 = \frac{\left(\dfrac{N_a}{N_{d,tub}}\right)(\phi_{o1} + V_{DD}) - \left(\dfrac{A_2}{A_1}\right)^2 \phi_{o2}}{\left(\dfrac{N_a}{N_{d,tub}}\right) + \left(\dfrac{A_2}{A_1}\right)^2}. \tag{9.1-61}$$

This is analogous to the asymptotic limit $V_2 = (G_1/G)V_{DD}$ obtained as eqn. (9.1-43) in the simplified RC approximation. It is interesting to note that if $A_1 = A_2$, $\phi_{o1} = \phi_{o2}$, and $N_a = N_{d,tub}$, then

$$V_2 \simeq \frac{1}{2}V_{DD}, \tag{9.6-62}$$

as expected. Of course, $(N_a/N_{d,tub}) < 1$ must be satisfied in order to create the n-tub in the p substrate so that V_2 will equilibrate at a different value. Also note that voltage-dependent subthreshold currents have been ignored in this analysis.

The differential equation (9.1-57) can also be simplified by using average values for the depletion capacitors. The total capacitance becomes

$$C_T \simeq C_L + C_{av,1} + C_{av,2}$$

$$= \text{Constant}. \tag{9.1-63}$$

Average values cannot be used for depletion widths since this would eliminate the point when the currents balance. The basic structure of the decay can be studied by assuming $A_1 = A_2$, $\phi_{o1} = \phi_o = \phi_{o2}$, and $x_{do1} = x_{do2}$. The equation then reduces to the form

$$\frac{dV_2}{dt} \simeq -\frac{qAn_i x_{do}}{2\tau_o C_T}\left(\sqrt{1 + \frac{V_2}{\phi_o}} - \sqrt{1 + \frac{V_{DD} - V_2}{\phi_o}}\right). \tag{9.1-64}$$

Although a numerical solution is still required, the behavior of $V_2(t)$ is more transparent. In particular, it is seen that $V_2(t)$ decreases with time such that the square-root terms eventually cancel each other. Obviously, the approximations give this point as $V_2 = (V_{DD}/2)$.

EXAMPLE 9.1.3-1

The charge leakage approximations developed in this section can be used to illustrate some of the design tradeoffs involved in using CMOS TG gates. The CMOS parameters are assumed to be

$$k'_n = 40\,[\mu A/V^2], \quad k'_p = 16\,[\mu A/V^2],$$

with pn junction parameters of (see Example 5.8-1)

nMOS: $N_a = 10^{15}\,[\text{cm}^{-3}]$, $\phi_o = 0.879\,[V]$ (n^+-to-substrate)

pMOS: $N_{d,\text{tub}} = 10^{16}\,[\text{cm}^{-3}]$, $\phi_o = 0.939\,[V]$ (p^+-to-n,tub).

Sidewall corrections will be ignored for simplicity.

The charge leakage properties are described by the derivative (dV_2/dt) in eqn. (9.1-57). As long as $(dV_2/dt) < 0$, charge leakage causes a change in the output voltage. The equilibrium voltage in eqn. (9.1-61) gives the point where $(dV_2/dt) = 0$. In a given CMOS process, the design variables are

A_1 = Area of p^+/n-tub junction in pMOS transistor

and

A_2 = Area of n^+/p-substrate junction in nMOS transistor.

Changing the areas affects the transmission properties of the gate and also establishes the leakage current levels since these are proportional to area.

Two TG designs are shown in Fig. E9.2. The layout in part (a) is designed

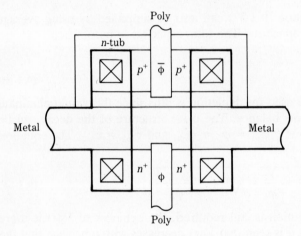

Symmetric transmission gate
(a)

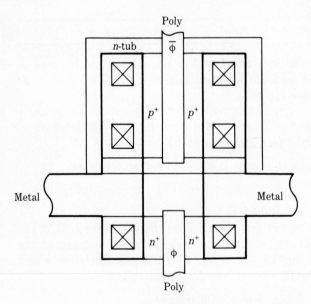

Increased pMOS TG layout
(b)

FIGURE E9.2

with equal geometry pass transistors with $(W/L)_n = (W/L)_p \simeq 2$. This gives $\beta_n \simeq 80\,[\mu A/V^2] \simeq 2.5\beta_p$, indicating nonsymmetrical electrical properties but simple layout. Using eqn. (9.1-61) with $A_1 = A_2$ and assuming $V_{DD} = 5\,[V]$ gives the equilibrium voltage as

$$V_2 \simeq \frac{(0.1)(0.939 + 5) - 0.879}{0.1 + 1} \simeq -0.26\,[V].$$

Since this value is negative, it says that $I_2 > I_1$ is always true, so charge leaks to ground with a final state of $V_2 = 0\,[V]$. This shows that the simple layout geometry will not hold charge in this technology.

The drawing in Fig. E9.2(b) shows a different TG geometry, where

$$(W/L)_n \simeq 2, \quad (W/L)_p \simeq 5,$$

so $\beta_n = \beta_p \simeq 80\,[\mu A/V^2]$. The electrical conduction properties are identical. However, now $A_1 \simeq (2.5)A_2$, so eqn. (9.1-61) gives

$$V_2 \simeq \frac{(0.1)(5.939) - (1/2.5)^2(0.879)}{0.26} \simeq +1.73\,[V].$$

The positive voltage indicates that an equilibrium point will be reached before all of the charge is lost. Physically this occurs because the junction area of the p^+/n-tub junction (which is connected to V_{DD}) has been increased to the point that compensates for the leakage to ground through the n^+/p-substrate junction.

The design tradeoff is in layout complexity: the identical MOSFET gate is easier to incorporate than the nonsymmetric TG.

(It should be noted that this example illustrates only the basic considerations involved in analyzing the electrical properties. Items such as the sidewall contributions to the leakage current should be included in a more detailed analysis.) ∎

9.2 Synchronized TG CMOS Logic

The first type of synchronous CMOS logic to be studied uses clocked transmission gates to control the data transfer. Circuits of this type are constructed in the same manner as depletion load nMOS pass-transistor configurations. However, since CMOS TGs are controlled by both a signal and its complement, different clocking schemes are possible in CMOS circuits.

9.2.1 Single-Clock Logic

Figure 9.12 shows a single-clock CMOS shift register circuit constructed by connecting static inverters with transmission gates. Each transmission gate is controlled by ϕ and $\bar{\phi}$. Synchronized data transfer is obtained by alternate

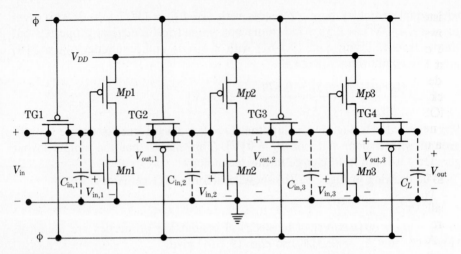

FIGURE 9.12 Single-clock CMOS transmission gate shift register.

phasing of the clock connections to the TGs. For example, TG1 has ϕ connected to the nMOS transistor and $\bar{\phi}$ connected to the pMOS device. TG2 is exactly opposite, with ϕ controlling the p-channel MOSFET and $\bar{\phi}$ controlling the n-channel MOSFET. Since both ϕ and $\bar{\phi}$ are used, this is generally classified as 2-phase logic even though only one clocking input is required.

The data transfer can be understood by referring to the ideal clock waveforms shown in Fig. 9.13. Suppose that the input voltage to the shift

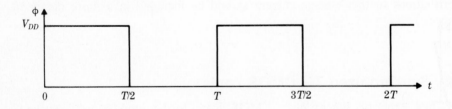

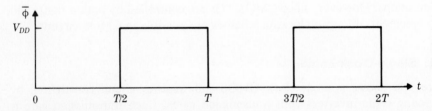

FIGURE 9.13 Ideal single-clock waveforms.

register is set at $V_{in} = V_{DD}$. When $\phi = V_{DD}$ and $\bar{\phi} = 0$ [V], TG1 conducts and allows the input capacitance $C_{in,1}$ to charge to a logic 1 voltage of $V_{in,1} = V_{DD}$. The output voltage of the first stage then goes to $V_{out,1} = 0$ [V]. During the next half clock cycle, $\phi = 0$ [V] and $\bar{\phi} = V_{DD}$. This places TG1 in a cutoff mode so that charge is dynamically stored on $C_{in,1}$. TG2 conducts with these clock signals, allowing charge on $C_{in,2}$ to escape to ground through TG2 and nMOS transistor $Mn1$. $V_{in,2}$ is thus established at 0 [V]. The behavior during the next clock cycle is easily extrapolated from this discussion. Note that the data move one stage per half bit time ($T/2$) as expected.

The clocking scheme employed in this circuit alternately isolates inverter stages in the chain. Inputs are permitted during one half cycle. The input TGs are in cutoff during the next half cycle, when the output transfer occurs. Ideally, this technique works quite well. However, problems can arise when more realistic clock waveform shapes are analyzed.

A simple example is shown in Fig. 9.14. The clock signals are no longer taken as ideal pulses but are assumed to have finite rise and fall times denoted by t_R and t_F. Obviously,

$$[\phi(t)][\bar{\phi}(t)] = 0 \tag{9.2-1}$$

is true only when the signals are at V_{DD} or 0 [V]; the two overlap during a transition. This can lead to problems in transfer timing if the transition time intervals are long compared with the gate delay.

Overlapping also arises from clock skew, in which the complementary clock $\bar{\phi}$ is time-shifted by an amount t_s as shown in Fig. 9.15. Linear ramp transitions have been used for simplicity. As shown by the plot of the overlap $\phi\bar{\phi}$, this results in an extended period of time during which $[\phi(t)][\bar{\phi}(t)] \neq 0$. Since both the input and output transmission gates are conducting, the stage

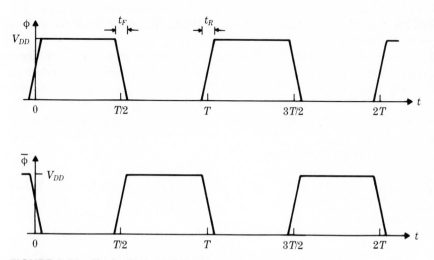

FIGURE 9.14 Single-clock waveforms with finite rise and fall times.

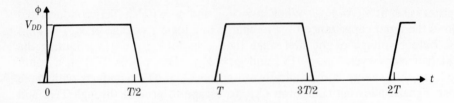

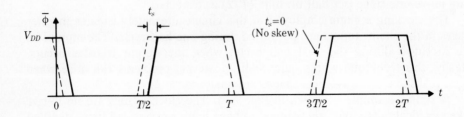

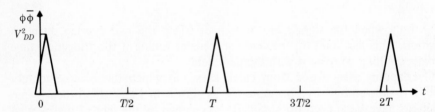

FIGURE 9.15   Clock skew.

isolation is degraded. The overlap time is dependent on t_s. Sources of clock skew include normal gate delay and unbalanced resistive and capacitive loading on the clock distribution lines. Controlling clock skew is always important in clocked systems.

The basic circuit in Fig. 9.16 provides a simple example of clock skew generation. The input clock waveform $\phi(t)$ is fed through an inverter to produce the complementary clock $\bar{\phi}(t)$. This introduces a propagation time delay of t_p, so that the skew time for this case is $t_s = t_p$ as referenced to the original clock. The amount of clock skew for this circuit is approximated using the results of the CMOS inverter transient analysis. Assuming that $\beta_n = \beta_p$ and $V_{Tn} = |V_{Tp}|$, eqn. (4.4-8) gives

$$t_s \simeq \tau_n \left\{ \frac{2V_{Tn}}{(V_{DD} - V_{Tn})} + \ln \left[\frac{4(V_{DD} - V_{Tn})}{V_{DD}} - 1 \right] \right\}. \qquad (9.2\text{-}2)$$

The time constant $\tau_n = \tau_p$ is

$$\tau_n = \frac{C_{\bar{\phi}}}{\beta_n(V_{DD} - V_{Tn})}. \qquad (9.2\text{-}3)$$

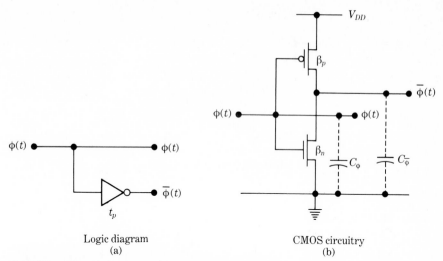

FIGURE 9.16 Clock skew from a simple inverter.

$C_{\bar{\phi}}$ is the capacitance of the $\bar{\phi}$ line driven by the inverter. In general, this will be different from the capacitance C_{ϕ} of the ϕ line.

One technique for reducing clock skew in the circuit of Fig. 9.16 is to add a delay element to equalize the timing. The circuit in Fig. 9.17 employs a CMOS transmission gate biased into conduction to provide the delay. As shown in the logic diagram of Fig. 9.17(a), the phase is referenced to a node that feeds both an inverter and a TG. The clock waveform $\phi(t)$ is obtained at the output of the inverter. The complementary clock $\bar{\phi}(t)$ is delayed through the transmission gate to compensate for the inverter propagation time. Figure 9.17(b) gives the circuit implementation for the logic.

The time delay introduced by the transmission gate may be studied using the first-stage output circuitry shown in Fig. 9.18. The output has been modeled as two parallel paths. One consists of the TG resistance R_{eq} connected to $C_{\bar{\phi}}$. The complementary clock $\bar{\phi}(t)$ is taken off this capacitor. The other path is through an output capacitance $C_{out,1}$. $C_{out,1}$ is the total output capacitance with contributions from both inverter stages and the transmission gate. It may be approximated by

$$C_{out,1} \simeq C_{GDn1} + C_{GDp1} + K(V_{DD}, 0)[C_{dbn1} + C_{dbp1}]$$
$$+ C_{line} + C_{G2} + C_{TG}, \tag{9.2-4}$$

where C_{G2} is the input gate capacitance seen looking into the second stage and C_{TG} represents the transmission gate contributions. Note that the voltage $V_{out,1}$ across $C_{out,1}$ constitutes the input voltage to the second stage.

$\bar{\phi}(t)$ is referenced to $V_{out,1}(t)$. To understand the behavior of $V_{out,1}(t)$, note that the first-stage MOSFETs $Mp1$ and $Mn1$ provide the charge and discharge paths for this node. A simple model for the switching is obtained by modifying

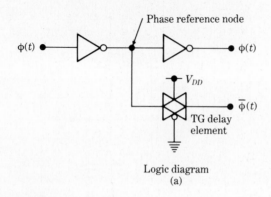

Logic diagram
(a)

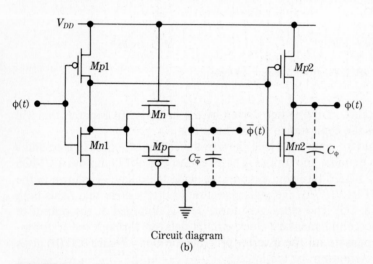

Circuit diagram
(b)

FIGURE 9.17 Transmission gate delay element for skew reduction.

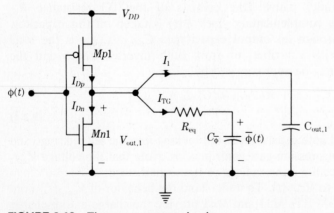

FIGURE 9.18 First-stage output circuit.

the time constants to account for the charging paths. Consider first the case where $\phi(t) \to 0\,[V]$, so $V_{out,1}$ charges toward V_{DD}. pMOS transistor $Mp1$ provides the charging path such that

$$I_{Dp} = I_{TG} + I_1. \tag{9.2-5}$$

For the special case $R_{eq} = 0$, the currents would be described by an equation of the form

$$I_{Dp}(V_{out,1}) = (C_{out,1} + C_{\bar{\phi}})\frac{dV_{out,1}}{dt}, \tag{9.2-6}$$

giving a rise time of

$$t_{LH} \simeq \tau_p \left\{ \frac{2\,|V_{Tp}|}{(V_1 - |V_{Tp}|)} + \ln\left[\frac{2(V_1 - |V_{Tp}|)}{V_0} - 1\right] \right\}. \tag{9.2-7}$$

The time constant for this case is given by

$$\tau_p = \frac{(C_{out,1} + C_{\bar{\phi}})}{\beta_p(V_1 - |V_{Tp}|)}. \tag{9.2-8}$$

$R_{eq} = 0$ represents the worst-case situation for charging C_{out} to a value of $V_{out,1} = V_{DD}$. This occurs because I_{TG} is maximized, reducing the value of I_1. Thus, for general values of R_{eq}, the actual charging time constant will be less than this maximum value, i.e.,

$$\tau_p < \frac{(C_{out,1} + C_{\bar{\phi}})}{\beta_p(V_1 - |V_{Tp}|)}. \tag{9.2-9}$$

The discharge time is obtained in a similar manner. t_{HL} assumes the form

$$t_{HL} \simeq \tau_n \left\{ \frac{2V_{Tn}}{(V_1 - V_{Tn})} + \ln\left[\frac{2(V_1 - V_{Tn})}{V_0} - 1\right] \right\}, \tag{9.2-10}$$

where

$$\tau_n < \frac{(C_{out,1} + C_{\bar{\phi}})}{\beta_n(V_1 - V_{Tn})} \tag{9.2-11}$$

establishes the limit on the discharge time constant. A symmetric CMOS design will give $\beta_n = \beta_p$ regardless of the actual value of R_{eq}.

The preceding discussion shows that $V_{out,1}(t)$ is described by the equations derived in the previous section for voltage transfer through a transmission gate. Only the time constants need to be modified. To find the time delay introduced by the transmission gate, note that the TG circuit shown in Fig. 9.18 uses $V_{out,1}(t)$ as a source in generating $\bar{\phi}(t)$. The current in this branch is

$$I_{TG} = \frac{V_{out,1} - \bar{\phi}}{R_{eq}} = C_{\bar{\phi}}\frac{d\bar{\phi}(t)}{dt}, \tag{9.2-12}$$

or

$$\frac{d\bar{\phi}}{dt} + \frac{\bar{\phi}(t)}{\tau_{\text{TG}}} = \frac{V_{\text{out},1}(t)}{\tau_{\text{TG}}},$$
(9.2-13)

where

$$\tau_{\text{TG}} = R_{\text{eq}}C_{\bar{\phi}}$$
(9.2-14)

is the TG time constant. This equation may be solved for the explicit behavior of $\bar{\phi}(t)$ when referenced to $V_{\text{out},1}(t)$.

Consider first the case where $C_{\bar{\phi}}$ is initially charged to V_{DD} and is then discharged to ground through $Mn1$. To simplify the mathematics, $V_{\text{out},1}(t)$ is approximated by the exponential

$$V_{\text{out},1}(t) \simeq V_{DD}e^{-t/\tau_n},$$
(9.2-15)

where τ_n is given by eqn. (9.2-11). Using this as the driving function on the right side of eqn. (9.2-13) yields a first-order differential equation that may be solved using standard techniques. The solution is

$$\bar{\phi}(t) \simeq V_{DD}e^{-t/\tau_{\text{TG}}} + \frac{V_{DD}}{\left(1 - \frac{\tau_{\text{TG}}}{\tau_n}\right)}(e^{-t/\tau_n} - e^{-t/\tau_{\text{TG}}}).$$
(9.2-16)

Figure 9.19 shows plots of $\bar{\phi}(t)$ as referenced to $V_{\text{out},1}(t)$ for various values of $(\tau_{\text{TG}}/\tau_n)$. It is seen that this ratio can be used as the design variable since it establishes the relative waveform delay. Increasing $(\tau_{\text{TG}}/\tau_n)$ gives longer delays, but the edge becomes less sharp.

The analysis of charging delay follows in a similar manner. For this case, a

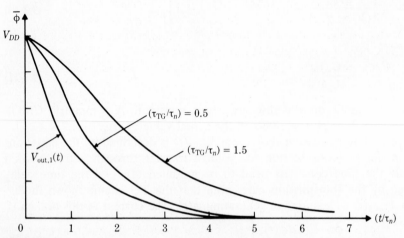

FIGURE 9.19 Fall time delay through transmission gate.

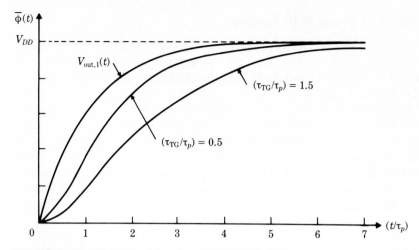

FIGURE 9.20   Rise time delay through transmission gate.

driving function of

$$V_{\text{out},1}(t) \simeq V_{DD}[1 - e^{-t/\tau_p}] \tag{9.2-17}$$

is used with τ_p given by eqn. (9.2-9). Substituting into eqn. (9.2-13) and solving with the initial condition $\bar{\phi}(0) = 0\,[\text{V}]$ gives

$$\bar{\phi}(t) \simeq V_{DD}(1 - e^{-t/\tau_{TG}}) + \frac{V_{DD}}{\left(\dfrac{\tau_{TG}}{\tau_p} - 1\right)}(e^{-t/\tau_p} - e^{-t/\tau_{TG}}). \tag{9.2-18}$$

This function is plotted in Fig. 9.20 for different values of (τ_{TG}/τ_p). It is again seen that increasing this ratio increases the signal delay through the transmission gate. Note that for a symmetric first-stage inverter, $\tau_n = \tau_p = \tau_1$ and (τ_{TG}/τ_1) is the important ratio for both rise and fall time delays.

The design factors come from an overall comparision of the $\phi(t)$ and $\bar{\phi}(t)$ line characteristics. Referring to the original circuit in Fig. 9.17, it is seen that $\phi(t)$ is obtained from the output of the second inverter made up of MOSFETs $Mn2$ and $Mp2$. This is delayed from the reference node by the inherent gate switching times. Assuming a symmetric design with $\beta_{n2} = \beta_{p2}$ and $V_{Tn} = |V_{Tp}|$ for simplicity, the time constant for this gate is

$$\tau_2 \simeq \frac{C_\phi}{\beta_{n2}(V_1 - V_{Tn})}, \tag{9.2-19}$$

with C_ϕ the capacitance of the ϕ line. The easiest design procedure is to first set $\beta_{n2} = \beta_{p2}$ to achieve the desired values of $t_{LH2} = t_{HL2}$ for $\phi(t)$. The next step is to compute the value of τ_{TG} needed to delay $\bar{\phi}(t)$ by an amount equal to that from the inverter. This then specifies the value of R_{eq} and thus gives the aspect ratios for the transmission gate MOSFETs.

9.2-2 Pseudo 2-Phase Shift Register

The use of two nonoverlapping clocks ϕ_1 and ϕ_2 simplifies the circuit timing. The easiest approach to this type of synchronization is to generate the complementary signals ϕ_1 and ϕ_2 so that the clocking waveforms appear as in Fig. 9.21. Since a pair of signals (e.g., ϕ_1 and $\bar{\phi}_1$) is used to control a transmission gate, the clock scheme is referred to as pseudo 2-phase logic.

Figure 9.22 shows the circuit for a pseudo 2-phase shift register. A quick comparison with the single-clock arrangement in Fig. 9.12 shows that the circuit structure is the same. The only difference is that every other TG is now controlled by the clock pair ϕ_2 and $\bar{\phi}_2$. Stage isolation is achieved so long as ϕ_1 and ϕ_2 do not overlap. This simplifies the circuit timing at the expense of adding another clock. A layout for the shift register circuit is shown in Fig. 9.22(b).

The maximum clock frequency $f_{\max} = (1/T_{\min})$ of the shift register can be estimated by a simple analysis. Symmetric inverters with $\beta_n = \beta_p$ and

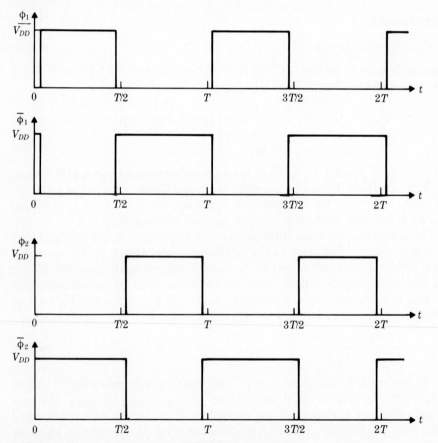

FIGURE 9.21 Pseudo 2-phase CMOS clock waveforms.

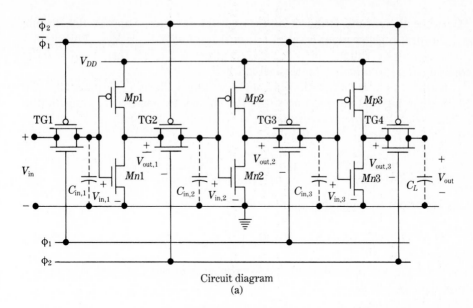

Circuit diagram
(a)

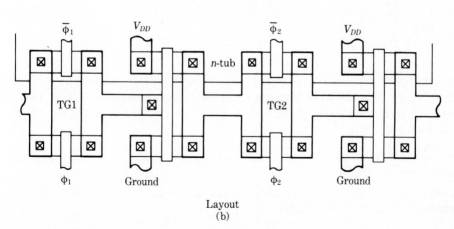

Layout
(b)

FIGURE 9.22   Pseudo 2-phase CMOS shift register.

$V_{Tn} = |V_{Tp}|$ will be assumed for simplicity. From Section 3.5, the critical input VTC points with this design are given by

$$V_{IL} = \frac{1}{4}\left(\frac{3}{2}V_{DD} + V_{Tn}\right), \qquad V_{IH} = \frac{1}{4}\left(\frac{5}{2}V_{DD} - V_{Tn}\right). \qquad \textbf{(9.2-20)}$$

Each transmission gate is characterized by an equivalent resistance R_{eq}. Also, C_{in} is taken to be the same for each stage.

A first-order approximation for f_{max} is obtained by adding two time intervals. The first is the time required to charge or discharge C_{in} through a

TG, while the second is the delay through the inverter. Suppose that during ϕ_1, the shift register input is set at a logic 1 level with $V_{in} = V_{DD}$. This gives a first-stage input voltage of $V_{in,1} = V_{DD}$ across $C_{in,1}$, so that $V_{out,1} = 0 \,[\mathrm{V}]$ corresponding to a logic 0 level is established.

During the next half clock cycle, $\phi_2 = 1$, which allows TG2 to conduct. To transfer the logic 0 to the second stage, the voltage $V_{in,2}$ across $C_{in,2}$ must fall to at least V_{IL}. The worst-case situation is where $V_{in,2}$ is initially at V_{DD}. Using eqn. (9.1-11), the discharge time is

$$t_{\mathrm{dis}} \simeq \tau_{n1}\left\{ \frac{2V_{Tn}}{(V_{DD} - V_{Tn})} + \ln\left[\frac{16(V_{DD} - V_{Tn})}{(3V_{DD} + 2V_{Tn})} - 1\right]\right\}. \tag{9.2-21}$$

It is important to remember that the time constant includes R_{eq} through

$$\tau_{n1} \simeq C_{in,1}\left(\frac{1}{\beta_{n1}(V_{DD} - V_{Tn})} + R_{\mathrm{eq}}\right). \tag{9.2-22}$$

The time needed for the input logic to propagate through the gate is estimated from eqn. (4.4-8) as

$$t_p \simeq \tau_{n2}\left\{ \frac{2V_{Tn}}{(V_{DD} - V_{Tn})} + \ln\left[\frac{4(V_{DD} - V_{Tn})}{V_{DD}} - 1\right]\right\}, \tag{9.2-23}$$

with

$$\tau_{n2} \simeq \frac{C_{\mathrm{out},2}}{\beta_{n2}(V_{DD} - V_{Tn})}, \tag{9.2-24}$$

where $C_{\mathrm{out},2}$ represents the capacitance seen at the second-stage output (with TG3 in cutoff). Summing these two time intervals gives the minimum clock period as

$$\left(\frac{T}{2}\right)_{\min} \simeq t_{\mathrm{dis}} + t_p. \tag{9.2-25}$$

Analysis of a logic 1 transfer gives the same results, as expected from the assumed symmetry of the circuit.

9.2.3 Complex Logic

Complex logic is easily implemented in pseudo 2-phase CMOS. Figure 9.23 provides a straightforward example. The logic diagram is shown in Fig. 9.23(a) and the CMOS circuit is drawn in Fig. 9.23(b). The first gate has complementary AOI structuring. Inputs are admitted during ϕ_1, with the gate producing a logic output of

$$F1 = \overline{AB + C}. \tag{9.2-26}$$

This is transferred to the input of the second gate during ϕ_2. The second gate is a 2-input NOR such that

$$G = \overline{(AB + C)F2} \tag{9.2-27}$$

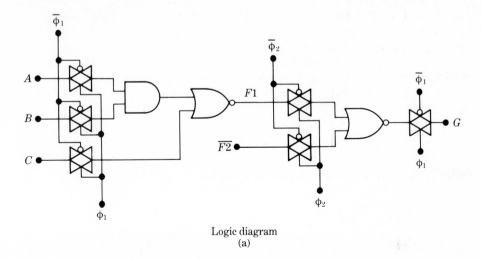

Logic diagram
(a)

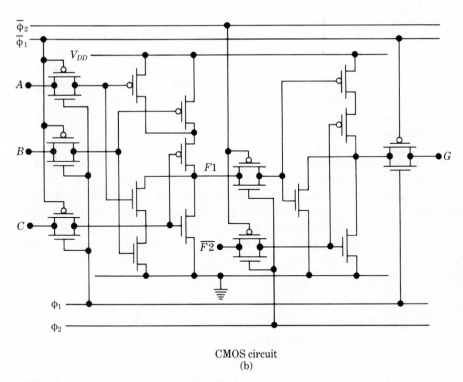

CMOS circuit
(b)

FIGURE 9.23   Combinational pseudo 2-phase CMOS logic.

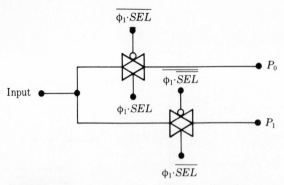

FIGURE 9.24   Path selector using transmission gates.

is transferred to node G during the next half clock cycle. It is seen that $f_{\max}$ is lower than that computed for the shift register. The main factor contributing to this reduction in clock frequency is the increased output capacitance of the complex gates. This gives longer propagation delay times, which must be accounted for in establishing the maximum clock frequency.

Composite transmission gate control signals are useful for many logic circuits. Figure 9.24 shows a simple path selector in which data transfer occurs during ϕ_1. The path is determined by the select signal SEL. Path P_0 is selected when $SEL = 1$, while $SEL = 0$ places the data onto P_1. Implementation of the logic requires four control signal combinations as shown on the drawing.

Figure 9.25 shows three methods for generating the TG control signals for the path selector. In Fig. 9.25(a), NAND logic is used to combine ϕ_1 and SEL. The equivalent NOR logic scheme in Fig. 9.25(b) has $\bar{\phi}_1$ and SEL as inputs. Both use inverters to generate complementary signals. This may lead to problems in clock skew because of the propagation delay. An alternate approach that combines the NAND and NOR circuits is shown in Fig. 9.25(c). The gates can be designed to minimize skew among the composite signals. The tradeoff is the increased real estate requirement.

CMOS transmission gates are capable of passing the full power supply level V_{DD}. Consequently, the simplified path selector shown in Fig. 9.26 constitutes a viable alternative. The data path is structured such that a clock signal of $\phi_1 = 1$ allows transmission to the branch point. If $SEL = 1$, the upper path (P_0) is selected, while a condition of $SEL = 0$ gates data to the lower path P_1. The only additional circuitry needed to implement this design is an inverter to produce the $\overline{SEL}$ signal.

The placement of the transmission gates has been chosen to avoid dynamic charge sharing problems. The criterion for possible charge sharing is the same as in nMOS, namely, it may arise when a soft node drives another soft node. The path selector circuit in Fig. 9.26 avoids this possibility because SEL is always equal to 0 or 1. When $\phi_1 = 1$ allows data transfer from the input, a

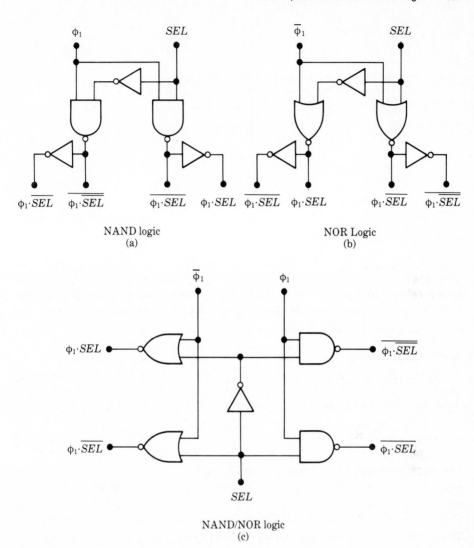

FIGURE 9.25 Control signal generation for path selector.

path is always preselected. The branch point is not a soft node, so no charge sharing occurs.

To illustrate the importance of this point, consider the path selector shown in Fig. 9.27(a) where the clock-controlled TGs and *SEL*-controlled TGs have been interchanged. This arrangement can be subject to charge sharing. An example is provided by the timing diagrams in Fig. 9.27(b).

Suppose that $V_{in}(t = 0) = V_{DD}$. During the time interval $[0, (T/2)]$, $SEL = 1$ but $\phi_1 = 0$. Capacitor C_A thus charges to a voltage of $V_A = V_{DD}$. When *SEL* goes to 0, the input TG turns off, which holds the charge on C_A. ϕ_1

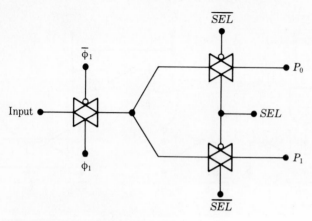

FIGURE 9.26 Simplified transmission gate path selector.

goes to 1 during the next half clock cycle, which induces charge sharing between C_A and C_B. To compute the final voltage on the pair, assume that initially $V_B(0) = V_o$. The total charge stored on the two capacitors is then

$$Q_T = C_A V_{DD} + C_B V_o. \tag{9.2-28}$$

After charge sharing, $V_A = V_B = V_f$, with

$$Q_T = (C_A + C_B)V_f. \tag{9.2-29}$$

Thus the final voltage across the pair is

$$V_f = \frac{C_A V_{DD} + C_B V_o}{C_A + C_B}. \tag{9.2-30}$$

The worst-case situation is when $V_o = 0$. Then

$$V_f = \frac{C_A}{C_A + C_B} V_{DD}. \tag{9.2-31}$$

The effects of charge sharing can be minimized by designing the circuit so that

$$C_A \gg C_B \tag{9.2-32}$$

is satisfied. This is chosen to enhance the forward charge transfer.

Selectively loadable dynamic registers can be constructed as shown in Fig. 9.28. An LD (load) signal is provided to control the input. A condition of $LD = 0$ holds the current state in the register, while setting $LD = 1$ permits new Data entry. The outputs have been chosen as $\overline{\text{Data}}$.

The circuit in Fig. 9.28(a) uses separate transmission gates for the LD and clock signals. The operation of the loading logic is straightforward. When $LD = 0$, the input transmission gate TG1 is in cutoff, prohibiting the entry of a new Data state. TG2 is conducting and allows for internal register feedback. Setting $LD = 1$ places TG1 into a conductive mode. Since TG2 is in cutoff,

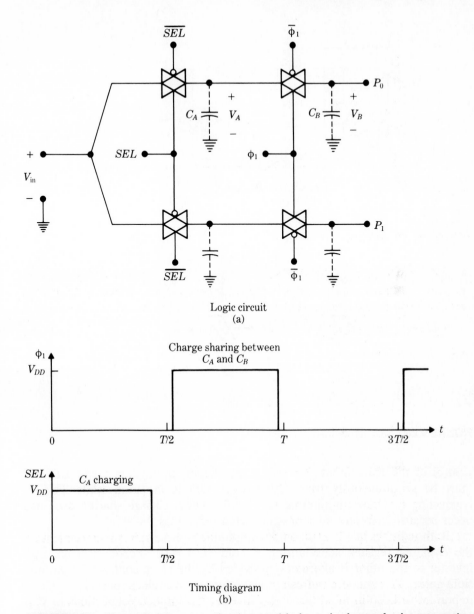

FIGURE 9.27 Transmission gate path selector with dynamic charge sharing properties.

the Data input is transferred to node X. If $\phi_1 = 0$ when this occurs, the new input state is dynamically held on this node until ϕ_1 goes to 1 and allows transfer to node X'. Charge sharing between X and X' occurs if $LD = 0$ during this transfer time interval.

An alternate arrangement that avoids charge sharing is shown in Fig. 9.28(b). The composite signals $\phi_1 \cdot LD$ and $\overline{\phi_1 \cdot LD}$ are applied to TGD for

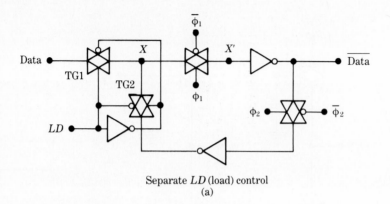

Separate LD (load) control
(a)

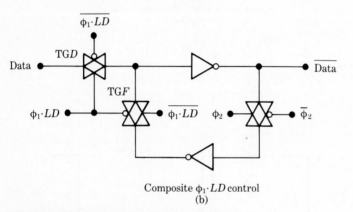

Composite $\phi_1 \cdot LD$ control
(b)

FIGURE 9.28   Selectively loadable register.

control of the Data input. To activate the input, both $\phi_1 = 1$ and $LD = 1$ must be simultaneously true. Otherwise, TGD is in cutoff while TGF is conducting to allow for internal register feedback. Charge sharing does not occur because there are no series-connected soft nodes.

Both registers in Fig. 9.28 employ (pseudo) 2-phase refreshing to maintain the dynamic register state during a hold ($LD = 0$). Feedback from one inverter to the other is alternately provided by the clock-controlled transmission gates. A complete refresh cycle requires one clock period T. (It is important to keep in mind that every inverter has implied connections to V_{DD} and ground.)

As a final example, consider the 2-to-4 decoder circuit shown in Fig. 9.29. The basic function table in Fig. 9.29(a) specifies the decoding of two inputs A_0, A_1 into four line outputs L_0, L_1, L_2, and L_3. The logic diagram in Fig. 9.29(b) implements this function using NAND gates. Selectively loadable input registers have also been provided.

The operation of the circuit is straightforward. Inputs A_0 and A_1 are admitted when $\phi_1 \cdot LD = 1$. A hold condition is attained by having $LD = 0$.

A_0	A_1	L_0	L_1	L_2	L_3
0	0	1	0	0	0
0	1	0	1	0	0
1	0	0	0	1	0
1	1	0	0	0	1

Basic function table
(a)

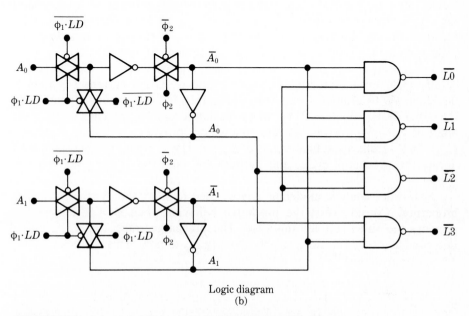

Logic diagram
(b)

FIGURE 9.29 2-to-4 decoder with selectively loadable inputs.

During ϕ_2, the register outputs become valid. The four signals A_0, $\bar{A}_0$, A_1, and $\bar{A}_1$ are decoded by the NAND gates, giving complementary outputs $\bar{L}_0$, $\bar{L}_1$, $\bar{L}_2$, and $\bar{L}_3$. These may be inverted if necessary. For example, output pairs L_i, $\bar{L}_i$ ($i = 0, 1, 2, 3$) could be used to control conduction through a transmission gate.

9.3 Single-Clock Dynamic Logic

Dynamic CMOS logic circuits rely on precharge and evaluate techniques to systematically provide logic functions. This section examines the basic concepts involved in single clock (2-phase) circuits.

9.3.1 Basic Operation

A dynamic CMOS inverter is shown in Fig. 9.30. The complementary MOSFETs *Mp* and *Mn* are clock-controlled and act as the precharge and evaluate transistors, respectively, in the dynamic arrangement. Logic inversion is obtained using the *n*-channel MOSFET *M*1. The inverter input voltage V_{in} is applied to the gate of *M*1, and the output voltage V_{out} is taken from the drain.

The operation of the inverter is similar to the dynamic nMOS circuit discussed in Section 8.7. A clock state of $\phi = 0$ initiates the precharge phase of a bit cycle by placing *Mp* in a conduction mode while keeping *Mn* in cutoff. Current flow I_{Dp} is established from the power supply through *Mp* to precharge the output node capacitance C_{out} to $V_{out} = V_{DD}$. A logic input voltage V_{in} is accepted during this half cycle. Note that if $V_{in} = V_{DD}$, corresponding to a logic 1 input, *M*1 conducts to precharge the *X*-node capacitance C_X.

Evaluation occurs during the next half cycle when $\phi = 1$. *Mp* is now in cutoff, while evaluation MOSFET *Mn* conducts. Drain current I_{Dn} flows to discharge C_X to ground. Conditional discharging of C_{out} occurs depending on the value of V_{in}. If $V_{in} = 0\,[\text{V}]$, *M*1 is in cutoff and V_{out} remains at V_{DD}. An input of $V_{in} = V_{DD}$ drives *M*1 into conduction, providing a discharge path for C_{out}. This results in a final value of $V_{out} = 0\,[\text{V}]$. Inputs are accepted only during the precharge cycle, and outputs are valid only during the evaluate phase.

First-order analytic modeling may be used to study the dynamic switching properties. Consider first the precharge behavior when $\phi = 0$. Figure 9.31 provides the subcircuit for this case. The precharge transistor *Mp* is biased into conduction by the clock signal, allowing I_{Dp} to flow. This charges C_{out} to a

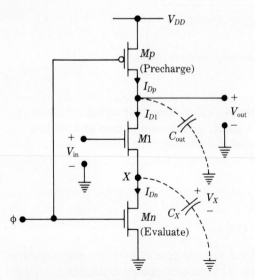

FIGURE 9.30 Dynamic CMOS inverter.

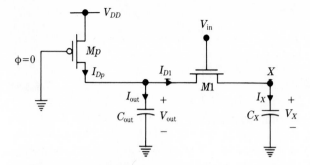

FIGURE 9.31 Precharge subcircuit.

voltage $V_{out} = V_{DD}$. C_X will also charge during this time if V_{in} is large enough to turn $M1$ on.

Suppose that $V_{in} = 0$. Logic transistor $M1$ is in cutoff with $I_{D1} = 0$. KCL gives $I_{out} = I_{Dp}$, so the analysis of Section 9.1.2 is valid. The charging time constant is thus given by

$$\tau_{ch} = \frac{C_{out}}{\beta_p(V_{DD} - |V_{Tp}|)}, \tag{9.3-1}$$

such that

$$t_{ch} = \tau_{ch}\left\{\frac{2|V_{Tp}|}{(V_{DD} - |V_{Tp}|)} + \ln\left[\frac{2(V_{DD} - |V_{Tp}|)}{0.1V_{DD}} - 1\right]\right\} \tag{9.3-2}$$

is the required precharge time.

The problem is more complicated when the inverter input is a logic 1 with $V_{in} = V_{DD}$. MOSFET $M1$ is biased into a conduction mode so that the current flow is described by

$$I_{Dp} = I_{out} + I_{D1}. \tag{9.3-3}$$

Since $I_X = I_{D1}$, current is diverted away from C_{out} to charge C_X. This increases the precharge time needed to achieve $V_{out} = V_{DD}$. An examination of the subcircuit shows that $M1$ acts as an nMOS pass transistor during the precharge. The final value for V_X is thus computed from

$$V_X = V_{DD} - V_{T1}(V_X). \tag{9.3-4}$$

The complicating factor in the analysis is that both V_X and V_{out} are functions of time t. A complete treatment requires solving a set of simultaneous differential equations for the pair of voltages. This is quite messy and will not be pursued here. Instead, approximations are introduced to simplify the problem to a point where the circuit operation can be understood.

The lowest-order approximation is obtained by assuming that $V_X(t) = V_{out}(t)$, which ignores the drain-source voltage (and hence the presence) of logic transistor $M1$. For this case,

$$I_{Dp} = C_{\text{out}} \frac{dV_{\text{out}}}{dt} + C_X \frac{dV_X}{dt}$$

$$\simeq (C_{\text{out}} + C_X) \frac{dV_{\text{out}}}{dt}, \tag{9.3-5}$$

which results in modifying the time constant to

$$\tau_{ch} \simeq \frac{(C_{\text{out}} + C_X)}{\beta_p (V_{DD} - |V_{Tp}|)}. \tag{9.3-6}$$

It should be noted that this also ignores the threshold voltage loss induced by $M1$, since it gives a final value of $V_X = V_{DD}$. This represents the worst-case value for t_{ch}.

The evaluation phase of the dynamic inverter is analyzed in a similar manner. Figure 9.32 gives the subcircuit when $\phi = V_{DD}$. The evaluate MOSFET Mn is biased into conduction, which allows for selective discharging of C_{out} depending on the value of V_{in}.

With $V_{\text{in}} = 0$, $M1$ is in cutoff and $I_{D1} = 0$. C_{out} thus remains charged at a voltage

$$V_{\text{out}} = V_{DD}, \tag{9.3-7}$$

corresponding to a logic 1 output. The only current flow in the circuit is I_{Dn}, which discharges C_X to $0\,[\text{V}]$. Mn acts as an nMOS pass transistor, so the analysis of Section 8.1.2 is valid. The discharge time constant for this case is (see eqn. 8.1-27)

$$\tau_{\text{dis}} = \frac{C_X}{\beta_n (V_{DD} - V_{Tn})}. \tag{9.3-8}$$

Although the discharge event does not affect the inverter output voltage, it is useful as a reference time interval.

C_{out} discharges to a final voltage $V_{\text{out}} = 0\,[\text{V}]$ when the inverter has a logic

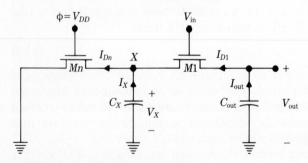

FIGURE 9.32 Evaluate subcircuit.

1 input of $V_{\text{in}} = V_{DD}$. Current flow in the circuit is described by

$$I_{Dn} = I_X + I_{D1}, \tag{9.3-9}$$

with

$$I_{D1} = I_{\text{out}} = -C_{\text{out}} \frac{dV_{\text{out}}}{dt}. \tag{9.3-10}$$

Note that an input voltage of $V_{\text{in}} = V_{DD}$ implies that V_X is precharged to $(V_{DD} - V_{T1})$. Mn must therefore sink discharge currents from both C_X and C_{out}.

The lowest-order approximation for the discharge is obtained by ignoring $M1$ to write $V_X(t) \simeq V_{\text{out}}(t)$. Then

$$I_{Dn} \simeq -(C_{\text{out}} + C_X) \frac{dV_{\text{out}}}{dt}, \tag{9.3-11}$$

so the effective time constant is modified to

$$\tau_{\text{dis}} \simeq \frac{(C_{\text{out}} + C_X)}{\beta_n (V_{DD} - V_{Tn})}. \tag{9.3-12}$$

This provides the worst-case value for the discharge time approximated by

$$t_{HL} \simeq \tau_{\text{dis}} \left\{ \frac{2V_{Tn}}{(V_{DD} - V_{Tn})} + \ln \left[\frac{2(V_{DD} - V_{Tn})}{0.1V_{DD}} - 1 \right] \right\}, \tag{9.3-13}$$

as derived previously in Section 9.1.1.

The preceding analysis represents only the lowest order of modeling, but it does serve to illustrate the basic behavior of the dynamic circuit. Two important comments concerning design can be extracted from the discussion. The first deals with the MOSFET aspect ratios. Large (W/L) values give faster switching, while minimum-size transistors with $(W/L)_{\text{min}}$ allow higher chip packing densities. Consequently, the dynamic tradeoff is between speed and real estate consumption. A layout example is provided in Fig. 9.33. The second point to note is that the output constitutes a (CMOS) soft node. When $V_{\text{in}} = 0$, charge is dynamically stored on C_{out} during the evaluate phase ($\phi = 1$). It is seen that charge leakage paths exist through Mp and $M1$. This establishes a minimum clocking frequency.

9.3.2 Generalized Combinational Logic

The dynamic inverter circuit is easily extended to provide gate-level logic functions. The idea is quite simple. Each gate is structured with a pMOS precharge transistor and an nMOS evaluate transistor. These provide conduction paths to V_{DD} and ground, respectively. Logic functions are implemented using nMOS configurations between the precharge and evaluate MOSFETs, as shown in Fig. 9.34. Precharging occurs when $\phi = 0$. During this time, C_{out} is

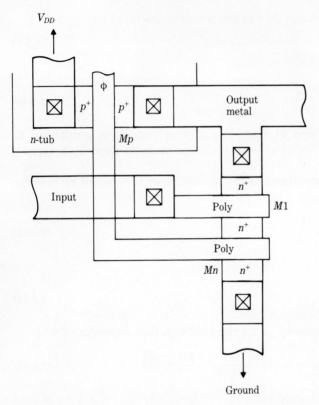

FIGURE 9.33 Dynamic inverter layout.

charged to a voltage $V_{\text{out}} = V_{DD}$ through Mp. Depending on the input combination, some internal MOS logic capacitances may also be charged. Input states are evaluated when the clock rises to $\phi = 1$. Conditional discharging of C_{out} to ground (through Mn) occurs during this time.

Figure 9.35 shows a 5-input NOR gate constructed using this approach. The logic is performed by the parallel-connected n-channel MOSFETs $M1$ through $M5$ such that the output is

$$F = \overline{X_1 + X_2 + X_3 + X_4 + X_5}. \tag{9.3-14}$$

The operation of the circuit is similar to that of the inverter. A clock signal of $\phi = 0$ places Mp in an active conducting mode and defines the precharge. C_{out} attains a voltage $V_{\text{out}} = V_{DD}$. If at least one of the inputs $(X_1, X_2, X_3, X_4, X_5)$ is at a logic high level, the associated nMOS logic transistor is conducting. This creates a current path to node X and charges C_X to a level $V_X = (V_{DD} - V_T)$ as in eqn. (9.3-4). Note that current is diverted away from C_{out}, thus lengthening the time required to charge C_{out}. This must be accounted for when determining the minimum precharge clock time. C_{out} itself can be approxi-

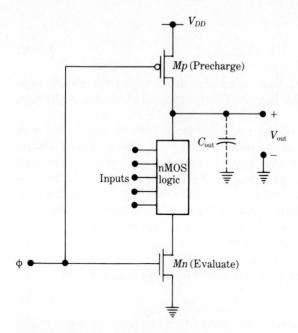

FIGURE 9.34 Generalized dynamic logic.

mated by

$$C_{\text{out}} \simeq (C_{GDp} + KC_{dbp}) + \sum_{i=1}^{5} (C_{GDi} + KC_{dbi}) + C_{\text{line}} + C_G, \qquad \textbf{(9.3-15)}$$

where the summation is over the nMOS logic transistors and C_G represents the input gate capacitance of the next stage. This is larger than the inverter C_{out} by

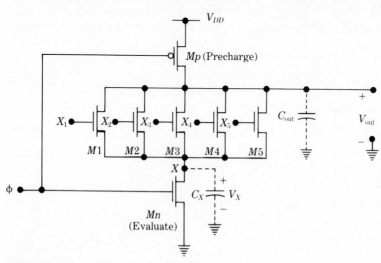

FIGURE 9.35 5-input dynamic NOR gate.

an amount

$$\Delta C_{\text{out}} \simeq \sum_{i=2}^{4} (C_{GDi} + KC_{dbi}) + \Delta C_{\text{line}}, \tag{9.3-16}$$

where ΔC_{line} is the additional line capacitance from wiring the nMOS logic transistors into the circuit. The increased output capacitance also contributes to longer precharge times.

Evaluation of the inputs occurs when the clock goes to $\phi = 1$. Mp is in cutoff, while Mn is placed in an active conducting mode. C_X is always discharged to ground during this portion of the logic cycle. Conditional discharging of C_{out} can occur depending on the inputs. If at least one input is high (at a logic 1 level), C_{out} discharges to ground through the conducting logic transistor(s) and Mn. This results in a final value of $V_{\text{out}} = 0\,[\text{V}]$. On the other hand, if all inputs are low (at a logic 0 level), then the charge is dynamically held on C_{out}. The output is thus maintained at $V_{\text{out}} = V_{DD}$, except for the decay introduced by soft node leakage.

A 3-input dynamic NAND gate that implements the logic function

$$F = \overline{X_0 X_1 X_2} \tag{9.3-17}$$

is shown in Fig. 9.36. The operational timing of this circuit is identical to the

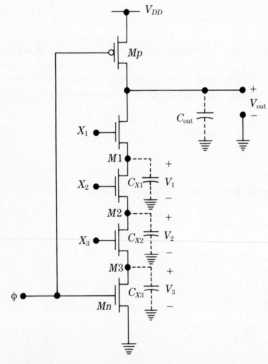

FIGURE 9.36   3-input dynamic NAND gate.

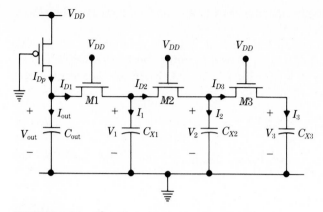

FIGURE 9.37 Charging circuit for NAND gate.

NOR gate: $\phi = 0$ defines the precharge while $\phi = 1$ defines the evaluation portions of the clock cycle. C_{out} constitutes the important capacitance that is precharged to $V_{out} = V_{DD}$. If $X_0 = X_1 = X_2 = 1$, then C_{out} discharges during the evaluation phase, giving $V_{out} = 0\,[V]$.

The minimum clock time intervals needed for the precharge and evaluate cycles are determined by the case where the inputs (X_1, X_2, X_3) are all at logic $1\,(= V_{DD})$ levels. The precharge time must be long enough to charge C_{out} to a level $V_{out} = V_{DD}$. The problem is illustrated by the subcircuit in Fig. 9.37, where it is seen that the pMOS current I_{Dp} splits between C_{out} and the nMOS logic transistors according to

$$I_{Dp} = I_{out} + I_{D1}. \tag{9.3-18}$$

Applying KCL to the remaining nodes gives

$$I_{D1} = I_1 + I_{D2},$$
$$I_{D2} = I_2 + I_{D3},$$
$$I_{D3} = I_3, \tag{9.3-19}$$

so the precharge current can be expressed as

$$I_{Dp} = I_{out} + I_1 + I_2 + I_3. \tag{9.3-20}$$

Noting that

$$I_i = C_{Xi}\frac{dV_i}{dt} \qquad (i = 1, 2, 3) \tag{9.3-21}$$

is the current into the ith capacitor,

$$I_{Dp} = C_{out}\frac{dV_{out}}{dt} + \sum_{i=1}^{3} C_{Xi}\frac{dV_i}{dt}. \tag{9.3-22}$$

The charging rate of C_{out} is described by (dV_{out}/dt). This equation shows that the current flow to the logic capacitances C_1, C_2, and C_3 increases the charge time needed for V_{out} to reach V_{DD}.

The worst-case value for the charge time constant is obtained by assuming that initially $V_i(0) = 0\,[V]$ ($i = 1, 2, 3$) and ignoring the logic MOSFETs $M1$, $M2$, $M3$ by writing $V_i(t) = V_{out}(t)$ for $i = 1, 2, 3$. Then all capacitors are in parallel, so the time constant is

$$\tau_{ch} \simeq \frac{(C_{out} + C_{X1} + C_{X2} + C_{X3})}{\beta_p(V_{DD} - |V_{Tp}|)}. \tag{9.3.23}$$

This represents the worst-case value. Similarly, the upper limit for the discharge time constant is

$$\tau_{dis} \simeq \frac{(C_{out} + C_{X1} + C_{X2} + C_{X3})}{\beta_n(V_{DD} - V_{Tn})}, \tag{9.2-24}$$

which views all capacitors in parallel and discharging through Mn.

AOI logic functions are easily implemented using this basic scheme. Figure 9.38 provides a circuit example for an output of

$$F = \overline{(A + B + C)X_1 + DX_2}. \tag{9.3-25}$$

Each branch may be analyzed using the techniques discussed in this section.

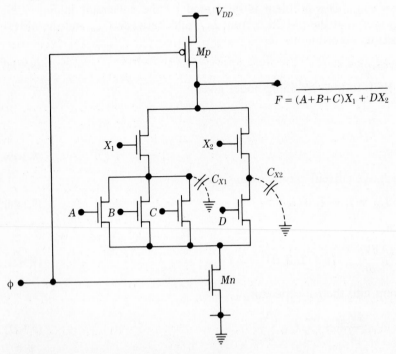

FIGURE 9.38   Dynamic AOI gate example.

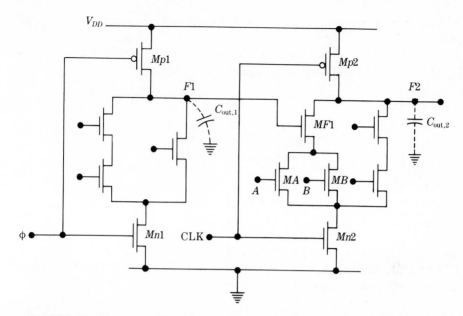

FIGURE 9.39   Timing problems in cascaded single-clock dynamic CMOS.

The capacitors C_{X1} and C_{X2} represent the internal logic nodes that are important for analyzing the switching times.

A significant problem that arises in single-clock CMOS circuits is that dynamic nMOS gates cannot be cascaded. To understand this limitation, refer to the circuit in Fig. 9.39. The first-stage logic is controlled by the clock signal ϕ. The second stage is timed by CLK, which will be either ϕ or $\bar{\phi}$ in a single-clock configuration.

Suppose that CLK $= \phi$, so that the precharge and evaluate times are the same for both stages. An error in logic transfer can occur during the evaluate portion of the clocking. To follow the events, first note that when $\phi =$ CLK $= 0$, the output capacitors $C_{out,1}$ and $C_{out,2}$ charge to a level V_{DD}. At this point, MOSFET $MF1$ is conducting. When the clocks switch to $\phi =$ CLK $= 1$, the circuit enters the evaluation mode and logic transistors $Mn1$ and $Mn2$ conduct. An error will occur if the first-stage output has a final value of $F1 = 0$ while $A = 1$ or $B = 1$ in the second stage. The problem lies in the fact that it requires a time interval $t_{HL,1}$ for $C_{out,1}$ to discharge and attain a valid $F1 = 0$ level. However, at the start of the evaluation cycle, capacitor $C_{out,2}$ will discharge through $MF1$ and either MA or MB to give $F2 = 0$. By the time the first-stage output $F1$ reaches a logic 0 level and drives $MF1$ into cutoff, the charge on $C_{out,2}$ has already been lost. This then gives an incorrect output of $F2 = 0$.

A more obvious problem occurs when CLK $= \bar{\phi}$. This clocking scheme alternates the precharge and evaluate cycles between the stages. Since the first stage is precharging $C_{out,1}$ while the second stage evaluates its inputs, $F1$ is

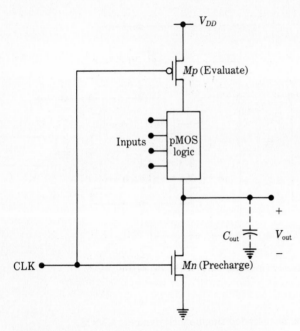

FIGURE 9.40   pMOS dynamic logic gate structure.

always at a logic 1 level regardless of the first stage inputs. In other words, it is not possible to pass the first stage logic through the gate! This situation, of course, must be avoided.

Dynamic cascades can be achieved by alternating nMOS and pMOS-based logic stages. A generalized dynamic pMOS gate is shown in Fig. 9.40. The structure is exactly the complement of that used for the nMOS gate. The output node is at the top of the nMOS evaluate transistor Mn. CLK = 1 defines a precharge event where Mn turns on and C_{out} is set to a voltage level of $V_{out} = 0$ [V]. (Although the term "precharge" may be confusing in this context, just remember that a pMOS gate behaves exactly opposite to the nMOS.) A clock state of CLK = 0 places Mp in a conducting state that evaluates the inputs. This results in a conditional charging of C_{out} to $V_{out} = V_{DD}$ if a conduction path is provided through the pMOS logic block.

Figure 9.41 provides an example of how an alternating cascade of dynamic nMOS and pMOS logic gates can be constructed. The clock signal ϕ is applied to the nMOS gate, while $\bar{\phi}$ acts to control the pMOS gate. This circuit avoids logic errors at the input to the second stage by using the fact that a high precharge voltage on the gate of p-channel MOSFET $M1$ places the transistor in cutoff. Unwanted conduction paths do not occur with this scheme.

During the time when $\phi = 0$, $Mp1$ is conducting to precharge C_{out} to a value of $V_{out,1} = V_{DD}$. This places $M1$ in cutoff with $I_{D1} = 0$. The pMOS gate has a clock signal of $\bar{\phi} = 1$ applied, so it is in a precharge interval with $Mn2$ conducting. This results in a value of $V_{out,2} = 0$ [V]. Logic evaluation takes

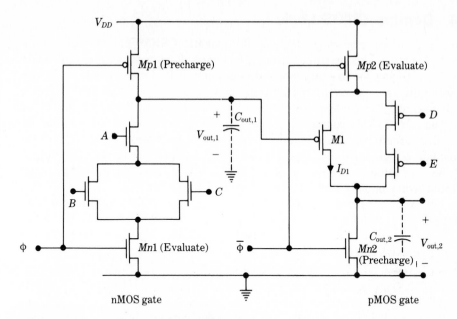

FIGURE 9.41 CMOS dynamic cascade.

place during the next half clock cycle when $\phi = 1$ and $\bar{\phi} = 0$. $C_{out,1}$ will discharge to $0\,[\mathrm{V}]$ if $A(B + C) = 1$. In this case, $M1$ will conduct current I_{D1}, which then charges $C_{out,2}$ to a logic 1 value of $V_{out,2} = V_{DD}$. On the other hand, if $A(B + C) = 0$, then $M1$ stays in cutoff. $C_{out,2}$ remains at a voltage $V_{out,2} = 0\,[\mathrm{V}]$ unless a conduction path from the power supply is formed with the input combination $DE = 0$.

The circuit design problem centers around setting the (W/L) values for the MOSFETs. Since the operation is purely dynamic, the logic is ratioless. In a critical design where the transient response times are important, (W/L) values must be chosen large enough to meet the timing specifications. Otherwise, the layout is simplified by using identical device geometries throughout. One drawback of the nMOS/pMOS logic cascade is the fact that

$$k_p' < k_n'. \tag{9.3-26}$$

Thus, $(W/L)_p > (W/L)_n$ will be required to equalize the stage response times.

A final point concerning this type of circuit is that the precharge in a pMOS block sets the output soft node to a level of $V_{out} = 0\,[\mathrm{V}]$. Charge leakage during an evaluate phase can be a problem because of the reverse current flow from V_{DD} to the output node through a pMOS logic transistor. For example, p-channel MOSFETs $M1$ and E in Fig. 9.41 both provide power supply leakage paths from the p^+/n-tub junctions. In a pMOS stage, the leakage path to ground through the precharge n-channel MOSFET aids in maintaining a logic 0 level. This is exactly opposite to the nMOS case.

9.4 Domino CMOS Logic

The problem of cascading single-clock dynamic CMOS circuits using only nMOS logic can also be solved using a technique known as *domino logic*. The basic structuring for one stage of a domino circuit is shown in Fig. 9.42. The logic inputs are applied to nMOS transistors arranged in a standard dynamic configuration. The dynamic output is buffered by the static inverter made up of *Mn2* and *Mp2*. This composite circuit is used as the basic building block for cascaded circuits. Domino circuits avoid logic transmission errors by always providing a logic 0 level at the output after an internal precharge. The output can be connected to the gate of an *n*-channel MOSFET in the next logic stage without worrying about an incorrect state interpretation.

The circuit operation may be understood by examining the clocking intervals. $\phi = 0$ defines the precharge cycle for the circuit. During this time, $Mp1$ conducts to charge $C_{out,1}$ to a level $V_{out,1} = V_{DD}$. This sets the minimum precharge clock interval to (see eqn. 4.4-3)

$$\left(\frac{T}{2}\right) > t_{LH,1} \simeq \tau_{p1}\left\{\frac{2\,|V_{Tp}|}{(V_1 - |V_{Tp}|)} + \ln\left[\frac{2(V_1 - |V_{Tp}|)}{V_0} - 1\right]\right\}, \qquad (9.4\text{-}1)$$

where the best-case value of the precharge time constant is

$$\tau_{p1} \simeq \frac{C_{out,1}}{\beta_{p1}(V_1 - |V_{Tp}|)}. \qquad (9.4\text{-}2)$$

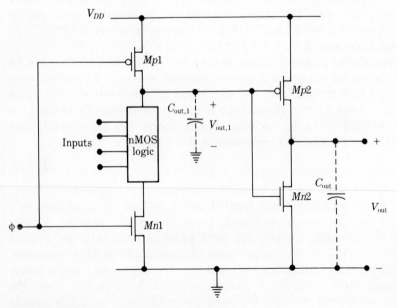

FIGURE 9.42 Basic domino logic circuit.

Since $V_{out,1}$ drives the static inverter, the precharge sets the overall circuit output at $V_{out} = 0 \, [V]$.

Evaluation takes place when the clock switches to $\phi = 1$. Transistor $Mn1$ conducts, providing a discharge path to ground if the logic sets up a conduction path through the nMOS logic block. V_{out} will remain at $0 \, [V]$ unless $C_{out,1}$ discharges. If this occurs, the inverter will switch when $V_{out,1}$ decays to the inverter V_{IL} value. In a symmetric inverter, this requires

$$V_{out,1} \simeq V_{IL} = \frac{3}{4}\left(V_{Tn} + \frac{3}{2}V_{DD}\right). \tag{9.4-3}$$

The domino output V_{out} reaches a logic 1 V_{DD} voltage after a propagation time of

$$t_p \simeq \tau_n\left\{\frac{2V_{Tn}}{(V_{DD} - V_{Tn})} + \ln\left[\frac{4(V_{DD} - V_{Tn})}{V_{DD}} - 1\right]\right\}. \tag{9.4.4}$$

Note that the inverter time constant is determined by the load capacitance C_{out}. The evaluate portion of the clock cycle must be long enough to allow for both the discharging of $C_{out,1}$ and the subsequent inverter propagation time. This requires

$$\left(\frac{T}{2}\right) > t_p + t_{HL,1}, \tag{9.4-5}$$

where $t_{HL,1}$ is the time needed for $V_{out,1}$ to decay to V_{IL}.

Domino cascades can exhibit fast logic propagation times. One reason for this is that the static inverter has a lower output capacitance than the dynamic gate. In addition, the inverter provides a charging path from the power supply that avoids soft node and charge sharing problems at the output. (Note, however, that the internal precharge node is still subject to these dynamic problems.)

Figure 9.43 illustrates the general structure of a domino cascade. Each gate is controlled by the clock signal ϕ. Precharging takes place when $\phi = 0$. During this time, the internal dynamic capacitance of each stage is charged to a voltage V_{DD}. Every static inverter thus has an output voltage of $0 \, [V]$ at the end of the precharge. When ϕ switches to 1, the inputs are evaluated and the dynamic capacitors undergo conditional discharging. The logic thus propagates through the circuit from left to right. The terminology "domino logic" is based on the observation that this resembles a chain of dominoes falling down on one another.

An example of a domino cascade is shown in Fig. 9.44. The input stage is connected to have a FanOut of 2. When precharging takes place with a clock state of $\phi = 0$, the first-stage internal dynamic capacitor $C_{out,1}$ charges to a voltage V_{DD}. This gives an output voltage of $V_1 = 0 \, [V]$ from the first stage, which in turn drives logic MOSFETs $M1$ and $M2$ into cutoff. $C_{out,2}$ and $C_{out,3}$ are then allowed to charge to V_{DD} during this time interval. Note that placing

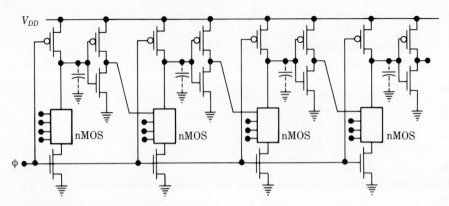

FIGURE 9.43 Cascaded domino logic.

$M1$ and $M2$ at the top of the series logic chain in each stage gives charging times that are independent of the other inputs. This simplifies the overall timing of the logic chain.

The domino logic propagation is triggered when the clock makes the transition from 0 to 1. The first-stage output is defined by

$$F_1 = X_1 X_2 X_3. \tag{9.4-6}$$

At the beginning of the evaluation cycle, $F_1 = 0$, so $F_2 = 0$ and $F_3 = 0$. These states will remain true unless all of the X inputs are at logic 1 levels. In this

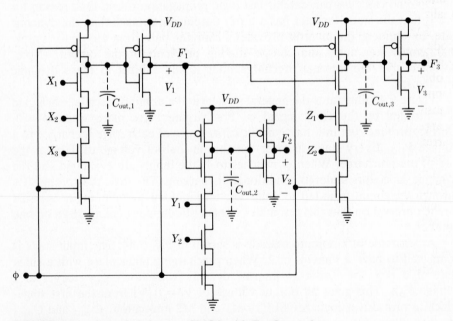

FIGURE 9.44 Cascaded domino CMOS with FanOut = 2.

case, F_1 makes a transition to a logic 1 level and

$$F_2 = F_1 Y_1 Y_2 = Y_1 Y_2,$$
$$F_3 = F_1 Z_1 Z_2 = Z_1 Z_2 \tag{9.4-7}$$

show how the outputs F_2 and F_3 will be determined by the Y and Z inputs. It is important to note that the logic propagates through all three stages during one evaluation time interval. Care must be taken to ensure that the worst-case propagation time through the chain does not exceed the $(T/2)$ limit.

Two comments should be made at this point. First, note that domino CMOS is ideally structured for AND-based logic cascades such as that shown in the example. This is because the output level of each stage is a logic 0 at the beginning of the evaluation phase. The only time this changes is when an AND condition is satisfied at the input stage. Then the logic propagates through the chain as long as subsequent AND operations produce a logic 1 output. The logic propagation is always stopped when an AND operation gives a logic 0 result. Generalized AOI structuring can still be used efficiently if it is placed toward the front of the chain.

The second comment is that domino logic is noninverting. Consequently, functions such as

$$F = A \oplus B = A\bar{B} + \bar{A}B \qquad \text{(XOR)}$$
$$F = A \odot B = AB + \bar{A}\bar{B} \qquad \text{(Equivalence)} \tag{9.4-8}$$

cannot be included in the domino chain. Inverting logic functions must be realized by using conventional static or other types of dynamic circuit sections.

The operation of the domino circuit depends on storing charge on the soft node between the dynamic input gate and the static output buffer. Static or low-frequency operation cannot be obtained because of normal charge leakage problems. However, this can be overcome by adding a pull-up MOSFET that connects the internal node to the power supply. In Fig. 9.45(a), a pMOS transistor Mp has been added as the pull-up device to supply $C_{\text{out},1}$ with charge. The device is biased with $V_{SGp} = V_{DD} = $ Constant, so it supplies current to keep $V_{\text{out},1} = V_{DD}$. To keep the DC power dissipation low, (W/L) is chosen to be small. An even simpler circuit is shown in Fig. 9.45(b). The pMOS precharge transistor has been eliminated, leaving only the pull-up MOSFET. The dynamic portion of the circuit thus resembles a clocked pseudo-nMOS arrangement. Since there is no controlled precharge path in the circuit, the clock period must be long enough to allow $C_{\text{out},1}$ to charge up to the required V_{DD} level. Note that the output becomes ratioed because of the path to the power supply during evaluation.

Standard domino logic cascades with series logic exhibit charge sharing characteristics that can induce logic glitches. The problem is illustrated by the circuit in Fig. 9.46(a). The total charge stored on this node is given by

$$Q_T = C_{\text{out},1} V_{DD}. \tag{9.4-9}$$

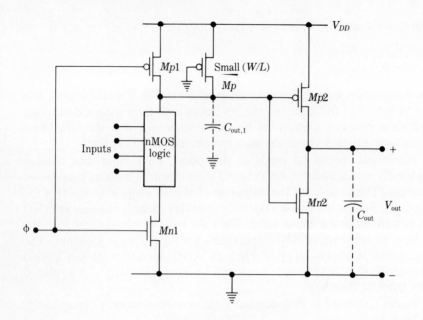

Basic configuration
(a)

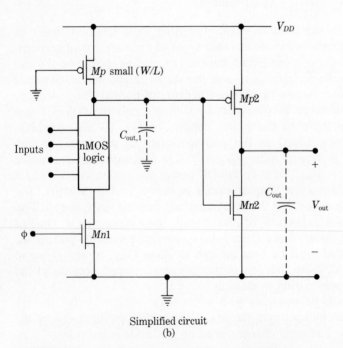

Simplified circuit
(b)

FIGURE 9.45 Domino stage with pull-up MOSFET.

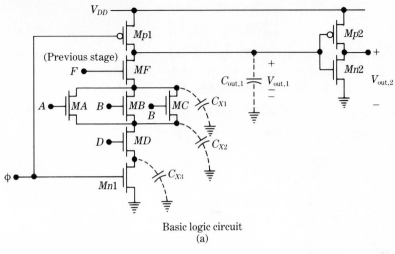

Basic logic circuit
(a)

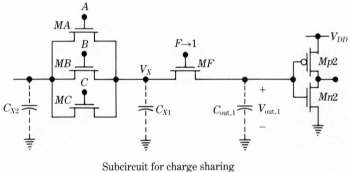

Subcircuit for charge sharing
(b)

FIGURE 9.46 Charge sharing in a domino chain.

F represents the output from the previous domino stage, so $F \to 0$ during this time. MOSFET MF is then driven into cutoff, isolating the dynamic output node.

Charge sharing can occur during the evaluation time ($\phi = 1$) if the previous stage in the chain induces a transition of $F \to 1$. This possibility is shown by the subcircuit in Fig. 9.45(b) for the case where it has been assumed that $D = 0$. This places MD in cutoff and prevents $C_{\text{out},1}$ from discharging through $Mn1$. Since $F = 1$, MF is conducting and charge must be shared with C_{X1}. In addition, if any of the inputs (A, B, C) are at logic 1 levels, then C_{X2} must be accounted for in the charge transfer. A logic error will occur if $V_{\text{out},1}$ falls below V_{IH} of the static CMOS inverter.

In the worst-case situation, both C_{X1} and C_{X2} will charge. This can be analyzed by first writing

$$V_X = V_{\text{out},1} - V_{TF}(V_X), \tag{9.4-10}$$

where

$$V_{TF}(V_X) = V_{T0n} + \gamma_n(\sqrt{2|\phi_{Fp}| + V_X} - \sqrt{2|\phi_{Fp}|}) \tag{9.4-11}$$

is the threshold voltage of MOSFET *MF*. Denoting the final voltage across $C_{out,1}$ after charge sharing by $V_{out,1} = V_f$ and assuming that $V_f > (V_{DD} - V_{TF})$, the total charge must distribute according to

$$Q_T = C_{out,1}V_f + (C_{X1} + C_{X2})(V_f - V_{TF}). \tag{9.4-12}$$

Equating the total charge to Q_T in eqn. (9.4-9) then gives

$$V_f = \frac{C_{out,1}}{(C_{out,1} + C_{X1} + C_{X2})}V_{DD} + \frac{(C_{X1} + C_{X2})}{(C_{out,1} + C_{X1} + C_{X2})}V_{TF}(V_X). \tag{9.4-13}$$

To solve for V_f, $V_{TF}(V_X)$ must be found. This may be computed by substituting V_f for $V_{out,1}$ in eqn. (9.4-10). Rearranging gives the self-iterating equation

$$V_X = \frac{C_{out,1}}{(C_{out,1} + C_{X1} + C_{X2})}[(V_{DD} - V_{T0n})$$
$$- \gamma_n(\sqrt{2|\phi_{Fp}| + V_X} - \sqrt{2|\phi_{Fp}|})] \tag{9.4-14}$$

for V_X. Solving then allows V_{TF} to be calculated for use in eqn. (9.4-13). If severe charge sharing takes place, then the final equilibrium voltage will be small and $V_f < (V_{DD} - V_{TF})$. In this case, there is no threshold voltage loss, so the simpler equation

$$V_f = \frac{C_{out,1}}{(C_{out,1} + C_{X1} + C_{X2})}V_{DD} \tag{9.4-15}$$

is valid.

The preceding analysis demonstrates that

$$C_{out,1} \gg (C_{X1} + C_{X2}) \tag{9.4-16}$$

should be satisfied to minimize charge sharing effects and ensure that $V_{out,1} > V_{IH}$ is maintained. This can be accomplished by designing the inverter MOSFETs *Mp2* and *Mn2* with large gate areas. Careful layout of the logic transistors is also important, since these contribute to C_{X1} and C_{X2}.

An alternate approach to treating the charge sharing problem is to add a feedback-controlled pull-up MOSFET as shown in Fig. 9.47. The device is connected such that

$$V_{SGp} = V_{DD} - V_{out,2}. \tag{9.4-17}$$

A precharge sets $V_{out,1} = V_{DD}$ and $V_{out,2} = 0$ [V]. Since

$$V_{SDp} = V_{DD} - V_{out,1}, \tag{9.4-18}$$

$V_{SDp} = 0$ after the precharge is complete. If $V_{out,1}$ should drop below V_{DD}, *Mp* turns on and allows I_{Dp} to flow to charge $C_{out,1}$. The device aspect ratio $(W/L)_p$ controls the level of current flow. The combination of *Mp* and the

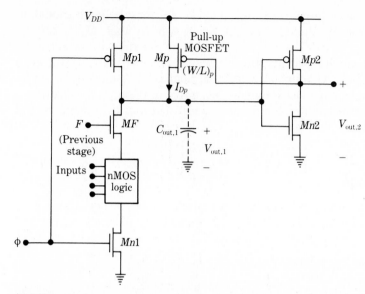

FIGURE 9.47 Use of feedback to control a pull-up MOSFET for charge sharing problem.

static inverter can be viewed as forming a simple latch because of the feedback loop.

Evaluating the inputs with $\phi = 1$ allows for two possibilities. First, suppose that charge sharing takes place. Mp will then supply current to $C_{out,1}$ to compensate for the charge lost to the internal capacitances of the nMOS logic block. Proper design of the feedback MOSFET Mp will maintain $V_{out,1} = V_{DD}$ as required. It should be mentioned that there is the possibility of a transient spike occurring as charge redistribution takes place. This can be controlled by the choice of $(W/L)_p$.

The second possibility is that the input logic evaluates to give $V_{out,1} = 0\,[\mathrm{V}]$ as the final state. In this case, $C_{out,1}$ will discharge through $Mn1$. The level of current I_{Dp} as determined by $(W/L)_p$ must be small enough to permit this discharge to take place. As soon as $V_{out,1}$ falls below V_{IL}, the inverter will switch to give $V_{out,2} = V_{DD}$. This reduces the source-gate voltage on the pull-up transistor to $V_{SGp} = 0\,[\mathrm{V}]$, which drives Mp into cutoff. The regenerative action maintains this state until the next clock cycle initiates a precharge.

The design criteria can be extracted from the preceding discussion. To this end, let $(W/L)_n$ represent the *effective* aspect ratio of the nMOS devices from the logic stage output to ground. In the present circuit, this is made up of MF, the nMOS logic block, and $Mn1$. A logic 0 level on $C_{out,1}$ is set by the ratioed behavior of Mp and the nMOS transistors, so it is convenient to use.

$$\beta_R = \frac{(W/L)_n}{(W/L)_p} \qquad (9.4\text{-}19)$$

as in pseudo-logic. The design tradeoff is that $(W/L)_p$ must be large enough to compensate for charge sharing and small enough so that $C_{\text{out},1}$ can be discharged to $V_{\text{out},1} = 0\,[\text{V}]$ when the input logic requires it. A large β_R indicates that Mp cannot provide much current, while a small β_R will keep $V_{\text{out},2}$ "stuck" at $0\,[\text{V}]$. Although a realistic calculation of β_R must be based on the process parameters and clocking intervals, a rule of thumb is to start with a value of β_R on the order of 1 for the first design cut and then simulate and adjust as necessary.

9.5   NORA Logic

NORA (NO RAce) logic structuring is a CMOS system design style that is particularly useful for *pipelined* circuits. An example of pipelining is provided in Fig. 9.48. In a central processing unit (CPU) of a computer, the instruction

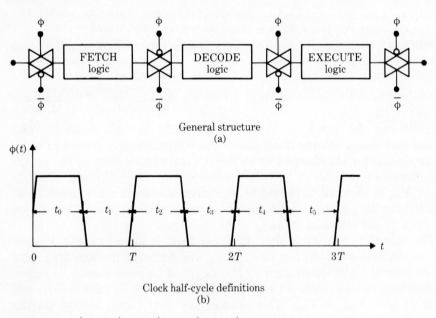

General structure
(a)

Clock half-cycle definitions
(b)

Logic block	t_0	t_1	t_2	t_3	t_4
FETCH	INST-1	INST-2	INST-3	INST-4	INST-5
DECODE		INST-1	INST-2	INST-3	INST-4
EXECUTE			INST-1	INST-2	INST-3

Instruction flow summary
(c)

FIGURE 9.48   Pipelined logic flow example.

sequencing may be broken down into three main operational logic groups:

1. FETCH: the instruction code is obtained by accessing the system memory;

2. DECODE: the code is "translated" into logic operations;

3. EXECUTE: the operations are executed by the system.

Pipelining this sequence enhances the overall system speed. The general structuring is obtained by cascading the logic circuit blocks as shown in Fig. 9.48(a). The logic sections are separated by TGs driven by clock signals ϕ and $\bar{\phi}$.

Pipeline logic can be understood by referring to half clock cycles t_1, t_2, t_3, etc., shown in Fig. 9.48(b). The instruction flow through the cascade is summarized by the table in Fig. 9.48(c). The term *pipeline* arises from the observation that instructions are continuously "piped" through the system, one after another. In terms of the notation used in the table, this implies that INST-1 is followed by INST-2, which is followed by INST-3, and so on. Since every logic stage always operates on an instruction, no clock cycles are wasted. This enhances the overall system speed over a nonpipelined structuring.

The "no race" characteristic of NORA logic arises from the internal circuit timing of the logic blocks when referenced to the input and output transmission gates. Clock and signal races can be understood using the logic block and timing diagram shown in Fig. 9.49. Assume that static circuits are used to implement the logic. A race problem can arise when ϕ and $\bar{\phi}$ change states. The clock overlap induced by the finite rise and fall times gives the "race

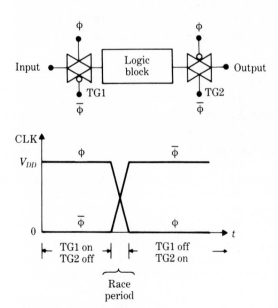

FIGURE 9.49 Signal race problem.

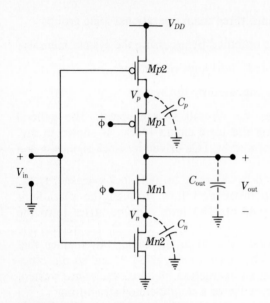

FIGURE 9.50   A C^{2}MOS latching circuit.

period", where it is possible that both TG1 and TG2 are in transmission modes. Changes in the input logic during this time can create a race condition between a new input combination and the previous input set. The "winner" will be the output state when TG2 goes into the nontransmitting mode. Obviously, this type of problem can lead to logic errors.

NORA logic blocks are constructed by alternately cascading dynamic nMOS and pMOS stages. A NORA output uses the C^{2}MOS (clocked CMOS) latch circuitry shown in Fig. 9.50 to avoid signal races. This circuit uses the complementary MOSFET pair $Mp1$ and $Mn1$ for clocking, while $Mp2$ and $Mn2$ are input devices. A clock signal of $\phi = 1$ is the precharge cycle, while $\phi = 0$ defines the hold state. C_p and C_n are the important precharge capacitors. The output state is maintained by the charge in C_{out}.

To analyze the operation of the latch, consider first the case where the input is a logic 0. During the precharge cycle ($\phi = 1$), both $Mp1$ and $Mn1$ are conducting. The input voltage $V_{in} = 0$ [V] allows current to flow through $Mp2$, which precharges C_p and C_{out} to V_{DD}. C_n is also charged during this time interval. The threshold voltage loss across $Mn1$ gives a maximum voltage of

$$V_n = (V_{DD} - V_{T0n}) - \gamma_n(\sqrt{V_n + 2|\phi_{Fp}|} - \sqrt{2|\phi_{Fp}|}) \qquad \text{(9.5-1)}$$

across C_n. The output node is isolated (latched) when the clock goes to $\phi = 0$ (a hold state). With $V_{in} = 0$ [V], C_{out} is initially charged to the power supply level V_{DD}, giving a logic 1 output. Note, however, that the output constitutes a CMOS soft node during this time.

The circuit operation is exactly opposite for a logic 1 input with $V_{in} = V_{DD}$. A precharge clock level of $\phi = 1$ implies that MOSFETs $Mp1$, $Mn1$, and $Mn2$

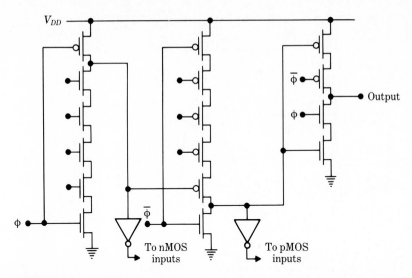

FIGURE 9.51   NORA ϕ-section.

are conducting to discharge the capacitors. In particular, C_n and C_{out} both drop to $0\,[\text{V}]$. A clock transition to a hold state ($\phi = 0$) maintains the output voltage at $V_{out} = 0\,[\text{V}]$.

The preceding discussions may how be combined to study the details of a NORA subblock. Figure 9.51 shows the basic circuitry for a NORA ϕ-*section*. It consists of alternating nMOS and pMOS stages with a C^2MOS output latch. Some important features deserve further comment.

First note that static inverters are provided at the output of each dynamic stage. These allow for driving output logic stages of the same polarity without false switching. The use of both nMOS and pMOS logic and inverters provides for inverted signals, which increase the logic flexibility of the system. Although only a single nMOS/pMOS combination is shown, it is also possible to have pMOS/nMOS cascades. A NORA section may thus be designed to implement a group of logical operations to an arbitrary degree of complexity.

The phasing of clock signals requires that a distinction be made between the behavior of the dynamic logic stages and that of the output latch circuitry. To clarify this statement, note that a clock of $\phi = 0$ precharges the logic circuits but places the output latch in hold. Since the output node is isolated from the input stages during this time, data transfer out of the ϕ-section can be accomplished. The condition of $\phi = 0$ will thus be termed a "precharge/transfer" state. A clock signal of value $\phi = 1$ provides for dynamic evaluation in the logic stages, with conditional discharging (charging) in the nMOS (pMOS) circuits. The overall results of the NORA logic chain are admitted to the output latch for transfer during the next half clock cycle. A clock state of $\phi = 1$ can still be termed as an evaluation phase without

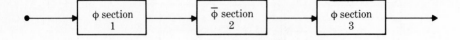

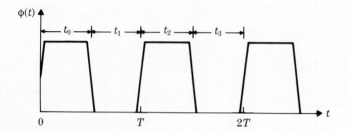

Logic section	t_0	t_1	t_2	t_3
1	PC/TR	EV OP1	PC/TR	EV OP2
2		PC/TR	EV OP1	PC/TR
3			PC/TR	EV/OP1

PC/TR = Precharge transfer
EV OP# = Evaluate operation number

FIGURE 9.52 NORA pipelined logic.

ambiguity. However, it is important to note that no data transfer out of the ϕ-section takes place during this time.

A NORA pipeline is shown in Fig. 9.52. It consists of alternating ϕ- and $\bar{\phi}$-sections, where a $\bar{\phi}$-section is created by simply interchanging ϕ and $\bar{\phi}$ in a ϕ-section. The pipelined operation is seen by noting that $\phi = 0$ places the ϕ-sections into precharge/transfer, while $\bar{\phi}$-sections are undergoing evaluation. When the clock changes to $\phi = 1$, data is transferred out of each ϕ-section and evaluated by the next $\bar{\phi}$-section in the chain. Pipelining is thus structured into the chain as indicated by the summary table. Race problems are avoided by the opposite internal phasing of the logic stages and output latch in each section.

NORA logic constitutes a structured approach to pipelined dynamic CMOS systems. The preceding discussion shows that the subsections are constructed using the operational circuit properties as applied to the system timing constraints. Problems such as charge sharing and leakage from soft nodes must still be considered when designing the circuitry. The overall integrity of the logic is thus determined by both circuit and system parameters. This, of course, is the realistic environment that a chip designer works in.

9.6 Zipper CMOS

Zipper CMOS logic is a dynamic nMOS/pMOS system design style that overcomes charge sharing and soft node leakage problems by using additional clocking signals. Figure 9.53 shows the basic clock generation scheme. The

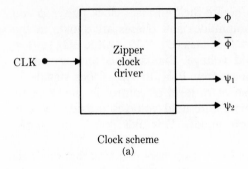

Clock scheme
(a)

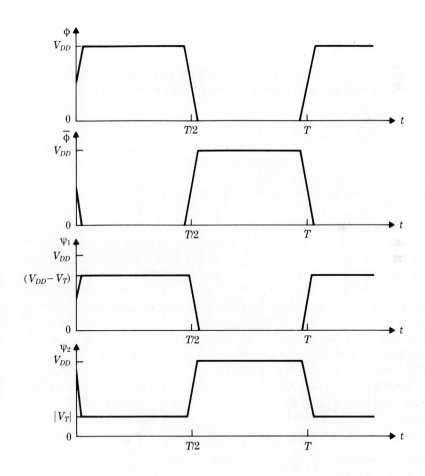

Timing diagrams
(b)

FIGURE 9.53 Zipper clocking signals.

master clock signal CLK is used to generate the "normal" clock pulses ϕ and $\bar{\phi}$, in addition to ψ_1 and ψ_2. The amplitudes and phases are shown in the timing diagram. ϕ and ψ_1 are in phase, but $\max(\phi) = V_{DD}$ while $\max(\psi_1) = (V_{DD} - V_T)$, with $V_T > 0$ a threshold voltage. Similarly, $\bar{\phi}$ and ψ_2 are in phase, but $\min(\psi_2) = V_T$, while $\bar{\phi}$ ranges from $0\,[\mathrm{V}]$ to V_{DD}. Clock signals ψ_1 and ψ_2 are introduced to allow for an extra level of control in the dynamic circuits. Offsetting ψ_1 and ψ_2 by device threshold voltages provides weakly conducting MOSFET channels to soft nodes. This aids soft node charge storage integrity.

Basic zipper structuring is shown in Fig. 9.54, where it is seen that the logic is formed using alternate nMOS and pMOS stages. nMOS stages are controlled by clock signals ϕ and ψ_1, while $\bar{\phi}$ and ψ_2 are used to time the pMOS stages. Precharging of the circuits occurs when the clocks are at voltages of

$$\phi = 0\,[\mathrm{V}], \quad \bar{\phi} = V_{DD}, \quad \psi_1 = 0\,[\mathrm{V}], \quad \psi_2 = V_{DD}. \tag{9.6-1}$$

During this time, output capacitors C_n of the nMOS stages are charged to V_{DD}, while the pMOS output capacitors C_p discharge to $0\,[\mathrm{V}]$. Logic evaluation occurs when

$$\phi = V_{DD}, \quad \bar{\phi} = 0\,[\mathrm{V}], \quad \psi_1 = (V_{DD} - V_T), \quad \psi_2 = V_T. \tag{9.6-2}$$

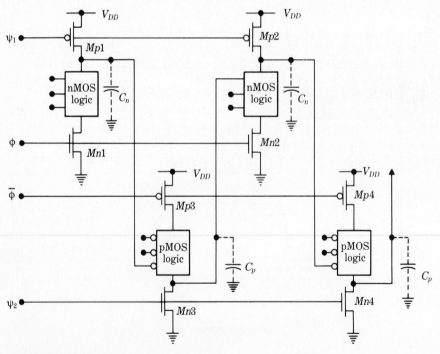

FIGURE 9.54 Zipper circuit structuring.

Capacitors C_n are conditionally discharged, while the C_p capacitors are conditionally charged. Evaluation takes place for all stages in the chain, so the logic is visualized as following a zigzag path, much like a zipper closing. This, of course, gives the basis for its name. The technique can be easily adapted to pipelined logic using this basic subsection structuring.

The circuit characteristics of zipper logic are seen by noting the placement of the clock signals. Consider the nMOS logic stages. ϕ controls the nMOS evaluate transistors $Mn1$ and $Mn2$, while ψ_1 is connected to the pMOS precharge FETs $Mp1$ and $Mp2$. Since ψ_1 has a maximum value of $(V_{DD} - |V_T|)$, the p-channel precharge transistors conduct even when $\phi = 1$. This provides charge flow to the soft node output capacitors, compensating for leakage and redistribution problems.

pMOS stages behave in a similar manner. $\bar{\phi}$ controls the pMOS evaluate transistors $Mp3$ and $Mp4$ while ψ_2 is used for the nMOS precharge devices $Mn3$ and $Mn4$. ψ_2 is always large enough to keep the n-channel transistors at least partially conducting so that the output nodes can sink excess charge to ground.

The zipper structuring automatically gives glitch-free logic because each output can make only a single transition during evaluation. For nMOS stages, only a 1 to 0 output change can occur, while pMOS outputs are characterized by having 0 to 1 as the only change. The existence of the single-output transition is sufficient to eliminate false charging or discharging of the soft nodes.

The design of the zipper clock driver network provides an interesting application of static circuit principles. Figure 9.55 shows a basic block diagram for the clock generator where the input clock CLK is used to generate the zipper clock signals. Standard inverter symbols (INV1 and INV2) are used for circuits that have full output voltage swing capabilities (from $0\,[\text{V}]$ to V_{DD}). These generate the ϕ and $\bar{\phi}$ signals. Inverters denoted with V_T (INV3 and INV4) indicate circuits that induce threshold voltage drops at the output. These circuits are introduced for ψ_1 and ψ_2. Note that the threshold-drop inverters use the ϕ or $\bar{\phi}$ in place of a power supply or ground connection.

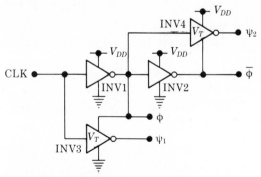

FIGURE 9.55 Zipper clock driver.

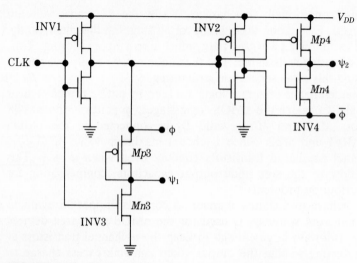

FIGURE 9.56   Zipper CMOS driver without body bias effects.

Zipper clock driver circuits may be designed to produce ψ_1 and ψ_2 voltage levels with or without body bias effects included. Figure 9.56 shows a circuit that does not have body bias effects and gives

$$\max(\psi_1) = V_{DD} - |V_{T0p}|, \qquad \min(\psi_2) = V_{T0n}. \tag{9.6-3}$$

Consider first "inverter" INV3. This circuit uses $Mn3$ as an nMOS switch while $Mp3$ is used to induce a threshold voltage drop with

$$V_{SG3} = \phi - \psi_1. \tag{9.6-4}$$

When CLK $= 0$ so that $\phi = V_{DD}$, $Mn3$ goes into cutoff. Thus, $MP3$ is described by

$$V_{SG3} = |V_{T0p}| = V_{DD} - \psi_1, \tag{9.6-5}$$

so

$$\psi_1 = V_{DD} - |V_{T0p}| \tag{9.6-6}$$

is the largest output value from this stage. An input of CLK $= V_{DD}$ gives $\phi = 0\,[V]$ and also turns $Mn3$ on. This results in a minimum value of $\psi_1 = 0\,[V]$.

Circuit INV4 works in the opposite manner, with $Mp4$ as the switching device and $Mn4$ the threshold voltage-inducing MOSFET. When the input is at CLK $= 0\,[V]$, $\phi = 0\,[V]$, $Mp4$ is in cutoff. Since $\phi = 0\,[V]$ during this time interval, $Mn4$ gives a minimum ψ_2 value of

$$\psi_2 = V_{T0n}. \tag{9.6-7}$$

This assumes that the ψ_2 output node was charged to V_{DD} during the previous half clock cycle.

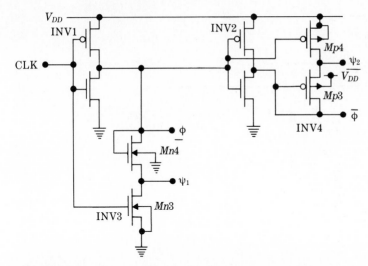

FIGURE 9.57 Zipper CMOS driver with body bias effects.

Figure 9.57 provides a driver circuit that includes body bias effects in the threshold voltage drop. To see this, consider the inverter circuit INV3 made up of n-channel MOSFETs $Mn3$ and $Mn4$. This is just a saturated enhancement-mode load inverter as discussed in Section 3.3.1. Since ϕ acts as the power supply in this circuit, the maximum output voltage is given by the self-iterating equation

$$\psi_1 = \phi - V_{Tn}(\psi_1)$$
$$= (V_{DD} - V_{T0n}) - \gamma_n(\sqrt{\psi_1 + 2|\phi_{Fp}|} - \sqrt{2|\phi_{Fp}|}), \qquad \textbf{(9.6-8)}$$

where $\phi = V_{DD}$ has been used. Similarly, INV4 made up of $Mp3$ and $Mp4$ forms the equivalent pMOS inverter with a minimum voltage of

$$\psi_2 = |V_{T0p} - \gamma_p(\sqrt{V_{DD} - \psi_2 + 2|\phi_{Fn}|} - \sqrt{2|\phi_{Fn}|})|, \qquad \textbf{(9.6-9)}$$

where $\bar{\phi} = 0\,[\text{V}]$.

The advantage of including body bias is seen by referring back to the zipper circuit structuring in Fig. 9.54. ψ_1 will fall to a lower level, which then turns on the nMOS logic evaluate transistors ($Mp1$ and $Mp2$) harder. In the same manner, the pMOS evaluate transistors ($Mn3$ and $Mn4$) are turned on slightly harder than for the case where body bias is not included. Thus, additional current is provided for charge restoration. One characteristic of the driver circuit in Fig. 9.57 is that the threshold drops are reversed, e.g., ψ_1 drives pMOS evaluate MOSFETs but attains a maximum value determined by V_{T0n}. In a process where $V_{T0n} = |V_{T0p}|$ is at least approximately true, this discrepancy will be negligible even though the body bias factors will be different.

The internal operation of a zipper stage can be seen by referring to the

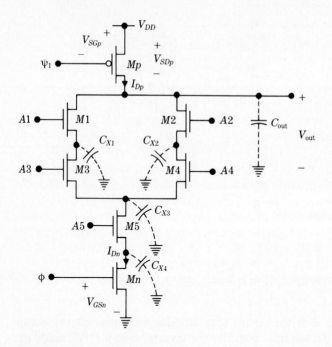

FIGURE 9.58   Charge compensation in a zipper nMOS logic stage.

nMOS logic circuit shown in Fig. 9.58. Precharging occurs when $\phi = 0\,[\text{V}] = \psi_1$. The output capacitance C_{out} is charged to V_{DD} through Mp since

$$V_{SGp} = V_{DD} \tag{9.6-10}$$

indicates that the device is conducting I_{Dp} at a high level. Note that at the end of the precharge cycle,

$$V_{SDp} = 0\,[\text{V}]. \tag{9.6-11}$$

Zipper structuring requires that the inputs $A1$ and $A2$ to the nMOS stage be outputs from preceding pMOS logic circuits. Thus, both $A1$ and $A2$ go to logic 0 states, which leaves the internal logic node capacitances C_{X1}, C_{X2}, C_{X3}, and C_{X4} at floating voltages.

   Evaluation takes place when the clocks change to

$$\phi = V_{DD}, \qquad \psi_1 = V_{DD} - V_T. \tag{9.6-12}$$

Mn enters a conducting stage since now

$$V_{GSn} = V_{DD} \tag{9.6-13}$$

indicates that I_{Dn} flows. This will discharge C_{X4} to $0\,[\text{V}]$; the remaining capacitors will conditionally discharge depending on the input combinations.

   The effect of having $\max(\psi_1) = (V_{DD} - V_T)$ is that a weak channel is formed in Mp during the evaluate period since

$$V_{SGp} = V_T > 0. \tag{9.6-14}$$

(The simplified MOSFET equations predict $I_{Dp} = 0$ at this source-gate voltage. However, a realistic device will admit small current flow levels.) As long as C_{out} remains charged at a level $V_{out} = V_{DD}$, $V_{SDp} = 0$ and $I_{Dp} = 0$. However, if V_{out} falls below V_{DD}, then Mp conducts. The amount of current supplied to the output node is dependent on $V_{SDp} = (V_{DD} - V_{out})$. For small values of source-drain voltage, Mp will be operating in nonsaturation with I_{Dp} established by $(W/L)_p$ for the device. The aspect ratio must be large enough to compensate for soft node leakage and charge sharing, but not so large that it interferes with a conditional discharge of C_{out} in the event that the input logic evaluates to give a logic 0 output. A logic design criteria can be formulated in terms of the nMOS/pMOS ratio (β_n/β_p), but optimizing this value for, say, charge sharing will probably degrade the transient response. Computer-aided simulations for different choices are thus in order for total design control.

Zipper CMOS provides certain advantages over basic domino or NORA structures. It is in itself an example of the design variations possible in CMOS logic circuits.

9.7 Dynamic Pseudo-2ϕ Logic

Dynamic CMOS circuits can be structured using pseudo-2 clocking where synchronization is provided by the pairs $(\phi_1, \bar{\phi}_1)$ and $(\phi_2, \bar{\phi}_2)$ such that

$$\phi_1(t)\phi_2(t) = 0 \tag{9.7-1}$$

is true for all times t. The presence of 4 clock signals allows for a number of different design styles. One approach is shown in Fig. 9.59. The circuit uses dynamic nMOS logic with transmission gates to control data inputs. Single-level logic is implemented by alternating the clock pairs from stage to stage.

Operation of the circuit can be understood by studying the phasing within a stage. The complementary clock signals ϕ_1 and ϕ_2 control the precharge and evaluate MOSFETs of the dynamic logic circuit. Precharging at the output node of a stage occurs when $\phi_i = 1$ ($i = 1, 2$), while evaluation takes place when $\phi_i = 0$. Data inputs are latched during the precharge phase, since the TGs conduct during this time. The nonoverlapping characteristic of ϕ_1 and ϕ_2 thus alternates the precharge/latch and evaluation cycles in the chain, synchronizing the logic propagation.

Some of the design problems involved in this CMOS style can be seen in the circuit of Fig. 9.60. C_{out} precharges to $V_{out} = V_{DD}$ when $\phi_1 = 1$ turns on Mp. The input TG is conducting during this time, so $A1$ can be latched. If $A1 = 0$, C_{out} is the only capacitor charged. On the other hand, an input of $A1 = 1$ charges the input capacitance C_{in} to V_{DD}, driving MOSFET $M1$ into

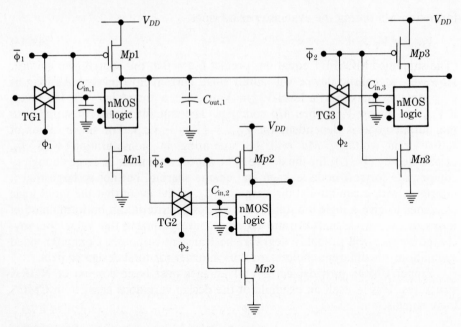

FIGURE 9.59   Single-level pseudo 2-phase circuit.

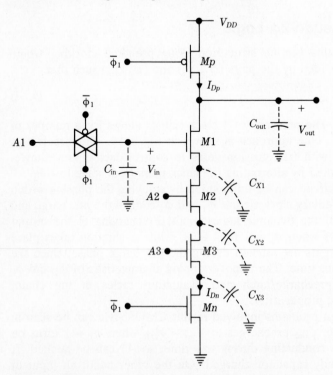

FIGURE 9.60   Pseudo 2-phase circuit design.

conduction. C_{X1} then precharges to a voltage V_{X1}, which has a final value of

$$V_{X1} = (V_{DD} - V_{T0n}) - \gamma_n(\sqrt{V_{X1} + 2|\phi_{Fp}|} - \sqrt{2|\phi_{Fp}|}). \qquad \text{(9.7-2)}$$

C_{X2} will also charge to this voltage if both $A1 = 1$ and $A2 = 1$, while $A1 = A2 = A3 = 1$ implies that all C_X capacitors reach this level.

Important design considerations revolve around the device aspect ratios. C_{in} is determined by the dimensions of the TG MOSFETs, the layout geometry, and the aspect ratio $(W/L)_1$ of logic transistor $M1$. C_{in} must charge to a level $V_{in} = V_{Tn1}$ to initiate conduction through $M1$. Since body bias effects are present,

$$V_{Tn1} = V_{T0n} + \gamma_n(\sqrt{V_{X1} + 2|\phi_{Fp}|} - \sqrt{2|\phi_{Fp}|}) \qquad \text{(9.7-3)}$$

shows that threshold voltage increases as C_{X1} charges. A large value of $(W/L)_1$ will increase the current flow, but it adds capacitance. The same comment holds for logic devices $M2$ and $M3$. The most critical device during the precharge is Mp, since this supplies I_{Dp} to charge C_{out} and internal node capacitances. The design tradeoff for the aspect ratio $(W/L)_p$ of this MOSFET is between current flow levels and capacitance, as expected.

Charge leakage and charge sharing are important during the evaluate phase ($\phi_1 = 0$). Since the TG is off, the charge on C_{in} will change because of leakage through the n^+p junctions in the TG MOSFETs. The voltage on C_{out} will also vary; the exact behavior depends on the input logic. If $A1 = 0$, the output constitutes a CMOS soft node with the usual leakage problems. An input of $A1 = 1$ turns $M1$ on. Charge sharing can occur if C_{X1} has not

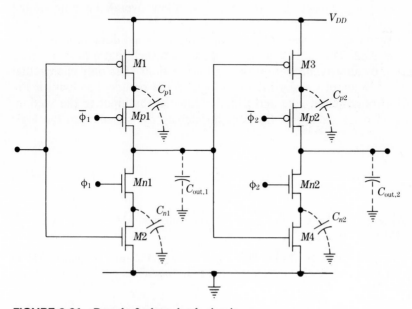

FIGURE 9.61 Pseudo 2-phase latch circuit.

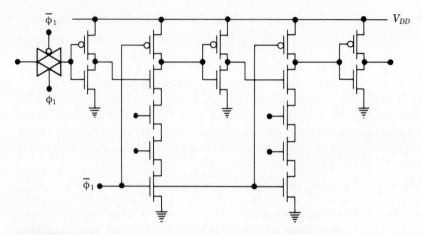

FIGURE 9.62   A ϕ_1-section of a pseudo-2ϕ domino system.

achieved the voltage level in eqn. (9.7-2). Charge redistribution takes place, and V_{out} falls below V_{DD} as described by the capacitance ratio equations. This situation can result if $(W/L)_p$ is kept small to minimize the capacitance, thus limiting the precharge current flow. Another possibility is where a high clock frequency gives a precharge interval that is too short to completely charge C_{X1}.

A pseudo-2ϕ latch based on C²MOS structuring is shown in Fig. 9.61 (compare with Fig. 9.50). The first stage is clocked such that $\phi_1 = 1$ defines the precharge/input cycle, while $\phi_1 = 0$ gives the hold/latch state. The second stage is identical, except that it is controlled by the clock pair ($\phi_2, \bar{\phi}_2$). Pseudo-2ϕ clocking phases the two stages to allow overall latching of the input.

A final example of a pseudo-2ϕ circuit is the modified domino structuring shown in Fig. 9.62. This can be termed a ϕ_1-section such that a domino chain is constructed by alternating ϕ_1-sections and ϕ_2-sections. The only change that has been used to adapt standard domino logic to the pseudo-2ϕ clocks is the addition of a transmission gate and a static buffer at the input to the section. This allows an extra degree of freedom in latching the signals into the logic sections.

REFERENCES

(1) M. Annaratone, *Digital CMOS Circuit Design*, Hingham, MA: Kluwer Academic Publishers, 1986.

(2) N. F. Gonclaves and H. J. De Man, "NORA: A Racefree Dynamic CMOS Technique for Pipelined Logic Structures," *IEEE J. Solid-State Circuits*, vol. SC-18, pp. 261–266, 1983.

(3) R. H. Krambeck, C. M. Lee, and H. H-F. S. Law, "High-Speed Compact Circuits with CMOS," *IEEE J. Solid-State Circuits*, vol. SC-17, pp. 614–619, 1982.

(4) C. M. Lee and E. W. Szeto, "Zipper CMOS," *IEEE Circuits and Dev. Mag.*, vol. 2, no. 3, pp. 10–17, May 1986.

(5) V. G. Oklobdzija and R. K. Montoye, "Design-Performance Trade-Offs in CMOS-Domino Logic," *IEEE J. Solid-State Circuits*, vol. SC-21, pp. 304–306, 1986.

(6) D. J. Meyers and P. A. Ivey, "A Design Style for VLSI CMOS," *IEEE J. Solid-State Circuits*, vol. SC-20, pp. 741–745, 1985.

(7) J. A. Pretorius, A. S. Shubat, and C. A. T. Salama, "Analysis and Design Optimization of Domino CMOS Logic with Application to Standard Cells," *IEEE J. Solid-State Circuits*, vol. SC-20, pp. 523–530, 1985.

(8) J. A. Pretorius, A. S. Shubat, and C. A. T. Salma, "Latched Domino CMOS Logic," *IEEE J. Solid-State Circuits*, vol. SC-21, pp. 514–522, 1986.

(9) N. Weste and K. Eshraghian, *Principles of CMOS VLSI Design*, Reading, MA: Addison-Wesley, 1985.

PROBLEMS

Assume that the CMOS process is characterized by the following parameters:

$$V_{T0n} = +0.8\,[\text{V}], \quad k_n' = 40\,[\mu\text{A/V}^2], \quad \gamma_n = 0.26\,[\text{V}^{1/2}],$$
$$2\,|\phi_{Fp}| = 0.579\,[\text{V}],$$
$$V_{T0p} = -0.8\,[\text{V}], \quad k_p' = 16\,[\mu\text{A/V}^2], \quad \gamma_p = 0.84\,[\text{V}^{1/2}],$$
$$2\phi_{Fn} = 0.699\,[\text{V}], \quad V_{DD} = 5\,[\text{V}],$$

unless otherwise stated. For capacitance levels use

$$C_{\text{ox}} = 0.691\,[\text{fF}/\mu\text{m}^2],$$

nMOS: $C_{j0} = 0.0975\,[\text{fF}/\mu\text{m}^2], \quad \phi_o = 0.879\,[\text{V}],$
$C_{jsw} = 0.107\,[\text{fF}/\mu\text{m}], \quad \phi_{osw} = 0.921\,[\text{V}];$ (p-substrate)

pMOS: $C_{j0} = 0.0298\,[\text{fF}/\mu\text{m}^2], \quad \phi_o = 0.939\,[\text{V}],$
$C_{jsw} = 0.362\,[\text{fF}/\mu\text{m}], \quad \phi_{osw} = 0.985\,[\text{V}].$ (n-tub)

Assume interconnect values of

$$C_{p-f} = 0.0576\,[\text{fF}/\mu\text{m}^2], \quad C_{m-p} = 0.0863\,[\text{fF}/\mu\text{m}^2],$$
$$C_{m-f} = 0.0345\,[\text{fF}/\mu\text{m}^2], \quad C_{m-n^+} = 0.0822\,[\text{fF}/\mu\text{m}^2].$$

9.1 The p-type substrate is doped with acceptors at a level of $N_a = 10^{15}\,[\text{cm}^{-3}]$. The n-tub donor doping is given as $N_{d,\text{tub}} = 10^{16}\,[\text{cm}^{-3}]$. Assume that the average carrier lifetime is given by $\tau_o = 1.1\,[\mu\text{s}]$. Verify that the zero-bias reverse leakage current levels for the process are given by:

pMOS: $J_{1o} = 0.369\,[\text{fA}/\mu\text{m}^2]$ $(p^+/n\text{-tub}),$
nMOS: $J_{2o} = 1.130\,[\text{fA}/\mu\text{m}^2]$ $(n^+/p\text{-substrate}),$

where sidewall current flow is ignored. Use these numbers when needed in working the problems in this chapter.

9.2 A CMOS inverter is designed with $(W/L)_n = 2$ and $(W/L)_p = 4$. The output drives the next stage through a CMOS transmission gate that has MOSFET aspect ratios of $(W/L)_n = 2 = (W/L)_p$. The circuit is modeled as shown in Fig. 9.2(b). Assume $C_2 = 80\,[fF]$.

(a) Estimate R_{eq} by assuming $R_{eq} = R_p(V_2 = 0)$. Use this value for the remainder of the problem.

(b) Suppose that $V_2(0) = 0\,[V]$ and that C_2 is charged through the TG and the p-channel MOSFET of the inverter. Calculate the time t_o (including the TG) when the charging transistor changes from saturated to nonsaturated behavior. Then plot $V_2(t)$ and calculate t_{LH}.

(c) Now suppose that a logic 0 is to be transferred through the TG. Plot $V_2(t)$, assuming that initially $V_2(0) = V_{DD}$. The find t_{HL} for the circuit.

9.3 Rework Problem 9.2 for the case where the TG transistor aspect ratios are increased to a value of $(W/L)_n = 10 = (W/L)_p$. (Assume that the inverter remains the same.) Discuss the results in terms of the time constants and transient time intervals.

9.4 Figure P9.1 provides the layout for a TG driving an inverter. All dimensions are in microns $[\mu m]$, and the gate overlap is assumed to be $L_o = 0.3\,[\mu m]$. Overlap is not shown explicitly in the layout.

(a) Calculate the maximum value of the TG depletion capacitance.

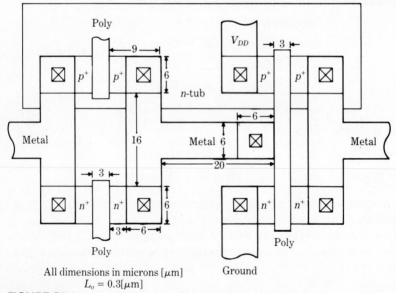

All dimensions in microns $[\mu m]$
$L_o = 0.3[\mu m]$

FIGURE P9.1

(b) Find the MOS interconnect and gate capacitance seen looking from the TG to the input of the inverter.

(c) Calculate the maximum charge Q held on the inverter input node when logic 1 input is stored there.

(d) Use the results of Problem 9.1 to estimate the maximum values of the leakage currents I_1 (from V_{DD} to the soft node) and I_2 (from the soft node to ground).

(e) Find the final value of V_2 assuming that a refresh does not occur. Does the logic state change from a 1 to a 0 in this case? Find the minimum value of V_2 that will still be interpreted as the logic 1 state. (The inverter MOSFETs have the same layout geometry as the TG transistors.)

* (f) Simulate the circuit using SPICE. Use SPICE depletion capacitance modeling, and simulate the interconnect capacitance by lumped-element models.

*(g) Write a program that solves the differential equation (9.1-57) for this case. The simplest form to start with is

$$\frac{dV_{in}}{dt} = -\frac{B}{C_T(V_2)}\left[\sqrt{\frac{\phi_{o2}}{N_a}}\left(1 + \frac{V_2}{\phi_{o2}}\right)^{1/2}\right.$$
$$\left. - \sqrt{\frac{\phi_{o1}}{N_{d,tub}}}\left(1 + \frac{V_{DD} - V_2}{\phi_{o1}}\right)^{1/2}\right],$$

where B is a constant and

$$C_T(V_2) = C_L + \frac{C_{j02}A_2}{\left(1 + \frac{V_2}{\phi_{o2}}\right)^{1/2}} + \frac{C_{j01}A_1}{\left(1 + \frac{V_{DD} - V_2}{\phi_{o1}}\right)^{1/2}}.$$

Different approaches are possible. Choose one that is straightforward to implement.

9.5 Consider the simple inverter shown in Fig. 9.16. The aspect ratios are set with $(W/L)_n = 3$, $(W/L)_p = 7.5$, while the capacitances are specified to be $C_{\bar{\phi}} = 125\,[\text{fF}]$ and $C_\phi = 140\,[\text{fF}]$. Calculate the amount of clock skew t_s introduced by this arrangement (assuming step input transitions).

9.6 The delay circuit of Fig. 9.17 is used to reduce clock skew. The inverter MOSFETs are designed with $(W/L)_n = 4$ and $(W/L)_p = 10$. C_ϕ is approximated as 650 [fF], while $C_{\bar{\phi}} \approx 700\,[\text{fF}]$. The output capacitance of the first stage is estimated to be $C_{out,1} = 95\,[\text{fF}]$.

(a) Calculate the time delay through the second stage.

(b) The exponential driving functions introduced to model $V_{out,1}(t)$ are difficult to use for design. It is simpler to solve eqn. (9.2-13) with a step function input.

Find expressions for $\bar{\phi}(t)$ valid for charging and discharging $C_{\bar{\phi}}$

through the TG by solving the homogeneous differential equation and then applying the initial conditions. Use the results to find a value of τ_{TG} that will (approximately) equalize the time delay.

(c) Now estimate $R_{eq} = R_p(\bar{\phi} = 0)$ and find the aspect ratio of the pMOS transistor in the TG. Choose the nMOS transistor to have the same dimensions.

(d) Use your design values in eqns. (9.2-16) and (9.2-18) to compare results.

*(e) Prepare a SPICE file that simulates the delay TG arrangement. Use the TG design obtained in part (c). Then vary the dimensions of the TG transistors and discuss the transient effects.

9.7 Design a full-adder in 2-phase CMOS logic using AOI structuring for the logic and TGs for data transmission.

9.8 Consider the control signal generation logic shown in Fig. 9.25(c).
(a) Draw the circuit implementation of the logic using static gates.
(b) The inverter in the *SEL* line introduces the most skew between *SEL* and the clock signals. Suppose that the inverter sees a capacitance of $C_{out} = 85\,[\text{fF}]$. Design the inverter such that the propagation time t_p is less than 0.1 [ns].

9.9 The selectively loadable register in Fig. 9.28(a) (which uses a separate load *LD* control) potentially exhibits charge sharing between nodes X and X' as mentioned in the text.

Suppose that $C_X = 155\,[\text{fF}]$ and that the inverter that uses X' as an input is designed with $(W/L)_n = 2 = (W/L)_p$. What is the largest node capacitance $C_{X'}$ at X' that will still allow transmission of a logic 1 voltage?

9.10 Design a 3-to-8 decoder with selectively loadable input registers. Denote the inputs by A_0, A_1, and A_2, which are controlled by pseudo-2 phase clocking and an *LD* signal (See Fig. 9.29.) Either NOR gates or NAND gates may be used.

9.11 The circuit diagram for a dynamic CMOS inverter was given in Fig. 9.30. A layout example for this circuit is shown in Fig. P9.2.
(a) Calculate the maximum value of the capacitance C_X for the circuit.
(b) Calculate the zero-bias depletion capacitances at the output node at both the nMOS and pMOS transistors (Mp and $M1$). Then average the contributions using $K(V_{OH}, V_{OL})$ for each region.
(c) Compute the average value of C_{out} that includes the metal-to-field contribution of the interconnect between Mp and $M1$ (only).
(d) Estimate the precharge time if the inverter input is $V_{in} = 0\,[\text{V}]$.
(e) Estimate the precharge time for the case where $V_{in} = V_{DD}$. Also compute the value of V_X for this case.

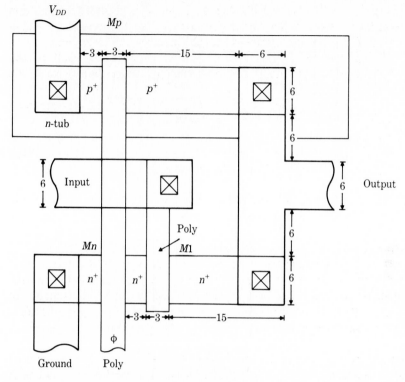

All dimensions in microns [μm]
$L_o = 0.3$ [μm]

FIGURE P9.2

9.12 The layout for a dynamic CMOS logic gate is shown in Fig. P9.3. The inputs are *A*, *B*, and *C*, and the timing is controlled by a single clock.

(a) What is the logic function performed by this gate?

(b) Find the average value of C_{out} with the layout shown (depletion only).

(c) Find the maximum capacitance between the *A*-input MOSFET and the *B*- and *C*-input transistors.

(d) Find the maximum capacitance between the *B*- and *C*-input MOSFETs and the evaluate nMOS transistor *Mn*.

(e) Examine the problem of charge sharing during an evaluation cycle. Look at the worst-case charge sharing problem, and indicate how this problem can be approached without changing the layout.

(f) Suppose that the *A*-input MOSFET is interchanged with the *B*- and *C*- input parallel transistors (so that the *B*- and *C*-inputs now have their drain connected to the output node). Redraw the circuit for this case, and examine the charge sharing problem with the new

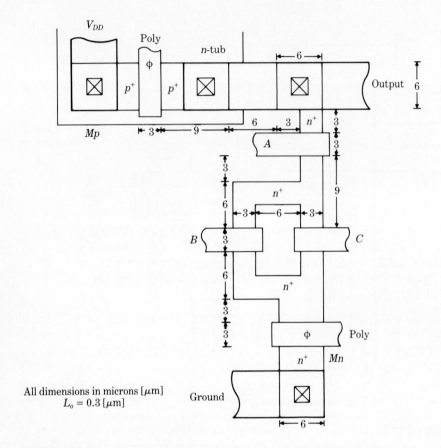

FIGURE P9.3

arrangement. What design tradeoff has been made in terms of the overall circuit performance?

9.13 Design a 2-to-4 decoder using a domino logic stage design with a dynamic NAND cascaded into a static inverter. Denote the inputs by A_0 and A_1, and provide static inverters where needed outside the domino stage. (Only a single stage is needed for each output.)

Extend your design so that a control signal $SEL = 1$ must be applied to make the decoder functional. Implement this modification so that the logic requires only one clock cycle to complete.

9.14 Consider the domino charge sharing circuit shown in Fig. 9.46. Assume that $A = 0 = B = C$ is maintained in the input logic but that F makes a transition from $0\,[V]$ to $V_{DD}\,[V]$. The capacitance levels are estimated to be $C_{out,1} = 185\,[fF]$ and $C_{X1} = 20\,[fF]$, and $C_{out,1}$ is initially charged to a voltage $V_{out,1} = V_{DD}$ during the precharge.

(a) Calculate the total charge Q_T that is on $C_{out,1}$ after the precharge is completed.

(b) Calculate the final voltage V_f after $F = 1$ is established.

(c) Recalculate V_f for the case where $C_{X1} = 78\,[fF]$.

*(d) Simulate the charge sharing problem using SPICE with an initial condition statement.

9.15 Calculate the values of max (ψ_1) and min (ψ_2) for the zipper clocking circuit shown in Fig. 9.57 using the parameter set given at the beginning of the problem section. Then calculate the amount of current admitted by a precharge pMOSFET in the zipper circuit if ψ_1 is applied to the gate and $(W/L)_p = 3$. (See Fig. 9.54; $Mp1$ is a typical MOSFET with this voltage applied.)

9.16 Consider the technique of compensating for charge leakage by using a pMOS transistor with a small $(W/L)_p$ value. An example of this is shown in Fig. 9.45(a). Analyze the device design requirements and establish an approach to choosing $(W/L)_p$ that will replenish lost charge but not interfere with conditional discharge.

9.17 A *tristate* output is a node that can produce logic 1 and logic 0 voltages in addition to an *infinite-impedance* level. The third output state is simply a soft node that has been isolated from the input circuitry. Different techniques are available to construct tristate outputs.

(a) Design a tristate inverter using two nMOS and two pMOS transistors that gives a high-impedance state when a control signal of $SEL = 0$ is applied. Analyze any obvious problems in the design.

(b) Construct a tristate inverter using one pMOS transistor and three nMOSFETs that gives a high-impedance state when $SEL = 0$. Discuss the problems involved with this type of design.

CHAPTER 10

Structured MOS Logic

Structured logic is particularly useful for VLSI system design. A structured approach to logic simplifies the implementation of complicated logic functions by requiring Boolean patterns of specified form. The patterns can usually be implemented on silicon by specifying the locations of transistors. This forms the direct bridge from basic digital MOS integrated circuits to VLSI system design.

This chapter centers on the basic techniques of structured logic, with emphasis on the circuit design problems of device layout and transient response.

10.1 Concepts in Structured Logic

Structured logic in digital chip design usually implies that

(a) the logic functions are in a structured form, or

(b) geometric arrays are used to implement layout.

In practice a combination of the two is used. Merging of the concepts gives substantial design variations, and designers tend to develop their own individual style.

Logic function structuring at the bit level is based on having *canonical* forms that follow regular algebraic patterns. The simplest such forms are the *sum-of-products* (SOP) and *product-of-sums* (POS) representations. Let A, B, and C denote input variables. An example of an SOP form is

$$F = ABC + \bar{A}B\bar{C} + \bar{A}BC + \bar{A}\bar{B}\bar{C}, \tag{10.1-1}$$

while a POS function looks like

$$G = (A + \bar{B})(\bar{A} + C)(B + C). \tag{10.1-2}$$

For gate-level implementations, AND-OR logic is the most natural for the SOP form while OR-AND structuring is appropriate for POS functions. Transistor arrays may be used for either form, and the topological regularity of the pattern often permits a more area-efficient solution to the design problem than that obtained using gate-level logic.

MOSFET arrays have device locations geometrically patterned such that logic formation is accomplished by routing interconnect lines. This allows for direct implementation of logic functions that are based on structured logic equations. Prewired gates consisting of groups of MOSFETs may be included in the array. *Gate arrays* are chips that may be user-defined by specifying interconnect patterns. These are lithographically translated to the chip layers and can be used for custom designs.

The techniques discussed in this chapter are some of the more common approaches to structured logic design in digital MOS ICs. Most tend to utilize both logic and device patterns in an effort to simplify the complexity of realistic design problems. In addition, structured techniques are adaptable to CAE (*computer-aided engineering*) system software.

10.2 Switch Logic

Switch logic uses transistors as basic switches that are either on (conducting) or off (nonconducting). Examples of switch logic have already been presented, e.g., the CMOS multiplexer circuit in Fig. 6.38 of Section 6.7.2. MOS technology admits three distinct switch types: nMOS, pMOS, and CMOS. These are shown in Fig. 10.1 along with their respective positive logic transmission properties. An X entry means that the switch is nonconducting,

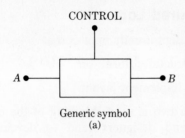

Generic symbol
(a)

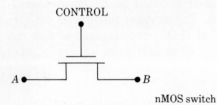

nMOS switch
(b)

A	CONTROL	B
0	0	X
1	0	X
0	1	0
1	1	1

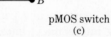

pMOS switch
(c)

A	CONTROL	B
0	0	0
1	0	1
0	1	X
1	1	X

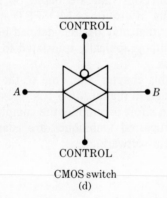

CMOS switch
(d)

FIGURE 10.1   Basic MOS switches.

so the state is not defined. To summarize, nMOS transistors conduct with a logic 1 control signal applied, while pMOS transistors require a logic 0 control signal to pass data. Single transistor switches exhibit threshold voltage drops. Assume that logic swings are from $0\,[\text{V}]$ to V_{DD}. Including body bias, the highest voltage transmitted through an nMOS switch is

$$V_N = (V_{DD} - V_{T0n}) - \gamma_n(\sqrt{V_N + 2\,|\phi_{Fp}|} - \sqrt{2\,|\phi_{Fp}|}), \qquad (10.2\text{-}1)$$

while the lowest voltage out of a pMOS switch is

$$V_P = |V_{T0p} - \gamma_p(\sqrt{V_{DD} - V_p + 2\,|\phi_{Fn}|} - \sqrt{2\,|\phi_{Fn}|})|. \qquad (10.2\text{-}2)$$

CMOS transmission gates can pass any voltage in the range $[0, V_{DD}]$ and logically act like nMOS switches.

10.2.1 nMOS Arrays

Logic formation is illustrated in Fig. 10.2. Conduction of the state P through the nMOS chain is controlled by the signals applied to the pass transistor gates. When a path is established through the chain, the output is described by

$$F = Y_0 Y_1 (Z_0 Z_1 + Z_2) P. \qquad (10.2\text{-}3)$$

However, this must be used with care since F remains undefined if there is a break in the path. Physically, an undefined value of F occurs because the voltage V_{out} on C_{out} floats according to the previous value as modified by charge leakage or redistribution among the other capacitances (C_0, C_1, C_2). An important rule is that the output from a chain using switch logic cannot be allowed to float in this manner. Also note that nMOS pass transistors induce threshold voltage losses according to eqn. (10.2-1).

Path selectors were discussed earlier in CMOS switch logic. The general block function is shown in Fig. 10.3. The logic is designed to select one of the n input lines using a binary combination on m control lines. This requires that $2^m \geq n$ must hold. An example is provided in Fig. 10.4 for a system with 6 input lines. Path selection requires $m = 3$ control signals, which are denoted

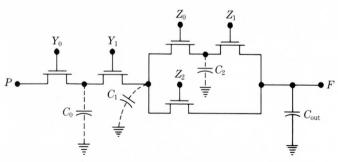

FIGURE 10.2 Basic logic formation using pass transistors.

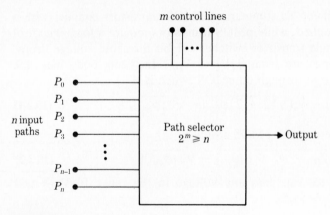

FIGURE 10.3   General path selector function.

by A, B, and C. Three control bits give a total of 8 possible combinations, so two will not be allowed. The choice shown correlates to the Boolean expression

$$F = ABCP_0 + \bar{A}BCP_1 + A\bar{B}CP_2 + AB\bar{C}P_3 + A\bar{B}\bar{C}P_4 + \bar{A}\bar{B}\bar{C}P_5, \quad (10.2\text{-}4)$$

as easily verified from the circuit. The control bit combinations $(ABC) = (001)$ and (010) give a floating output node and must be avoided. Obviously, a path selector that does not have this problem is one where $n = 2^m$.

The path selector example illustrates that switch logic is well structured for the canonical sum-of-products form. The circuit in Fig. 10.5 implements the function

$$F = ABC + A\bar{B}C + \bar{A}B\bar{C} + \bar{A}\bar{B}\bar{C} \quad (10.2\text{-}5)$$

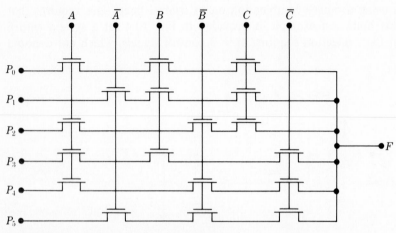

FIGURE 10.4   nMOS 6-to-1 path selector.

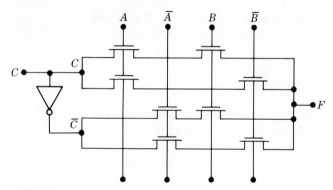

FIGURE 10.5 Sum-of-products switch logic.

by using A and B to control the variable C. Since the pair (A, B) constitutes the total control parameter set, all 4 paths are accounted for. The output is well defined with no possibility of a floating output. Simplification of the chip layout is aided by the patterned structure of the transistor array.

Arbitrary logic functions can be rewritten in SOP form using the rules of Boolean algebra or general theorems. To illustrate the process, consider

$$F = A + BC. \tag{10.2-6}$$

As a first step, multiply the first term by $(C + \bar{C}) = 1$ to get

$$F = AC + A\bar{C} + BC. \tag{10.2-7}$$

Although this form can be used to construct a workable switch logic circuit, implementing the circuit requires some care because each term has only 2 of the 3 variables. Figure 10.6(a) shows the problems involved. Suppose that $A = 1$, $B = 0$, and $C = 1$. Pass transistors $MP2$ and $MP3$ are in conducting modes, so both the A and B branches are connected to the output. Since $V_A = V_{DD}$ and $V_B = 0\,[\text{V}]$, a conduction path is established between A and B.

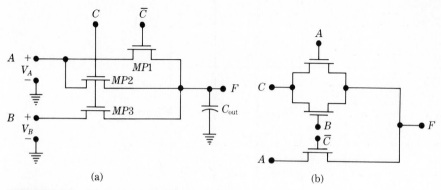

(a)   (b)

FIGURE 10.6 Improper switch logic designs.

Hence, the voltage across the output capacitor C_{out} is not well defined. This problem is intrinsic to the interpretation used to design the circuit. Simply restructuring the expression may not alleviate the problems. To see this, suppose that the equation is rewritten as

$$F = (A + B)C + A\bar{C}. \tag{10.2-8}$$

The term-by-term switch logic circuit for this form is shown in Fig. 10.6(b). It is easily seen that this circuit suffers from the same type of problems.

Canonical structuring can be obtained from eqn. (10.2-6) by multiplying the first term by $(B + \bar{B})(C + \bar{C})$. Rearranging and reducing gives

$$F = BC + A\bar{B}C + AB\bar{C} + A\bar{B}\bar{C} \tag{10.2-9}$$

for the SOP representation. A network designed with this expression is shown in Fig. 10.7. The power supply voltage V_{DD} has been used as a logic 1 value for the first term by writing $BC = (1)BC$. Since B and C constitute the control variables for the pass transistors, the 4 combinations indicate that the switching will be well defined and unique.

Other design variations are possible. For example, using the *absorption theorem* of Boolean algebra allows eqn. (10.2-6) to be written as

$$F = A(\overline{BC}) + BC$$
$$= A(\bar{B} + \bar{C}) + BC. \tag{10.2-10}$$

This simplifies the logic down to the network shown in Fig. 10.8. Comparing with Fig. 10.7 shows that a significant reduction in complexity has been achieved.

Another possibility is obtained by using both the power supply V_{DD} (logic 1) and ground (logic 0) lines to structure logic paths to the output node. Figure 10.9 shows this type of scheme. The upper half of the network is designed to

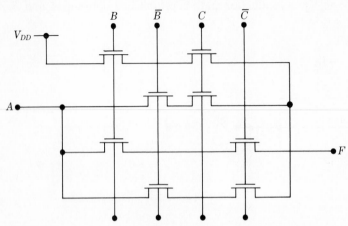

FIGURE 10.7 Sum-of-products switch network.

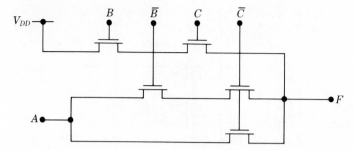

FIGURE 10.8 Network reduction using the absorption theorem.

provide $F = 1$ output values, while the lower half is the complementary network for $F = 0$ values. This type of logic formation is straightforward, but much of the MOSFET array structuring has been lost. Layout problems are similar to those encountered in random logic designs.

10.2.2 CMOS Arrays

CMOS switch logic is identical in form to nMOS. Figure 10.10 shows an 8-to-1 path selector described by

$$F = ABDP_0 + \bar{A}BCP_1 + A\bar{B}CP_2 + AB\bar{C}P_3 + A\bar{B}\bar{C}P_4$$
$$+ \bar{A}\bar{B}\bar{C}P_5 + \bar{A}B\bar{C}P_6 + \bar{A}\bar{B}CP_7. \quad \textbf{(10.2-11)}$$

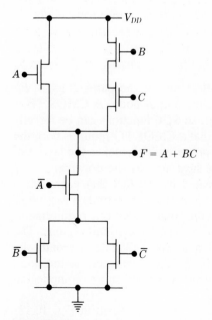

FIGURE 10.9 Switch network using logic 1 and logic 0 voltages.

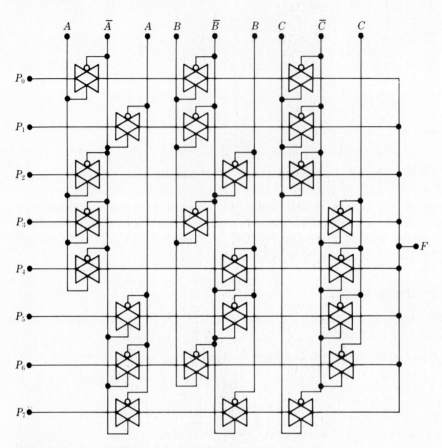

FIGURE 10.10   CMOS 8-to-1 path selector.

The control variables (A, B, C) uniquely select one of the paths $P_0, \ldots, P_7$. Since $2^3 = 8$, there is no possibility of a floating output node. A CMOS TG is logically identical to an nMOS pass transistor, so SOP functions can be directly implemented in TG arrays. Note, however, that a CMOS TG requires both the control signal and its complement to switch properly. This makes the layout of the control signal lines more complicated, as suggested by the drawing.

CMOS TG array layout is also complicated by the fact that each switch consists of 2 opposite polarity MOSFETs. Figure 10.11 represents a section of the 8-to-1 selector and illustrates two important points about this requirement. First, the pMOS transistors must be placed in n-type background regions. The drawing shows these as n-wells located in a p-type substrate. Second, the pMOS and nMOS transistors in a TG must be parallel-connected, as indicated by the heavy lines in the schematic. Layout becomes more complicated because of the interconnect routing required.

An alternate approach to using CMOS TG arrays is shown in Fig. 10.12. The paths P_0, P_1, and P_2 of the 8-to-1 selector have been implemented using

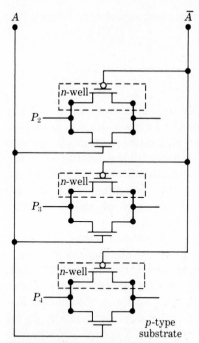

FIGURE 10.11 CMOS transmission gate interconnect and *n*-well layout problem.

separate pMOS and nMOS chains. Selection of a given path enables two possible paths. If the data is a logic 0, the nMOS chain provides the conduction; a logic 1 state is transmitted through the pMOS transistors. This avoids the threshold voltage loss incurred by using only a single polarity transistor and also provides a structured MOSFET array to aid in layout. One

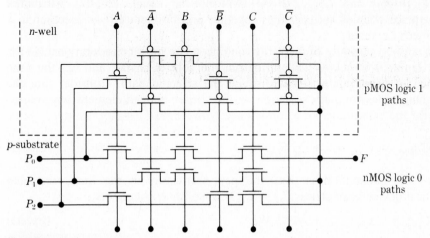

FIGURE 10.12 Partial implementation of an 8-to-1 selector using separate nMOS and pMOS switching paths.

important drawback of this technique is that the input capacitance to each line increases.

10.2.3   Circuit Considerations

In general, switch logic admits to topologically regular arrays, which aids the chip layout. This is especially useful in complicated system designs. From the circuit viewpoint, the DC characteristics are straightforward to design and evaluate, even with threshold voltage drops through single polarity switches. The transient switching characteristics are more complicated because of the line capacitances in the array interconnect wiring. Switching times increase with increasing capacitance, so the transient response of the array may constitute a limiting factor in the system performance.

General capacitance considerations can be illustrated using the 4-to-1 multiplexer in Fig. 10.13(a). The control signals (A, B) are applied to vertical lines in the array, while data is transferred along the horizontal paths $P_0, \ldots, P_3$. The lumped-element MOSFET gate capacitances C_G associated with the switching transistors are shown in Fig. 10.13(b). Similarly, Fig. 10.13(c) shows the drain-bulk and source-bulk capacitances (denoted by C_b for simplicity) along the horizontal data paths. Cross-coupling capacitances between the vertical and horizontal interconnects are given in Fig. 10.13(d). Two distinct capacitor types are shown. C_P represents the normal gate-n^+ pass transistor capacitances (C_{GS} or C_{GD}). Overlap areas between the horizontal and vertical interconnect lines gives the contributions denoted by C_{h-v}. Line-to-substrate capacitance is not shown explicitly in the drawings but can account for a significant fraction of the total value. Oxide thickness and layering will be different for the vertical and horizontal interconnects, so that $C_{\text{line,hor}}$ [F/cm] and $C_{\text{line,ver}}$ [F/cm] will not be equal. Finally, distributed capacitance couples vertical lines together and horizontal lines together in a continuous manner.

Transient response of the array centers on two main considerations: (1) the time required to set up the path from the control signals and (2) the time needed to propagate a signal through the selected path. Consider first the switching times for path selection. A first-order lumped-element approximation for the capacitance of a control line with length L_v is

$$C_{\text{control}} \simeq C_{\text{line},v}L_v + \sum_i C_{Gi}, \tag{10.2-12}$$

where the summation is over the pass transistors in the line. Assuming identical device geometries, $C_G = C_{\text{ox}}WL$, so for N transistors,

$$C_{\text{control}} \simeq C_{\text{line},v}L_v + N(C_{\text{ox}}WL). \tag{10.2-13}$$

Since the array was designed with all possible (A, B) combinations, each control line has $N = 2$ in the present example. To this order of approximation, the selection time is determined by the drive capability of the control signal buffers relative to the total vertical capacitance.

Logic propagation through the horizontal data paths is more complicated owing to the transistor characteristics and distributed capacitances. An accurate analysis requires the use of computer simulations. However, it is possible to illustrate some important aspects of the problem by introducing models that simplify the mathematics.

Figure 10.14 shows a lumped-equivalent circuit for a pass chain with n transistors. MOSFETs are modeled as simple linear resistors R_i, giving the RC-ladder network structure. Assume that all capacitors are initially uncharged and that a logic 1 voltage level $V_{in} = V_1$ is applied at the input. Transferral of the input state requires that the output capacitor C_n charge to a

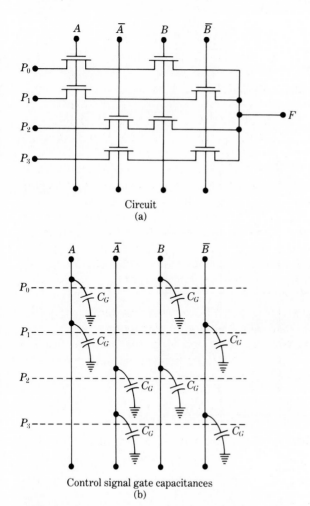

Circuit
(a)

Control signal gate capacitances
(b)

FIGURE 10.13 Switch logic array capacitances in the lumped-element approximation.

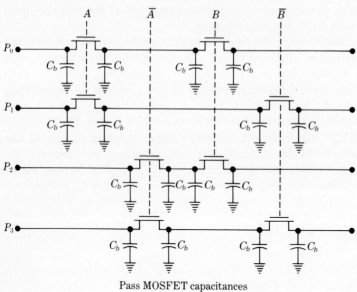

Pass MOSFET capacitances
in horizontal paths
(c)

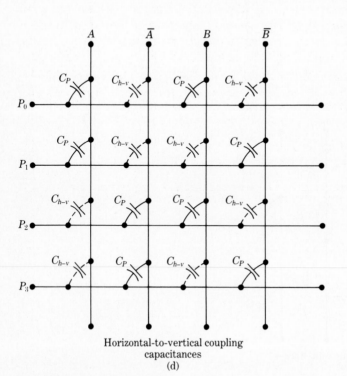

Horizontal-to-vertical coupling
capacitances
(d)

FIGURE 10.13 (Continued)

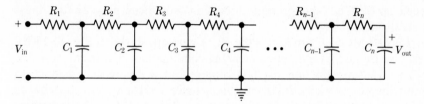

FIGURE 10.14 Lumped-element model for switch transistor chain.

value of $V_{out} = V_n$, which can be interpreted as a logic 1 voltage. This problem has been analyzed in the literature (see [4]). The simplest case is for identical RC branches where $R_1 = \cdots = R_n = R$ and $C_1 = \cdots = C_n = C$. Let t_{prop} be the propagation time needed to charge C_n to a logic 1 voltage V_n. Then

$$t_{prop} \leq \left(\frac{V_1}{V_1 - V_n} \right) \frac{1}{2} RCn(n + 1) \tag{10.2-14}$$

establishes the upper limit on the propagation time through the chain. Since t_{prop} increases as $n(n + 1)$, there is a practical limit to the number of transistors used in a data path.

Distributed interconnect parameters can be included by using the transmission line network shown in Fig. 10.15(a). Transmission lines account for the finite time interval required to propagate a voltage wave between devices. A complete analysis will also include resistive losses in the interconnect material and reflections at line discontinuities. Assume that all capacitors are initially uncharged and that a voltage V_{in} is applied to the line. The voltage wavefront propagates down the chain, eventually charging the load capacitor C_L. Transmission line effects can be included in circuit designs using computer

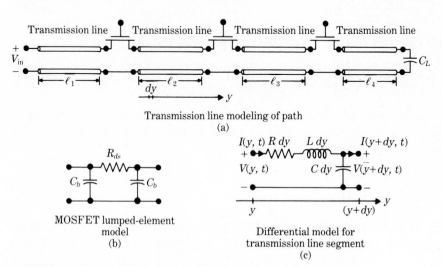

FIGURE 10.15 Path modeling using transmission line models.

codes such as SPICE. The present discussion will focus only on the basic physics involved.

The simplest MOSFET model is the *RC n*-network shown in Fig. 10.15(b), which can be treated as a lumped-element combination in transmission line theory. The transmission line itself is modeled in a differential manner to include distributed resistance R [Ω/cm], inductance L [H/cm], and capacitance C [F/cm]. A segment of length dy is illustrated in Fig. 10.15(c). The important quantities are the voltage $V(y, t)$ and current $I(y, t)$ as the wavefronts propagate along the transmission line.

Applying the Kirchhoff laws to the differential segment gives the relations

$$\frac{\partial V(y, t)}{\partial y} = -RI(y, t) - L\frac{\partial I(y, t)}{\partial t}$$

$$\frac{\partial I(y, t)}{\partial y} = -C\frac{\partial V(y, t)}{\partial t}.$$

(10.2-15)

Eliminating the current yields

$$\frac{\partial^2 V}{\partial y^2} = RC\frac{\partial V}{\partial t} + LC\frac{\partial^2 V}{\partial t^2}$$

(10.2-16)

as the differential equation describing the voltage propagation. Proper boundary conditions must be used at the endpoints of each line in the chain. The full analysis is beyond the scope of this text. Instead, interest will be directed toward the important properties of the wavefronts.

Two special cases can be analyzed. Suppose first that $R = 0$, indicating a *lossless line*. For this case, eqn. (10.2-16) reduces to wave equation

$$\frac{\partial^2 V}{\partial y^2} - \frac{1}{v^2}\frac{\partial^2 V}{\partial t^2} = 0,$$

(10.2-17)

which describes voltage wavefronts traveling with velocity

$$v = \frac{1}{\sqrt{LC}} = \frac{1}{\sqrt{\mu_0 \varepsilon}},$$

(10.2-18)

i.e., the speed of light in the dielectric. For SiO_2 with a relative permittivity of about 3.9, $v \simeq 152$ [μm/ps]. The voltage and current waves are related by the *characteristic impedance*

$$Z_o = \sqrt{\frac{L}{C}} = \frac{V(y, t)}{I(y, t)},$$

(10.2-19)

where Z_o is determined by the geometry of the interconnect line. At this order of approximation, the only effect of the transmission line is to introduce a delay into the voltage.

Propagation can be understood using the single-segment transmission line

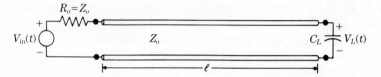

FIGURE 10.16   Charging of a load capacitor through an ideal transmission line.

example as shown in Fig. 10.16. The input voltage is described by

$$V_{in}(t) = V_1 u(t),$$ (10.2-20)

where $u(t)$ is the unit step function

$$u(t) = \begin{cases} 0 & (t < 0), \\ 1 & (t \geq 0), \end{cases}$$ (10.2-21)

introduced to model an input transition from a logic 0 to a logic 1 state at time $t = 0$. The source is assumed to be matched to the line with an internal impedance $R_o = Z_o$; this eliminates reflections at the source end. Analyzing the problem gives the voltage across C_L as

$$V_L(t) = \begin{cases} 0 & (t < T), \\ V_1[1 - e^{-(t-T)/R_o C_L}] & (t \geq T), \end{cases}$$ (10.2-22)

where

$$T = \frac{\ell}{v}$$ (10.2-23)

is the time required for the voltage wavefront to travel from the source to the load. For this simple example, the capacitor voltage is just a delayed RC charge event.

The second special case is that where L is negligible so R and C dominate the line parameters. Equation (10.2-16) now reads

$$\frac{\partial^2 V}{\partial y^2} = RC \frac{\partial V}{\partial t},$$ (10.2-24)

which is recognized as the diffusion equation introduced earlier in the fabrication discussion of Section 5.1.2. For an infinite line with an input voltage $V_{in}(t) = V_1 u(t)$, the mathematics is identical to a constant surface concentration impurity diffusion (predeposit). The voltage is thus described by

$$V(y, t) = V_1 \, \text{erfc} \left(\sqrt{\frac{RC}{4t}} \, y \right) u(t).$$ (10.2-25)

Using the first equation in (10.2-16) gives the current as

$$I(y, t) = V_1 \sqrt{\frac{C}{\pi R t}} e^{-RCy^2/4t} u(t),$$ (10.2-26)

which is easier to interpret. Since L is ignored, the line model of Fig. 10.15(c) degenerates into a differential RC ladder network. The current $I(y, t)$ can be viewed as diffusing down the line, decaying exponentially with y^2. The line does not support wave propagation, so the length must be kept small to transmit logic 1 voltage levels.

A transmission line with arbitrary R, L, and C supports wave propagation with amplitude attenuation because of the resistance. A step function input

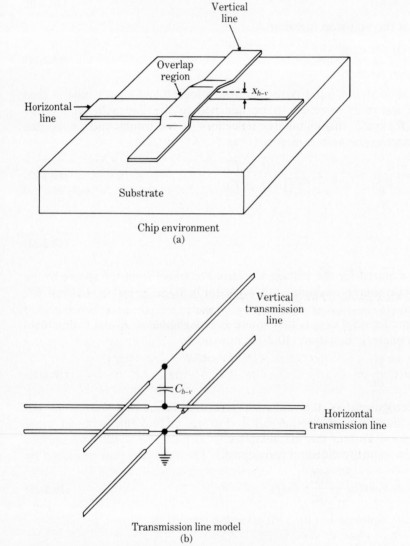

Chip environment
(a)

Transmission line model
(b)

FIGURE 10.17   Horizontal-to-vertical line coupling.

results in a current wavefront of the form [7]

$$I(y, t) = \frac{V_1}{Z_o} e^{-Rt/2L} I_o\left(\frac{R}{2L}\sqrt{t^2 - \left(\frac{y}{v}\right)^2}\right) u\left(t - \frac{y}{v}\right),$$ (10.2-27)

where I_o is the modified Bessel function of the first kind of order zero. The function $u[t - (y/v)]$ indicates wave propagation at velocity v while the exponential and Bessel function combine to give amplitude decay.

Transmission lines can also be used to model the coupling between horizontal and vertical interconnects in the transistor array. Figure 10.17(a) shows the situation where a vertical interconnect crosses over a horizontal path line. The oxide thickness between the two is shown as x_{h-v}. Assuming an overlap area of A_{h-v}, this gives a lumped capacitance of

$$C_{h-v} = \frac{\varepsilon_{ox} A_{h-v}}{x_{h-v}}.$$ (10.2-28)

The transmission line equivalent circuit is constructed in Fig. 10.17(b). C_{h-v} couples voltage and current waves between the lines. This causes changes in the signal propagations because of reflections and charging.

This discussion demonstrates the complexity of MOSFET array response calculations. If timing is critical, accurate modeling and computer simulations become mandatory, particularly in large arrays where the propagation and intermediate charging effects slow logic transmission.

10.3 Weinberger Structuring

Combinational logic circuits can be designed using a structured approach that aids layout and increases packing density. Transient response times can also be improved by minimizing transmission path lengths and capacitances. Many current approaches to this problem in both static and dynamic circuits originate from the concept of LSI patterning established by Weinberger [15].

Weinberger arrays are created by placing transistors on the chip in a geometrically regular manner. Horizontal and vertical interconnect patterns are used to wire the devices together. Efficient logic design of this type can be accomplished using one type of gate; for example, NOR gates form a complete logic set for nMOS circuits. The discussion here will use nMOS NOR logic for examples. However, the techniques are easily generalized.

The basic concepts involved in Weinberger structuring are illustrated in Fig. 10.17. A 3-input NOR gate with an output of

$$F = \overline{(A + B + C)} = \bar{A}\bar{B}\bar{C}$$ (10.3-1)

is shown in Fig. 10.18(a). The circuit has been drawn so that both input and output lines run horizontally while the power supply and ground connections are vertical. While this circuit can be used without modification, different

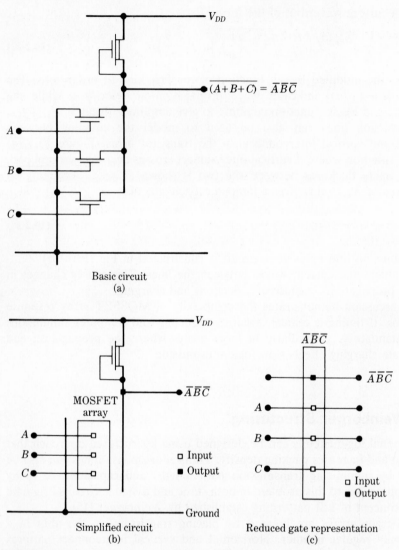

FIGURE 10.18 NOR gate reduction for Weinberger array structuring.

levels of abstraction help to describe more complicated arrays. Figure 10.18(b) shows an equivalent representation where the driver logic MOSFETs are not shown explicitly. Empty squares are used to denote input connections to transistor gates, while the output contact is shown as a filled square. An even simpler view is provided in Fig. 10.18(c). The gate circuit has been reduced to a vertical box. Input and output connections are created using parallel horizontal lines running perpendicular to the gate circuit. Logic formation is viewed as the inputs (A, B, C), creating a vertical output of $\bar{A}\bar{B}\bar{C}$, which is routed to the output by the contact (filled square).

Figure 10.19 illustrates the steps used to implement a 3-to-8 decoder using Weinberger structuring. The NOR-based logic diagram is shown in Fig. 10.19(a). A Weinberger array can be based on the equivalent gate placement shown in Fig. 10.19(b). The important point to note is that input and output lines are horizontal while the gate circuitry is vertical. This directly gives the circuit implementation of Fig. 10.19. Other layout patterns are possible, and the one shown is not optimized for compactness. However, the basic points remain valid.

Different layout techniques are possible within a Weinberger structure. Imaging is extremely useful, and variations on a *Weinberger image* are readily found in chip designs. The concept of a mirror image can be seen in the circuit of Fig. 10.20. The power supply rail V_{DD} is used as the "pivot" line. The circuit above the V_{DD} rail is the mirror image of the circuit below the power supply. This automatically introduces another degree of regularity into the layout patterns. Additional images can be added as needed. For example, Weinberger

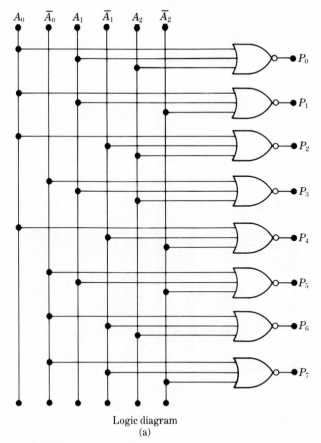

Logic diagram
(a)

FIGURE 10.19   Example of Weinberger structuring for a 3-to-8 decoder.

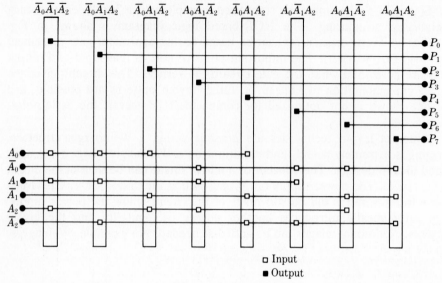

Basic gate structuring
(b)

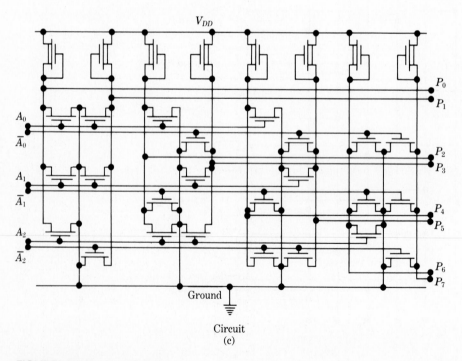

Circuit
(c)

FIGURE 10.19   (Continued)

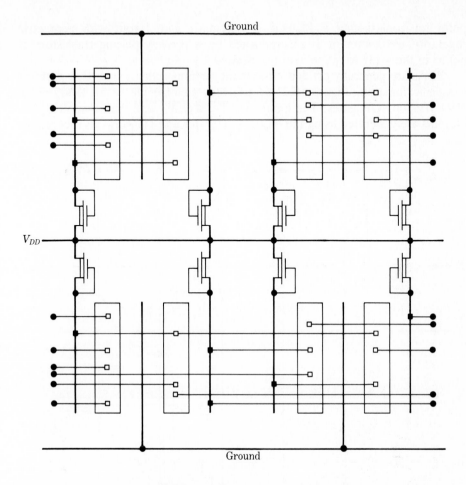

□ Input
■ Output

FIGURE 10.20 Weinberger imaging about the power supply rail.

images can be created above the top ground line and below the bottom ground line. These would be bounded vertically by V_{DD} rails.

Weinberger structuring and variations on the technique can be useful for complex combinational logic implementations in large system designs. These approaches can be readily adapted to CAD (computer-aided design) algorithms and can conceivably form a basis for expert systems.

10.4 Programmable Logic Arrays

Programmable logic arrays (PLAs) constitute an important class of structured logic circuits. Many PLAs are based on Weinberger patterning. The starting point for constructing a PLA is a systematic gate layout with geometric regularity. Programming is accomplished by selectively placing transistors at points in the array to implement the desired logic function.

PLAs are generally divided into circuit sections called *planes* that perform a specific function. Figure 10.21(a) provides an example of an AND-plane using nMOS NOR gates. The inputs are X_0, X_1, X_2, and their complements. Outputs are generated using the NOR/NOT gates and DeMorgan's theorem to

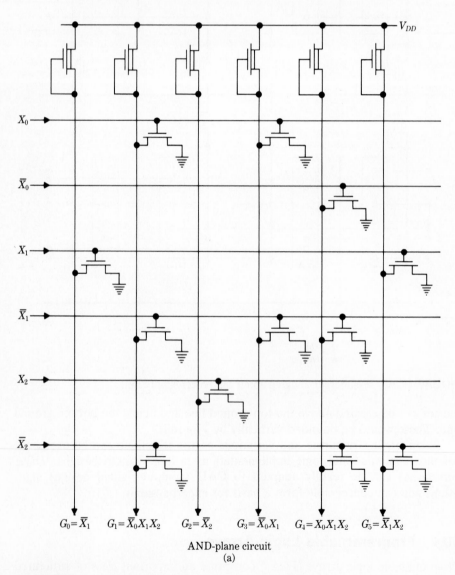

AND-plane circuit
(a)

FIGURE 10.21   nMOS AND-OR programmable logic array.

write

$$G_0 = \bar{X}_1, \quad G_1 = \bar{X}_0 X_1 X_2, \quad G_2 = \bar{X}_2,$$

$$G_3 = \bar{X}_0 X_1, \quad G_4 = X_0 X_1 X_2, \quad G_5 = \bar{X}_1 X_2. \tag{10.4-1}$$

An OR-plane circuit example is shown in Fig. 10.21(b). The inputs $F_0, \ldots, F_5$ (from the AND-plane) feed the NOR gate array. Output inverters then give OR-plane functions of

$$F_0 = \bar{X}_1 + \bar{X}_2,$$

$$F_1 = \bar{X}_2 + \bar{X}_0 X_1,$$

$$F_2 = \bar{X}_1 + X_0 X_1 X_2, \tag{10.4-2}$$

$$F_3 = \bar{X}_0 X_1 + \bar{X}_1 X_2,$$

$$F_4 = \bar{X}_0 X_1 X_2 + \bar{X}_2,$$

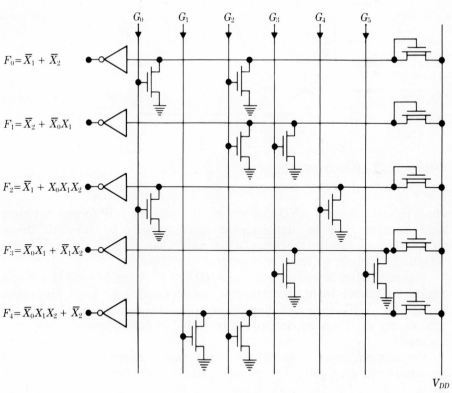

OR-plane circuit

(b)

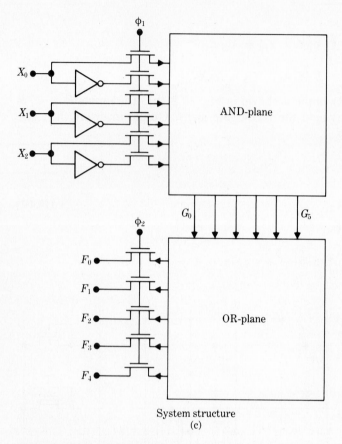

System structure
(c)

FIGURE 10.21   (Continued)

which indicate that the AND-OR scheme is useful for SOP forms. Arbitrary logic functions can be implemented (programmed) by selective driver MOSFET placement in the array. One mask change (e.g., the active area definition) is usually sufficient to reprogram the chip.

Figure 10.21(c) illustrates how the AND-OR PLA can be used in a simple clocked 2ϕ system. Input pass transistors conduct during ϕ_1. Logic propagates through the AND-plane and then through the OR-plane gates. Outputs are valid during ϕ_2. Latching circuits can be added to both inputs and outputs if required.

Propagation times through the logic planes depend on the output capacitances at each gate. The problem can be understood by the cascaded circuit shown in Fig. 10.22, which represents the logic path G_4-F_2 in the PLA example. The inputs $X_0, X_1,$ and X_2 are fed into a 3-input NOR gate. Propagation through the first stage is set by the value of $C_{\text{out},G4}$. A simple

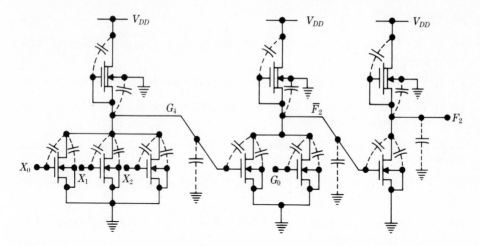

FIGURE 10.22 Capacitance contributions for the $G_4 - F_2$ PLA data path.

approximation is

$$C_{\text{out},G4} \simeq [K(V_{OH}, V_{OL})C_{sbL} + \dot{C}_{GDL}]$$

$$+ \sum_i [K(V_{OH}, V_{OL})G_{dbi} + C_{GDi}]$$

$$+ C_{\text{line},G4} + \sum_j C_{Gj}, \qquad\qquad \textbf{(10.4-3)}$$

where the first sum (i) is over the NOR drivers while the second sum (j) counts the input capacitance into the second stage. Since the array structuring is used, the line capacitance $C_{\text{line},G4}$ can represent a significant, if not dominant, contribution. The AND-plane circuit for this logic path gives an output F_2 whose response depends on

$$C_{\text{out},\bar{F}_2} \simeq [K(V_{OH}, V_{OL})C_{sbL} + C_{GDL}] + \sum_j [K(V_{OH}, V_{OL})C_{dbj} + C_{GDj}]$$

$$+ C_{\text{line},\bar{F}_2} + C_{G,\text{inv}}. \quad \textbf{(10.4-4)}$$

Again, the line capacitance in the array layout may dominate. Finally, the logic must propagate through the inverter, which has the normal circuit delays.

Logic propagation times are proportional to output capacitances, so speed through the array may become critical. In particular, the half-clock cycle time ($T/2$) in the 2ϕ input-output design must be long enough to allow worst-case logic formation. This constraint can limit the size of the array. Alternately, the topology can be "folded" in different ways to allow for denser logic without excessively increasing the PLA real estate requirements.

REFERENCES

(1) D. K. Cheng, *Field and Wave Electromagnetics,* Chapt. 9, Reading, MA: Addison-Wesley, 1983.

(2) M. I. Elmasry (ed.), *Digital VLSI Systems,* New York: IEEE Press (Wiley), 1985.

(3) O. F. Folberth and W. D. Grobman (eds.), *VLSI: Technology and Design,* New York: IEEE Press (Wiley), 1984.

(4) L. A. Glasser and D. W. Dobberpuhl, *The Design and Analysis of VLSI Circuits,* Reading, MA: Addison-Wesley, 1985.

(5) E. E. Hollis, *Design of VLSI Gate Array ICs,* Englewood Cliffs, NJ: Prentice-Hall, 1987.

(6) T. C. Hu and E. S. Kuh (eds.), *VLSI Circuit Layout: Theory and Design,* New York: IEEE Press (Wiley), 1985.

(7) P. C. Magnusson, *Transmission Lines and Wave Propagation,* Boston: Allyn and Bacon, 1970.

(8) C. A. Mead and L. Conway, *Introduction to VLSI Systems,* Reading, MA: Addison-Wesley, 1980.

(9) A. Mukerjee, *Introduction to nMOS and CMOS VLSI Systems Design,* Englewood Cliffs, NJ: Prentice-Hall, 1986.

(10) D. A. Pucknell and M. Eshraghian, *Basic VLSI Design,* Sydney: Prentice-Hall, 1985.

(11) D. Radhakrishnan, S. R. Whitaker, and G. K. Maki, "Formal Design Procedures for Pass Transistor Switching Circuits," *IEEE J. Solid-State Circuits,* vol. SC-20, pp. 531–536, 1985.

(12) S. Ramo, J. R. Whinnery, and T. Van Duzer, *Fields and Waves in Communication Electronics,* 2nd ed., Chapt. 5, New York: Wiley, 1984.

(13) B. Randell and P. C. Treleaven, *VLSI Architecture,* London: Prentice-Hall, 1983.

(14) J. D. Ullman, *Computational Aspects of VLSI,* Rockville, MD: Computer Science Press, 1984.

(15) A. Weinberger, "Large Scale Integration of MOS Complex Logic: A Layout Method," *IEEE J. Solid-State Circuits,* vol. SC-2, pp. 182-190, 1967.

(16) N. Weste and K. Eshraghian, *Principles of CMOS VLSI Design,* Reading, MA: Addison-Wesley, 1985.

PROBLEMS

Assume interconnect capacitance levels of

$$C_{p-f} = 0.0576\,[\text{fF}/\mu\text{m}^2],$$

$C_{m-p} = 0.0863\,[\text{fF}/\mu\text{m}^2],$

$C_{m-f} = 0.0345\,[\text{fF}/\mu\text{m}^2],$

$C_{m-n^+} = 0.0822\,[\text{fF}/\mu\text{m}^2],$

and $C_{ox} = 0.691\,[\text{fF}/\mu\text{m}^2]$ for the gate capacitances. Consult the problem sections in Chapter 8 for nMOS parameters and in Chapter 9 for CMOS parameters if needed.

10.1 Consider the logic function

$$F = AB + BC + AC.$$

(a) Use nMOS switch logic to implement this function in canonical SOP form.

(b) Now use CMOS TG logic structuring instead. Compare the transistor count and examine the interconnect problems for both.

10.2 Stick diagrams (introduced in Problem 5.13) are quite useful for structuring arrayed logic. An example is shown in Fig. P10.1, which gives both the layout and logic for a 4-to-1 path selector. A transistor site with dimensions is also shown. In working this problem, assume a metal-to-metal spacing of $18\,[\mu\text{m}]$ (horizontal, edge to edge) and an n^+-to-n^+ spacing of $21\,[\mu\text{m}]$ (vertical, edge to edge).

(a) Calculate the poly gate input capacitance seen from a metal line looking into a MOSFET site.

(b) Calculate the amount of metal-to-n^+ capacitance seen at each site. What is the main effect of this capacitance?

(c) Calculate the capacitance of metal line as seen from the top (P_0) to the bottom (P_3) that contains two MOSFET sites. (The $\bar{B}$ line is the easiest to visualize.)

(d) Calculate the maximum depletion capacitance seen along the P_0 line from the A-control to the left edge of the B-control line (so that this length contains one metal line with a MOSFET and one metal line as just an overcrossing interconnect). Then estimate the capacitance of each horizontal line.

10.3 The drawing in Fig. P10.2 is used to model the interaction between crossing lines in an array. The capacitance values are given as $C_1 = 4.10\,[\text{fF}]$, $C_2 = 19.60\,[\text{fF}]$, and $C_3 = 27.6\,[\text{fF}]$. The input voltage is specified to be

$$V_{in}(t) = V_{DD}[1 - e^{-t/\tau}],$$

where τ is a time constant characterizing the voltage buildup on the line. Assume that all capacitors are initially uncharged.

(a) Find the current $i(t)$ through C_2.

(b) Calculate the final voltage $V_{out}(t \to \infty)$ across C_2. Then discuss what this result implies about designing the array layout.

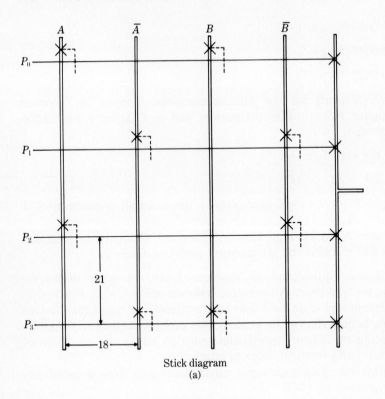

Stick diagram
(a)

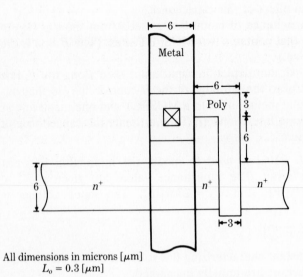

All dimensions in microns [μm]
$L_o = 0.3$ [μm]

MOSFET location
(b)

FIGURE P10.1

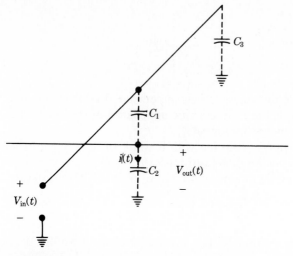

FIGURE P10.2

10.4 Consider a CMOS TG 4-to-1 path selector that is described by an output function

$$F = P_0 A \bar{B} + P_1 \bar{A} \bar{B} + P_2 AB + P_3 \bar{A} B,$$

with A and B the control signals and P_i the input lines. Use normal TG structuring, such as that shown in Fig. 10.10 for the 8-to-1 selector.
(a) Draw the circuit schematic for the circuit, keeping pMOS transistors in one n-tub. Discuss the interconnect problem with this scheme. What problems arise when separate n-tubs are used for each TG?
(b) Restructure the logic using separate pMOS and nMOS paths (see Fig. 10.12). Look at the interconnect problem for this case.

10.5 The characteristic impedance Z_o of an interconnect can be approximated by using the *microstrip* geometry shown in Fig. P10.3. This consists of an interconnect strip of width w at a distance d above a ground plane. For use in MOS ICs, d may be approximated (to lowest order) as the oxide

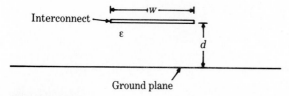

FIGURE P10.3

thickness. For $(w/d) \geq 1$,

$$Z_o \simeq \frac{(\eta_o/\sqrt{\varepsilon_r})}{\left[\dfrac{w}{d} + 2.42 - 0.44\dfrac{d}{w} + \left(1 - \dfrac{d}{w}\right)^6\right]} \, [\Omega]$$

may be <u>used</u> to calculate the characteristic impedance. In this expression, $\eta_o = \sqrt{\mu_o/\varepsilon_o} \simeq 377 \, [\Omega]$ is the characteristic impedance of free space and ε_r is the relative permittivity (3.9 for SiO_2).

(a) Calculate Z_o for a polysilicon line with a width of $w = 3 \, [\mu m]$ that travels over a FOX region $0.60 \, [\mu m]$ thick.

(b) Calculate Z_o for a metal line of width $8 \, [\mu m]$ over a $1.0 \, [\mu m]$ oxide.

10.6 Design a full-adder using NOR-based logic and Weinberger structuring in nMOS.

10.7 Extend Weinberger structuring to the case of dynamic CMOS. What design tradeoffs exist between NAND and NOR logic basis?

Appendix: Summary of *pn* Junction Properties

A step-profile *pn* junction is shown in Fig. A.1. The *p*-type region ($x < 0$) has a constant acceptor doping N_a [cm^{-3}], while the *n*-type region ($x > 0$) has a constant donor doping N_d [cm^{-3}]. The depletion edges are specified as $-x_p$ on the *p*-side and x_n on the *n*-side such that $W = (x_p + x_n)$ is the total depletion width. The equilibrium ($V = 0$) analysis is summarized first.

The depletion electric field $\mathscr{E}(x)$ in $[-x_p, x_n]$ is calculated using the Poisson equation for the potential $\phi(x)$:

$$-\frac{d^2\phi(x)}{dx^2} = \frac{d\mathscr{E}(x)}{dx} = \frac{\rho(x)}{\varepsilon_{Si}}. \tag{A.1}$$

$\rho(x)$ is the volume charge density of the ionized dopants such that

$$\rho(x) = \begin{cases} +qN_d & [0, x_n], \\ -qN_a & [-x_p, 0]. \end{cases} \tag{A.2}$$

Integrating with the boundary conditions $\mathscr{E}(x_n) = 0 = \mathscr{E}(-x_p)$ gives

$$\mathscr{E}(x) = \begin{cases} -\dfrac{qN_d}{\varepsilon_{Si}}(x_n - x) & [0, x_n], \\[2mm] -\dfrac{qN_a}{\varepsilon_{Si}}(x + x_p) & [-x_p, 0]. \end{cases} \tag{A.3}$$

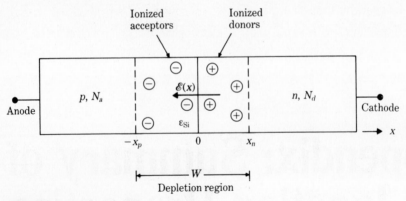

FIGURE A.1   Step-profile *pn* junction.

The maximum electric field is at $x = 0$ and has a magnitude

$$\mathcal{E}_{max} = \frac{qN_d x_n}{\varepsilon_{Si}} = \frac{qN_a x_p}{\varepsilon_{Si}}, \tag{A.4}$$

which shows conservation of depletion charge.

The potential is obtained by integrating the electric field. The built-in potential ϕ_o is a characteristic of the doping and is found to be

$$\phi_o = \int_{-x_p}^{+x_n} \mathcal{E}(x)\, dx,$$

$$\simeq \left(\frac{kT}{q}\right) \ln\left(\frac{N_d N_a}{n_i^2}\right). \tag{A.5}$$

$\phi(x)$ can be obtained directly from the electric field expressions. The depletion widths are found to be

$$W = \sqrt{\frac{2\varepsilon_{Si}\phi_o}{q}\left(\frac{1}{N_a} + \frac{1}{N_d}\right)},$$

$$x_n = \frac{N_a}{(N_a + N_d)}W, \qquad x_p = \frac{N_d}{(N_a + N_d)}W. \tag{A.6}$$

A *one-sided* junction is obtained if $N_d \gg N_a$ or vice versa. In this case,

$$W \simeq \sqrt{\frac{2\varepsilon_{Si}\phi_o}{qN}}, \tag{A.7}$$

where $N =$ smaller (N_a, N_d). Most of the depletion is forced to the lighter doped side.

Applying an external voltage V changes the internal fields. Assuming that the positive side of the voltage is connected to the p-side specifies $V > 0$ as the condition for forward bias, while $V = -V_r < 0$ is needed for reverse bias. The equilibrium quantities may be modified to account for the applied voltage by the transformation $\phi_o \rightarrow (\phi_o - V)$. For example, the total depletion width becomes

$$W = \sqrt{\frac{2\varepsilon_{Si}}{q}(\phi_o - V)\left(\frac{1}{N_a} + \frac{1}{N_d}\right)}. \tag{A.8}$$

Junction capacitance orginates from the depletion charge. It is important in reverse bias, where it is given by

$$C_j(V_r) = \frac{\varepsilon_{Si}}{W(V_r)}[F/cm^2]. \tag{A.9}$$

C_j is nonlinear since it changes with the reverse voltage V_r.

Current flow through the junction is established by tracking the minority carriers. The three main components are electron current I_n on the p-side, hole current I_p on the n-side, and recombination-generation current originating from the depletion region. I_n and I_p combine to give the ideal diode equation

$$I = I_o(e^{qV/kT} - 1), \tag{A.10}$$

where

$$I_o = qA\left(\frac{D_n n_{po}}{L_n} + \frac{D_p p_{no}}{L_p}\right) \tag{A.11}$$

is the reverse saturation current. The reverse generation current is found as

$$I_{gen} \simeq -\frac{qAn_i}{2\tau_o}W(V_r), \tag{A.12}$$

while the forward recombination current assumes the form

$$I_{rec} \simeq \frac{qAn_iW}{2\tau_o}e^{qV/2kT}, \tag{A.13}$$

where τ_o is the average carrier lifetime. These contributions must be added to the ideal diode current.

Index

Acceptor 2
Accumulation 4
Acetic acid 260
Activation energy 234
Active area 263, 286
Active load 97, 101
Adder 341, 370
Alignment 286
Alkali charge 232
Aluminum 3
Annealing 246
Anomalous phosphorus diffusion 242
AOI logic 325, 358, 534
Array logic 585
Arsenic 235, 239
Arsine 249
Aspect ratio 13

Bandgap narrowing 241
Barrier potential, MOSFET 73
Bessel function, modified 601
Biasing parameter (m) 119
Bipolar junction transistor (BJT) 74, 297
  current gain 299
Bird's beak 63, 267, 274, 307
Bistable circuits 386
Bit time 453
Body-bias effects 29
Body-bias factor 11, 140
Boltzmann's constant (k) 8
Boolean logic levels 80, 84
Bootstrapping 484
  isolation network 489

Boron 2, 235, 239
Breakdown voltage 56
Built-in voltage 38, 203
Bulk depletion charge 7, 10, 140
Bulk external voltage 2, 102
Buried contact 278, 445
Butting contact 277

Cantilever structure 262
Capacitance
  bootstrap 485
  coupling 86, 285
  depletion 202, 440, 519
  field oxide (FOX) 63, 265
  gate 3, 44
  interconnect 281
  line 209
  memory cell (DRAM) 502
  MOSFET 199
  output, inverter 208
  overlap 199, 288
  oxide 3
  sidewall 202
  trench 270
Carbon tetrachloride 261
Carbon tetrafluoride 261
Carrier density
  intrinsic (n_i) 2
  majority 2
  minority 2
Carrier diffusion lengths 440
Carrier lifetimes 440
Central processing unit (CPU) 562
Channel charge 37

Channel electric field 19
Channel length 13
    physical and drawn 198, 288
Channel length modulation 25
Channel stop 13, 36, 266, 270, 274
Channel transit time 160
Channel voltage 19, 45
Channel width 13
Characteristic impedance 598, 613
Charge
    bulk depletion 7, 10, 140
    inversion 10, 20
    oxide 9
Charge leakage 438, 456, 465, 513
Charge sharing 467, 482, 494, 536, 557
Charge storage 438
Charge time (t_{ch}) 165
Charge transfer 430
Charge-state vacancies 239
Chemical vapor deposition (CVD) 247
Chip layout 284
Chip real estate 50
Chlorinated oxide 232
Clock period 392
Clock skew 523
Clocked CMOS (C^2MOS) 564
Clocked static registers CMOS 539
Clocked static registers nMOS 465
CMOS 137
CMOS circuit symbols 346
CMOS TG symbols 359
Combinational logic 311, 349, 545
Complementary error function 236,
        599
Complex logic 325, 358, 534
Computer-aided engineering
        (CAE) 585
Concentration dependent diffusion 237
Conditional discharge 492
Constant surface concentration
        diffusion 234
Constant voltage scaling 54
Contact cut 252
Contact printing 251
Critical inverter voltages 83
Crystalites 3
Current integrator 244
Current-voltage $(I-V)$
        characteristics 1, 14

Cutoff in a MOSFET 14
CVD processes 247

Decoder 461, 540, 603
Degree of anisotropy 259
Depletion-load inverter 120, 275
Depletion-mode MOSFETs 35, 119
Design equations 22
Design rule checker (DRC) 292
Design rules 284
    CMOS 303
    nMOS 287
Destructive read 502
Device parameters
    nominal values 293
    working values 293
Device transconductance (β) 23
Diborane 249
Diffraction effects 253
Diffusion coefficient 233, 237, 440
Diffusion equation 234, 599
Diffusion, impurity 233
Diffusion current 22
Diode
    capacitance 441, 513
    leakage current 439, 513
    modeling 513
Discharge time 169
Domino logic 554, 576
Donors 35, 119
Dosimetry 244
Double poly process 278
Drain current 14
Drain induced barrier lowering
        (DIBL) 74
Drain, MOSFET 12
Drift current 21
Drive-in 237
Driver-load ratio 99, 106, 118, 123
Dry etching 260
Dynamic CMOS logic
Dynamic memory cell 501, 502
Dynamic nMOS logic 430, 473, 487
Dynamic pMOS logic 552

Effective channel length 64
Effective diffusion coefficient 237
Effective intrinsic density (n_{ie}) 241
Einstein relations 449

Electromagnetic interference (EMI)   87
Electron motion   14, 160
Electron-beam lithography   256
Electric field   5
Electron mobility   21, 42
Encroachment   267, 286
Energy of switching   188
Enhancement-mode MOSFET   13
Equality function   334
Equivalence function   334
Error function   236
Etch rates   258
Etch selectivity   260
Etching   258
Evaluate MOSFETs   492, 542
Exclusive NOR (XNOR)   331, 367
Exclusive OR (XOR)   331, 367

Fall time   159
Fermi potential   8, 140
Fick's law   233
Field enhancement factor   239
Field implant   63
Field oxide (FOX)   264
Field regions   263
Field-effect   1
Fixed-point iteration   107
Flatband voltage   9
Flip-flop CMOS TG   413
Flip-flop clocked   392, 413
Flip-flop
    JK   393, 420
    master-slave   397, 418
    SR
        NAND   412
        NOR   387, 408
    D-type   393, 417
Floating polysilicon filler   271
Form factor, *f* (SCE)   61
Form factor, *g* (NWE)   65
Frequency
    clock   392
    maximum switching   173, 219, 457
    minimum switching   456
Full-adder   342, 371
Fully-recessed LOCOS

Gate arrays   585
Gate overhang   287

Gate oxide   3
Gate voltage   4
Gaussian distribution
    current in a transmission line   599
    ion implant model   244
    process modeling   294
    diffusion profile   237
Generation current   440, 518
Gold   249, 256
Gradual channel approximation
        (GCA)   18
Guard rings   139, 299

Half adder   341, 370
High-to-low time (t_{HL})   89, 159
Hole mobility   137
Hydrofluoric acid   260
Hysteresis   399

Inductance   86
Input-high voltage (V_{IH})   84
Input-low voltage (V_{IL})   83
Integration techniques   336
Interconnect   3, 209, 263, 278, 284
Inversion charge   10, 20
Inverter, general properties   80
Ion energy loss   246
Ion implantation   243
    activation   246
    dose   10, 35, 244
Ion milling   261
Ion stopping power   245
Isolation   263

Junction field-effect transistor
        (JFET)   36

Large-scale integration (LSI)   137
Laser annealing   247
Latch   386
Latchup   138, 297
Lateral doping   256
Lateral etching   258
Layout mask drawings   288
Leakage currents   439, 513
Light sensitizers   255
Limited-source diffusion   237
Line inductance   86
Line resistance   163, 185

Lithography
  e-beam 256
  optical 250
  X-ray 255
Load device, inverter 80
Load line 91, 101, 113, 121
Loading effects in plasma etching 261
LOCOS 63, 267
Logic swing 89
Low-to-high time (t_{LH}) 89, 165

Mask drawings 288
Masks 250, 256
Maximum switching frequency
    (f_{max}) 173, 457
Metallization 249
Microstrip transmission line 613
Midpoint voltage 88
Minimum linewidth 284
Minimum size effects (MSE) 69
Minimum spacing 285
Minimum-area MOSFETs 357, 437,
    480
Minimum switching frequency
    (f_{min}) 456
Misalignment errors 286
Mobility 21, 38, 42, 47, 53, 137
Molybdenum 249
MOS
  system 1
  accumulation mode 4
  depletion mode 6
  inversion mode 7
MOSFET
  circuit equations 22, 82
  operation 12
  resistance 362
  structure 12
MOSFET scaling 49
Multiplexer 366

n-tub CMOS 297
n-type 36
nMOS 2, 12
NAND gate
  CMOS 346
  nMOS 317
Native oxide 229
Narrow-width effects (NWE) 63

Nitric acid 260
Noise immunity (NI) 89
Noise margins (NM) 84
Noise sensitivities (NS) 88
Non-overlapping clocks 451
Nonsaturated current flow 16, 22
Nonstandard CMOS 371
NOR gate
  CMOS 352
  nMOS 311
NORA logic 562
Number of squares (n) 210

On condutance 109
On resistance 98
Output capacitance (C_{out}) 208
Output-high voltage (V_{OH}) 83
Output-low voltage (V_{OL}) 83
Oxidation
  high pressure 231
  rate constants 230
  CVD 248
  thermal 249
Oxide charge 9
Oxide pinholes 55

p-tub CMOS 297
p-type 2
pMOS 137
pMOS inverter 141
pMOS positive logic 344
Pass gate 429
Pass transistor 428
Passivation 248
Path selector 461, 536, 587
Pattern transfer 250
Pellicle 255
Permittivity 4, 7
Phosphine 248
Phosphoric acid 260
Phosphorus 235, 239
Phosphosilicate glass (PSG) 232
Photoresist 252
Pipelined logic 562
Plasma etching 261
Platinum 249
pn junction 439, 513
  depletion charge 58
  depletion depth 7, 440
  junction saturation current 439

Poisson's equation   50, 72
Polycides   88, 250
Polysilicon   3, 248
Portable design rules   289
Positive logic   80
Power dissipation   54, 138, 188, 492
Power supply voltage (V_{DD})   81
Power-delay product (PDP)   188
Precharging   491, 543
Precharging in pMOS   552
Precharge capacitors (C^2MOS)   564
Predeposit   235
Probability density $p(x)$ 293
Probability normalization   293
Process flow   273
Process transconductance (k')   23
Process variations   292
Product-of-sum (POS) logic   585
Programmable logic arrays (PLAs)   605
Projected range   244
Propagation time (t_p)   89, 175
Proximity printing   253
Pseudo-nMOS logic   372
Pseudo-pMOS logic   372
Pseudo 2-phase clocking   532, 573
Pull-up MOSFET   577
Punch-through   74

Race condition   397, 562
Random-access memory (RAM)   501
Random logic   311
Rapid anneal techniques   247
Ratioed logic   99, 108, 118, 128, 313,
        320, 327, 340, 374, 457, 477
Ratioless logic   480
RC-ladder   595
Reactive ion etching (RIE)   261
Recessed oxide isolation (ROI)   267
Recipe   232
Refractory metals   249
Refresh   465
Register   454
Registration errors   286
Resin, photoresist   253
Resistivity (ρ)   2
Resistor model for inverter   97
Reticle   250
Retrofit CMOS   297
Reverse bias   203

Reverse current, diode   513
Ring oscillator   175
Room temperature   3

S curve   80
Saturated current flow   17, 24, 30
Saturation voltage   24, 31
Scaled inverter performance   218
Scaling factor (S)   50
Scaling theory   49
Schmitt trigger
    CMOS   420
    nMOS   398
Second-gate effect   74
Selectively loadable registers   461, 538
Self-aligned MOSFET   265
Self-iterations   104, 107
Sequential logic   451
Sheet resistance   86, 210, 284
Shift register   454
Short-channel effects (SCE)   58
Sidewall doping   202
Silane   248
Silicides   249
Silicon dioxide   3, 55, 229, 248
Silicon nitride   248
Silicon-controlled rectifier (SCR)   297
Silicon-on-sapphire (SOS)   301
Single-clock CMOS logic   523, 541
Single-phase nMOS   451
Sodium ions   232
Soft bake   252
Soft node   430, 467, 513, 545
Solid solubility limit (N_0)   235
Source, MOSFET   12
Spot size, e-beam   256
Sputtering   249
Standard CMOS structuring   344
Standard deviation   293
Static induction   74
Static logic   311
Statistical process analysis   292
Stick diagrams   289, 309, 384
Straggle   244
Stress-relief oxide   248
Structured logic   584
Substrate bulk   2, 14
Subthreshold current (I_{sub})   74, 439, 513
Successive diffusion   237

Sum-of-products (SOP) logic   585
Super buffer   221
Surface carrier density   6
Surface electric field   6
Surface potential   5
Surface state charge   9
Switch logic   585
Symmetric CMOS design   147
Synchronous logic   428, 454
System response   163

Thermal voltage   8
Threshold voltage   9
Threshold voltage adjustment   10, 35, 274, 283
Threshold voltage loss   102, 432, 587
Threshold voltage, field (FOX)   63, 267
Threshold voltage, inverter (V_{th}) 88
Throughput   255
Time constant
  charge sharing   471
  CMOS TG   504
  nMOS pass gate   432, 437
Titanium   249
Transition width (TW)   89
Transmission gate (CMOS)   359, 532
  equivalent resistance   360
  layout   521, 592
  logic formation   359
Transmission line effects   597
Trench isolation   269
Tristate gate   583

Twin-tub CMOS   297
Two-dimensional MOSFET effects   71
Two-level logic   478
Two-phase logic   451, 473
Two-step diffusion   237

Unity gain line   88

Vertical etching   258
Very-large-scale integration (VLSI)   48
Velocity, electron   160
Velocity saturation   160
Voltage transfer characteristics
    (VTC)   80
Voltages, MOSFET   13
VTC critical voltages   83

Wafer stepper   254
Wave equation   598
Weak-inversion   73
Weinberger structuring   601
Wet etching   260
White ribbon effect   269
Work function   9, 43, 139

X-ray lithography   255

Yield   286

Zero-threshold MOSFET   283, 435
Zipper CMOS   566